United Nations,
Economic Commission for Europe,
Geneva

Food and Agriculture Organization
of the United Nations
Rome

European Timber Trends and Prospects to the year 2000 and beyond

Volume I

UNITED NATIONS
New York, 1986

NOTE

The designations employed and the presentation of material in this publication do not imply the expression of any opinion whatsoever on the part of the Secretariat of the United Nations concerning the legal status of any country, territory, city or area or of its authorities, or concerning the delimitation of its frontiers or boundaries.

ECE/TIM/30
Volume I

UNITED NATIONS PUBLICATION

Sales No. E.86.II.E.19

ISBN 92-1-116368-4
(complete set of two volumes)
ISBN 92-1-116369-2 (Vol. I)

12000P, Vols. I and II
Not to be sold separately

Preface

Mandate

This study has been carried out on behalf of two inter-governmental bodies with responsibility in the field of forestry and forest products, the FAO European Forestry Commission and the Timber Committee of the UN Economic Commission for Europe. A central part of the work programmes of both bodies, which have a joint secretariat and work in close co-operation with each other, has been the preparation of studies of the long-term outlook for the forest and forest products sector. (The first such study was published in 1953.) The work programme of the Timber Committee includes the following:

The Committee will keep under review and up-date projections and forecasts of long-term trends in production, trade, consumption and prices of forest products, comparing them with projections made in previous long-term studies. The studies will provide Governments with the latest projections and forecasts of long-term prospects at the regional and subregional level as a framework for national studies and as a basis for drawing up national policies for the forest and forest products sector. Amongst other factors, the Committee's analyses will take into account the impact of air pollution on the forestry and forest products sector, including the prospects for future wood supply. (ECE/TIM/28, annex IV, project 14.1.1.1.)

The study has been prepared in accordance with this mandate, and a similar one in the programme of the European Forestry Commission.

Objectives

This study, referred to hereafter as ETTS IV (it is the fourth in the series of European timber trends studies), is intended above all as a tool for decision-makers and their advisers, governments, forest owners, forest industries, traders in forest products and others who have an interest in the outlook for the forest and forest products sector. Its objectives are therefore:

(a) To present and analyse past trends in the forest and forest products sector which may affect the outlook for the future;

(b) To prepare forecasts, wherever possible in quantitative terms, for consumption and supply of forest products, and of non-wood goods and services, and to examine the outlook for the sector as a whole by comparing the various forecasts.

Great importance has been attached at every stage to presenting the source of data, evaluating their quality and presenting the methods of analysis used, so that readers will be able:

(a) To evaluate the quality of the forecasts;

(b) In some cases to prepare alternative forecasts.

Most of the comment is at the regional or country group level. However, in order to facilitate analysis and help decision-makers at the national level, data are provided by country whenever possible.

The study itself (volume I) is accompanied by a companion volume mostly consisting of statistical material with national data (volume II). There are also a number of publications connected with ETTS IV, listed in the explanatory notes.

Preparation of the study

The preparation of the study was directed and coordinated, and the major part of the work on it was carried out by members of the Timber Section, ECE/FAO Agriculture and Timber Division in Geneva. However, the invaluable contribution made by others must be recorded, in particular the staff of the Statistics and Economic Analysis Group of the Forestry Department of FAO in Rome.

Tribute should also be paid to the numerous experts in the countries participating in the work of the Timber Committee and the European Forestry Commission. In particular, appreciation is expressed for the support given by the correspondents nominated by countries. It is no exaggeration to say that, without the active support of experts at the national level and the considerable interest shown in the project by members from the time of its inception, the problems of preparing and completing the study would have been almost overwhelming. The secretariat expresses its sincere thanks to all those who have contributed in one way or another to the study. A list of the national correspondents is presented in volume II.

In addition, a number of governments made resources available for the preparation of the study by contributing to a special trust fund, by putting experts at the disposal of the secretariat, or by funding one of a series of "mini-seminars" on the situation and outlook in particular countries. The secretariat expresses its sincere thanks to these governments who have contributed greatly to improving the quality and widening the scope of the study.

The draft of the study was discussed at a review meeting in November 1985, and benefited greatly from the detailed and positive suggestions made by the participants. (The meeting was not however asked officially to approve the draft.)

Responsibility

The responsibility for the study, including the views expressed and the conclusions reached, rests entirely with the secretariat.

EXPLANATORY NOTES

Sources of data

Unless otherwise indicated, all data are taken from the ECE/FAO data base, maintained in Geneva, which draws data from official replies to regular questionnaires. Forest products statistics are published regularly in the ECE/FAO *Timber Bulletin* and the FAO *Yearbook of Forest Products*. The data base is updated continuously.

Country groupings

It is recognized that no country grouping is perfect. For this reason, wherever possible, data are provided by country so that readers may construct their own grouping if they wish. For the sake of continuity the same country groups as in the previous study (ETTS III) are used:

Nordic countries (NC): Finland; Iceland; Norway; Sweden;

European Economic Community (EEC(9)): Belgium; Denmark; France; Germany, Federal Republic of; Ireland; Luxembourg; Netherlands; United Kingdom;

Central Europe (CE): Austria; Switzerland;

Southern Europe (SE): Greece; Portugal; Spain; Turkey; Yugoslavia;

Mediterranean countries: Cyprus; Israel; Malta;

Eastern Europe (EE): Bulgaria; Czechoslovakia; German Democratic Republic; Hungary; Poland; Romania; *USSR*;

North America (NA): Canada; United States of America.

Frequently data are not shown separately for "Mediterranean countries", as the amounts concerned are small. When separate data are not shown for "Mediterranean countries", they are included in "Southern Europe".

The term "Western Europe" in this study refers to all the countries in Europe, except Eastern Europe.

Data are usually not available for Albania. Where such data are available (e.g. for the forest resource), Albania is included in "Mediterranean countries" and this is indicated in the text or footnotes to tables.

Greece, Portugal and Spain, although now full members of the EEC, were retained in Southern Europe, for the sake of continuity.

All countries covered by the study are either members of the ECE Timber Committee or of the FAO European Forestry Commission or, for most countries, of both.

Symbols used

m³	cubic metre, solid volume
m³ o.b.	cubic metre, overbark
m³ u.b.	cubic metre, underbark
m³ EQ	cubic metre, equivalent wood in the rough
ha	hectare

kg	kilogram
m.t.	metric ton
–	nil or negligible
..	unknown or not available
*	unofficial figure or secretariat estimate
Billion	thousand million

Most frequently used abbreviations

ECE	United Nations Economic Commission for Europe
FAO	Food and Agriculture Organization of the United Nations
GATT	General Agreement on Tariffs and Trade
CMEA	Council for Mutual Economic Assistance
EEC	European Economic Community
OECD	Organisation for Economic Co-operation and Development
USSR	Union of Soviet Socialist Republics
USA	United States of America
GDP	gross domestic product
NMP	net material product
GAI	gross annual increment
NAI	net annual increment
OWL	other wooded land
R and D	research and development
ISIC	International Standard Industrial Classification

Terminology

The terminology and definitions used are the standard ones used in FAO and FAO/ECE publications, defined in the FAO/ECE *Timber Bulletin* and the FAO *Yearbook of Forest Products*.

The term "pulpwood" includes all wood raw material used for the manufacture of pulp, particle board and fibreboard, i.e. pulpwood, round and split, coniferous and non-coniferous, as well as wood residues, chips and particles.

The term "logs" is often used for sawlogs and veneer logs.

Earlier timber trends studies

Full title	Abbreviation	Published in:	Forecasts to:
European Timber Trends and Prospects	ETTS I	1953	1960
European Timber Trends and Prospects, A new Appraisal, 1950-1975	ETTS II	1964	1975[a]
European Timber Trends and Prospects, 1950-1980, An Interim Review	Interim Review	1969	1980[a]
European Timber Trends and Prospects, 1950 to 2000	ETTS III	1976	2000

[a] With tentative extrapolations to the year 2000.

Rounding of the data

In many of the tables, it will be found that, when adding up the detail, the result does not correspond exactly with the total shown. This is due to the fact that the data in the tables are based on the official statistics received from countries, which are generally reported to the nearest 100 units. The totals as shown have been aggregated from the individual elements in their original degree of detail. Both the detail and totals shown have been rounded from the original data, leading to apparent inconsistencies.

Connected publications and data sources

LONG-TERM SERIES

Long-term series for removals, production and trade of forest products, with annual data from 1969 to 1984, by country and country group, for all ECE member countries, Israel and Japan are being issued in a separate document which also presents summaries of the TIMTRADE data base on trade flows in forest products (see annex 16.19) for 1970, 1975, 1980 and 1984. This publication, entitled *Forest products statistics: production, trade and apparent consumption, 1969 to 1984* (symbol ECE/TIM/31) is available free on request from the secretariat.

FOREST RESOURCE AROUND 1980

The results of the enquiry summarized in chapter 3 are published as *The Forest Resources of the ECE Region (Europe, the USSR, North America)*, ECE/FAO, Geneva, 1985, ECE/TIM/27.

REGULAR PUBLICATIONS

The ECE/FAO and FAO data bases are both updated on a continuing basis and the results regularly published. The main publications are:

— ECE/FAO *Timber Bulletin* (production, trade and prices of forest products, ECE member countries, plus Israel and Japan, ten numbers per year, of which the first, the *ECE Timber Committee Yearbook*, provides information on activities and a complete bibliography for relevant publications over the past year);

— The *FAO Yearbook of Forest Products* provides data for all countries in the world on production and trade of forest products.

There are also occasional publications on specific subjects, notably the structure, capacity and raw material consumption of various industry sectors.

Further information on these publications and how to obtain them may be obtained on request from the secretariat (ECE/FAO Agriculture and Timber Division, Palais des Nations, CH-1211 Geneva 10, Switzerland).

CONTENTS

Summary

S.1 INTRODUCTION

S.1.1 Scope of the study

The study, known as ETTS IV, deals essentially with supply and consumption of wood and forest products. It also describes and analyses the forest industry and trade in wood and forest products, and non-wood benefits of the forest. It covers present conditions as well as trends. Forecasts are derived, based on assumptions which are clearly spelt out.

S.1.2 Aim

The study is intended to be a working tool for people who have to make long-term planning decisions in the forest and forest products sector and their advisers: forest owners, public and private, forest industries and those responsible for national or local forest and forest industry policy, as well as policies for trade, employment, rural development and the environment.

The description and analysis of the present situation and of past trends can be seen as an aim in itself. Here the study presents (mainly in its annexes) a data base which has a wide potential use.

The forecasts serve a further purpose. They are intended to give some guidance to decision-makers as to possible developments. These are not predictions but show the consequences of the present situation under assumptions which are realistic in the view of the people who worked on the study. The data presented should make it possible for other users to work with other assumptions and thus derive their own forecasts, and in every case to form their own judgement of the validity of the forecasts.

S.1.3 Country grouping

The following country grouping has been used throughout the study:

Nordic countries: Finland; Iceland; Norway; Sweden;

EEC(9): Belgium; Denmark; France; Germany, Federal Republic of; Ireland; Italy; Luxembourg; Netherlands; United Kingdom;

Central Europe: Austria; Switzerland;

Southern Europe: Cyprus; Greece; Israel; Malta; Portugal; Spain; Turkey; Yugoslavia;

Eastern Europe: Bulgaria; Czechoslovakia; German Democratic Republic; Hungary; Poland; Romania.

In order to facilitate comparison with ETTS III, the EEC(9) which was used in ETTS III, has been retained, rather than the EEC(12) which came into existence only on 1 January 1986. However, the data for the EEC(12) have been calculated and are reproduced in the annexes.

No country grouping is perfect. However, country data are also presented so that analysts may construct their own country groupings as they wish.

S.1.4 Structure of the study

The study itself (volume I) contains the following seven parts, which will be summarized below.

1. Economic, demographic and social background (chapter 1);
2. The European forest (chapters 2-6);
3. Demand for forest products (chapters 7-13);
4. The European forest industries (chapters 14-15);
5. Trade in forest products (chapters 16-17);
6. Wood and energy (chapters 18-19)
7. Synthesis and conclusions (chapters 20-22).

Volume II contains annexes, mostly of a statistical nature, which provide supplementary information, notably data by country.

S.2 ECONOMIC, DEMOGRAPHIC AND SOCIAL BACKGROUND

The population of Europe is expected to grow more slowly than foreseen in ETTS III (583 million people by the year 2000 instead of 601). The fastest growth will be in Southern Europe.

1

TABLE S.1

Europe: population trends and prospects, 1950 to 2025

	1950	1965	1980	2000 [a]	2025 [a]
Total population (millions)					
Nordic countries	14.4	16.2	17.4	17.6	16.9
EEC(9).......................	215.4	44.0	260.3	265.9	260.4
Central Europe................	11.6	13.2	13.9	13.4	12.2
Southern Europe	83.1	103.8	128.5	165.4	207.2
Eastern Europe	88.5	100.1	109.8	121.1	131.2
Europe....................	413.0	477.3	529.9	583.4	627.9
Average annual percentage change over previous period					
Nordic countries	..	+0.8	+0.5	+0.1	−0.2
EEC(9).......................	..	+0.8	+0.4	+0.1	−0.1
Central Europe................	..	+0.9	+0.3	−0.2	−0.5
Southern Europe	..	+1.5	+1.4	+1.3	+1.1
Eastern Europe	..	+0.8	+0.6	+0.5	+0.4
Europe....................	..	+1.0	+0.7	+0.5	+0.4

[a] Projections are the UN medium variant.

Gross domestic product (GDP) is an important variable as it affects, directly or indirectly, supply and consumption of wood and wood products. GDP scenarios were developed by FAO in co-operation with the World Bank, OECD and several other international organizations. Two scenarios were used (table S.2).

TABLE S.2

Past and projected growth in GDP

(*Annual percentages*)

	Past trends		Scenarios 1980-2000	
	1965-1973	1973-1980	Low	High
Nordic countries.....	3.7	3.3	2.8	3.6
EEC(9).............	4.5	2.2	2.4	2.9
Central Europe......	4.4	1.5	2.2	3.0
Southern Europe	6.5	3.9	3.5	4.3
Eastern Europe......	6.5	4.5	3.7	5.1
Europe...........	4.7	2.6	2.6	3.3

The low scenario departs from the low growth of recent years and leads to a gradual recovery. During the late 1990s the growth rate would be similar to that of the 1970s. In the high scenario growth rates higher than those of the 1970s, but lower than those of the 1960s, are expected.

Developments until the end of 1985 seemed to indicate that even the growth rates of the low scenario were rather optimistic. The drop in oil prices since then has again changed the situation so that the scenarios now (1986) look more realistic. This may show the difficulties involved and encourage caution in the use of the projections.

Construction is one of the major sectors affecting wood consumption. Activity in Western Europe, in particular in the housing sector, has been stable or, in some countries, declining, since the mid-1970s. The outlook for residential investment is central to the outlook for consumption of sawnwood and panels, but is almost impossible to predict, as factors are apparent which might raise or lower residential investment. Positive factors include rising real incomes, desire for better quality housing, and specialized housing while negative factors include high real interest rates, decline in inflationary expectations, restrictions on government spending and in some areas, competition for land.

An interesting feature is that the share of timber frame (TF) houses among new low rise houses has increased markedly in many countries during the 1970s and early 1980s, although it is still well below 20% outside the Nordic countries, where it is usually well over 90%. In the United Kingdom, however, strong promotion of TF led to a significant increase in the percentage, but it dropped steeply again after some adverse media comment. This shows that markets can change quickly, drastically and in different directions and that the forest sector itself has a role to play in expanding or defending markets.

The increase in the productivity of *agriculture* is expected to continue. As most European countries are broadly self-sufficient in food, this will lead either to production of the same amount of food on less land, or the production of costly agricultural surpluses. In both cases, there would be pressure to convert agricultural land to other uses, including forestry.

S.3 THE FOREST RESOURCES OF EUROPE IN 1980 AND OUTLOOK TO THE YEAR 2000 AND BEYOND

S.3.1 The resource

A forest resource enquiry was undertaken in 1982. It presented the situation around 1980 and contained three parts:

1. General forest inventory data;
2. Woody biomass;
3. Non-wood goods and services from the forest.

The results were published as "The Forest Resources

TABLE S.3

Population, land area and forest resources in Europe around 1980,
by country groups

| | | | Forest and other wooded land (million ha) | Exploitable closed forest | | |
	Population (millions)	Land area (million ha)		Area (million ha)	Growing stock (million m³ o.b.)	Net annual increment (million m³ o.b.)	Total removals [a] (million m³ u.b.)
Nordic countries	17.4	112.4	59.9	48.3	4 407	146	103
EEC(9)	260.3	150.7	34.5	27.8	3 565	128	82
Central Europe	13.9	12.2	4.9	4.0	1 109	25	19
Southern Europe [b]	131.2	179.8	66.6	27.1	2 582	93	60
Eastern Europe	109.8	96.8	28.0	25.8	4 278	113	78
Europe	532.6	551.9	193.8	132.9	15 941	505	341

[a] Recorded removals from all sources (not only exploitable closed forest). Not exactly comparable with forecasts (see section 5.1 and table S.4).
[b] Including Albania.

of the ECE Region (Europe, the USSR, North America)",
ECE/FAO, Geneva, July 1985.

Certain main results on population, land area and forest
resources are summarized in table S.3. The study stresses
the wide variation between (and within) conditions in different countries not only as regards climate and site conditions but as regards ownership structure, fiscality and
forest management methods and objectives.

S.3.2 Factors affecting removals

Factors affecting removals are discussed in detail in the
study. The following are of particular interest:

(a) Conflicts between different uses of the forest such
as wood production, recreation, etc. seem to be a
concern in many countries;

(b) The ownership structure in some countries (prepon-

derance of small-scale owners, often with management objectives not devoted to wood) sometimes
hinders maximum mobilization of wood;

(c) Fiscal factors can also affect forest management;

(d) The difficulties for most countries to find a market
for small sized wood from thinnings are mentioned. This negatively affects the level of silvicultural activities.

S.3.3 Supply scenarios

Member countries made forecasts of future removals
and the future forestry situation. The scenarios were made
for ten year periods up to the year 2020. The forest equations were applied strictly by most countries (growing stock
at beginning of period plus increment minus drain equals
growing stock at end of period).

TABLE S.4

Europe: forecasts of wood removals in 2000
by country group and species group

| | Base period [a] (million m³ u.b.) | 2000 (million m³ u.b.) | | Percentage change | |
		Low	High	Low	High
Nordic countries	103.8	109.8	130.1	+ 5.7	+25.3
EEC(9)	91.1	101.1	114.7	+11.0	+25.9
Central Europe	16.6	19.9	21.8	+19.9	+31.3
Southern Europe	59.6	77.1	82.3	+29.4	+38.1
Eastern Europe	79.5	82.8	89.2	+ 9.7	+12.2
Europe	350.5	390.8	438.1	+11.5	+25.0
of which:					
Coniferous	227.9	256.4	289.2	+12.5	+26.9
Non-coniferous	121.8	134.4	148.9	+10.3	+22.2

[a] A year or period around 1980, varying from country to country, which was used as a basis for the forecasts.

FIGURE S.1

Europe: past and forecast levels of removals and net annual increment

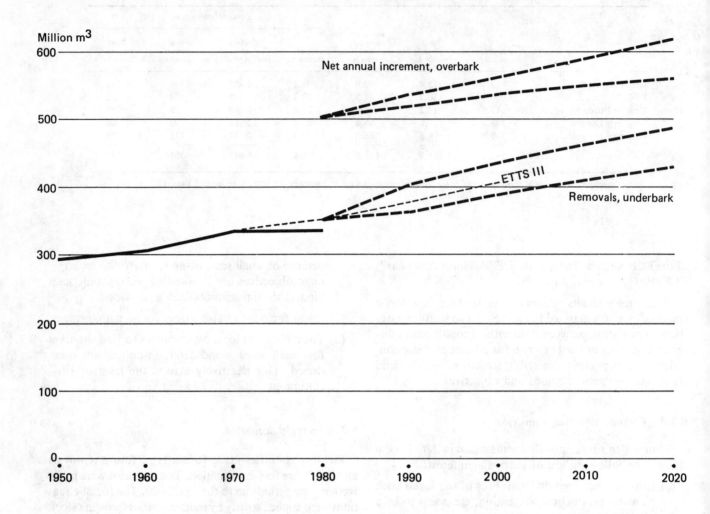

The forecasts imply an increase in removals between 1980 and 2000 of between 11% and 25%. Net increment (gross increment less natural mortality) is also expected to increase substantially during the period. The development of removals underbark and of net increment overbark is shown in figure S.1 (please observe the difference in the unit of measurement). It can be seen that the outlook for removals does not differ very much between ETTS III and ETTS IV, although there are large regional deviations. Even the growing stock is expected to increase up to the year 2000 by 12% in the low and 10% in the high removals scenario.

The following is of particular interest:

(a) The westernmost countries of Europe, especially Ireland, the United Kingdom, France, Portugal and Spain foresee a far above average increase of their forest resources. Large new plantations, mostly of coniferous species, will mature;

(b) In a number of countries in the middle of Europe, however, the forecasts are for stagnation or even decline (e.g. Czechoslovakia, Federal Republic of Germany in the low scenarios);

(c) Hardwood removals have been decreasing since the 1950s, but have turned upwards again during recent years. The use of wood for fuel seems to be an important factor behind this development.

S.3.4 Damage to forests

The present extent and possible future effects of forest damage, particularly that attributed to air pollution, are discussed in chapter 6.

At present there is neither the statistical basis nor any other experience that could be used to prepare forecasts of the extent of damage attributed to air pollution. As a result, a number of non-quantified scenarios are proposed concerning the interaction of forest damage, fellings, increment, growing stock and management decisions. Up to the mid-1980s, it was possible to compensate any sanitation fellings necessary by reductions in planned fellings.

It was therefore decided not to modify in any way countries' forecasts for the forest resource (of which two specifically take forest damage into account).

S.3.5 Non-wood goods and services

Chapter 4 reviews the long-term outlook for demand and supply of non-wood benefits of the forest. Although very little quantitative analysis can be done, it can still be predicted that the non-wood benefits in total, especially the provision of leisure facilities, will account for an increasing proportion of the total goods and services supplied by the forest, even if wood production will remain by far the largest single function overall (but not in some forest areas e.g. near large centres of population where tourism is important, or in vulnerable terrain).

S.4 THE DEMAND FOR FOREST PRODUCTS

S.4.1 Past trends in consumption

The fastest growth in consumption has been for wood-based panels, notably particle board (+8.7% p.a. between 1950 and 1980), followed by paper and paperboard (+5.3% p.a.), and then sawnwood (1.7% p.a.). For all products, the 1970s saw violent cyclical fluctuations and a slowdown in the long-term growth rates.

An examination of recent trends in the consumption of different forest products reveals that there are big differences in the consumption *per caput* in different regions including Canada, the USA, USSR and Japan (figures 7.8 to 7.10). Southern Europe shows the lowest or nearly lowest consumption figures for sawnwood, wood-based panels and paper. The highest consumption figures *per caput* (USA, Canada, the Nordic countries) are usually five to seven times higher than the lowest. It can also be seen from the figures that the growth rate in consumption *per caput* for Southern Europe is bigger than in any other region even though the region has the fastest growth in population in Europe.

For fuelwood consumption the downward trend since the 1950s was reversed in the late 1970s. All statistics about fuelwood consumption are very uncertain however and it is likely that the real consumption is higher than recorded statistics show.

S.4.2 The uses of forest products

Chapter 8 examines how the consumption of sawnwood and wood-based panels is distributed into end-use sectors. The main finding is that the use of wood per unit produced in the most important sector, the construction sector, varies very widely and that there are big differences between countries. Such a situation carries both possibilities and threats. The market is sensitive to such factors as costs, convenience in use, durability, availability and not least variations in fashion. The state of knowledge about the situation in the different end-use sectors (notably the volumes consumed and the factors affecting competition between products) is poor and should be improved. In several areas, there appear to be substantial possibilities to raise levels of consumption of sawnwood and panels.

Chapter 9 reviews the end-uses for the different grades of paper and paperboard and the main factors which will affect the outlook. The major uncertainty concerns the effect of developments in electronic communications technology on consumption of graphic papers. The consensus of experts is that consumption will not be significantly affected before the mid-1990s, but that thereafter the outlook is very uncertain.

S.4.3 Price trends

Price trends for forest products are presented (chapter 10) in such a way as:

— To allow comparisons with the general producer price index;

— To study price fluctuations;

— To allow price comparisons within a country between competing forest products or between raw material and products (for instance pulpwood and chemical pulp).

The possibilities of interpretation are limited by changes in quality and in composition over time within a product group. It is however possible to say that there is no upward or downward trend for most products, despite major fluctuations, especially in the 1970s. An exception is particle board, and, to a lesser extent, the other panels, which showed a clear downward trend in prices. There is thus no basis in past trends to assume either upward or downward price movements in the future.

TABLE S.5

Consumption scenarios for ETTS III and ETTS IV

(Million units)

	Units	1970	1980	ETTS III 1980 [a]	ETTS III 2000 [b]	ETTS IV 2000 Low	ETTS IV 2000 High	Difference in 2000 between ETTS III and IV Low	Difference in 2000 between ETTS III and IV High
Sawnwood	m³	93	102	99	113	119	141	+ 6	+ 28
Wood-based panels	m³	23	36	46	113	50	58	− 63	− 55
Paper and paperboard	m.t.	38	49	56	100	67	92	− 33	− 8
Fuelwood[c]	m³	69	72	54	35	86	109	+ 33	+ 56
Other industrial wood	m³	30	23	26	20	21	..	+ 1	..

^a Low scenario.

^b Adjusted forecasts (table 9/12).

^c Fuelwood: Data for 1969-71 from official sources, for 1979-81 according to natio-

nal estimates. ETTS III scenarios are comparable with data for 1970 and ETTS IV scenarios with data for 1980. To achieve comparability, figures for the difference have been adjusted by the difference between statistics and estimates for 1980, i.e. 18 million m³.

S.4.4 Models for demand projections

Models for demand play a central role in the study. Projections by models of any type are based on the following assumptions:

— That the model correctly mirrors historical correlations between the consumption of forest products and the explanatory variables;

— That the scenarios used for the development of the explanatory variables are reasonable;

— That the correlations between consumption and explanatory variables remain unchanged.

Basically two models were used:

— The "GDP elasticities model" was used for almost all countries in the world, for broad product groups (sawnwood, wood-based panels and paper and paperboard);

— The "end-use elasticities model" was used for twelve important consumer countries in Western Europe and for five categories of sawnwood and wood-based panels. It requires more detailed input data.

The GDP elasticities model uses GDP and border unit values (import or export unit values, whichever is more important for the country in question) as explanatory variables. In addition a linear trend variable is introduced, which adjusts for technological changes, such as market penetration by a new product.

The end-use elasticities model describes more precisely end-use activity and prices. It applies weighted activity indices for those sectors in which a certain forest product is used and it works as far as possible with prices on domestic markets. A trend variable is also used.

For Eastern Europe, forecasts were based on expert opinion and long-term trend analysis.

S.4.5 Scenarios for the consumption of forest products

Scenarios for consumption of various products to 2000 either derived by one of the two models described above,

or from an enquiry to correspondents (for fuelwood) or estimated (for less important products) are shown in table S.5. In this table, a comparison is also made between the results of ETTS III and ETTS IV. It can be seen that in the new study a rather slower rate of growth is foreseen as compared to ETTS III, in particular for wood-based panels, but also for paper and paperboard. The forecast for sawnwood is for a rather slow increase in consumption, although slightly faster than the ETTS III forecast. For fuelwood, however, ETTS IV forecasts a rise, while ETTS III expected a fall.

The following rates of growth are forecast by ETTS IV, for the period between 1980 and 2000 (annual percentages):

	Low	High
Sawnwood	+ 0.8	+ 1.6
Wood-based panels	+ 1.7	+ 2.5
Paper and paperboard	+ 1.6	+ 3.2
Fuelwood	+ 0.9	+ 2.1

The study also examines the outlook for demand in the first quarter of the twenty-first century. Naturally, for a period 40 years in the future, the uncertainty of any quantitative estimates is stressed, as well as the possibility of major changes in trend. "Base" scenarios are proposed, which assume continuation of the *per caput* consumption levels of 2000 as well as the "extreme" scenarios, for a continued rise after 2000 to new record levels by 2025 or, on the other hand, for a drop to the levels of the 1980s or below.

The major factors affecting consumption of forest products before or after 2000 may be summarized as follows:

(*a*) Population growth;

(*b*) Macro-economic growth;

(*c*) Level of residential investment;

(*d*) Technical and economic competitivity of sawnwood and panels in the building sector;

(*e*) Developments in electronic data processing and transmission and their acceptance by society;

FIGURE S.2

Europe: ETTS IV consumption scenarios

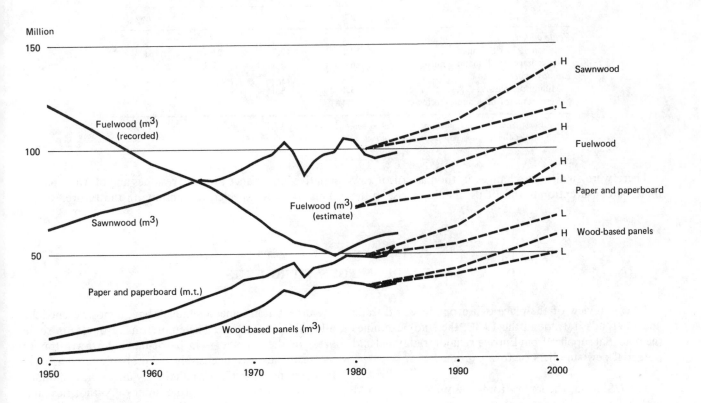

(f) Technical and economic compititivity of paper and paperboard in the wrapping and packaging sector;

(g) The energy situation, especially the relative price and availability of different fuels;

(h) Changes in consumer habits.

While many of these − (a), (b), (c), (e) and (g) − are not susceptible to influence by the forest and forest prod-ucts sector, which is of relatively minor importance in most countries, it should be stressed that the sector does have a strong influence over some others, notably the technical and economic competitivity of its products. It is to a large extent for the sector itself to decide whether consumption levels are to be allowed to drift downwards or whether the necessary measures will be taken to defend and develop markets — control of costs, fundamental research, prod-uct development, standardization, marketing, etc.

S.5 THE EUROPEAN FOREST INDUSTRY

These chapters describe the role of the forest industry in society, its structure and capacity and its raw material consumption.

Table S.6 shows value added and employment in the forest industries, in total and in relation to all manufacturing industry.

Chapter 14 presents the trends in the structure and capacity of the forest industries, as well as for production and self-sufficiency in the various products. There is also a preliminary review of the economic health of these industries, especially trends in productivity and profitability, although further research is needed in this area.

TABLE S.6

Value added and employment in forest industry (ISIC 331 + 332 + 341)
Western Europe and North America

	Western Europe [a]		North America	
	1970	1980	1970	1980
Value added				
Billion 1970 US dollars	17.1	17.7	23.4	31.4
Percentage of all manufacturing	6.7	6.3	7.3	7.5
Employment[b]				
Million people.	3.1	3.0	1.7	1.9
Percentage of all manufacturing	7.4	7.3	8.8	9.0

[a] Except Luxembourg and Switzerland, for which no data were available.

[b] Includes data for Czechoslovakia, Hungary and Poland.

There were significant changes in the pattern of raw material consumption in the 1970s, notably a rise in the relative importance of wood residues and of waste paper. This trend is expected to continue in the future.

S.6 TRADE IN FOREST PRODUCTS

After a review of the place of Europe in world trade and of trends in trade in the 1970s, the study examines the potential supplies from Europe's major traditional and potential new suppliers of forest products.

The *USSR* has the largest forest resource in the world with millions of hectares as yet untouched. However, removals have fallen since the mid-1970s as forest activities have moved east and north, into Siberia, with resulting increased costs and other constraints (e.g. climate and terrain, limited labour resources). Furthermore the forest sector has to compete with other sectors for limited resources for investment. There is also a large potential for increased domestic consumption, especially for panels and paper. Thus, despite the large difference between allowable cut and removals, and the effort to use available resources more intensively, there is little prospect of any significant increase in exports from the USSR in the foreseeable future. Exports are expected to maintain roughly their present levels, unless unforeseen events necessitate a significant change in policy priorities.

Canada also has a very large, mostly coniferous, forest resource, but in Canada removals have almost reached the allowable annual cut. The coniferous resources in some areas of the country — in particular old-growth timber — are already used intensively and probably could not sustain present harvest levels indefinitely under the conditions of recent management programmes. Authorities and forest-based industries have become increasingly aware of the developing imbalances in Canada and of the uncertainty about their extent. Measures proposed to raise production in the long term include improved protection against insects and fire, improved regeneration, more use of hardwoods, of resources at present considered economically inaccessible and of lower quality raw material, and improved raw material yields. There is undoubtedly a potential to expand exports further, although there are many problems. There is at present a major public debate, as yet incomplete, in Canada on the future of forestry, which makes the outlook difficult to foretell. However it is clear that without the expansion of the raw material base in the ways outlined above, it would hardly be possible for Canada's exports of forest products to rise much beyond present levels, at least in the longer term and on a sustained basis.

TABLE S.7

Europe's place in world trade in forest products, 1980
(Million m³ EQ)

To:	From:	World	Europe	USSR	Canada	USA	Other
World		474	167	36	119	63	90
of which:							
Europe		225	141	24	21	18	20
USA		84	1	–	79	–	4
Japan		71	1	7	9	24	30
Other		94	24	5	10	21	36

Although trade is relatively less important for the *USA*, the size of this country's domestic market, the world's largest by far, means that developments there will affect world trade patterns. An official study, published in 1982, foresaw increased demand for forest products, higher domestic supply and higher net imports (despite increased exports). Doubts have since been raised about the forecast levels of demand, and the resource potential, notably in the southern states. At the same time there has been an increase in the interest paid by government and industry to expanding exports. Studies are in hand on many of these questions, but the outlook at present, both for US imports and exports must be considered uncertain, especially as it will be strongly affected by the level of the US dollar, which has been very volatile in the mid-1980s.

About 12 million m³ EQ of European imports (2.5% of total supply) originate in the *natural tropical hardwood forest*, which covered about 1,160 million ha in 1980 (closed forest only). Of this 860 million ha is productive, of which 38 million ha managed, 154 million ha logged over and 668 million ha undisturbed. The proportion of logged over forest is higher in many of the traditional tropical hardwood exporters, including some who are no longer able to export, due to depletion of the resource. Deforestation of closed broadleaved forests is estimated at 6.9 million ha (0.6%) per year. There were in the late 1970s 11.5 million ha of tropical plantations, of which 7.1 million ha industrial plantations. Since then more have been established. According to FAO/UNEP, recent trends, notably that for deforestation and the establishment of plantations, are expected to continue in the near- to medium-term future. In the medium- to long-term, if present trends were to continue, the total area of undisturbed productive closed forests would have dropped to 540 million ha in 2000. However, the area of plantations is expected to increase faster than in the recent past and programmes for reforestation and sustained management of forest plantations to be implemented. The result of these developments would be the partial replacement of the natural tropical resource by plantation silviculture, on a smaller area, but with much greater productivity and producing in most cases a totally different type of wood. The supply potential of the natural tropical hardwood forest, however, exists well beyond the year 2000, although this would contribute to the further depletion of the resource. The determining factor seems to be the policies adopted by the producer countries for their forests. A shift to a more restrictive stance than in the past could lead to severe imbalances between supply and demand of natural tropical hardwoods.

There are also *new potential sources of supply*. The source with the largest potential, as well as the greatest uncertainty, is tropical and sub-tropical plantations usually of fast growing species oriented to export markets. Very high rates of increment may be achieved, but there are enormous silvicultural, social and economic (e.g. infrastructure) problems. Some recent success stories indicate that these can be overcome. If there is strong demand for forest products on international markets, the economic possibilities to overcome the problems will be greater.

Two countries, *Chile* and *New Zealand*, have developed forest resources based largely on Radiata pine, and have set ambitious, but apparently realistic, goals for increasing exports, notably of logs, sawnwood and pulp, to world markets. Chile expects removals of 25-40 million m³ in 2000 and New Zealand about 20 million m³.

The outlook for imports by non-European countries is also briefly reviewed, as it could affect availability of forest products for Europe. The *Japanese* authorities expect a small increase in imports by the mid-1990s and a larger increase in domestic supply. It is possible however that the increase in imports could be higher than forecast, as previous targets for increased domestic supply have not been met.

There is very great uncertainty about future levels of consumption in *developing countries*, which depend on these countries' success in carrying out development programmes, and, *a fortiori*, about future levels of imports which will also be influenced by domestic removals and availability of funds for imports. Import demand from developing countries would, in certain circumstances, be very strong. There is particular uncertainty over the outlook for China.

Given the supply potential of traditional and new exporters, this rough assessment does not appear to confirm the opinion that a world-wide shortage of forest products is imminent, as exporters appear to have the potential to satisfy even high import demand (if developing country imports are not at the top of the range).

The possibility of world-wide tension between forest products supply and demand would arise however if a strong effective import demand materialized from developing countries.

In short, the world forest products markets seem to have the potential to be quite flexible in the medium- to long-term and to adapt even to significantly changed situations. Whether this potential flexibility is realized depends on many economic and institutional factors, including the speed with which signals of impending change are understood and acted on.

S.7 WOOD AND ENERGY

Two chapters of the study are devoted to the outlook for wood and energy, because of the fundamental changes in this sector which have taken place since the mid-1970s.

Around 1978, under the influence of rising energy prices and increased interest in renewable sources of energy, fuelwood consumption, which had been declining since the 1950s, reversed its trend and started to rise again. However recorded removals of fuelwood are only one part of the picture. A special enquiry revealed that around 1980 the following volumes of wood were used as a source of energy:

	Million m³
Fuelwood	72
Wood and bark residues of forest industries	40
Forest products used for energy after their original use	11
Estimated wood equivalent of pulping liquors burnt in chemical pulping....................................	44
TOTAL	165

This implies that over 40% of the volume of wood and bark removed is used as a source of energy, and that the provision of energy remains the single most important use of wood, in volume terms.

Energy wood is mostly burnt in households (about 60% of the total) and in the forest industries (under 30%) and accounted for about 2% of European energy consumption, but for a significantly higher share in a few countries (Turkey, Finland, Yugoslavia, Sweden, Portugal).

Future levels of consumption of wood for energy will be determined by a wide range of factors, including notably the overall supply/demand balance for energy, the organization of energy wood markets, the development of simple and reliable wood-burning equipment, harvesting costs and the results of research undertaken on new forms of wood-derived fuels (e.g. synthetic liquid or gaseous fuels) and on energy plantations.

National correspondents, in response to an enquiry, estimated that the consumption of wood for energy would increase by 2.0-2.5% per year, almost exclusively in the conventional uses (households, forest industries, medium-size consumers such as district heating units). Little increase was expected in the new forms of energy.

Demand for energy wood is not expected to have significant adverse effects on availability of wood raw material for the forest industries, although there may be some price effects for low quality pulpwood.

S.8   SYNTHESIS AND CONCLUSIONS

S.8.1   The consistency analysis model

Earlier chapters of this study dealt separately with the trends and prospects for supply, notably removals, and for consumption of forest products. The final part (chapters 20-22) compares the forecasts for consumption and supply and draws conclusions from the results.

A model has been designed to check the consistency of the forecasts at the level of individual countries. The basic principle of the model is that from forecasts for consumption, "derived removals" are calculated. These are the removals necessary to satisfy the forecast consumption after consideration of trade, yields, residue transfer and recycling of waste paper. The derived removals are then compared with national, independently prepared, forecasts for removals. Conclusions can be drawn from the difference between the two, the "gap". It should be noted that, for 1979-81, data are adjusted to remove distortions due, for instance, to faulty conversion factors, so that the gap is zero, i.e. derived and forecast removals are equal.

The procedure can be summarized graphically as follows (residues and waste paper omitted for simplicity):

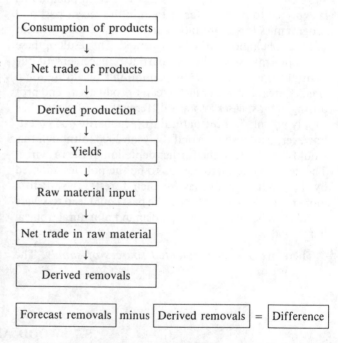

A more detailed figure appears in chapter 20 (fig. 20.1)

A special difficulty in constructing the balances is that it is almost impossible to make reasonable assumptions about future trade. In order to overcome this difficulty the level of foreign trade has been "frozen" in this study by the theoretical construction of keeping imports and exports constant at the level of 1979-81. This implies that the calculated gap needs a rather special interpretation which will be commented on below.

This approach is not intended to provide a forecast of the future situation of the forest and forest products sector, but as a tool to analyse the interrelationship of the forecasts obtained in other parts of the study.

S.8.2 "The gap" and its interpretation

What is the significance of the "difference" or "gap" calculated by this model? Clearly, there will be no "gap" in the future as consumption will exactly equal supply. The "difference" should rather be seen as an indication of the degree of consistency between the forecasts made in other chapters. No attempt has been made to reduce the differences by adjusting the forecasts. The differences, along with the secretariat comments on them, are intended as the starting point for a discussion at the national level of the implications of these forecasts for the outlook for the national forest policy.

The "difference" could be adjusted by changing the scenarios for one or more of the following:
- Consumption;
- Removals;
- Trade;
- Conversion factors (yields from raw material);
- Recovery rates for wood residues and waste paper.

The factor to be adjusted will depend on national circumstances.

Whereas quantitative forecasts are available for three of the above factors (consumption, removals and recovery rates), the situation is different for trade and conversion factors. For both of these, it was not possible to provide scenarios for future developments (see chapters 15 and 17). It was therefore necessary for the consistency analysis model to assume that they will remain at the level of 1979-81. In no way should this procedure be misunderstood as indicating that the authors of ETTS IV believe that trade and conversion factors will not change. These assumptions are merely the first step in an analytical procedure, and the outlook for trade and raw material yields is discussed at greater length below.

S.8.3 Consistency analysis of the outlook

The "differences" are presented and analysed by country in section 20.3. Their significance varies widely according to national circumstances and cannot be summarized here.

Is it possible to make any general statements, applicable to Europe, rather than to individual countries? In particular, can anything be said on the basis of this consistency analysis about the outlook for the supply/demand balance as a whole?

The difference between derived and forecast removals for Europe as a whole in table 20.3 is quite small for 1990

(a difference of about 6 million m³). No difficulty is foreseen in adjusting this difference, especially as the negative difference for Eastern Europe (– 4.7 million m³), arises from internal circumstances and will certainly not result in significantly increased imports to this region. It appears therefore that, for the period around 1990, supply and demand may be roughly in balance, without increased imports from other regions. If however extra-European suppliers were able to increase their exports to Europe, by successful marketing or competitive prices, there might even be some over-supply. In any case, fairly competitive market conditions could be expected.

For the period around 2000, a rather different situation can be deduced from the consistency analysis. Demand (converted into derived removals) is expected to be between 39 and 60 million m³ higher than supply (forecast removals). This should not however be interpreted as a forecast of a "shortage" of wood on international markets, for the following reasons:

- A number of countries, in Eastern and Southern Europe, are unlikely to increase their imports significantly, notably because of currency restraints. Adjustments in these countries, which account for 9-16 million m³ of the negative difference, will probably not affect the *international* supply/demand balance;

- It will probably be possible to improve raw material yields, notably through an increasing share of "high-yield" pulps in the paper furnish. This might reduce the "gap" by about 10 million m³, assuming an average 5% improvement of yields.

These factors together could reduce the "gap" at the European level by 20-25 million m³. It does not appear realistic to propose a higher rate of residue transfer or waste paper recycling, as the high assumptions for these seem to represent a maximum. The remaining gap, of 20-35 million m³, could be adjusted by one or more of the following:

- Higher imports from outside Europe;

- Lower consumption of forest products (including fuelwood);

- Higher European removals.

The scenarios for consumption and removals were based on analysis of trends and potentials, discussed at length elsewhere in the study. The scenario for trade however – that it remained constant at the 1979-81 level – was merely an assumption made to further the preparation of the consistency analysis model and is clearly *not* a likely future development. It appears reasonable, therefore, to assume that most of the remaining "gap" will be filled by increases in net imports. Thus, if the scenarios of the rest of this study, notably those for consumption of forest products and for European removals, are accepted, Europe's imports from other regions may be expected to increase by 20-35 million m³ over their 1979-81 level, or Europe's exports to other regions to decline by a similar amount, or a combination of these two developments.

An increase in imports from other regions does not imply a tight supply/demand balance or "shortages", unless potential exporting regions have difficulty in supplying the quantity required. Will potential exporters to Europe be able to expand their exports by 20-35 million m³? The (non-quantitative) analysis in chapter 17 would seem to indicate that suppliers outside Europe would be physically able to supply this volume. It is not possible however, with the analytical tools used in the study, to say whether any economic scarcity, leading to higher prices, is likely or whether the low costs of new producers like Brazil, Chile or New Zealand will exert a downward pressure on prices.

These conclusions are rather different from those in earlier timber trends studies, which stressed the need to develop the European forest resource to avoid possible future supply problems. This change in the outlook is at least partly due to the fact that decision-makers in the field of forestry have accepted the results of earlier studies. Consequently they "believed in a future for wood", and have continued the process of intensifying forest management and of creating new forests in Europe.

S.8.4   Outlook for the twenty-first century

The removals forecasts for 2020 prepared by national experts are compared in chapter 21 with some rough estimates of consumption. For a period so far into the future, no precise comparisons are possible.

There are however two main conclusions:

(*a*) If *per caput* consumption in the 2020s were to remain at the level projected for 2000, forecast European removals would be able to satisfy the increased demand with no increase in net imports over the levels forecast for 2000;

(*b*) On the other hand, consumption could be either significantly higher or lower than the conservative basic assumptions proposed, so neither of the extreme scenarios can be entirely ruled out.

It is also stressed that the forest sector itself has a major role to play in determining the level of consumption of forest products in the future.

S.8.5   Conclusions

The final chapter identifies a number of broad conclusions which may be drawn from the study, in particular a number of areas where there is a need for improved data and analysis:

– The behaviour and management objectives of small scale forest owners;

– The situation and outlook with respect to forest damage;

– Structure and capacity of the sawmilling and wood-based panels industries;

– The economic health of the forest industries;

– The outlook for construction;

– Factors affecting end-uses for sawnwood and panels;

– Possible effects of new electronic media;

– Outlook for the supplying and importing regions outside Europe;

– Global energy trends and their consequences for wood-derived energy.

It will also be necessary to monitor trends on a continuing basis and to compare them with the outlook forecast by ETTS IV.

In addition, a few major policy questions arising out of the study are also identified, concerning the following:

– Consequences of the increasing demand for non-wood goods and services (especially recreation);

– Responses in forest policy to the changes in the supply/demand balance forecast by ETTS IV;

– Problems connected with the fragmentation of ownership and management of forest and some forest industries, notably sawmilling;

– Possible consequences for forestry of changes in the agricultural sector;

– The possible need for strategies to revitalize the forest industries;

– Means to improve the competitivity of forest products in their markets;

– The desirability, or otherwise, of diversifying the outlets for the products of the forest sector;

– Is there a need for changed wood energy policies?

– Possibilities of optimising the organization of the "wood chain" (*filière bois*) from the tree to the final product.

Attention is drawn to the problem of forest damage, whether from forest fires or attributed to air pollution and to the challenge of forest conservation and protection facing European society.

Finally attention is drawn again to the uncertainty of the outlook for the twenty-first century, and to the unavoidable necessity of taking forestry decisions on the basis of imperfect information. Forestry will remain, as always, an expression of confidence in the future.

CHAPTER 1

The economic, demographic and social environment to the year 2000

1.1 INTRODUCTION

This study is concerned with the outlook for the forest and forest products sector. Yet developments outside the sector will undoubtedly have a determining effect on the sector − developments which in most cases the sector cannot influence in any way. The authors of this study are not, of course, competent to prepare a global outlook study, an ambitious but controversial project attempted by, among others, the Club of Rome, the OECD and the Administration of the USA. Nor is a complete, coherent global outlook study necessary in order to study the outlook for the forest and forest products sector, as many global aspects are not at all relevant to the sector.

It is however necessary to make a certain number of assumptions about future developments for those features which do directly affect the forest and forest products sector, for instance the rate of economic growth, the size and distribution of population and the level and type of construction activity, as well as some social aspects, such as the use of leisure time, which affects the recreational functions of the forest.

This chapter will present the assumptions regarding the economic, demographic and social environment which the secretariat has used as an input to the analysis in other chapters. On questions of such complexity, no certainty is possible. Estimates from official international organizations have been used whenever possible, but where necessary the secretariat has made its own estimates. Frequently a range of possibilities is presented, which it is hoped will include all likely developments.

If readers find themselves in fundamental disagreement with the assumptions presented in this chapter, it is clear that they will not be able to accept the conclusions reached in the rest of the study. In some cases, notably the demand modelling, the analysis is presented in such a way that if other assumptions are preferred, the results can be recalculated to produce modified projections. In other cases, however, notably where the analysis is not quantitative, the outlook is less easy to modify.

Finally, it is assumed that no natural or man-made disaster will occur of such proportions that the whole economic and social environment is modified. It is not possible to analyse in any meaningful way the consequences for the forest and forest sector of such a catastrophe. (The problem of the consequences of air pollution damage to forest which, although extremely serious, is of a rather different nature is, however, treated explicitly in chapter 6.)

1.2 DEMOGRAPHIC TRENDS

The data in this section are taken from *World Population Prospects as Assessed in 1982*, prepared by the Population Division of the UN Secretariat, issued in 1985, and its predecessor, *Demographic Indicators of Countries: Estimates and Projections as assessed in 1980*, UN, New York, 1982.

Table 1.1 and annex table 1.1 show that European population growth has been slow and is expected to continue so. Indeed population growth in Europe is slower than in any other region of the world. In a few countries, fertility rates have fallen below replacement levels, but the "medium variant" of the UN projections foresees a slow rise for Europe as a whole, from 413 million in 1950, to 530 million in 1980, to 583 million in 2000 and to 628 million in 2025, with an average annual rate of increase falling from 1.0% in the period 1955-1965 to 0.4% in 2000-2025.

TABLE 1.1

Europe: population trends and prospects, 1950 to 2025

	1950	1965	1980	2000 [a]	2025 [a]
Total population (millions)					
Nordic countries	14.4	16.2	17.4	17.6	16.9
EEC(9)	215.4	244.0	260.3	265.9	260.4
Central Europe	11.6	13.2	13.9	13.4	12.2
Southern Europe	83.1	103.8	128.5	165.4	207.2
Eastern Europe	88.5	100.1	109.8	121.1	131.2
Europe	413.0	477.3	529.9	583.4	627.9
Average annual percentage change over previous period					
Nordic countries	..	+ 0.8	+ 0.5	+ 0.1	− 0.2
EEC(9)	..	+ 0.8	+ 0.4	+ 0.1	− 0.1
Central Europe	..	+ 0.9	+ 0.3	− 0.2	− 0.5
Southern Europe	..	+ 1.5	+ 1.4	+ 1.3	+ 1.1
Eastern Europe	..	+ 0.8	+ 0.6	+ 0.5	+ 0.4
Europe	..	+ 1.0	+ 0.7	+ 0.5	+ 0.4

[a] Projections are the UN medium variant.

Southern Europe shows the fastest rates of population increase, followed by Eastern Europe. In the Nordic countries, the EEC and Central Europe slow growth before 2000 is expected to give way to declines in the first decades of the next century. On the country level the population growth foreseen for Turkey is particularly dramatic, passing from 20.8 million in 1950 (the sixth largest in Europe) to 99.3 million in 2025, by far the largest population in Europe. (The next largest, according to the projections, will be France, with 58.5 million.)

The population of Europe is aging. Over the 75 years from 1950 to 2025, the median age of the population of Europe (without Cyprus, Israel and Turkey, as country groupings used for population studies differ from those in ETTS IV) is expected to rise from 30.5 years to 40.4 years. As a consequence, the size and characteristics of the dependent population (children and old people, who must be supported by the working population) will change. The fluctuation foreseen in the total age dependency ratio (see table 1.2 for data and definitions) is the resultant of two clear trends. The age dependency ratio for children has been falling and is expected to continue to fall until 2000 and that for old people has been rising steadily and will rise until at least 2020. This will have consequences for planners of the economy as a whole, notably for the social services. For the forest products sector, this trend might be of interest in as much as it concerns demand for construction (e.g. fewer schools and more specialized housing for old people).

Another strong trend evident from table 1.2 is the steadily growing proportion of the population which lives in towns. This percentage was already 70.5% in 1980 and

TABLE 1.2

**Europe:[a] population trends and prospects,
by age, and by area of residence (rural/urban)**

	1950	1980	2000	2025
Median age [b]	30.5	33.0	37.3	40.4
Age dependency ratio (%)[c]				
Under 15 years	38.5	34.4	29.1	29.3
65 years and over	13.2	20.1	22.0	28.7
Total	51.7	54.5	51.1	58.1
Percentage of urban population	55.4	70.5	78.4	85.6
Average rate of annual change:				
Urban	+ 1.6	+ 1.1	+ 0.7	+ 0.3
Rural	− 0.3	− 1.1	− 1.3	− 1.6

[a] In this table, Europe is as defined by the Population Division i.e. *without* Cyprus, Israel and Turkey.

[b] The age which divides the population into two groups of equal size, one of which is younger and the other of which is older than this median.

[c] The ratio of the dependent population, defined as children under 15 years and/or people 65 years and over, to the population of intermediate age (i.e. 15-64), in per cent.

is expected to reach 85.6% by 2025, which must be quite near a practical maximum. It is expected that most of the internal migration will be directed towards medium or small towns in the suburbs rather than the largest urban centres, many of which are actually losing population. Rural depopulation might have negative consequences for labour availability for forestry and the forest industries, both of which are essentially rural in nature.

Within the general context of the move from rural to urban living, there may be moves from the larger towns to small country towns and villages, especially in industrialized societies. Second-home ownership which mostly implies a "country cottage" may also mean that part of the urban population lives in rural areas for part of the year and regards it as its holiday, i.e. recreational, area. Pressures on the forest could, therefore, increase in the vicinity of large towns and in favoured holiday areas (which may be distant from towns).

1.3 ECONOMIC GROWTH

From the early 1950s to the early 1970s the world experienced sustained economic growth at hitherto unheard of rates, which profoundly changed not only nearly all national economies but the organization of society as a whole. This economic growth, with the social and technological changes which made it possible and resulted from it, changed standards of living, social structures, life styles, and expectations of the future. Indeed, in the long-term historical perspective, the growth during the quarter century 1950 to 1975 is probably the exception and not the rule for economic change.

From 1973 on, a change in the structure of the world economy appears to have taken place. At first this change was only perceived as a rather pronounced downswing in the business cycle, caused essentially by the political and economic events in 1973 known as the "oil shock". At the same time there was increased concern about pollution of the environment and limits to natural resources. Some analysts constructed scenarios for economic and physical catastrophe, or at least for zero growth (in favourable or unfavourable circumstances) in perpetuity. The first report of the Club of Rome *The Limits to Growth* played a special role in this movement of thought and came to symbolize its major concerns.

The late 1970s and early 1980s have seen intense investigations of the causes and characteristics of the structural changes which are under way or have taken place. Among the subjects examined are the following:

– Growth in total Gross Domestic Product (GDP). Since the mid-1970s there appear to have been stronger cyclical fluctuations around a lower growth rate. Most market economies seem to move more in parallel, rather than counteracting each other's movements;

– Investment levels. Gross fixed capital formation seems to be falling as a proportion of GDP (in western Europe at least);

– At one stage, permanent high inflation rates, with consequent economic distortions and changes in expectations were feared. Recently, however, inflation rates have fallen to more "normal" levels in most countries;

– "Real" interest rates (i.e. nominal rates minus inflation) have been at high levels since the early 1980s;

– Unemployment in nearly all market economy countries is at very high levels. Some analysts fear that the level of structural unemployment has risen in recent years and that the economies will not be able to regain the lower unemployment levels of the 1960s;

– The growth in labour productivity may be slowing down;

– Since the collapse of the Bretton Woods system, international financial arrangements have been in a state of flux, characterized notably by large and fast changes in exchange rates;

– Concern has been expressed about the future availability of international liquidity to finance expansion.

There appears, as yet, to be no consensus among economists as to the fundamental nature of these structural changes and the direction of likely future developments. There does appear to be agreement, however, that growth rates will not return to the high 1960s levels ("Such an eventuality appears well nigh impossible", according to Interfutures) nor will there be a prolonged period of complete stagnation ("zero growth"), as Governments would find such a situation untenable due to the public discontent and increased unemployment which would inevitably follow.

For the purposes of this study, the quantitative GDP scenarios used are those prepared by FAO for use in its outlook work, based on consultations with the World Bank, OECD and a number of other international organizations. These scenarios or "projection assumptions" are described by FAO as follows:

All GDP series are expressed in US dollars (at 1980 prices and exchange rates). However, official exchange rates do not necessarily represent

TABLE 1.3

GDP growth rates and scenarios by country group

(*Average annual percentage growth*)

	Real		Scenarios 1980-2000	
	1965-1973	*1973-1980*	*Low*	*High*
Nordic countries	3.7	3.3	2.8	3.6
EEC(9)	4.5	2.2	2.4	2.9
Central Europe	4.4	1.5	2.2	3.0
Southern Europe	6.5	3.9	3.5	4.3
Eastern Europe	6.5	4.5	3.7	5.1
Europe	4.7	2.6	2.6	3.3

Note: For absolute values and country detail see annex tables 1.2 and 1.3.

the relationship between price levels in different countries. It is widely appreciated that the exchange rate conversion of GDPs of different countries to a common currency such as the US dollar do not yield a reliable basis for international comparisons.

Provisional rates of growth of total GDP were selected for five yearly periods up to the year 2000. Two alternative assumptions as to GDP growth in real terms were selected for each country, based on projections prepared by the UN for the strategy of the Third Development Decade, the UN Economic Commission for Europe, the World Bank, national development plans and various unofficial sources.

(*a*) A *"low"* or *"adjusted trend"* scenario broadly in line with historical trends, due account being taken of the most recent changes affecting these trends. Under this assumption, the developing countries taken as a group would gradually recover from the low growth experienced in 1980-85 so that growth in the latter half of the 1990s would be similar to the rates achieved in the 1970s. The developed countries as a group would show a similar pattern of recovery under this scenario.

(*b*) A *"high"* or *"normative"* scenario which would involve progressive achievement of accelerated economic growth in the developing countries reflecting the fulfilment of national development plans and targets. For developed countries, growth under this scenario would be more rapid than in the 1970s but would fall short of the rates achieved in the 1960s.

Both of these assumptions would imply that growth in developing countries would in general be more rapid than in the developed countries. However, in view of the present widely expressed uncertainties as to the medium- and long-term prospects of the world economy, the two assumptions should be mainly interpreted as the limits of a range of likely economic growth possibilities.

It should be noted that the national accounts systems of the centrally planned economies of Eastern Europe are based on different principles than those of market economies. National income in Eastern Europe is measured by Net Material Product (NMP) rather than Gross Domestic Product (GDP). There are also technical problems in converting NMP data in national currencies to US dollars. The data on trends and projections in GDP for east European countries may not therefore be fully comparable with those for other countries.

The scenarios for Europe are summarized in table 1.3, which shows that the "low" scenario assumes average annual growth rates from 1980 to 2000 quite close to the low rates of the middle and late 1970s – 2.6% for Europe as a whole. The "high" scenario is still below the 1960s/early 1970s rates – 3.3% for Europe, as compared to 4.7%. As in the past, higher growth rates than elsewhere are foreseen for Southern and Eastern Europe.

In terms of GDP *per caput* the growth rates are almost identical to those for total GDP as only minimal population growth is foreseen.

1.4   CONSTRUCTION

Construction is the major end-use sector for sawnwood and wood-based panels (see chapter 8) and changes in the level and type of construction activity strongly influence consumption of these products. This section discusses the outlook for construction, and proposes quantified scenarios for residential investment, chosen as an independent variable in the "end-use elasticities" model presented in chapter 11. This discussion is partly based on the ECE publication "Human Settlements – Key Factor in Economic and Social Development", *Economic Bulletin for Europe*, vol. 35, No. 1, March 1983, published by Pergamon Press for the United Nations, notably chapters 3 and 9.

FIGURE 1.1

Past and projected levels of population
(*UN projections, medium variant*)

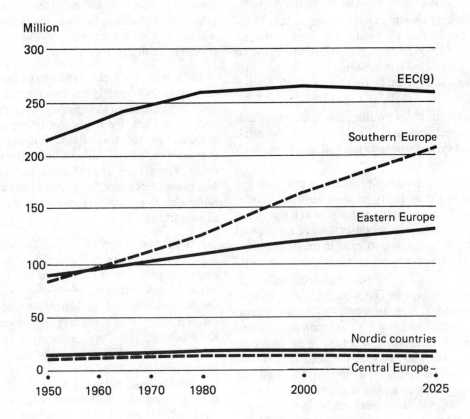

FIGURE 1.2

Europe: past and projected levels of GDP
(*FAO projections*)

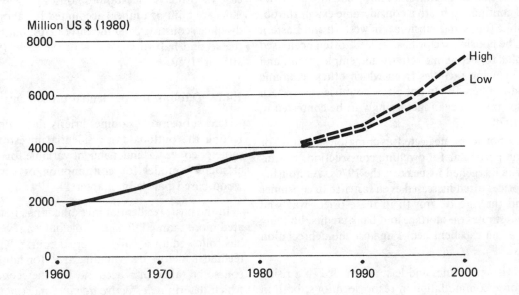

1.4.1 General trends

Most of Europe has seen strong building activity since the 1950s, reflecting first the reconstruction needs and subsequently the effects of expanding economic growth with rising incomes and also, in turn, increasing leisure time.

This is particularly evident in the housing sector. European housing production rose gradually from some 3.5 new dwellings per 1,000 inhabitants in the early 1950s, to a peak level of 8.6 per 1,000 inhabitants in 1973. In several countries, it was as high as 11 to 12 new dwellings per 1,000 inhabitants.

In a number of countries housing stock and standards had by then reached a relatively high level, and the social, institutional and other infrastructure had also been much developed.

From then on new housing production has shown a declining trend in the majority of west European countries but has continued generally to grow in east European countries, albeit at relatively modest rates. In some cases the floor area of dwellings, in particular in one-family houses, also declined, in order to counter rising building costs.

The overall decline in new housing construction was to some extent compensated by the growing importance of renovation and modernization work, for which several reasons can be discerned. Firstly, renovation and modernization of existing buildings had tended to be neglected in favour of new building, so that the need for renovation accumulated. Secondly, new designs and building methods, as well as building quality did not always meet expected standards over the longer term. A further reason was renewed interest in, and appreciation of, the building design and style of earlier periods.

This strong building activity was the result of a number of factors. First, there was a trend to increase housing space *per caput* notably on account of rising incomes – although views differ as to the degree of elasticity of demand for housing. Second, a demand for housing was generated by the conversion of city centres from being primarily housing areas to primarily office areas and by the extensive labour migration. Third, housing activities were, and continue to be, to a considerable extent the object of public funds and subsidies, in western and Eastern Europe. The housing sector has, indeed, often been used as a stimulus for economic activity and employment, and enjoyed financing facilities from which other economic sectors did not benefit. In not a few countries in western Europe this policy began increasingly to be contested in the later 1970s.

A factor not to be neglected as a major stimulus for building, in particular for dwelling construction, was inflation, which reached its peak in the 1970s. As housing and land prices often increased even faster than consumer prices, and the rate of growth in these prices was well above interest rates on mortage and bank credits, building appeared as an excellent hedge against the depreciation of savings.

Finally, rising incomes and leisure time led to a rising demand for accommodation in resort locations, both in the form of hotels and of secondary and seasonal residences. The extent of the latter, is often not well known statistically and therefore difficult to assess accurately but is clearly large. It should be noted that statistical coverage of housing (and other construction) may in general not always be complete, and may thus underestimate the true extent of production, for instance due to clandestine work which has shown a markedly rising trend in many countries, especially in recent years.

The qualitative aspects of housing (and building in general) were increasingly reviewed in the light of new factors, in particular a sharpening environmental awareness and, after 1973, the rising cost of energy which not only affected directly costs related to living comfort, but also indirectly the cost of building and other materials and products. The effects of both these factors on demand for, and the utilization of forest products has already become noticeable in several regards and is likely to grow further in importance.

From the point of view of forest products, a particularly important development was the increasing use of the timber frame building method for low-rise residential construction in several countries (in some countries, notably the Netherlands and the United Kingdom, often with an exterior brick wall cladding). In the case of the United Kingdom, timber frame construction increased quite markedly in the 1970s from below 20,000 completions per year to over 40,000, or between 15% and 20% of all housing completions, although a sharp reduction, to around 12% was recorded in 1984 following adverse media comment, which was strongly contested. In Belgium, timber frame construction accounted in 1969 for 2% of all one-family houses built in that year, but for 14% in 1980.

It is interesting to note that also in countries with an historical tradition of timber frame construction (e.g. Finland, Norway and Sweden) this building type showed a rising trend after it had tended to decline in the 1960s. This movement was partly a result of the reversal of the trend to large-scale housing developments. Thus, in the case of Sweden, one-family houses (of which well over 90% are built of timber) accounted for about 30% of all dwellings constructed in the early 1970s, but for 70% from the second half of the 1970s to the early 1980s and for 50% in 1983.

1.4.2 Outlook for residential investment

This subsection examines briefly the major factors affecting the outlook for residential investment, the parameter chosen as independent variable for the "end-use elasticity" model for consumption of sawnwood and wood-based panels in chapter 11.

In the past, residential investment has not only fluctuated more than GDP as a whole but, in several countries has followed a radically different course. The social and technical factors determining *need* for housing must of course be taken into account, but the economic factors which determine *effective demand* must not be ignored.

Some of the social or technical factor are:

(a) The state of the housing stock. In this context, it should be noted that in most ECE countries there is a crude surplus of dwellings over households, and little substandard housing remains (although the definition of "substandard" changes with time and with expectations as regards quality);

(b) Plans for new towns and similar enterprises have been undertaken in the past for a variety of social and economic reasons. The present tendency in human settlements planning is away from such large scale undertaking and towards more piecemeal adaptation and upgrading of neighbourhoods, for instance in decaying city centres;

(c) There are changes in the number and type of households. This is not merely a question of the number of people in a certain age group or even of the marriage rate: increasingly, young people wish to live away from their families before establishing permanent households, and specialized housing is necessary for the increasing number of older people who want to and can live alone. A rising divorce rate also increases demand for separate dwellings. These growing desires for housing are not, however, in many cases, imperative needs: for instance, if no suitable housing is available at a reasonable price, young people stay longer with their parents;

(d) There is probably an upper limit to the floor space which each person can reasonably use. There is however very great latitude for improving the quality of housing, even to standards which appear very high today;

(e) Around most large population centres, there is strong competition for all available land between different potential uses including housing, industry, communications, infrastructures, agriculture, sports and recreation, etc. This is likely to encourage fairly compact housing and discourage "urban sprawl". There is also a growing preference among many people for older houses, which are seen as having more "character", even if considerable amounts of money have to be invested on their renovation;

(f) There may be a preference for single family housing, if land and finance are available. This trend may be in contradiction with that described under (e).

In addition, there is the question of choice in the expenditure of household income: here housing competes with other types of expenditure (e.g. for recreation, or consumer goods) and saving. Priorities in this respect may change over time.

The economic factor influencing residential investment most frequently mentioned is family income, but this cannot be seen separately from housing prices or rents (affected by both building costs and land prices) and real interest rates which have risen very sharply since the late 1970s. For many years, low or even negative real interest rates and inflationary expectations made an investment in a house, paid for by a long term mortgage a very rational decision in economic terms. In recent years this situation has changed as real interest rates have risen sharply and inflationary expectations fallen.

Finally demand for new housing, publicly or privately owned, is strongly affected by the decisions of governments, including local and regional authorities, which traditionally have provided a wide range of incentives for housebuilding, and in many countries have themselves accounted for a significant part of new housing construction. Some of these incentives have been part of a long term programme aimed at providing decent housing for all but others have been closely linked with management of the economy (e.g. counter-cyclical investment) or with more political objectives. It may be however that, in recent years, the perception that most people are decently housed, along with increasing concern about budget deficits and fears that funds used to finance housing are thereby diverted from investment in industry, have led governments to attach lower priority to housing subsidies and programmes.

It is possible to model all these factors but, in the words of the above mentioned ECE publication "housing market models are ambitious experiments. They require, in order to function properly, a great deal of data collected through sampling studies, as well as advanced technical and professional skill, which is not always available." In addition separate models may well be needed for different regions within countries. Such models are clearly not possible in the context of ETTS IV.

As an indication of relationships in the past between GDP, residential investment and manufacturing production (discussed in section 1.5 below), annex figure 1.1 shows indices (1964 = 100) for these in most western European countries between 1964 and 1981. The figures show that, whatever the trend in the 1960s and early 1970s, residential investment does not show any upward trend after the mid 1970s in the 15 countries for which complete series are available. Even in Norway, whose economy is strongly affected by oil revenues, residential investment appears to level off after 1978. In several countries, the last few years even show a downward trend. This would seem to indicate that, of the factors outlined above, those which tend to restrain growth in residential investment, such as higher prices and interest rates, have counterbalanced or even outweighed those which would encourage expansion, notably the increase in households and the desire for better quality housing.

The following two scenarios for future residential investment are therefore proposed:

(a) *Low scenario*

Real interest rates remain high. Land prices and building costs continue to rise faster than inflation. Little new building land is released. Governments, conscious of their budget deficits and considering their population is already decently housed, do not increase housing subsidies and even reduce them in some cases. Nevertheless, households still attach importance to improving the quality of their housing, even if this implies spending a greater percentage of their income on this and/or carrying out some or all of the work themselves (do-it-yourself). Residential investment is increasingly concentrated on maintenance and transformation of existing housing. *Residential investment stays constant at around the 1980 levels.*

(b) *High scenario*

Real interest rates return to levels equivalent to those of the 1960s and 1970s. Although costs still rise and land is fairly scarce, Governments attach priority to raising housing standards, and therefore maintain or even increase present levels of subsidy, and make special efforts to provide housing for special groups, notably the old and the young. They ensure sufficient land is available, although a considerable part of the investment goes to ambitious upgrading of older property. Nevertheless, given the weight of the constraining factors, and the realization that almost all people already have adequate housing, it does not appear possible that residential investment could grow at the same rate as GDP. *Residential investment grows at half the rate of GDP* (taking the high GDP scenario).

1.5 MANUFACTURING PRODUCTION

Manufacturing production is also one of the independent variables in the "end-use elasticity" model. It has traditionally been seen as the "motor" of the economy and for many years grew rather faster than GDP as a whole.

In recent years, however, services have grown faster than manufacturing and increased their share of both value added and employment.

The structural changes mentioned above have also brought into question the relation between developments for manufacturing production and for GDP. Some analysts consider the economies are moving to a "post industrial" era and that the rate of growth of manufacturing production will slow down or stop. A study for the Senior Economic Advisers to ECE Governments concluded in 1981 that "the slower growth of industrial production than of GDP since 1973 is linked with the special circumstances of the period − circumstances which may well continue into the 1980s and cannot yet be taken as evidence of a long-term trend in patterns of output or demand of anything like the magnitude of the shift in pattern shown by the figures for the few years after 1973. All the same, the net slowing down of what was and will still be, regarded as the "motor" of the economy calls for reflection among those who endeavour to peer into the future" (EC.AD(XVIII)/R.2, para. 30).

In the majority of western European countries between 1964 and 1981, manufacturing production rose at a rate very close to that of GDP, although with slightly stronger fluctuations. In Italy and the Netherlands, it rose rather faster (elasticity of about 1.15) and in Finland considerably faster (elasticity of 1.6 over the period, but of 1.3 for 1974 to 1981). In Norway and the United Kingdom, manufacturing production and GDP rose together until 1974 when manufacturing production stabilized while GDP continued to rise. This development is presumably due to the large-scale devotion of the national resources of these two countries to the oil sector with the coming on stream of the North Sea oil fields in the 1970s.

In almost all countries of Southern Europe however, manufacturing production rose considerably faster than GDP. This is consistent with the finding of the document cited above that "the share of manufacturing is negatively associated − among the industrial western countries − with the level of real income per head" (EC.AD (XVIII)/R.2, para. 27), i.e. that manufacturing production plays a proportionally larger role in those countries which are developing from the economic point of view, but that as incomes rise, the growth rates of GDP and manufacturing production will converge. The GDP elasticities for manufacturing production in Southern European countries for the period 1964 to 1981 were as follows:

Greece 1.56
Portugal 1.43
Spain . . . . 1.85 (1964 to 1977 : 2.24
                                              1977 to 1981 : negative)
Turkey. 1.09
Yugoslavia 1.60

From the above it would seem that the most useful concept in order to construct scenarios for manufacturing production is that of GDP elasticity (e.g. if GDP elasticity is 1, manufacturing production will rise at the same rate as GDP).

The following two scenarios are proposed:

(a) *Low scenario*

There is a limited degree of de-industrialization, so that manufacturing production does not rise quite as fast as GDP (in the low scenario) in western European countries. Elasticities remain lower than the average in Norway and the United Kingdom and higher than the average in Finland and Southern Europe. GDP elasticities diminish in these countries over the forecasting period, notably as income levels in Southern Europe get closer to the European average.

(b) *High scenario*

Manufacturing production expands in line with GDP (FAO high scenario), for western Europe including Norway and the United Kingdom. Elasticities for Southern Europe do drop, but not by as much as in the low scenario.

The elasticities used are set out in annex table 1.5.

The "end-use elasticities" model also requires explicit scenarios for production of furniture. It appeared difficult however to construct such scenarios in view of the cyclical and highly competitive nature of the industry and its regional, even local, character. It has therefore been assumed that furniture production in each country will grow at the same rate as manufacturing production as a whole, although it is recognized that such an assumption could be misleading, for instance if one country's furniture industry becomes uncompetitive and suffers serious losses of market share.

TABLE 1.4

Scenarios for GDP/manufacturing production elasticities

	Low scenario (elasticities relative to low GDP scenario)		High scenario (elasticities relative to high GDP scenario)	
	1980-1990	1990-2000	1980-1990	1990-2000
Finland	1.40	1.20	1.40	1.20
Norway	0.80	0.80	1.00	1.00
Sweden..........................	0.95	0.90	1.00	1.00
Belgium-Luxembourg	0.95	0.90	1.00	1.00
Denmark	0.95	0.90	1.00	1.00
France	0.95	0.90	1.00	1.00
Germany, Fed. Rep. of	0.95	0.90	1.00	1.00
Ireland	0.95	0.90	1.00	1.00
Italy	1.0	0.95	1.10	1.05
Netherlands.....................	1.0	0.95	1.10	1.05
United Kingdom..................	0.80	0.80	1.00	1.00
Austria.........................	0.95	0.90	1.00	1.00
Switzerland	0.95	0.90	1.00	1.00
Greece	1.45	1.30	1.50	1.40
Portugal........................	1.35	1.20	1.40	1.30
Spain	1.20	1.05	1.30	1.15
Turkey	1.10	1.10	1.20	1.20
Yugoslavia......................	1.50	1.35	1.60	1.40

1.6 OUTLOOK FOR AGRICULTURE

Agriculture and forestry are the two major users of rural land (44% and 35% of total land respectively). Developments in agriculture therefore may affect the availability of land for forestry.

The tendency throughout Europe has been for agricultural productivity to increase. There is nothing to indicate a general reversal of this trend although in particular regions the pace of change may slow down. Thus the use of chemical fertilizers may be reduced or abandoned for economical and ecological reasons or a combination of both and less intensive farming methods may be re-adopted, but overall the likelihood is that it will be possible to produce the same amount of food on a smaller area. Grazing in particular is likely to be more concentrated, not only because it is technically more efficient, but because of the difficulty of getting flockmasters (shepherds) and their families to continue to live and work in isolated conditions in remote areas.

Many governments have attached great importance not only to ensuring adequate food supplies, but to preserving the rural economy and the rural landscape. These policy goals have often been approached through price support mechanisms for many agricultural products. These have in many cases achieved their aim of increasing food supplies and supporting the rural economy, but at the cost of seriously distorting the agricultural economy and building up enormous and expensive agricultural surpluses, which have been the object of increasing criticism. Many governments, notably in the European Economic Community are now reassessing the fundamentals of their agricultural policies. It is not yet possible however to predict what changes could occur.

As a consequence, the use of marginal agricultural land may be seen as depending essentially on government policy, rather than on economic processes or food supply imperatives. It will therefore be government policy, taking into account not only agricultural questions but employment and social considerations, as well as land-use and budgetary aspects, which will determine whether marginal, and even some not so marginal, agricultural lands remain in their present use or are converted to other uses − possibly forestry.

1.7 SOCIAL DEVELOPMENTS

This section mentions briefly some social developments which may have an influence on the forest and forest products sector, notably on the demand for the non-wood benefits of the forest (chapter 4). In this area, the element of speculation is of necessity greater than in areas where an analysis in more quantitative terms is possible.

A few likely or possible developments are set out below. It should be pointed out that these are not universally applicable: there are very large regional and country differences.

If GDP continues to grow as foreseen in section 1.3, it is likely that personal disposable income will also grow — although it is possible that a larger part of the increase will go to investment or savings. Individuals will decide how this increase in their disposable income will be spent — on recreational activities (of what type?), on housing, on consumer goods, etc.

The amount of leisure time is likely to increase further. In addition to physical tasks, many of which have already been mechanized, computers and other equipment will further reduce manpower requirements for many labour-intensive service functions (e.g. accounting, banking, stock control). Job sharing and a shorter working week may be seen as palliatives to unemployment.

More widespread car ownership and improved road infrastructure has greatly increased the mobility of the population, making it possible for town dwellers to visit even remote rural areas quite easily. In several countries, car ownership must be approaching saturation levels, but the continuing rapid expansion of the travel business seems to indicate that those who are able to travel will continue to seek out new areas for recreation (for vacation or for shorter, one-day trips). Forest areas will be among the objects of this travel.

The public as a whole, and especially the urban population, has become more aware of issues concerned with forestry and, to a lesser extent, with the forest industries. Educational programmes for adults as well as children are now much more widely based and cover biological and ecological aspects of forests in a comprehensive and interesting way. The economic side of forestry is much less well covered. This is one of the causes of the far greater importance being attached by the public to the non-wood aspects of the forest — notably landscape, nature conservation and recreational values — than to its economic role. A higher degree of public awareness of forest issues is to be welcomed, provided that the professionals of the sector ensure that it is as well informed as possible. It is not at all impossible that, because of this increased public awareness, decisions on the forest and forest industries sector may come to be taken at higher, more political levels. Indeed, this is already apparent in a number of countries, particularly in connection with the damages resulting from air pollution but also with regard to policies on land use, recreation etc.

1.8 OVERVIEW

The preceding sections may be summarized as follows:

— Population is expected to grow slowly or stagnate in most countries, except in a few southern European countries;

— Economic growth is expected to be relatively slow, between 2.6% and 3.3%, with higher rates in Eastern and Southern Europe. Structural changes also appear to be taking place in the economies;

— There is great uncertainty about the future of construction, especially residential construction, with conflicting influences from different factors. Two scenarios were chosen: stability at the 1980 level and slow growth, although a decline cannot be ruled out;

— Manufacturing production is expected to grow more slowly than GDP in most countries as European economies move into the "post-industrial" era;

— Rural land-use patterns will be affected by changes in the agricultural sector where productivity is expected to continue to increase. Agricultural policies are being reviewed and the options chosen may affect forestry, which is one of the most plausible alternative uses for agricultural land.

Where possible, quantitative scenarios for the above aspects have been prepared which will be used as input in other parts of the study, notably the models of demand for forest products.

CHAPTER 2

Overview of the European forest resource and of factors affecting wood supply

2.1 INTRODUCTION

The present chapter serves as an introduction to chapters 3 to 6 which cover the European forest:

— Chapter 3 presents the statistical data on the European forest around 1980, drawing mostly on the assessment of the forests of the ECE region published in 1985; [1]

— Chapter 4 presents the situation and outlook for the non-wood benefits of the forest;

---

[1] *The Forest Resources of the ECE Region (Europe, the USSR, North America)*, ECE/TIM/27, Geneva 1985.

— Chapter 5 contains national forecasts of the outlook for the forest resource, especially as regards wood supply (removals);

— Chapter 6 deals with the problems raised by forest damage, especially that attributed to airborne pollution.

The chapter is divided into two parts:

(*a*) A brief overview of the European forest resource, which attempts to put it in its social, economic and environmental context;

(*b*) A review of the main factors affecting wood supply from the European forest.

2.2 OVERVIEW OF THE EUROPEAN FOREST RESOURCE IN ITS ECONOMIC AND SOCIAL CONTEXT

Although data on the forest resource are presented in ETTS IV, as in other studies, in harmonized form and comparable units (e.g. hectares or cubic metres), the European forest is not at all homogeneous. The following are all examples of parts of the European forest resource (the list is by no means complete):

— Slow growing, scattered trees near the tree-line near the Arctic circle;

— Intensively managed stands of pine and/or spruce in Scandinavia;

— Production plantations, newly established on bare land;

— Forests of 200 to 300-year-old oaks used essentially for recreation;

— Hardwood coppice, originally managed for fuelwood, now often degraded and unmanaged;

— Large uniform stands of fast-growing maritime pine;

— Mediterranean scrub ("maquis" or "matorral") classified as "other wooded land";

— Uneven aged, mixed species stands in mountainous regions, intended primarily for soil protection and prevention of erosion and avalanches;

— Eucalyptus or poplar, managed intensively on short rotations, usually for pulpwood;

— Middle altitude stands of spruce or fir interspersed with agriculture, in northern and Central Europe;

— Old oak stands managed on a 200-300 year rotation, for very high quality sawnwood and veneers;

— Plantations, often of pine, in Southern Europe, intended to combat erosion;

— Subalpine forests of mixed beech, spruce and fir;

— Forests whose layout and species composition are conceived primarily with game management in mind.

This incomplete list serves to demonstrate the extreme diversity of the "European forest", which any statistical presentation is bound to over-simplify and distort. This diversity affects the environment, notably the climate and site conditions, of a forest, the ownership, the objectives of the owner, the forest policy of the country concerned, the economic context (notably as regards management costs and the demand for wood), the legal and institutional framework and, last but not least, the history of the forest in question, especially earlier management choices and actions.

It is possible to characterize in a very crude and summary way the main forest regions of Europe as follows.

In the *Nordic countries*, especially Finland and Sweden, forests are the dominant ground cover; forestry, with the forest industries which depend on it, forms a central part of the economy, and makes a major contribution to the balance of payments. There is limited species diversity (spruce, pine, and birch are the major species) and growing conditions are often harsh leading to slow growth, but intense management and rationalization enable Finland and Sweden to remain competitive exporters on world markets. In Norway, the forest sector has become relatively less significant, economically, since the development of the oil industry and the consequent structural adjustments.

The *EEC(9)* account for 49% of Europe's population, but for only 27% of its land area and 18% of its forest area: the priorities and objectives of forest management are therefore very different from those in the Nordic countries. For many forests near population centres, the provision of recreational facilities is a major management

TABLE 2.1

Population, land and forest resources in Europe around 1980

	Population (millions)	Land (excl. water) (million ha)	Forest and OWL [b] (million ha)	Hectares per caput Land	Hectares per caput Forest and OWL [b]	Forest OWL [b] as percentage of land	Exploitable closed forest Growing stock/ha (m³ o.b./ha)	Exploitable closed forest NAI [c]/ha (m³ o.b./ha)
Finland	4.8	30.5	23.2	6.36	4.84	76.0	81	3.18
Iceland	0.2	10.0	0.1	50.00	0.50	1.0	..	..
Norway	4.1	30.8	8.7	7.50	2.12	28.3	87	2.62
Sweden	8.3	41.1	27.8	4.96	3.35	67.7	99	3.01
Nordic countries	17.4	112.4	59.9	6.46	3.44	53.2	90	3.03
Belgium	9.8	3.1	0.7	0.31	0.07	22.7	122	7.50
Denmark	5.1	4.2	0.5	0.83	0.09	11.4	115	8.51
France	53.8	54.3	15.1	1.01	0.28	27.8	116	4.05
Germany, Fed. Rep. of	61.7	24.3	7.2	0.39	0.12	29.6	155	5.63
Ireland	3.4	6.9	0.4	2.03	0.11	5.5	92	7.29
Italy	56.2	30.1	8.1	0.54	0.14	26.8	144*	3.07
Luxembourg	0.4	0.3	0.1	0.72	0.23	31.5	161	4.10
Netherlands	14.2	3.4	0.4	0.24	0.03	11.7	99	4.22
United Kingdom	55.7	24.1	2.2	0.43	0.04	9.0	101	5.56
EEC(9)	260.3	150.7	34.5	0.58	0.13	22.9	128	4.59
Austria	7.5	8.3	3.8	1.10	0.50	45.5	252	6.19
Switzerland	6.4	4.0	1.1	0.62	0.18	28.3	392	6.54
Central Europe	13.9	12.2	4.9	0.88	0.35	39.8	280	6.26
Albania	2.7	2.7	1.2	1.01	0.46	45.3	86	3.16
Cyprus	0.6	0.9	0.2	1.54	0.29	18.7	33	1.01
Greece	9.6	12.9	5.8	1.34	0.60	44.6	74	2.05
Israel	3.9	2.0	0.1	0.52	0.03	5.4	44	3.21
Portugal	9.7	8.6	3.0	0.88	0.31	34.8	73	4.42
Spain	37.5	49.9	25.6	1.33	0.68	51.3	70	4.28
Turkey	44.5	77.1	20.2	1.73	0.45	26.2	96	2.89
Yugoslavia	22.3	25.6	10.5	1.15	0.47	41.0	128	3.27
Southern Europe	131.2[a]	179.8	65.1	1.37	0.50	37.4	95	3.43
Bulgaria	9.0	11.1	3.8	1.23	0.42	34.2	90	1.82
Czechoslovakia	15.3	12.6	4.6	0.82	0.30	36.4	221	5.38
German Dem. Rep.	16.7	10.6	3.0	0.63	0.18	27.3	170	5.79
Hungary	10.7	9.1	1.6	0.85	0.15	18.0	162	6.15
Poland	35.8	30.5	8.7	0.85	0.24	28.7	138	3.38
Romania	22.2	23.0	6.3	1.04	0.28	27.3	210	5.52
Eastern Europe	109.8	96.8	28.0	0.88	0.26	28.7	163	4.39
Europe	532.6[a]	551.9	193.8	1.04	0.36	35.1	120	3.80

[a] Includes Malta (population 0.4 million) which has no significant forest resource.

[b] OWL = other wooded land.

[c] NAI = net annual increment.

objective, or a least a significant constraint on management for wood production. Even away from population centres, recreation can be a major management objective (e.g. in holiday areas or where hunting is important). Elsewhere, however, forests, at least those owned by public bodies or large private owners, are intensively managed for wood production. Forests belonging to small-scale forest owners are however often badly managed or not managed at all (see section 2.3.3 below). Within the EEC(9) there is a very wide variety of growing conditions and forest types.

Forestry in *Central Europe* is shaped by the mountainous landscape which fixes the pre-eminent objective – maintenance of a satisfactory forest cover to prevent erosion and avalanches – and presents numerous problems and extra costs to the forest owner. High operating costs cause economic problems. In lower areas, however, forestry can be very competitive. Of the two countries, Austria is a major exporter on world markets, while Switzerland is mostly an importer, except for some logs.

With some major exceptions, (e.g. pine and eucalyptus in Portugal and Spain), many forests of *Southern Europe* have been degraded by a very long history of over-cutting (for fuelwood and industrial wood), by over-grazing and by forest fires, which destroy hundreds of thousands of hectares every year. There also exist millions of hectares of Mediterranean scrub and low quality woodland. Some countries (e.g. Spain, Turkey) have been carrying out long-term programmes to upgrade the forest resource and establish new plantations (principally aimed at erosion control) but much remains to be done. Most of the same problems (notably fire) face the forests of Italy and southern France, which are included in the EEC(9).

In *Eastern Europe*, most of the forest belongs to the State, which usually assigns the highest priority to wood production (except near urban areas, and in mountainous areas). The three northerly countries (Czechoslovakia, German Democratic Republic and Poland) have a predominantly coniferous resource, while the others (Bulgaria, Hungary, Romania) have a large proportion of non-coniferous species. Several countries in the region are engaged in long-term programmes to develop their forest resource, which was in some cases severely degraded.

Table 2.1 compares each country's forest area to both its total land area (excluding water) and to its population, as these are perhaps the most fundamental factors providing the context within which the goals of a national forest policy may be formulated.

The size of the forest resource ranges from 27.8 million ha in Sweden, to very small or negligible areas in Malta, Iceland and Israel. Six countries have more than 10 million ha of forest and other wooded land, three in Southern Europe (Turkey, Spain, Yugoslavia), two in the Nordic countries (Sweden, Finland) and one in the EEC(9) (France). The southern European countries' resource includes large areas of rather unproductive "other wooded land".

There are equally wide variations as regards the percentage of forest cover. Finland and Sweden have by far the highest share of forest and other wooded land (76% and 67% respectively) Spain has 51% and four other countries, of which three in Southern Europe, between 40% and 45%. Countries with a very low percentage of forest cover include, in addition to countries with small forest resources four EEC countries which have largely lost their natural forest cover over past centuries (Belgium, Ireland, Netherlands, United Kingdom).

Another indication of the relative importance of forests is the area of forest and other wooded land for each inhabitant. No less than seven EEC countries have less than 0.15 ha/caput because of the combination of relatively small forest areas and high population densities. In contrast, the forest area per head in Finland (4.84 ha) is over 150 times greater than in the Netherlands. Finland, Sweden and Norway all have more than 2 ha/caput of forest and other wooded land.

The nature of the different forests also varies widely. Tables 2.1 and 2.3 permit a summary comparison of two

TABLE 2.2

Variations between European countries in area of forest and other wooded land in absolute terms and relative to total land and population 1980

A. AREA OF FOREST AND OTHER WOODED LAND
(Europe 193.8 million ha)

10 million ha or more	*0.5 million ha or less*
Sweden (27.8)	Malta (–)
Spain (25.6)	Iceland (0.1)
Finland (23.2)	Israel (0.1)
Turkey (20.2)	Ireland (0.4)
France (15.1)	Netherlands (0.4)
Yugoslavia (10.5)	Denmark (0.5)

B. FOREST AND OTHER WOODED LAND AS PERCENTAGE OF TOTAL LAND
(Europe 35.1%)

Over 40%	*Under 20%*
Finland (76.0%)	Malta (–)
Sweden (67.7%)	Iceland (1.0%)
Spain (51.3%)	Israel (5.4%)
Austria (45.5%)	Ireland (5.5%)
Albania (45.3%)	United Kingdom (9.0%)
Greece (44.6%)	Denmark (11.4%)
Yugoslavia (41.0%)	Netherlands (11.7%)
	Hungary (18.0%)
	Cyprus (18.7%)

C. FOREST AND OTHER WOODED LAND PER INHABITANT
(Europe 0.36 ha/cap)

0.5 ha/cap or more	*0.15 ha/cap or less*
Finland (4.84)	Malta (–)
Sweden (3.35)	Israel (0.03)
Norway (2.12)	Netherlands (0.03)
Spain (0.68)	United Kingdom (0.04)
Greece (0.60)	Belgium (0.07)
Austria (0.50)	Denmark (0.09)
Iceland (0.50)	Ireland (0.11)
	Fed. Rep. of Germany (0.12)
	Italy (0.14)
	Hungary (0.15)

of the main parameters, growing stock and net annual increment, compared to the area of exploitable closed forest. It should be borne in mind that these data are national averages which themselves conceal a wide variation within countries.

Seven countries have exploitable closed forests with an average growing stock of more than 150 m³ o.b./ha. These countries form a contiguous group in central/northern continental Europe, which might be said to share a common "forestry culture" laying emphasis on long rotations and the production of high quality large-sized wood. Switzerland is an extreme case with growing stock per hectare more than three times the European average. Whereas high levels of growing stock per hectare often result from a silvicultural choice, low levels are often the result of natural constraints, e.g. low rainfall and infertile soil in the five southern European countries and cold climate with a short growing season in Finland and Norway. In some countries a high proportion of young stands also contributes to the low figure of growing stock per hectare.

There are also wide differences in species structure. The proportion of coniferous species in total growing stock exceeds 75% in the Nordic countries, Austria and two northerly countries of Eastern Europe (German Democratic Republic, Poland), as well as in Cyprus and Ireland, where development of exploitable closed forest has concentrated on coniferous plantations. Most countries with less than 50% coniferous species are in the south-eastern part of Europe.

Net annual increment (NAI) per hectare, is also the result of a combination of climatic and site conditions and silvicultural choices. Most of the countries with NAI/ha under 3.3 m³/ha/year are effected by some type of climatic constraint (even if silviculture is intensive, as in Finland and Sweden). Those with NAI over 5 m³ o.b./ha/year do have relatively favourable growing conditions and relatively intense forest management. In the latter group there are five EEC countries, four from Eastern Europe and both central European countries.

TABLE 2.3

Variations between European countries in average growing stock and net annual increment per hectare of exploitable closed forest 1980

A. GROWING STOCK PER HECTARE (M³ O.B./HA)
(Europe 120)

150 m³ o.b./ha or more	90 m³ o.b./ha or less
Switzerland (392)	Cyprus (33)
Austria (252)	Israel (44)
Czechoslovakia (221)	Spain (70)
Romania (210)	Portugal (73)
German Dem. Rep. (170)	Greece (74)
Hungary (162)	Finland (81)
Fed. Rep. of Germany (155)	Norway (87)
	Bulgaria (90)

B. PERCENTAGE OF CONIFEROUS SPECIES IN GROWING STOCK ON EXPLOITABLE CLOSED FOREST (%)
(Europe 63.3)

75% or more	50% or less
Cyprus (100)	Albania (12)
Austria (85)	Hungary (15)
Sweden (85)	Luxembourg (15)
Finland (82)	Yugoslavia (28)
Norway (80)	Bulgaria (34)
Ireland (78)	France (39)
Poland (77)	Italy (36)
German Dem. Rep. (77)	Romania (40)

C. NET ANNUAL INCREMENT PER HECTARE (M³ O.B./HA)
(Europe 4.10)

5 m³ o.b./ha or more	3.3 m³ o.b./ha or less
Denmark (8.51)	Cyprus (1.01)
Belgium (7.50)	Bulgaria (1.82)
Ireland (7.29)	Greece (2.05)
Switzerland (6.54)	Norway (2.62)
Austria (6.19)	Turkey (2.89)
Hungary (6.15)	Sweden (3.01)
German Dem. Rep. (5.79)	Italy (3.07)
Fed. Rep. of Germany (5.63)	Albania (3.16)
United Kingdom (5.56)	Finland (3.18)
Romania (5.52)	Israel (3.21)
Czechoslovakia (5.38)	Yugoslavia (3.27)

2.3 FACTORS AFFECTING ROUNDWOOD SUPPLY

There are several factors, other than of a purely silvicultural nature, which strongly affect the level of roundwood supply from the European forests. It is usually impossible either to quantify their effects in the past or to take them explicitly into account, when assessing the outlook for the future. The more important of these factors are briefly presented below.

2.3.1 Physical accessibility

In many parts of the world, large areas of forest are classified as "inaccessible" because remoteness, lack of roads or bad working conditions make harvesting or silviculture difficult, if not impossible, to carry out. This is not a major problem in Europe, and in those few areas where the transport and communications network is not very dense (e.g. parts of Norway or Yugoslavia) or where terrain is difficult (e.g. in the Alps), the question of "accessibility" is better presented in economic terms or as connected with the protective functions of the forest.

2.3.2 Factors connected with the provision of non-wood goods and services

Forests provide a wide range of goods and services other than wood. The situation and outlook for these non-wood goods and services is presented at length in chapter 4: here it is sufficient to note that, although in almost all cases wood is produced alongside these benefits, the forest management practices (or regulations) necessary to provide them may cause the supply of wood from the forest concerned to be lower than would be the case if maximum wood production were the only management objective.

As chapter 4 will show, the variety of circumstances is very wide; few, if any, generalizations are possible. Nevertheless, a few types of interaction between wood supply and the supply of non-wood benefits may be mentioned.

An extreme example is when areas are set aside for scientific research, or for wilderness (relatively rare in Europe), when harvesting may be completely forbidden. In many nature reserves, wood is removed only when this is desirable to maintain or improve the natural habitat. Forests with an important protection function can usually not be clearcut and the volume of wood to be removed may be strictly limited. The management of areas with an important recreational function will also be different from those which are devoted primarily to wood production.

Furthermore the cost of providing non-wood benefits is often borne from the income from wood supply, with negative effects on the economics of the latter.

2.3.3 Ownership structure

Forest owners in Europe can be very roughly divided into four groups:

— Public authorities (within this group there are differences, as regards management objectives and size of holding, between State, regional, communal and other public institutions);
— Forest industries;
— Large-scale private forest owners (often individuals or families, but also enterprises such as pension funds or insurance companies);
— Small-scale private forest owners (who may be further subdivided into those who live close to their forest and may exploit it themselves, and absentee owners, many of whom live in towns).

The ownership structure of the European forest is summarized in table 3.7, which breaks down private forest into "forest industry", "farm forest" and "other". Just over half the European forest is publicly owned with a high proportion of public ownership in Eastern and Southern Europe, and a low proportion in the Nordic countries. In many countries, the main public forest owner is the State, in others communal authorities (e.g. Switzerland) or regional governments (e.g. Federal Republic of Germany). Forest industries only own a significant part (15%) of the forest in the Nordic countries, especially Sweden (24%). In the Nordic countries and Central Europe 50-60% of the forest is farm forest, and in the EEC(9) "farm forest"

and "other privately-owned" forest each account for just over 30% of the forest area.

As regards size of holding, a crucial factor for forest management, the partial data in chapter 3 (table 3.10) indicate that small holdings (i.e. less than 10 ha) account for 15-25% of the forest area in some countries, notably in the EEC(9) and Central Europe. There are probably well over 6 million forest holdings of under 10 ha in Europe − an estimated 3 million in France alone.

The larger-scale owners, public, forest industry and private, in almost all cases have the resources and skills to apply a satisfactory level of management to their forest holdings, which are big enough to enable costs to be kept at reasonable levels relative to income.

On the other hand, in many countries the intensity and the standard of management of small-scale forests is very low; some holdings are not managed at all; and in extreme cases, where the provisions of the Napoleonic law on inheritance have resulted in the splitting of forest holdings over successive generations, country dwellers that have moved into cities are not even aware that they are forest owners. Furthermore, the 1970s and 1980s have seen changes in the ownership pattern, notably an increase in non-farmer private forest owners.

In the Nordic countries and notably in Finland, the standard of management by small-scale forest owners is considered satisfactory so that, whilst small-size woodlands may have some intrinsic disadvantages, these can be turned to advantage in some circumstances. To assess the future role of small-scale forests as producers, particularly of wood, it is necessary to look briefly at some of the factors that affect their management.

The intrinsic disadvantage which militates most against the management of small-scale woods is their size; the difficulty of mechanizing operations in woodlands that at most measure only a few hectares and often less than one hectare needs no elaboration here. Even where many small ownerships taken together add up to a considerable area of contiguous forest, and where the terrain is suited to the use of machinery, the difficulties − legal, fiscal, psychological − of getting owners to co-ordinate the management of their individual holdings and the marketing of their timber are formidable.

In countries where the owners are farmers living near their forest holdings and the winter climate makes agricultural activity highly seasonal, the very smallness of their forest holding may encourage them to use labour intensive methods or to use their farm tractors (suitably equipped for forestry). But to manage the forest successfully in this way, the owner must have

— Adequate knowledge of management techniques;
— Adequate capital, e.g. to equip his farm tractor for forest operations;
— Markets for the wood he harvests.

He probably also needs to have an adequate financial reward − either in the short run or perhaps more distantly in, say, capital appreciation. Above all he must have an interest in his forests.

To try to overcome the problems inherent in small-scale forest ownership, several countries with significant numbers of such ownerships have tried to stimulate a more effective forest management by encouraging co-operatives of forest owners (e.g. for equipment purchase or marketing). At the same time they have provided extension services and training and fiscal stimuli to good management.

It must be recognized that many, if not all, small-scale forest owners have other sources of income. Their market behaviour is not the same as that of owners − and especially larger-scale owners − who to a greater or lesser extent depend on their forests to provide a regular income or to meet periodic peaks in expenditure. The small-scale owner's forest income will in any case be episodic so that in bad market conditions he will often prefer not to harvest his crop and, if he can afford to do so, to await a better price. He will also be influenced by the availability (or lack of availability) of opportunities to invest the income from harvesting or by the need to do so − e.g. to replace farm machinery or fixed equipment. In some cases (e.g. in periods of inflation) it is more rational economically to keep capital in a growing forest than to invest in financial instruments.

The problems which can be raised by the unwillingness of private forest owners to harvest were highlighted by events in Finland and Sweden in the late 1970s, when raw material shortages occurred, which were explained by the type of behaviour described above. The shortages were ended by a combination of market reorganization and higher roundwood prices. It is not clear yet whether the solutions found were permanent, as cyclical factors played a role in the situation.

It has also to be recognized that for some small-scale owners the forest is not perceived as a productive economic enterprise. It is not uncommon to find owners who have sufficient income from other sources to be able to regard their forest as a site for recreation, e.g. as a site for a holiday house or for hunting. They will not undertake any forest operations which would reduce the site's recreational and amenity value to the owner.

Different larger-scale forest owners also have different objectives: public bodies do usually seek income from the forest holdings, but may choose to forgo maximum income for the sake of achieving other objectives of national or local forest policies (see below); forest industries may attach higher priority to providing raw material for processing operations than to the economics of the forest holding seen as an isolated unit. Large scale non-industry private forest owners may also have differing objectives, e.g. maximum income or maximum capital appreciation, according to their situation (some may even seek a "loss", in accounting terms at least, for tax purposes). In some countries forest holdings have also played a role in minimizing the effect of redistributive taxation on large private fortunes.

2.3.4 Fiscal factors

Forestry has many characteristics which make it unusual from a fiscal point of view, notably the extremely long period between investment and return on investment and the very episodic nature of the income from one forest stand. In some cases fiscal systems do not take these special characteristics into account and thereby can unintentionally discourage good forest management. On the other hand, in a few countries, special systems have been devised to encourage good forest management, and a steady wood supply for the forest industries.

These questions were examined by Dr. Cyril Hart in a study entitled "Effect of Taxation on Forest Management and Roundwood Supply", prepared under the auspices of FAO and ECE and published in 1980 as supplement 4 to volume XXXIII of the *Timber Bulletin for Europe*. There are enormous differences between national fiscal systems and between the fiscal situations of forest owners, so it is difficult to make generalizations about the effects of forest fiscality, still less to quantify these effects.

The following extracts from Dr. Hart's study (based on the fiscal situation of the mid-1970s) give some idea of the main problems encountered:

12.9 Fiscal incidence: Forest taxation in Europe differs widely among countries in its provisions, the level at which the taxes are raised, and the tax rate. The systems do not always fully accord with the principles of sound silviculture and proper forest management.

The incidence and effect of some taxes is a burden, of others a relief, concession or incentive, and in some cases neutral. Thus forest policy through taxation measures can have the effect on forestry of being negative or positive or neutral to it. Some provisions hinder management, and have an adverse effect on roundwood supply. Some encourage one or more aspects of forestry, while others have little or negligible effect. There are sometimes anomalies whereby two types of taxation may have contradictory effects on forest management and harvesting.

Often there is a significant difference between the intended and the actual effects of taxation. For example, a provision meant to encourage forestry by giving exemption from income tax may have the unintended effect of reducing desirable silvicultural expenditure, such as tending, because it can neither be set against other taxable income, nor carried forward for setting against (i.e. recouped from) future net income from the forest.

12.9.1 *General fiscal incidences*: The effect of taxation provisions relating to *regeneration* and *afforestation* are often incentives in as much that they exempt young stands until they can produce a net income. However, where the net expenditure during the unproductive years cannot be offset against other taxable income, the effect becomes a disincentive.

There is not always recognition of the variations in the productive capacity of forest soils − the same level of taxes is payable on forest land of the highest productivity-class as on land capable of producing only the lowest yield. (A notable exception is Finland.)

Taxation incentives are generally oriented to *wood production*, but sometimes also to *landscape/amenity* (e.g. the Netherlands and Portugal) and *public access*. Some countries prefer instead to subsidize landscape and public access by monetary grants (e.g. the United Kingdom under Basis III).

As to *continuity of ownership*, despite the desirability of sustained management and production, continuity has not always been achieved. Among the reasons are inheritance taxes, particularly where they are assessed not on capitalized-income value but on full market value ...

Investment in forestry: Taxation policies can have a major impact, negative as well as positive, on flows of private funds into and out of forestry. Because of the lengthy periods during which capital can be locked up in forest growing stock, inheritance taxes and taxes on capital gains in particular can have an important impact on private investment in forestry. More favourable tax treatment (perhaps with other practical incentives) would encourage greater investment in forestry.

Dr. Hart also examined specific fiscal incidences under the following headings:

(a) Valuation of forest assets for tax purposes;

(b) Taxation of capital as an addition to income tax;

(c) Inheritance and gift taxes;

(d) Capital gains on forest land;

(e) Land and other real property taxes;

(f) Assessment of forestry income for income tax purposes;

(g) Taxes on turnover;

(h) Consideration of changes in forest taxation systems.

2.3.5 Availability and cost of labour

In a few forest areas, there may be difficulties in obtaining sufficient labour for forestry work which can still be strenuous, dangerous, remote from urban attractions and badly paid. The conditions of forest workers generally have improved, however: much more work is mechanized, machines are being better designed ergonomically, with considerable stress on safety features and training. Wages paid by the hour, week or month have replaced piece work rates in some countries. The effect of these changes has been to make forest workers more qualified, full-time (previously there was much more seasonal work) and better paid — with a corresponding increase in labour costs and consequent necessity to increase productivity. There are still wide variations in conditions of work within the region. Shortages of labour in absolute terms may be a constraint on wood supply in a few places but in most cases, labour questions are important for their effect on the economy of silviculture and harvesting.

Two studies have been carried out by Finnish experts under the auspices of the Joint FAO/ECE Working Party on Forest Economics and Statistics:

"Trends in Forestry Employment in Europe and North America, 1965 to 1977" (supplement 3 to volume XXXIII of the Timber Bulletin for Europe; Geneva 1980);

"Labour Productivity in Forestry" (draft issued as TIM/EFC/WP.2/R.76, May 1985).

Table 2.5 suggests that there are well over 1 million people employed in forestry in Europe. The data in the table, however, are not completely comparable between countries (e.g. as regards assessment of part-time work). As the 1980 study pointed out, there are particular uncertainties in countries with a large proportion of private forests, especially small woodlands. As the forest owners and their families often do their own logging as well as the silvicultural work, the true forest labour force and labour input includes self-employment and is therefore very difficult to estimate.

Most countries are faced with a general upward movement in wages and some countries with limited availability of labour in rural areas (despite widespread unemployment). In these circumstances gains in productivity are of great importance. The 1985 draft study was able to bring together reasonably comparable information on the development of productivity in a number of countries and made the following comments.

A general look at the differences of productivity levels and their development in the reporting countries gives the impression that the relative differences have increased from the 1950s to the 1980s. If, however, we disregard the short series for France, Switzerland and Norwegian farm forests and the lespromkhoz of the USSR and also Bulgaria, which may not be comparable with other countries, the remaining curves indicate a clear decline in productivity variation, especially from the 1960s.

The study period is mainly characterized by the mechanization of the wood harvesting process, which started early in Canada and the USSR, due to the possibility of large clearcuttings. In Scandinavian countries and Central Europe the conditions are less favourable for extensive clearcutting, even in State forests. Therefore mechanization started later — mainly in the 1960s — but speeded up thereafter. The relatively slow progress of the USSR lespromkhoz after a rapid start in the 1950s might be due partly to features specific to the USSR, notably inclusion of non-harvesting activities, including primary conversion.

Not too much weight should be given to the differences in productivity between countries and the data should be treated as indicative. The comparability is probably hampered in many cases by differences in reliability and the coverage of the data, in the definitions of, for example, hauling and road transport, in converting labour input to comparable man-days, and above all in harvesting conditions (forest ownership, cutting methods, size of trees, distances, terrain, etc.).

The complexity of these differences is demonstrated by the fact that the highest levels in logging productivity have apparently been reached under the mountainous conditions experienced in British Columbia and Norway (State forestry).

Bulgaria has an even but relatively slow growth of labour productivity of 2.5% per annum, which seems to indicate modest mechanization during the period covered. The reasons are explained in the pilot study (TIM/EFC/WP.2/R.58).

TABLE 2.5

Employment in forestry, 1977

	Total (1 000 persons)		Share of permanent (%)
		Of which permanent	
Finland	49.0	..	..
Norway	9.0	..	..
Sweden[a]	191.9	18.6	9.7
Belgium[b,c]	2.2	1.0	45.5
Denmark[a]	5.6	1.1	19.6
France[c]	78.0	41.0	52.6
Germany, Fed. Rep.of[d]	92.4	44.9	48.6
Luxembourg[a]	0.5	0.1	20.0
Netherlands[b]	7.7	3.5	45.5
United Kingdom[d]	19.7	..	..
Austria	11.8	..	..
Switzerland	21.3	4.6	21.6
Cyprus	1.1	0.2	18.2
Greece[b]	53.4	4.4	8.2
Spain	121.0	46.0	38.0
Yugoslavia	64.2	..	..
Bulgaria	40.5	27.0	66.7
Czechoslovakia	94.4	75.0	79.4
Hungary[b]	60.5	45.8	75.7
Poland	121.1	93.9	77.5
Total (20 countries)	1 045.3	..	..

Source: Trends in Forestry Employment, 1965 to 1977, Annex table 1.

[a] 1976.

[b] 1975.

[c] Logging only.

[d] 1974.

Baden-Württemberg state forestry in the Federal Republic of Germany also had a remarkably even growth for a still longer period but at some 6% p.a.

Austria, represented by one big enterprise only and Finland (all forests) both represent a type of modest increase in 1951-1965, then a rapidly advancing mechanization of both cutting and hauling, reaching 10% per annum growth rate and possibly showing some signs of a levelling off in the last few years.

Sweden (large-scale forestry), Canada, Denmark, Norwegian state forestry and Switzerland (plateau region) show clear evidence of a slow-down of the mechanization progress. Sweden showed a 10% p.a. growth up to the mid-1960s and then a remarkable 15% p.a. growth to the mid-1970s, followed by stagnation. Norway had a largely similar trend. The productivity in Canada (all forests) grew at 6% p.a. in the 1960s and up to 1975, but then stagnated.

Part of the declining growth of mechanization and hence productivity might be due to a growing opposition of the population in many countries to large clearcuttings and, especially, the use of heavy machines in forests. Accumulation of neglected thinnings again aroused growing concern among professional foresters and led to the necessity of developing logging machinery with less weight and often lower productivity. The continuing steep 10% p.a. rise of the productivity in Finland compared with Sweden's stagnation is at least partly explained by the fact that Sweden in the late 1970s changed from a mainly piece rate wage system to a time rate based one. The productivity in cutting decreased considerably.

2.3.6 Economic factors

The supply of roundwood is, of course, determined or at least strongly influenced by economic factors. Most of the above-mentioned factors have economic effects, typically on the costs of forest management and harvesting (e.g. ownership structure, labour costs, fiscality). There are, however, some other economic aspects concerned with wood supply which should also be briefly mentioned.

On the demand side, the volume and price of round-wood supplies depend on the economic health of wood consuming industries within an economic transport distance.

There are several cases in Europe of forests which are not harvested because no market exists at present for their products, usually because there is no appropriate processing industry in the region. In this way the health of the forest depends in part on the success of the local forest industries. If they are not successful in producing a technically suitable, competitively priced product, on the basis of locally available raw material, the forest will be under-exploited, uneconomic and ultimately of lower quality.

A particular problem is that of stands, notably of hard-woods, which are considered "low quality" because of species or the technical characteristics of the wood. Increasingly the owners of this type of stand are having difficulties in finding outlets for this material (other than as energy wood, often at low prices), especially as some outlets (e.g. railway sleepers, pit wood) have been losing importance. Examples of forest resources, for which difficulty has been experienced in finding outlets, are the oak of southern Sweden, the degraded hardwood coppice in France and Belgium, low quality beech in Switzerland and the Federal Republic of Germany, and some hardwoods in Hungary.

There are other factors, such as the organization of roundwood markets (which sometimes lack transparency or are dominated by a few participants) or the level of interest rates, which, along with length of rotation, play a crucial role in determining the economics of forestry. It is unfortunately the case that for many forests the true economic situation, and in particular the economics of round-wood supply, is not known or is seriously distorted, either because the return on capital is not correctly calculated (forestry is a very capital intensive enterprise) or because the costs of providing non-wood benefits — to individuals or to the collectivity — are not correctly assessed or covered by the income from wood production.

2.4 OBJECTIVES OF FOREST POLICY

One of the most important influences on wood supply, along with the soil and climatic potential and the existing forest resource, is national forest policy. Governments have a wide range of instruments with which to further their forest policies, which vary widely according to the economic and social system of the country concerned and its forest resource traditions. It is clearly beyond the scope of ETTS IV to examine these in any detail.

It is however worth pointing out that many types of objective are taken into account when forming forest policy and that some of these objectives have little or no connection with wood supply, although they may, directly or indirectly, affect wood supply. Among the objectives which may have to be reconciled when forming forest policy in Europe are the following:

(a) Supplying the country's needs for forest products;

(b) Supplying the country's needs for energy;

(c) Maintaining or improving the trade balance;

(d) Maintaining and developing the rural economic and social structure;

(e) Maintaining or increasing employment;

(f) Encouraging decentralization;

(g) Improving the income of individuals and public bodies;

(h) Soil protection;

(i) Ensuring a harmonious countryside;

(j) Preserving sites of special aesthetic, cultural, historical or scientific interest;

(k) Encouraging nature conservation;

(*l*) Supplying recreational opportunities (especially for city dwellers);

(*m*) Ensuring preparedness for a national emergency;

This list is, of course, partial and in no order of priority.

Each country (or region within a country) formulates the objectives of its forest policy and, explicitly or impli-citly, assigns priorities to the different objectives. It is not for the authors of ETTS IV to analyse these decisions which are usually arrived at through normal national decision-making processes. However, when assessing the outlook for wood supply, it must be borne in mind that the maximization of wood supply is not the only objec-tive of forest policy.

2.5 METHODS OF FORECASTING ROUNDWOOD SUPPLY IN ETTS IV

It is clear from the above overview of factors that to provide a reliable forecast of roundwood supply, it is necessary to take into account not only forest inventory data, supplemented by data on costs and prices, but many other factors, whose future development would be diffi-cult to predict — most of all at an international level.

National correspondents were therefore requested to prepare forecasts for roundwood supply, using their own experience and "feel" for their country's situation taking into account all these factors, as well as national forest policies. These forecasts are presented in chap-ter 5.

CHAPTER 3

The European forest resource around 1980

3.1 INTRODUCTION

The main objective of this chapter is to present the forest resource situation in Europe in quantitative terms. The chapter covers the following aspects of the forest resource:

— Land classification and land-use categories;

— Distribution of stocked exploitable closed forest by age-classes;

— Forestry situation on exploitable closed forest with regard to growing stock, net annual increment and removals;

— Forest ownership, management status and structure of forest holdings;

— Recognized major functions of the forest and other wooded land;

— Changes in the forest resource over time;

— Trends in removals since 1950, concentrating on the 1970s.

The chapter concentrates especially on exploitable closed forest — the most important part of the forest sector from the point of view of wood production. It provides the background for other chapters of the study, especially for chapter 5, which presents forecasts for the forestry situation and wood supply prospects to the year 2020.

The basic information on forest resources presented in this chapter is taken mainly from Part I of the 1985 ECE/FAO publication *The Forest Resources of the ECE Region (Europe, the USSR, North America)* which contains detailed information on the above-mentioned aspects by country and country groups. This publication is referred to below as the "1980 Assessment" (it was published in 1985 but the data in it mostly refer to a period around 1980). A few countries have corrected or supplemented their data since the publication of the 1980 Assessment. This chapter shows the modified data and therefore does not always correspond exactly to the published version. (The major change is for Spain where 11.6 million ha of Mediterranean scrub, previously classified "non-forest land" are now classified as "other wooded land".)

While analysing the data presented in chapter 3 and attempting to reach conclusions, readers are asked to bear in mind the following factors:

(a) The national forest inventories were carried out at different periods in different countries, which created problems in trying to compare the data between the countries and country groups; but most of the latest inventories took place in the late 1970s or early 1980s;

(b) Despite the existence of common definitions,[1] different approaches were adopted by different countries to land classification, definition of forest land-use, sub-division of forest into coniferous and non-coniferous and tree measurement during inventories, which reduced the comparability of data in some cases;

(c) Most of the information was provided by the countries themselves through their correspondents, who took the data from national forest inventories or other equally authoritative sources, but the secretariat had to use other sources in some cases to complete the tables.

Great care should be taken therefore in making comparisons between countries or over time.

_____

[1] The main terms and definitions used in the study are set out in appendix 1 of the 1980 Assessment.

3.2 LAND CLASSIFICATION AND LAND-USE CATEGORIES

Figure 3.1 helps to explain the relationship between the terms defined in the annex 3.1 to the 1980 Assessment and the way in which the statistical information on forest resources is presented.

"Stocked exploitable closed forest" is broken down by:

A. SPECIES GROUPS: (i) Coniferous; (ii) Non-coniferous.

B. AGE CLASSES.

"Forest and other wooded land" is broken down by:

A. OWNERSHIP

  (i) *Publicly owned*
     – State
     – other

  (ii) *Privately owned*
     – forest industries
     – farm forest
     – other

B. MANAGEMENT STATUS

  (i) Land managed according to a plan;
  (ii) Land not managed according to a plan but having controls relating to management or use;
  (iii) Land having no controls relating to management or use.

C. STRUCTURE OF HOLDINGS

  (i) Size of holdings;
  (ii) Number of holdings

D. RECOGNIZED MAJOR FUNCTIONS

  (i) With wood production the recognized major function;
  (ii) With protection, conservation and biological uses the recognized major functions;
  (iii) With recreation the recognized major function.

Table 3.1 presents a general picture of total land use in Europe around 1980. According to the latest data more than 190 million hectares or 35% of the total land is classified as forest and other wooded land. The highest percentage of land accounted for by forest and other wooded land is in the Nordic countries (53%), the lowest in the countries of the EEC(9) (23%).

Exploitable closed forest covers some 133 million hectares or 69% of the total of forest and other wooded land.

In Eastern Europe, 93% of forest and other wooded land is exploitable closed forest, the highest percentage in Europe, the lowest being in Southern Europe (41%). The other three country groups have practically the same proportion of exploitable closed forest – about 81%.

Unexploitable closed forest in Europe occupies about 14 million hectares or nearly 7% of total forest and other wooded land. Nearly two thirds of Europe's unexploitable closed forest is in Southern Europe and the Nordic countries (5.5 million ha or 38% and 3.7 million ha or 25% of the European total respectively). There are several reasons why this area is classified as unexploitable, including physical inaccessibility, legal restrictions on commercial cuttings because of protection, conservation, biological and recreation functions of the forests, as well as economic criteria (low stand productivity, excessive harvesting or transport costs, etc.).

Other wooded land, defined as having a tree cover of less than 20%, occupies about 46 million hectares in Europe or about 24% of the total area of forest and other wooded land. Over 72% of this is in Southern Europe.

Non-forest land amounts to some 358 million hectares in Europe or 65% of the total land area. Agricultural land covers 242 million hectares (44%) while about 116 million hectares (21%) comes into the "other" category which includes built-up and related land, including land under transport and communication facilities, as well as land without a recognized economic use, such as mountainous areas, tundra and bare land. A substantial part of this category of land is in the Nordic countries and Southern Europe – 37% and 35% of the European total respectively.

TABLE 3.1

Europe: main land-use categories by country groups, around 1980

	Total land (excl. water) (million ha)	Forest and other wooded land (OWL)		Exploitable closed forest		Unexploitable closed forest (million ha)	Other wooded land (million ha)	Non-forest land (million ha)	
		Total (million ha)	% of total land	Total (million ha)	% of forest and OWL			Agricultural land	Other
Nordic countries	112.4	59.9	53.2	48.3	80.6	3.7	7.9	9.8	42.8
EEC(9)	150.7	34.5	22.9	27.8	80.6	3.3	3.5	96.2	20.0
Central Europe	12.2	4.9	39.8	4.0	81.2	0.7	0.2	5.6	1.8
Southern Europe	179.8	66.6	37.0	27.1	40.7	5.5	33.9	72.0	41.2
Eastern Europe	96.8	28.0	28.7	25.8	92.8	1.2	1.0	58.5	10.3
Total Europe	551.9	193.8	35.1	132.9	68.6	14.4	46.5	242.0	116.0

Source: 1980 Assessment.

TABLE 3.2

Main land-use categories, including forest and other wooded land (OWL), by country and country groups

	Years	Total land (excl.) water (1 000 ha)	Forest and other wooded land		Exploitable closed forest		Unexploitable closed forest (1 000 ha)	Other wooded land (1 000 ha)	Non-forest land (1 000 ha)		
			Total (1 000 ha)	% of total land	Total (1 000 ha)	% of forest and OWL			Total	Agricultural land	Other land
Finland	1975-81	30 547	23 225	76.0	19 445	83.7	440	3 340	7 322	3 122	4 200
Iceland	1981	10 000	100	1.0	..	..	..	100	9 900	2 300	7 600
Norway	1967-79	30 754	8 701	28.3	6 600	75.6	1 035	1 066	22 053	1 039	21 014
Sweden	1973-77	41 148	27 842	67.7	22 230	79.8	2 170	3 442	13 306	3 300	10 006
Nordic countries		112 449	59 868	53.2	48 275	80.6	3 645	7 948	52 581	9 761	42 820
Belgium	1970	3 080	680	22.7	600	88.2	–	80	2 400	1 500	900
Denmark	1981	4 230	484	11.4	400	82.6	66	18	3 746	2 920	826
France	1981	54 319	15 075	27.8	13 340	88.5	535	1 200	39 244	32 466	6 778
Germany, Fed. Rep. of	1961	24 322	7 207	29.6	6 838	94.9	151	218	17 115	14 221	2 894
Ireland	1980	6 889	380	5.5	347	91.3	–	33	6 509	4 690	1 819
Italy	1980	30 126	8 063*	26.8	3 868*	48.0	2 495*	1 700*	22 063	18 919	3 144
Luxembourg	1983	258	82	31.5	80	97.6	2*	–	176	141	35
Netherlands	1982	3 394	397	11.7	294*	74.1	37	66	2 997	2 395	602*
United Kingdom	1980	24 098	2 178	9.0	2 017	92.6	10	151	21 920	18 966	2 954
EEC(9)		150 716	34 546	22.9	27 784	80.5	3 296	3 466	116 170	96 218	19 952
Austria	1971-80	8 256	3 754	45.5	3 165	84.3	589	–	4 502	3 632	870
Switzerland	1972-78	3 976	1 124	28.3	795	70.7	140	189	2 852	1 947	905
Central Europe		12 232	4 878	39.8	3 960	81.2	729	189	7 354	5 579	1 775

TABLE 3.2 (continued)

	Years	Total land (excl.) water (1 000 ha)	Forest and other wooded land		Exploitable closed forest		Unexploitable closed forest (1 000 ha)	Other wooded land (1 000 ha)	Non-forest land (1 000 ha)		
			Total (1 000 ha)	% of total land	Total (1 000 ha)	% of forest and OWL			Total	Agricultural land	Other land
Greece	1983	12 894	5 754	44.6	1 793	31.2	719	3 242	7 140	3 960	3 180
Portugal	1980	8 562	2 976	34.8	2 590	87.0	37	34	5 586	4 096	1 490
Spain	1980	49 940	25 622	51.3	6 506	25.4	1 882	17 234	24 318	20 508	3 810
Turkey	1963-72	77 079	20 199	26.2	6 642	32.9	2 214	11 343	56 880	27 280	29 600
Yugoslavia	1979	25 600	10 500	41.0	8 500	81.0	600	1 400	15 100	14 400	700
Southern Europe		174 075	65 051	37.4	26 031	40.0	5 452	33 568	109 024	70 244	38 780
Albania	1981	2 740	1 242*	45.3	930*	74.9	..	312*	1 498	731	767
Cyprus	1980	923	173	18.7	100	57.8	53	20	750	605	145
Israel	1981	2 031	109	5.4	66	60.6	9	34	1 922	421	1 501
Mediterranean countries		5 694	1 524	26.7	1 096	71.9	62	366	4 170	1 757	2 413
Bulgaria	1980	11 100	3 800	34.2	3 300	86.8	100	400	7 300	5 800	1 500
Czechoslovakia	1980	12 552	4 578	36.4	4 185	91.4	308	85	7 974	6 924	1 050
German Dem. Rep.	1980	10 600	2 955	27.3	2 590	87.6	110	255	7 645	5 015	2 630
Hungary	1981	9 094	1 637	18.0	1 563	95.5	49	25	7 457	6 585	872
Poland	1978	30 457	8 726	28.7	8 410	96.4	178	138	21 731	19 200	2 531
Romania	1985	22 954	6 265	27.3	5 723	91.3	467*	75	16 689	14 963	1 726
Eastern Europe		96 757	27 961	28.7	25 771	92.8	1 212	978	68 796	58 487	10 309
TOTAL EUROPE		551 923	193 828	35.1	132 917	68.6	14 396	46 515	358 095	242 046	116 049

Source: 1980 Assessment, table 1.1.

TABLE 3.3

Europe: distribution of stocked exploitable closed forest by age classes in selected countries (percentage of total area)

	Total (1 000 ha)	Uneven age stands (percentage)	Predominantly even-age stands (high forest) (percentages)									Average age of rotation (years)
			<10 years	11-20 years	21-40 years	41-60 years	61-80 years	81-100 years	101-120 years	121-140 years	>140 years	
Finland												
Coniferous	17 797	—	8.3	7.4	12.6	15.5	18.4	12.8	8.0	5.2	11.8	80/120, 110-180[a]
Non-coniferous	1 470	—	4.0	5.5	18.0	25.4	25.9	13.0	5.4	1.0	1.8	60/80, 70-90[a]
Sweden												
Coniferous	19 641	—	8.7	9.3	9.3	13.2	16.5	18.0	14.3	5.9	4.8	80-100, 100-120[a]
Non-coniferous	1 018	—	5.1	10.5	26.0	31.8	16.3	6.9	2.5	0.7	0.2	80-100
Belgium												
Coniferous	272	—	18.4	23.9	29.0	21.0	5.9	1.8	—	—	—	80-100
Non-coniferous	109	100.0	—	—	—	—	—	—	—	—	—	180-200
Denmark												
Coniferous	325	—	17.8	20.0	30.2	15.4	6.5	5.5	4.6[b]	:	:	45-120, 150
Non-coniferous	141	—	9.9	17.0	15.6	20.6	12.1	7.8	17.0[b]	:	:	81-150
France												
Coniferous	3 314	—	12.4	15.5	26.0	16.5	11.5	6.2	4.6	3.1	4.2	..
Non-coniferous	5 484	0.7	7.1	14.2	59.0	14.5	3.2	0.8	0.4	0.1	—	
Germany, Fed. Rep. of												
Coniferous	4 395	—	2.8	26.0	21.1	20.6	13.5	8.8	7.2[b]	:	:	80-120
Non-coniferous	1 999	3.3	1.5	14.4	13.1	13.2	14.1	12.2	28.2[b]	:	:	120-180
Ireland												
Coniferous	289	—	30.1	32.5	27.0	8.3	2.1[c]	:	:	:	:	50
Non-coniferous	44	—	9.1	4.5	13.6	15.9	56.9[c]	:	:	:	:	150
Luxembourg												
Coniferous	25	:	56.0[d]	24.0	12.0	8.0	—	—	—	—	—	50-100
Non-coniferous	40	:	5.0[d]	2.5	2.5	2.5	5.0	35.0	47.5	—	—	140-160
Netherlands												
Coniferous	167	—	8.7	11.2	36.0	27.2	11.7	4.0	0.8	0.4[e]	—	60-80
Non-coniferous	66	—	21.5	13.9	24.4	17.7	10.3	6.2	2.9	3.1[e]	:	25-40,[f] 100-130
United Kingdom												
Coniferous	1 379	—	25.2	28.0	32.1	10.9	1.9	(1.4	..	0.5[e]	:	50-60
Non-coniferous	559	—	3.4	6.4	25.4	17.9	14.7	(21.3)	..	10.9[e]	:	80-100, 120

TABLE 3.3 (continued)

	Total (1 000 ha)	Uneven age stands (percentage)	Predominantly even-age stands (high forest) (percentages)									Average age of rotation (years)
			<10 years	11-20 years	21-40 years	41-60 years	61-80 years	81-100 years	101-120 years	121-140 years	>140 years	
*Switzerland**												
Coniferous.................	922	7.1	6.0	(..	(12.0)	(..	36.0)	(..	25.9)	(..	13.0)	80-140, 250-350
Non-coniferous.............	202	6.9	6.0	(..	(11.9)	(..	36.1)	(..	26.2)	(..	12.9)	120-140, 180-300
Portugal												
Coniferous.................	1 222	32.3	3.6	10.5	43.6	9.0	1.0	–	–	–	–	50-100
Non-coniferous.............	1 231	81.4	11.0	4.8	2.8	–	–	–	–	–	–	8-12, 12, 50-200^g
Spain												
Coniferous.................	6 085	60.0	12.8	15.2	11.4	0.4	0.2	–	–	–	–	60-80, 100-120
Bulgaria												
Coniferous.................	1 100	..	31.8	27.3	9.1	9.1	9.1	9.1	2.7	0.9	0.9	100-120
Non-coniferous.............	2 100	..	9.5	14.3	38.1	14.1	9.5	4.8	4.8	3.3	1.4	120-140
Czechoslovakia												
Coniferous.................	2 646	–	7.9	8.4	14.4	20.2	22.5	15.3	7.7	2.7	0.9	100-120
Non-coniferous.............	1 352	–	6.9	9.4	24.1	21.7	15.6	10.9	6.6	3.4	1.4	100-150
German Dem. Rep.												
Coniferous.................	2 275	5.0	–	20.3	20.8	17.0	16.6	12.7	5..	2.5	–	80-120
Non-coniferous.............	680	47.8*	–	3.4	5.6	6.5	7.2	7.9	7.8	13.8	–	..
Hungary												
Coniferous.................	222	–	33.8	24.3	27.0	8.1	5.4	1.4	-	–	–	60-75
Non-coniferous.............	802	–	14.9	20.2	24.9	16.7	13.7	7.1	2.5^b	..	..	75-120
Poland												
Coniferous.................	6494	1.8	10.3	14.7	22.3	21.7	15.9	9.4	2.6	1.3	..	100-110
Non-coniferous.............	1648	5.7	7.9	10.0	29.9	20.8	11.9	8.2	3.4	2.2	..	45-75, 110-135
Romania												
Coniferous.................	1 681	4.9	13.5	16.4	12.7	12.9	16.5	10.4	(12.7	..	..)	110
Non-coniferous.............	3 496	3.4	7.4	8.0	21.5	20.4	13.3	9.0	(17.0	..	..)	100-120

Source: 1980 Assessment, table 7.

a Longer rotations refer to the northern parts of the country.

b Over 100 years.

c Over 60 years.

d 20 years and less.

e Over 120 years.

f Poplar.

g 8-12 Eucalyptus, 12 Chestnut, 150-200 Cork oak.

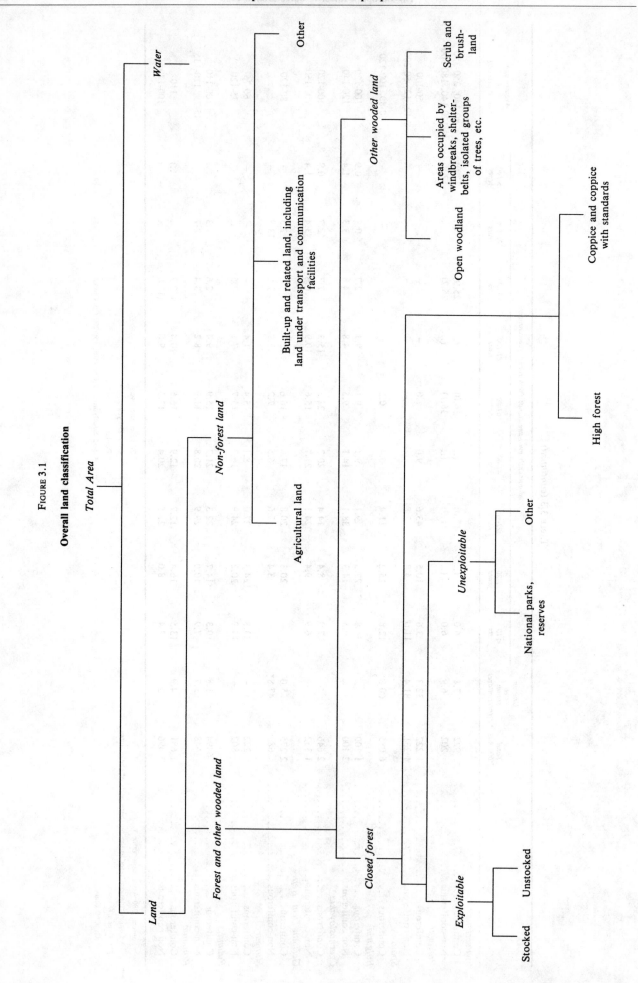

FIGURE 3.1

Overall land classification

3.3 DISTRIBUTION OF STOCKED EXPLOITABLE CLOSED FOREST BY AGE CLASSES

Table 3.3 (pages 36 and 37) shows the available information on age-class distribution of "high forest" country by country. The age structure of the forest differs considerably from country to country as does the average rotation for coniferous and non-coniferous species which varies even within a single country.

3.3.1 Coniferous

Several countries have a significant percentage of young and middle-aged coniferous stands. The percentage of these stands reaches almost 60% in France, Ireland, Luxembourg, United Kingdom, Spain, Bulgaria and Poland. This is in most cases the result of a sustained programme to develop the forest resource by new plantations

or conversion of existing stands. Among European countries, Finland, Sweden, and Czechoslovakia have the most even age distribution as regards coniferous stands.

3.3.2 Non-coniferous

The distribution by age-classes of non-coniferous stands in the European countries is not the same as for coniferous. Many countries have rather large areas occupied by mature and over-mature non-coniferous stands, that is, at the age of cutting or close to it (Luxembourg, about 47%, United Kingdom — more than 30%). On the other hand some countries have a considerable proportion of non-coniferous stands between 20 and 40 years old (France 59%, Bulgaria 38%, Poland 30%).

TABLE 3.4

**Europe: Forestry situation on exploitable closed forest
by country groups around 1980**

	Exploit-able closed forest (million ha)	Growing stock (million m³ o.b.)	Gross annual increment (million m³ o.b.)	Natural losses (million m³ o.b.)	Net annual increment With bark (million m³ o.b.)	Net annual increment Without bark (million m³ u.b.)	Removals [a] (million m³ u.b.)
Nordic countries	48.3	4 407	155.0	8.9	146.1	127.4	99.68
EEC(9)	27.8	3 565	137.3	9.8	127.5	112.7	78.7
Central Europe	4.0	1 109	24.9	0.1	24.8	22.4	16.6
Southern Europe	27.1	2 582	96.8	3.6	93.2	76.6	51.5
Eastern Europe	25.8	4 278	123.9	10.8	113.1	99.7	78.0
Total	132.9	15 941	537.9	33.1	504.8	438.8	323.9

[a] As reported for the 1980 Assessment. See chapter 5 for discussion of differences between these data, those reported annually to ECE/FAO and the "base period" for the removals forecasts.

TABLE 3.5

**Country groups' share of European total of exploitable closed forest,
growing stock, net annual increment and fellings, around 1980**

(Percentage of European total)

	Exploitable closed forest	Growing stock (overbark)	Net annual increment (overbark)	Fellings (overbark)
Nordic countries	36.3	27.6	28.9	31.2
EEC(9)	20.9	22.4	25.3	24.3
Central Europe	3.0	7.0	4.9	5.0
Southern Europe[a]	20.4	16.2	18.5	16.4
Eastern Europe	19.4	26.8	22.4	23.1
	100.0	100.0	100.0	100.0

[a] Including Albania.

TABLE 3.6

Europe: forestry situation on exploitable closed forest by country and country group around 1980

	Exploitable closed forest (total) (1 000 ha)	Growing stock (living trees)				Net annual increment (NAI)				NAI as % of growing stock		
		Total (million m³ o.b.)	Volume per ha (m³/ha)	Coniferous (million m³ o.b.)	Non-coniferous (million m³ o.b.)	Total (million m³ o.b.)	NAI per hectare (m² o.b./ha)	Coniferous (1 000 m³ o.b.)	Non-coniferous (1 000 m³ o.b.)	Total	Coniferous	Non-coniferous
Finland	19 445	1 568	81	1 290	278	61 930	3.18	48 119	13 811	3.95	3.73	4.97
Norway	6 600	575	87	459	116	17 310	2.62	13 710	3 600	3.01	2.99	3.10
Sweden	22 230	2 264	99	1 934	330	66 941	3.01	55 311	11 630	2.96	2.86	3.52
Nordic countries	48 275	4 407	90	3 683	724	146 181	3.03	117 140	29 041	3.32	3.18	4.01
Belgium	600	73	122	40	33	4 500	7.50	2 438	2 062	6.16	6.10	6.25
Denmark	400	46	115	26	20	3 405	8.51	2 520	885	7.52	9.88	4.47
France	13 340	1 550	116	605	945	54 000	4.05	23 500	30 500	3.48	3.88	3.23
Germany, Fed. Rep. of	6 838	1 062	155	744	318	38 000	5.63	29 200	9 300	3.62	3.92	2.92
Ireland	347	32	92	25	7	2 528	7.29	2 451	77	7.93	9.74	1.14
Italy	3 868	557*	144*	201*	356*	11 880*	3.07	4 277*	7 603*	2.13	2.13	2.14
Luxembourg	80	13	161	2	11	330*	4.10	120*	210*	2.54	5.45	1.94
Netherlands	294	29	99	21	8*	1 240	4.22	910*	330*	4.28	4.33	4.12
United Kingdom	2 017	203	101	111	92	11 200	5.56	8 600	2 600	5.52	7.75	2.83
EEC(9)	27 784	3 565	128	1 775	1 790	127 583	4.59	74 016	53 567	3.58	4.17	2.99
Austria	3 165	797	252	674	123	19 581	6.19	16 657	2 924	2.46	2.47	2.38
Switzerland	795	312	392	210*	102*	5 200*	6.54*	3 600*	1 600*	1.67*	1.71*	1.57*
Central Europe	3 960	1 109	280	884	225	24 781	6.26	20 257	4 524	2.23	2.29	2.01
Greece	1 793	133	74	77	56	3 683	2.05	1 941	1 742	2.77	2.52	3.11
Portugal	2 590	189	73	116	73	11 447*	4.42	8 326*	3 121*	6.06	7.18	4.28
Spain	6 506	453	70	280	173	27 833	4.28	19 532	8 301	6.14	6.98	4.80
Turkey	6 642	637	96	412	225	19 204	2.89	11 695	7 509	3.01	2.83	3.34
Yugoslavia	8 500	1 084	128	298	786	27 754	3.27	7 334	20 420	2.56	2.46	2.60
Southern Europe	26 031	2 496	96	1 183	1 313	89 921	3.45	48 828	41 093	3.60	4.13	3.13
Albania	930*	80*	86*	10*	70*	2 940*	3.16	550*	2 390*	3.68	5.50	3.41
Cyprus	100	3	33	3	–	101	1.01	100	1	3.07	3.20	0.71
Israel	66	3	44	2	1	212	3.21	98	114	7.24	5.66	9.50
Mediterranean countries	1 096	86	78	15	71	3 253	2.97	748	2 505	3.78	5.00	3.53
Bulgaria	3 300	298	90	101	197	6 000	1.82	2 600	3 400	2.01	2.57	1.73
Czechoslovakia	4 185	923	221	686	237	22 503	5.38	17 055	5 448	2.44	2.49	2.30
German Dem. Rep.	2 590	440	170	339	101	15 000	5.79	11 500	3 500	3.41	3.39	3.47
Hungary	1 563	253	162	38	215	9 612	6.15	1 305	8 307	3.80	3.43	3.86
Poland	8 410	1 162	138	897	265	28 454	3.38	23 430	5 024	2.45	2.61	1.90
Romania	5 723	1 202	210	482	720	31 594	5.52	10 992	20 602	2.63	2.28	2.86
Eastern Europe	25 771	4 278	163	2 543	1 735	113 163	4.39	66 882	46 281	2.65	2.63	2.67
TOTAL EUROPE	132 917	15 941	120	10 083	5 858	504 882	3.80	327 871	177 011	3.17	3.25	3.02

Tables 3.4 and 3.5 present the general situation as regards exploitable closed forest (area, growing stock, increment and fellings). Around 1980 the Nordic countries had 36% of Europe's total exploitable closed forest, 28% of growing stock, 29% of net annual increment and 31% of fellings, with the EEC(9) in second place. Country groups' shares of growing stock and increment differ from those of exploitable closed forest, because of differences in growing conditions, policies relating to rotation age, stocking, and so on.

Thus, in Central Europe, where exploitable closed forest occupies an area of only 3% of the European total, the shares of growing stock and net annual increment are of 7% and 5% respectively. The figures of table 3.5 are practically the same as those in ETTS III, indicating that the country groups' shares of the European total around 1980 have not changed very much since 1970.

The figures in table 3.6 on the situation on exploitable closed forest show considerable variations between the countries and country groups. The volume of growing stock per hectare in Central Europe is more than three

times higher than in the Nordic countries and Southern Europe, which may be explained partly by differences in growing conditions, but also by management traditions, intensity and objectives. The average volume of growing stock per hectare in Europe around 1980 was 120 m³ o.b./ha.

The net annual increment (NAI) in 1980 in Europe of 505 million m³ o.b. represents an average NAI of about 3.80 m³ o.b. per hectare of exploitable closed forest or 3.2% of the growing stock. There are quite wide variations in net annual increment per hectare between the countries and country groups. It ranges from 8.51 m³/ha in Belgium to 1.01 m³/ha in Cyprus and from 6.26 m³/ha in Central Europe to 3.00 m³/ha in the Nordic countries. In some cases NAI per hectare depends on the age-class structure of the forest stands. In countries with large areas of young stands and plantations with still low growing stock volume per hectare, the percentage rate of increment is relatively high (e.g. 7.18% in Portugal and 6.98% in Spain for coniferous forests) although relatively low in terms of volume per hectare.

3.4 FOREST AREA BY OWNERSHIP AND MANAGEMENT STATUS

Tables 3.7 and 3.8 (table 3.7 is a summary of table 3.8) show the distribution of forest and other wooded land between the main categories of public and private ownership in Europe by country and country group. More than half of Europe's closed forests are publicly owned (51%) of which 40% belong to the State. The rest of the European forest is in private hands (49%) of which 37% belongs to farmers and only 5% to the forest industries. The proportion of forests in public ownership ranges from over 90% in Eastern Europe to one-third in the EEC(9) and

Central Europe and only one quarter in the Nordic countries. In general, in Europe, industry ownership has little importance, only the Nordic countries having this form of ownership to a significant extent (15%). The most common type of private ownership of forests is by farmers, the proportion of which ranges from some 60% in Central Europe and the Nordic countries to 9% in Eastern Europe. Some consequences of this ownership structure for roundwood supply were discussed in chapter 2.

TABLE 3.7

Europe: forest area by main ownership groups

	Area covered [a]	Publicly owned			Privately owned			
		Total	State	Other	Total	Forest industries	Farm forest	Other
				(Million hectares)				
Nordic countries	59.8	15.6	12.7	2.9	44.2	9.0	33.4	1.8
EEC(9)	32.8	12.4	5.6	6.8	20.4	0.1	9.8	9.4
Central Europe	4.9	1.6	0.7	0.9	3.3	–	3.0	0.3
Southern Europe [b]	64.7	42.2	31.6	10.6	22.5	0.4	21.1	1.0
Eastern Europe	28.0	25.5	25.2	0.3	2.4	..	2.4	–
Total	190.1	97.2	75.8	21.4	92.9	9.5	69.6	12.5
				(Percentage of area covered)				
Nordic countries	100.0	26.0	21.2	4.8	74.0	15.1	55.9	3.0
EEC(9)	100.0	37.8	17.0	20.7	62.2	0.3	29.9	28.7
Central Europe	100.0	32.7	14.3	18.4	67.3	–	61.2	6.1
Southern Europe [b]	100.0	65.2	48.8	16.4	34.8	0.6	32.6	1.5
Eastern Europe	100.0	91.1	90.0	81.1	8.9	..	8.6	–
Total	100.0	51.1	39.9	11.3	48.9	5.0	36.6	6.6

[a] May differ from total in other tables.　　[b] Excluding Albania, Cyprus and Israel.

TABLE 3.8

Forest area by main ownership groups

	Area covered [a] (1 000 ha)	Publicly owned				Privately owned				
		Total (1 000 ha)	Percentage of total area covered	State (1 000 ha)	Other (1 000 ha)	Total (1 000 ha)	Percentage of total area covered	Forest industries (1 000 ha)	Farm forests (1 000 ha)	Other (1 000 ha)
Finland	23 225	6 793	29.2	6 359	434	16 432	70.8	1 751	14 150	531
Iceland	..	..	..		..	..	..	..	..	..
Norway	8 701	1 343	15.4	1 104	239	7 358	84.6	450	5 626	1 282*
Sweden	27 842	7 426	26.7	5 200	2 226	20 416	73.3	6 794	13 622	–
Nordic countries	59 768	15 562	26.0	12 663	2 899	44 206	74.0	8 995	33 398	1 813
Belgium	680	285	41.9	70	215	395	58.1	–	95	300
Denmark	460[a]	180	39.1	148	32	280	60.9	8	272	–
France	15 075	3 881	25.7	1 428	2 453	11 194	74.3	–	3 130	8 064
Germany, Fed. Rep. of	7 207	4 049	56.2	2 238	1 811	3 158	43.8	–	2 455	703
Ireland	380	298	78.4	298	–	82	21.6	–	–	82
Italy	6 363	2 527	39.7	373	2 154	3 836	60.3	48*	3 788*[b]	–
Luxembourg	82	38	46.3	7	31	44	53.7	–	12	32
Netherlands	331	159	48.0	104	55	172	52.0	–	–	172
United Kingdom	2 178	945	43.4	945	–	1 233	56.6	..	..	..
EEC(9)	32 756	12 362	37.7	5 611	6 751	20 394	62.2	56	9 752	9 353
Austria	3 754	750	20.0	618	132	3 004	80.0	–	2 861	143
Switzerland	1 124	823	73.2	64	759	301	26.8	2	121	178
Central Europe	4 878	1 573	32.2	682	891	3 305	67.8	2	2 982	321
Greece	5 754	5 372	93.4	4 212	1 160	382	6.6	–	382	-
Portugal	2 627	630	24.0	75	555	1 997	76.0	57	970	970
Spain	25 622	8 592	33.5	1 123	7 469	17 030	66.5	400	16 630	–
Turkey	20 199	20 173	99.9	20 173	–	26	0.1	–	–	26
Yugoslavia	10 500	7 400	70.5	6 000	1 400*	3 100	29.5	–	3 100	–
Southern Europe	64 702	42 167	65.2	31 583	10 584	22 535	34.8	457	21 082	996
Cyprus	173	160	92.5	159	1	13	7.5	–	13	-
Israel	109	105	96.3	34	71	4	3.7	–	4	-
Med. countries	282	265	94.0	193	72	17	6.0	–	17	-
Bulgaria	3 800	3 800	97.4	3 700	100	–	–	–	–	-
Czechoslovakia	4 578	4 575	99.9	4 444	131	3	0.1	..	..	3
German Dem. Rep.	2 955	2 601	88.0	2 601	–	354	12.0	..	351	3
Hungary	1 637	1 170	71.5	1 157	13	467	28.5	..	457	10
Poland	8 726	7 105	81.4	7 057	48	1 621	18.6	–	1 621	–
Romania	6 265	6 265	100.0	6 265	–	–	–	–	–	–
Eastern Europe	27 961	25 516	91.3	25 224	292	2 445	8.7	..	2 429	16
TOTAL EUROPE	190 347	97 445	51.2	75 956	21 489	92 902	48.8	9 510	69 660	12 499

Source: 1980 Assessment, table 2.1.

[a] May differ from totals shown elsewhere. See source for details.
[b] Farm forests and other privately-owned forests.

Table 3.9 summarizes the information in the 1980 Assessment on the management status of closed forests in Europe. Nearly 60% of the closed forests in Europe are worked under management plans, and although the rest are not managed according to a plan, they do mostly have controls relating to management or use. Nowadays, there are hardly any closed forests in Europe which are not under some form of working plan or legal control. Nearly 90% of publicly owned forest is managed according to a plan: the proportion is higher for State-owned forests (95%) than for other publicly owned forests (62%). Eighty-six per cent of forests owned by the forest indus-

tries are managed according to a plan, but the proportion is much lower for other types of privately owned forest: 24% for farm forests and only 10% for other private forests. This is presumably one result of absentee ownerships and of the small size of holdings.

Table 3.10 shows the relative importance of small (less than 10 ha) forest holdings in selected European countries. It should be noted that it is practically impossible to obtain accurate data on the total number of private and even of public forest holdings or owners (the two are not necessarily the same: one owner may have one or several holdings or several owners may share one holding).

TABLE 3.9

Europe: area of closed forest under some form of legal control, management plan and other regulations

(Million hectares)

	Total	Managed according to plan						Under other regulations [a]					
		Publicly owned			Privately owned				Publicly owned		Privately owned		
	Total	Total	State	Other	Forest industries	Farm forest	Other	Total	State	Other	Forest industries	Farm forest	Other
Nordic countries	51.9	26.0	9.2	2.1	7.3	6.9	0.5	25.9	0.9	0.5	0.8	22.6	1.1
EEC(9)	31.0	10.3	5.4	3.8	–	0.4	0.7	20.7	–	2.9	–	6.3	11.5
Central Europe	4.9	2.1	0.7	0.7	–	0.6	0.1	2.8	–	0.2	–	2.4	0.2
Southern Europe [b]	27.3	18.7	15.3	1.9	..	1.5	–	8.6	1.8	1.3	0.4	5.1	..
Eastern European	27.4	26.7	24.7	–	..	2.0	–	0.7	0.2	0.2	..	0.3	–
Total Europe	142.5	83.8	55.3	8.5	7.3	11.4	1.3	58.7	2.9	5.1	1.2	36.7	12.8

(Percentage of closed forest under some form of legal control, management plan or other regulation)

	Total	Managed according to plan						Under other regulations [a]					
		Publicly owned			Privately owned				Publicly owned		Privately owned		
	Total	Total	State	Other	Forest industries	Farm forest	Other	Total	State	Other	Forest industries	Farm forest	Other
Nordic countries	100.0	50.1	17.7	4.0	14.1	13.3	1.0	49.9	1.7	1.0	1.5	43.6	2.1
EEC(9)	100.0	33.2	17.4	12.3	–	1.3	2.2	66.8	–	9.4	–	20.3	37.1
Central Europe	100.0	42.9	14.3	14.3	–	12.2	2.1	57.1	–	4.1	–	49.0	4.0
Southern Europe [b]	100.0	68.5	56.0	7.0	..	5.5	–	31.5	6.6	4.8	1.5	18.6	..
Eastern Europe	100.0	97.4	90.1	–	..	7.3	–	2.6	0.8	0.7	..	1.1	–
Total Europe	100.0	58.8	38.8	6.0	5.1	8.0	0.9	41.2	2.0	3.6	0.8	25.8	9.0

[a] Not managed according to a plan, but having controls relating to management and use.   [b] Excluding Albania, Cyprus, Israel and Portugal.

TABLE 3.10

**Relative importance of forest holdings under 10 ha in selected
European countries, around 1980**

	Total					
				Of which: holdings of 10 ha or less		
					Percentage of total	
	Area covered (1 000 ha)	Number of holdings	Area (1 000 ha)	Number	Area	Number
Finland [a]	13 164	322 800	640	95 700	4.9	29.6
Norway	8 701	142 019	..	59 004	..	41.5
Sweden.....................	27 842	236 073	453	60 199	1.6	25.5
Belgium	600	105 878	101	100 525	16.8	94.9
France	15 347	3 269 171	4 300[a]	3 000 000	28.0[a]	91.8
Germany, Fed. Rep. of........	6 989	485 538	1 995[b]	474 930[b]	28.5[b]	97 8[b]
Luxembourg	..	12 328	..	11 200	..	91.0
Netherlands	331	5 278[c]	53	2 123[c]	16.0[c]	40.2
Austria	3 211	227 774	540	189 424	16.8	83.2
Switzerland	1 124	258 146	251	252 775	22.4	97.9
Poland	8 588	1 431 884	1 376	1 425 558	16.0	99.6

Source: 1980 Assessment, table 6.1.

[a] Farm forest and other private (non-forest industry owned) forest only.

[b] 50 ha and less.

[c] Excluding private forest holdings of less than 5 ha.

The table shows clearly that in many countries, especially in the EEC(9) and Central Europe as well as in Poland, 90% or more of the holdings are of less than 10 hectares, with all the management problems that often accompany small holdings i.e. low investment, lack of expertise or interest by the owner, irregularity of income, difficulty of obtaining larger items of equipment, etc. (see chapter 2). Furthermore, these holdings can occupy roughly 15-25% of the forest area: as a result, a significant part of the forest may well have difficulties in achieving high levels of productivity (if this is what the owner wishes, which often may not be the case, as in some cases wood production has relatively low priority for the small-scale forest owner).

According to table 3.11 the largest forest holdings are State-owned. They range considerably in size between

TABLE 3.11

**Average size of holdings of closed forest and other wooded land
according to the structure of ownership in selected countries**

(ha/holding)

Country	State owned	Other publicly owned	Owned by forest industries	Farm forests etc.
Austria	..	..	..	14[a]
Belgium	362[b]	338[b]	–	4[a]
Cyprus	22 714	..	..	..
Denmark	2 145	123	400	9
Finland.............................	79 500	..	438	458
France.............................	843	173	–	3
Germany, Fed. Rep. of..............	2 160	140	–	7
Greece.............................	801	290	–	302
Hungary...........................	6 968	..	–	318
Ireland	1 221	–	–	..
Luxembourg	133	131	–	4
Norway............................	1 498	523	–	52
Poland	2 856	26	–	1
Spain.............................	498	597	500	6
Sweden............................	126 800	..	113 200*	58*
Switzerland	210	..	..	1
Turkey	12 600*	–	–	202*
United Kingdom....................	2 625	..	..	..

[a] Average size of holdings of closed forest only.   [b] Managed according to forestry regulations.

European countries: from 126,800 hectares in Sweden to 210 hectares in Switzerland and 133 hectares in Luxembourg (the differences may partly result from different definitions of an individual State holding). Farm forests in all European countries are very much smaller, which reflects, in most cases, traditional or historical systems of forest management and tenure in different countries. In the Nordic countries, for instance, the average size of farm forest

holdings is about 50 hectares; in the EEC(9) it ranges from 3 ha (France) to 9 ha (Denmark). The smallest average sizes of farm forest holdings are in Switzerland and Poland. In the case of Hungary, the average size of all State forest holdings is nearly 7,000 hectares, but 93% of the State forests is concentrated in twenty large-scale enterprises whose average size of forest holding is about 49,000 hectares.

3.5 RECOGNIZED MAJOR FUNCTIONS OF THE FOREST AND OTHER WOODED LAND

The present section contains the information available on the recognized major functions of the forest and other wooded land in Europe. Table 3.12 gives a broad indication of the sub-division of forest and other wooded land according to three main recognized functions (wood production, protection and recreation) as reported by countries. This categorization into three major recognized functions was a new element introduced in the 1980 assessment and caused difficulties for some countries. Because of the partly subjective nature of the estimates, the degree of comparability between countries may not be high.

These estimates refer to *major* functions. Most forests in fact have two or more simultaneous functions and only occasionally does one function exclude others (e.g. some

forms of nature conservation or forests of special scientific interest). The allocation of forests to "major functions" is therefore bound to be to a greater or lesser extent arbitrary, although it does give a broad overview of the functions of the forest. The main function of forest and other wooded land in Europe in the opinion of countries remains "wood production" (72%), "protection forest" is in second place (26%) and the remaining 2% of the area is allocated to the "recreation" function. Among the country groups, the Nordic countries have the highest share of forests for wood production (90%). The countries of Southern Europe consider that only 44% of their total forest and other wooded land is mainly for wood production, while 55% of it was reported to have "protection" as its major function.

TABLE 3.12

Europe: estimates of forest and other wooded land (OWL) by recognized major functions by country groups

	Forest and OWL (million ha)				Percentage of total forest and OWL			
	Total[a]	Wood production	Protection	Recreation	Total[a]	Wood production	Protection	Recreation
Nordic countries	59.8	54.1	4.5	1.2	100.0	90.5	7.5	2.0
EEC(9)	32.8	29.2	3.5	0.1	100.0	89.0	10.7	0.3
Central Europe[b]	3.7	2.9	0.8	–	100.0	78.4	21.6	–
Southern Europe	61.9	27.2	34.2	0.5	100.0	43.9	55.3	0.8
Eastern Europe	28.0	20.9	4.8	2.3	100.0	74.7	17.1	8.2
Total	186.2	134.3	47.8	4.1	100.0	72.1	25.7	2.2

[a] Total forest and other wooded land figures may differ from those shown in table 3.2.

[b] Austria only. All forests of Switzerland are considered multipurpose forests.

3.6 CHANGES IN THE EUROPEAN FOREST SINCE THE 1970s

The preceding sections have described the European forest around 1980: but what changes have occurred over time? Successive inventories, at the national and international level have tended to show increased forest area, growing stock and increment. However these results are

usually not comparable as a significant part of the increase is undoubtedly due to improvements in inventory methods, and some of the apparent changes are due to changes in definitions.

Correspondents were therefore requested to provide

data for a few key parameters which were comparable for a period around 1970 and a period around 1980. Data were received for fifteen countries, which between them accounted for 67% of exploitable closed forest in Europe. This may be considered a satisfactory indication of general trends. The data received for these countries are presented in annex tables 3.1 to 3.4.

The definitions used by correspondents were not always the same as those used elsewhere in this study, as estimations or changes in definitions were sometimes necessary to obtain comparability over time. The data presented in annex tables 3.1 to 3.4 should therefore be used *only* as an indication of changes over time; they are not fully comparable with other data in this study or indeed between countries. For this reason only the changes over time are shown in table 3.13. The data have been adjusted to show changes over a 10-year period.

The broad trend is clear: while the area of exploitable closed forest has increased slightly (+2.5% over 10 years), growing stock and increment have increased more rapidly (+12.3% and +8.9% respectively). This would indicate that within a relatively stable pattern of rural land use (with some major exceptions, discussed below) the forest has been developed, partly by silvicultural investment and partly as drain has remained below increment.

Table 3.13 also shows the results of action taken in some countries significantly to increase the area of productive forest. In ten years the area of exploitable closed forest in Ireland and the United Kingdom grew by about 30%, by over 9% in Spain, nearly 8% in the Netherlands and over 7% in Hungary. The largest increase in absolute terms was in Finland where the area of exploitable closed forest

increased by 677,000 ha (3.6%), also mostly in the context of a broad policy of enlarging the forest resource although there has been some natural extension. The declines in forest area in a few other countries may be attributed to pressure from other land uses, notably infrastructure and urbanization, and to forest damage (e.g. fire).

All responding countries reported an increase of growing stock on exploitable closed forest, with particularly rapid increases over the ten-year period in Ireland (+80%) and the United Kingdom (+50%). In volume terms, however, the largest increase was in France (+342 million m³ o.b., or 29%), despite the relatively minor increase in the area of exploitable closed forest. This results from the specific characteristics of the French forest improvement plan which concentrates on up-grading the quality of existing forest land rather than on planting on bare land, (e.g. in Ireland, and the United Kingdom) or other wooded land (e.g. "matorral" in Spain). Finland and Sweden also recorded a large increase in growing stock (over 110 million m³ o.b. each), which may be attributed partly to investment in forest improvement (e.g. drainage in Finland) and partly to the fact that because of low demand, removals in the 1970s were significantly below the level of drain, allowing an accumulation of growing stock.

A similar pattern is observable for increment as for growing stock with increases in most countries, especially the United Kingdom (+78%), Ireland (+39%), France (+15%), Denmark and Sweden (+13% each). The 16% drop recorded in Poland is due to infestation of that country's forests by the pine nun moth (*Limantria monacha*), the worst of which now appears to be past. The increment of Polish forests is expected to recover, if other negative influences also recede.

TABLE 3.13

Changes for exploitable closed forest between the period around 1970 and the period around 1980

	Area		Growing stock		Net annual increment	
	1 000 ha	Percent-age	Million m³ o.b.	Percent-age	Million m³ o.b.	Percent-age
Finland......................	+677	+3.6	+112	+7.8	+5.5	+9.9
Norway	+100	+1.5	+59	+11.4	+1.5	+9.5
Sweden......................	+42	+0.2	+116	+5.1	+9.1	+13.1
Nordic countries.............	+819	+1.6	+287	+6.8	+16.1	+11.4
Denmark	−5	−1.3	+3	+7.9	+0.3	+13.0
France	+227	+1.7	+342	+29.1	+6.8	+14.6
Ireland	+79	+29.5	+12	+80.0	+0.7	+38.9
Netherlands..................	+23	+7.8	+2è	+10.0	−	+2.2
United Kingdom..............	+496	+32.6	+68	+50.4	+4.9	+77.8
Austria	−65	−2.0	+46	+6.1	+1.1	+5.9
Switzerland	−2[a]	−[a]	+75	+18.4	−	−
Cyprus	−17*	−14.2*	−	+1.3	−	+0.8
Spain	+548	+9.2	+26	+6.1	+2.3	+9.0
Czechoslovakia	−64	−1.5	+122	+15.3	−	−
Hungary.....................	+130	+8.9	+38	+17.7	+0.7	+7.9
Poland	+35	+0.4	+103	+9.8	−5.5	−16.0
Total (15 countries).........	+2 204	+2.5	+1 124	+12.3	+27.4	+8.9

Note: Original data for different periods adjusted to a comparable 10 year period. The original data are presented in annex tables 3.1 to 3.4.

[a] Exploitable and unexploitable closed forest.

3.7 ESTIMATES OF WOODY BIOMASS

The traditional forest inventory measures only a part of the forest, that which was considered usable by traditional forest industries, notably sawmilling. In effect, this is usually the stems and large branches of trees over a certain minimum diameter. Other parts of the tree and trees outside the forest have always been used for fuelwood. However, technical advances in the pulpwood-using industries have made it possible to use some of this material also as raw material for pulp and wood-based panels. In times of strong pulpwood demand (e.g. in 1974) or in countries where domestic raw material availability is a constraint on the expansion of the industries, widespread interest has been expressed in expanding the availability of raw material by using parts of the biomass which had not previously been used — branches, stumps, foliage, small trees, etc.

A pre-requisite for planning in this area is the estimation of the volume and mass of woody biomass. As complete biomass inventories are extremely expensive, in most cases the biomass has been estimated on the basis of traditional inventory data. The 1980 Assessment was the first attempt to bring together at an international level, on a comparable basis, national biomass estimates. These are presented in part II of the Assessment, and summarized in table 3.14.

For Europe as a whole, total tree biomass on exploitable closed forest is about 50% higher than inventoried biomass. A further 10% is estimated to exist outside the exploitable closed forest. There are however wide differences in the technical and economic characteristics of the different types of biomass. Stumps and roots for instance, which are estimated to be equivalent to 20% of the inventoried biomass, are expensive to extract and heavily contaminated with soil and stones, which makes processing more difficult. In addition, doubts have been expressed about the possible long-term effect on forest productivity of some stands of removing stumps, or of diminishing the soil protection capacity of the forest (e.g. on steep slopes).

Nevertheless the research carried out in several countries and the estimates summarized in table 3.14, show that the potential does exist for expanding the supply of raw material for the forest industries by using a larger part of the forest biomass. Such an expansion would only take place however if demand was high enough to compensate the extra costs of using this type of raw material.

TABLE 3.14

Tree biomass in the ECE region: estimates by national correspondents and the secretariat

			Exploitable closed forest					All lands.[a]
				Above-ground tree biomass				
					Inventoried material			
	Total tree biomass	Stumps and roots	Total	Total	Wood	Bark	Other above ground	Total tree biomass
			(Million m.t. oven dry)					
Nordic countries	2 750	424	2 326	1 803	1 562	241	253	2 992
EEC(9)	2 666	348	2 317	1 762	1 619	143	555	2 846
Central Europe	715	110	605	498	449	49	107	728
Southern Europe.................	1 828	227	1 600	1 291	1 171	121	308	2 252
Eastern Europe	3 023	373	2 651	2 017	1 774	242	634	3 066
Europe	10 982	1 482	9 499	7 371	6 575	796	2 127	11 884
		(Percentage of inventoried material on exploitable closed forest)						
Nordic countries	153	24	129	100	87	13	14	166
EEC(9)	151	20	131	100	92	8	31	161
Central Europe	144	22	121	100	90	10	21	146
Southern Europe.................	142	18	124	100	91	9	24	174
Eastern Europe	150	18	131	100	88	12	31	152
Europe	149	20	129	100	89	11	29	161

Source: 1980 Assessment, table 11.26. See Part II of the assessment for description of both national and secretariat methods of assessment.

[a] Exploitable and unexploitable closed forest, other wooded land and trees outside the forest.

3.8 TRENDS IN REMOVALS

Data are available on roundwood removals in a considerable degree of detail. The trends between 1950 and the early 1970s were analysed in ETTS III, section 5.3. The present section is intended to update this analysis and therefore uses essentially the same format for the tables. The data presented for removals do not include Albania.

The data analysed in this section are those provided annually to ECE/FAO and stored in the ECE/FAO data base. They are not exactly the same as data on removals in the 1980 Assessment or those used as base period for the forestry forecasts, although the differences are not so great as to distort the analysis. This question is discussed further in chapter 5 (see especially annex table 5.1). Long-term series for removals are included in the ETTS IV companion publication ECE/TIM/31.

Total European removals rose by nearly 16% between 1949-51 and 1979-81, to reach 341 million m^3. Most of this growth took place in the 1960s while the 1970s were marked more by cyclical fluctuations than by a trend to growth. The rise in total European removals, by decade is as follows:

1949-51 to 1959-61: +11.9 million m^3 (4%);
1959-61 to 1969-71: +30.9 million m^3 (10%);
1969-71 to 1979-81: +4.0 million m^3 (1%).

FIGURE 3.2

Europe: removals 1965-1984, by assortment

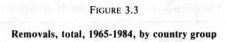

FIGURE 3.3

Removals, total, 1965-1984, by country group

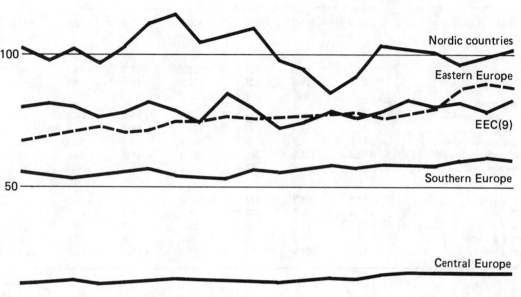
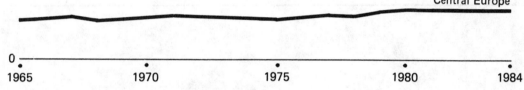

This overall trend, however, concealed two conflicting movements: while removals of sawlogs, veneer logs and pulpwood increased strongly (by 64% for sawlogs and veneer logs and by 178% for pulpwood over the 30 year period), removals of other assortments, notably fuelwood, fell. Fuelwood, pitprops and other industrial wood, taken as a group, accounted for 54% of European removals in 1949-51, but for only 22% in 1979-81. Fuelwood removals fell from 121.7 million m^3 to 53.4 million m^3, a drop of 56%. As fuelwood is mostly of non-coniferous species, removals of these species also fell in absolute terms and accounted for a significantly smaller share of the European total.

If the trend for more recent years is examined in more detail (see graphs based on annual data in figs. 3.2 and 3.3), the importance of cyclical fluctuations becomes apparent, especially for the Nordic countries, and to a lesser degree the EEC(9). These swings are mostly due to demand factors, notably the recession of the late 1970s, but part of the decline at that time in the Nordic countries was due to problems in mobilizing the supply of roundwood from private forest owners (see discussion in chapter 2).

In 1978 recorded[2] fuelwood removals stopped their long term downward trend and started to rise again. If this change of direction is not a temporary aberration, there are profound implications for the pattern of future roundwood removals. The outlook for fuelwood removals is discussed in chapter 19.

On the country group level, the largest changes (in volume terms) between 1949-51 and 1979-81 were for Eastern Europe (+ 16 million m^3) and the Nordic countries (+ 15 million m^3) followed by Southern Europe (+9 million m^3) and Central Europe (+6 million m^3). The greatest increase, in percentage terms, was for Central Europe, where removals nearly doubled over the 30 years. In the EEC(9) on the other hand, removals stagnated at around 80 million m^3. As a consequence the EEC(9)'s share of European removals fell from 27.7% to 23.9%. In the 1970s the strongest increase was for Eastern Europe. In 1979-81 the Nordic countries had not yet recovered their record levels of the early 1970s.

---

[2] Real fuelwood removals are undoubtedly greater, but no year-by-year data exist. See chapter 18.

TABLE 3.15

Europe: recorded removals, 1949-51 to 1979-81

	Volume (million m³ underbark)				Change 1949-51 to 1979-81		Percentage of total	
	1949-51	1959-61	1969-71	1979-81	Volume (million m³ u.b.)	Per-cent	1949-51	1979-81
TOTAL	293.9	305.8	336.7	340.7	+ 46.8	+ 15.9	100.0	100.0
of which:								
A. *By species*								
Coniferous...................	170.7	186.0	211.2	227.9	+ 57.2	+ 33.5	58.1	66.9
Broadleaved	123.2	119.8	125.5	112.8	− 10.4	− 8.4	41.9	33.1
B. *By assortment*								
Sawlogs and veneer logs	97.8	117.7	143.7	160.4	+ 62.6	+ 64.0	33.3	47.1
Pulpwood	37.3	59.9	93.7	103.5	+ 66.2	+ 177.5	12.7	30.4
Other industrial wood	37.1	34.7	30.8	23.4	− 13.7	− 36.9	12.6	6.9
Total industrial wood	172.2	212.3	268.3	287.3	+ 115.1	+ 66.8	58.6	84.3
Fuelwood.................	121.7	93.5	68.4	53.4	− 68.3	− 56.1	41.4	15.7
C. *By country groups*								
Nordic countries..............	87.5	98.7	111.8	102.9	+ 15.4	+ 17.6	29.8	30.2
EEC(9)	81.4	78.2	78.3	81.5	+ 0.1	+ 0.1	27.7	23.9
Central Europe..............	12.6	15.5	16.1	18.8	+ 6.2	+ 49.2	4.3	5.5
Southern Europe	50.7	48.6	57.7	59.6	+ 8.9	+ 17.6	17.2	17.5
Eastern Europe...............	61.7	64.9	72.8	77.9	+ 16.2	+ 26.3	21.0	22.9

The specifications and precision of the data for removals and harvesting losses are not sufficient to enable a precise comparison of growth (as measured by net annual increment) with drain (removals plus harvesting losses). It is clear however from the data in this chapter that drain has remained below, often well below, increment in almost all countries. A few countries over-cut in the exceptional circumstances of the immediate post-war years. In Finland and Sweden, drain approached or even temporarily surpassed increment in the 1960s and early 1970s. Overall, however, the drain on European forests has remained well below increment, leading to an expansion in growing stock.

TABLE 3.16

Europe: share of main assortment groups
in roundwood removals, 1949-51 to 1979-81

(*Percentage of total*)

	1949-51	1959-61	1969-71	1979-81
Total				
Sawlogs and veneer logs	33.3	38.5	42.7	47.1
Pulpwood........................	12.7	19.6	27.8	30.4
Other (including fuelwood)	54.0	41.9	29.5	22.5
Coniferous				
Sawlogs and veneer logs	46.7	48.7	51.1	54.2
Pulpwood........................	20.5	28.1	32.6	31.5
Other (including fuelwood)	32.8	23.2	16.3	14.3
Broadleaved				
Sawlogs and veneer logs	14.7	22.6	28.5	32.5
Pulpwood........................	1.9	6.3	19.8	27.6
Other (including fuelwood)	83.4	71.1	51.7	39.9

TABLE 3.17

Europe: roundwood removals by country, main assortment groups and species, 1949-51 to 1979-81

(1 000 m³ u.b.)

	Sawlogs and veneer logs		Pulpwood		Other (including fuelwood)		Total		Total coniferous		Total broadleaved	
	1949-51	1979-81	1949-51	1979-81	1949-51	1979-81	1949-51	1979-81	1949-51	1979-81	1949-51	1979-81
Finland	10 728	20 457	8 920*	19 563	19 719	4 967	39 367	44 987	28 653	37 577	10 714	7 420
Norway	3 754	4 671	3 747	3 899	2 826	804	10 327	9 374	9 026	8 619	1 301	755
Sweden	12 967	22 058	14 052	24 242	19 814	2 218	37 833	48 518	32 900	42 608	4 933	5 910
Nordic countries	27 449	47 186	26 719	47 704	33 359	7 989	87 527	102 879	70 579	88 794	16 948	14 085
Belgium-Luxembourg	902	1 437	67	582	1 382	659	2 351	2 678	1 268	1 548	1 083	1 131
Denmark	1 013*	1 191	4*	287	956	598	1 973	2 075	910	1 233	1 063	843
France	8 207	18 485	631	8 767	21 583*	4 023	30 421	31 275	9 984*	14 811	20 437*	16 464
Germany, Fed. Rep. of	14 999	16 959	1 982	9 518	11 348	4 985	28 329	31 461	19 737	21 538	8 592	9 923
Ireland	76	242	1	87	92	50	169	379	54	349	115	30
Italy	2 567	2 623	419	955	11 066	5 079	14 052	8 657	2 516	1 477	11 536	7 181
Netherlands	149	242	20*	345	431	298	600	884	308	600	292	284
United Kingdom	2 639	2 647	127ê	972	702	510	3 468	4 128	1 037	2 808	2 431	1 321
EEC(9)	30 552	43 825	3 251	21 513	47 560	16 201	81 363	81 539	35 814	44 363	45 549	37 176
Austria	5 266	8 945	1 595	3 229	2 241	2 212	9 102	14 386	7 968	12 066	1 134	2 320
Switzerland	1 365	2 613	328	817	1 799	963	3 492	4 393	2 720	2 960	772	1 433
Central Europe	6 631	11 558	1 923	4 046	4 040	3 175	12 594	18 779	10 688	15 026	1 906	3 753
Greece	129	482	–	143	3 126	1 965	3 255	2 590	616*	664	2 639*	1 926
Portugal	646*	4 550	114*	3 264	3 757*	606	4 517*	8 420	2 610*	5 544	1 907*	2 876
Spain	1 036*	3 552	31*	6 492	10 183*	2 331	11 250*	12 376	4 106*	7 637	7 144*	4 738
Turkey	573	4 944	42	792	5 941	16 640	6 556	22 376	919*	14 016	5 637*	8 360
Yugoslavia	5 047	7 627	668	1 545	19 355	4 450	25 070	13 622	6 370	4 492	18 700	9 130
Southern Europe	7 431	21 155	855	12 237	42 362	25 993	50 648	59 384	14 621	32 355	36 027	27 030
Cyprus	10	41	–	17	22	18	38	76	25	73	13	2
Israel	–	26	–	49	8	43	8	118	1	63	7	55
Mediterranean countries	10	67	–	66	2 436	61	46	194	26	136	20	57
Bulgaria	1 498	1 478	37*	1 102	3 527*	1 858	5 062	4 438	1 457	1 197	3 605	3 241
Czechoslovakia	5 890*	9 893	1 287*	4 754	4 185	3 716	11 362	18 364	8 737	14 062	2 625	4 302
German Dem. Rep.	6 290	4 046	1 849	2 565	4 904	3 378	13 043	9 989	11 322*	7 933	1 721	2 057
Hungary	417	1 839	7*	1 231	2 048	3 096	2 472	6 166	89*	400	2 383*	5 766
Poland	6 450	10 787	1 206*	4 531	5 490	5 246	13 146	20 564	11 987	16 700	1 159	3 865
Romania	5 177*	8 577	144*	3 776	11 333*	6 065	16 654*	18 417	5 433	6 942	11 221*	11 475
Eastern Europe	25 722	36 621	4 530	17 959	31 487	23 360	61 739	77 940	39 025	47 235	22 714	30 705
EUROPE	97 795	160 411	37 278	103 525	158 844	76 779	293 917	340 715	170 753	227 909	123 164	112 806

Note: Data in the same format for 1969-71 are in ETTS III, table 17/5.

TABLE 3.18

Europe: recorded removals by main assortments, 1949-51 to 1979-81

	Volume (million m³ underbark)				Percentage change		
	1949-51	1959-61	1969-71	1979-81	1949-51 to 1959-61	1959-61 to 1969-71	1979-71 to 1979-81
Sawlogs and veneer							
logs	97.8	117.7	143.7	160.4	+ 20	+ 22	+ 12
Coniferous	79.7	90.6	107.9	123.6	+ 14	+ 19	+ 15
Broadleaved	18.1	27.1	35.8	36.8	+ 50	+ 32	+ 3
Pulpwood	37.3	59.9	93.7	103.5	+ 61	+ 56	+ 10
Coniferous	35.0*	52.3*	68.9	72.0	+ 49*	+ 32*	+ 5
Broadleaved	2.3*	7.6*	24.8	31.5	+ 230*	+ 226*	+ 27
Other industrial							
wood (incl. pitprops)	37.1	34.7	30.8	223.4	− 6	− 11	− 24
Coniferous*	25.4	22.8	17.6	14.9	− 10	− 23	− 15
Broadleaved*	11.7	11.9	13.2	8.4	+ 2	+ 11	− 36
Total industrial							
wood	172.2	212.3	268.3	287.3	+ 23	+ 26	+ 7
Coniferous	140.2	165.7	194.4	210.5	+ 18	+ 17	+ 8
Broadleaved	32.0	46.6	73.8	76.8	+ 46	+ 58	+ 4
Fuelwood	121.7	93.5	68.4	53.4	− 23	− 27	− 22
Coniferous	30.6	20.3	16.8	17.4	− 34	− 17	+ 4
Broadleaved	91.2	73.2	51.7	36.0	− 20	− 29	− 30
TOTAL	293.9	305.8	336.7	340.7	+ 4	+ 10	+ 1
Coniferous	170.7	186.0	211.2	227.9	+ 9	+ 14	+ 8
Broadleaved	123.2	119.8	125.5	112.8	− 3	+ 5	− 10

CHAPTER 4

Non-wood benefits of the forest*

4.1 INTRODUCTION

One of the major global issues of the second half of the present century has been man's alteration of the natural environment on which he depends absolutely for his being.

The power to change has never been greater and the pace of change has never been greater. To many, the price that is being paid for changes in land use outweighs the benefits that accrue from them.

The conservationists or environmentalists point to examples of new techniques in the forest sector which have had unforeseen adverse consequences; for example, the planting of large areas with a single exotic species of conifer with the attendant risk of a natural disaster such as devastating insect attack or the pollution of water courses and even underground water with fertiliser run-off. Some go so far as to refuse to recognise the harvesting of wood as a benefit.

Previous studies of the demand for and supply of wood and wood products in Europe as well as part III of the forest resource assessment have confirmed that the greater part of Europe's forest area is managed with the primary aim of producing timber for industry. The policies of individual countries have nevertheless emphasised the need to use the forest resources wisely and to have regard to the other functions of the forest.

A change in people's attitude to the forest and in their demands upon it has become very evident since the early 1970s. More and more informed opinion, and not just so-called pressure groups, is becoming concerned with the protection of the environment and the demand for the non-wood functions of the forest.

Decision-makers are put in a difficult position. Seen from the point of view of the adverse balance of European wood supplies, the argument is all in favour of increasing wood production. But the growing demand for environmental and recreational services from the forest requires policies to take full account of alternative, perhaps conflicting land uses. Indeed any discussion of non-wood benefits raises issues and introduces subjects which may be within the competence of government departments and organizations unconnected with forestry and the timber trade.

It was therefore decided to devote a chapter of ETTS IV specifically to the outlook for non-wood benefits, not seen purely as influences on the level of wood production. This chapter will:
- Present a simple classification of the benefits of the forest;
- Describe the situation and outlook for each of these benefits, concentrating on the outlook for demand and for the forest's ability to supply the benefit in question;
- Draw attention to some broader policy questions arising from the presentation.

It should be mentioned that there are severe methodological obstacles to a quantitative analysis of these questions. It is relatively easy to assign a market value to most of the non-wood *products* although, in some countries, the production or gathering of particular products may be subsidised for social reasons and the market may be correspondingly distorted. More difficult to measure and to quantify are the benefits that are conferred by the *environmental* functions of the forest. Sometimes it may be easier to assess a value – say, of protection forest – by assessing the value of physical damage to the locality or region that would occur if the forest were removed. There have been many attempts to devise methods of measuring in monetary terms the costs and benefits of *recreational* facilities and there have been some successes in assessing these costs and benefits for particular situations. In some countries the costs of providing recreation in the forest has been assessed without it having been possible to measure the corresponding benefits in monetary terms.

This chapter does not attempt to examine the methodologies that have been used or proposed for the monetary evaluation of the non-wood functions of the forest.

Some of the factors which will influence demand for and supply of the non-wood functions have already been discussed in chapter 1.

The problem of formulating acceptable forest policies cannot be bypassed; decisions have to be taken. And guidelines for decision-making can be established, if not in strictly economic terms then on the basis of a consensus of views among experts from different disciplines with weight also being given to informed public opinion.

* This chapter has been prepared with the invaluable help of Mr. E. G. Richards, United Kingdom. It also benefited greatly from the comments of a small meeting of experts convened to discuss the first draft (Messrs. P. Bartelheimer (Federal Republic of Germany), A. Madas (Hungary), R. Morandini (Italy), P. Bakker (Netherlands) and K. Janz (Sweden)).

4.2 CLASSIFICATION OF BENEFITS OF THE FOREST

The non-wood benefits of the forest are many and their relative values change with time and place.

For the purpose of discussion they are considered under two main headings.

There are the *service functions* of the forest which are here considered in two groups:
- The environmental functions;
- The recreational functions.

The choice of functions and their allocation to environment or recreation is somewhat arbitrary e.g. landscape might be considered under the environmental functions.

The second heading concerns the goods produced by the forest: these are divided into wood and what used to be referred to in forestry literature as *"minor forest products"*; they are here referred to as *non-wood products* of the forest.

The choice of non-wood products has also been made on an arbitrary basis. Some are by-products of the mainstream wood industries (e.g. naval stores derived from pulping liquors) but where a product is harvested from the forest in its own right (e.g. naval stores from resin tapping) it has been included and as the text makes clear, considered mainly in this sense.

Figure 4.1 presents a brief classification of these benefits; this chapter is organized according to the classification in figure 4.1 which should not be taken as a final classification:

FIGURE 4.1

Classification of the benefits of the forest

Goods	Services
A. *Wood* (Logs, pulpwood, fuelwood, etc.)	A. *Environmental functions* Protection (e.g. from erosion, avalanches, floods, winds)
B. *Non-wood products* Food (e.g. berries, mushrooms, nuts, honey, meat) Fodder (for domestic animals) Wool and skins Tannins Christmas trees, foliage, mosses lichens, other decorative materials Cork	Environment Global (CO_2/O_2 exchange) Local (screening, absorption of noise and pollutants) Water regulation and quality Nature conservation B. *Recreational functions* Leisure pursuits and tourism (e.g. walking, riding, orienteering, skiing) Hunting and fishing Landscape

4.3 PRESENT SITUATION AND FUTURE OUTLOOK FOR NON-WOOD FUNCTIONS OF THE FOREST AND NON-WOOD PRODUCTS

4.3.1 Introduction

In this section the various goods and services provided by the forest are first described and discussed from a current point of view. An attempt is then made to forecast how the situation may have changed by the year 2000 to 2010 under the influence of the various factors affecting change in demand and supply, which are discussed in chapter 1.

Any attempt to summarize the current situation on the European level requires the simplification of many complex issues. Yet it is possible to detect a few threads that are common to most if not all countries included in this study.

Thus the forest policies of most countries now reflect, to a greater or lesser extent, the increasing demand for a better regard for nature conservation and the landscape. At the same time, other demands, often older and more traditional (e.g. protection, water conservation, food gathering) have not diminished. And there is everywhere an increasing use of the forest for recreation by populations who are moving from the rural countryside into towns and villages but who are becoming increasingly mobile as car ownership increases.

4.3.2 Environmental functions

4.3.2.1 PROTECTION

The area of protection forests maintained and managed to help control soil erosion, flooding and avalanches appears to be adequate to meet most needs for this type of forest in all but the Mediterranean countries. There are afforestation and reforestation programmes that have soil restoration and stabilisation as their primary aim.

Protection against damage by wind is another matter. Shelter belts are still being created on an extensive scale in the Soviet Union to prevent the blowing of soil and to ameliorate the microclimate for agriculture. On a very much smaller scale other countries, often with a rather small area of forest relative to their land surface, continue to encourage the planting of small shelter belts on agricultural land.

The specialised forests planted to prevent the often serious drifting of sand dunes along the coast line are not being added to on any large scale, although their maintenance may be a continuing commitment as, for example, in the Landes region of France.

Where shelter belts and hedges have been cleared in order to increase the size of individual fields and so facilitate the use of large agricultural machines, there are sometimes moves to restore some of the shelter lost. This means recreating shelter belts, not necessarily in their original locations.

There are plenty of examples to demonstrate the beneficial effects of forests on the water regime: without tree cover in catchment areas the run-off water carries large quantities of silt and reservoirs quickly become silted up. In the Mediterranean region, for example, forest cover (including coppice) may help prevent the undesirable effects of flash flooding following heavy rainfall on dry, steep slopes. Without adequate forest cover, loss of soil can reach disastrous proportions as, for example, is reported by Southern European countries.

It is considered unlikely that in general there will be any significant reduction in the area or change in the location of existing protection forests, since the need to control erosion or avalanches or floods or to mitigate the effect of the wind, will still be of paramount importance in various parts of Europe. Fears have been expressed however about the possible effect on the protection functions of damage attributed to air pollution. Where protective forests are being exploited commercially, future demand prospects in Europe point to their continuing to be managed as wood producing as well as protective forests (although it is not technically necessary from the protection point of view to manage for wood production).

The present protective forest area will however become increasingly costly to manage, compared to other forests where cost limiting measures can be applied more easily. The main reasons are:

- Forest operations cannot be fully mechanized because many protection forests are on very steep slopes;
- The selection system under which they are mostly managed is not amenable to mechanised management and is more intensive than most other systems of management;
- Ageing crops which will need to be regenerated if the stands are to remain effective in the future in their protective role;

The explosion in skiing with the creation of new resorts, the expansion of existing resorts, the installation of ski-lifts and the creation of downhill pistes, cause serious damage to protection forests in several ways e.g. the forests may be divided up into relatively narrow belts; natural regeneration is damaged by the sharp ski edges.

It is likely that the area of protection forests will be increased significantly by new planting in the Mediterranean and Southern Europe, especially on sites where heavy rains follow long periods of drought and carry away the dry soil in "flash floods". These forests are likely to be created on abandoned agricultural land including land currently grazed by local or nomadic herds of sheep or goats.

The value of the assets which these forests protect will continue to be seen as far outweighing the cost of establishing and maintaining them.

4.3.2.2 GLOBAL ENVIRONMENTAL FUNCTION

The argument is often advanced that the world's forests make a vital contribution to the continuing balance of the earth's atmosphere. The hypothesis is that with the burning of more fossil fuels such as coal and oil, the carbon dioxide (CO_2) layer of the atmosphere is increased and the oxygen ratio diminished; the additional CO_2 will prevent the reflected heat from the earth's surface escaping and temperatures will rise. The most gloomy prediction is that eventually the polar ice caps will melt, the seas rise and the inhabitable land surface will diminish. Not all scientists take such a dismal view, and some believe that there is "cause for concern but not panic."

There does however seem to be some common ground among scientists, namely that to have any significant adverse effect on the situation would require the destruction of the world's forests on a massive scale. Likewise to have any significant beneficial effect − by absorbing large amounts of carbon dioxide − would require massive new plantations "the size of Europe".

No future scenario for Europe's forests can reasonably contemplate such extremes.

4.3.2.3 LOCAL ENVIRONMENTAL FUNCTION

The ability of even a relatively small area of trees to screen a new factory, reduce the noise from a motorway, or to absorb people, has led to a large amount of planting in industrial or urban situations. Planting with one object in view will often confer other benefits as well.

Planting on the scale envisaged under this head may more properly be called planting "trees outside the forest" but in aggregate it can amount to a considerable area and make a small but significant contribution to total wood resources in wood-poor countries.

In other situations the woodlands may have another role – that of separating open areas from one another and thus giving these open areas a greater feeling of privacy that would not be the case if they were not screened from each other.

The planting of trees around industrial sites, motorways and around towns generally will be a continuing process. The cost of these "environmental" woodland areas will be borne by the State or the local community. The production of wood will be secondary to the protection of the environment by the reduction of noise and screening of buildings, increased wildlife and the improvement of air quality by filtering out obnoxious particles.

The choice of species will be made with these main objectives in view. These plantations will contain a variety of species, and they will be managed intensively on a selection system so that the tree cover is maintained permanently. Near towns, forests which were originally planted for wood production will more and more be managed as environmental forests, with wood production as a relatively low priority.

Advances in silvicultural knowledge and in tree breeding techniques will mean that new varieties or species (e.g. hybrids) of tree will be better able to withstand the adverse conditions in which they are planted to capture high concentrations of dust and pollutant particles. It is indeed conceivable that environmental planting and maintenance will become a rather more specialized and more important branch of forest science than it is today.

4.3.2.4 WATER

In most European situations, the beneficial role of trees in controlling run-off is not in dispute. Nor is the role of trees in improving water quality. Slowing down the run-off increases the chances that rain water will percolate through the soil where surface-induced impurities will be removed. Coniferous forests may however tend to increase soil acidity, for reasons which are not wholly understood.

Trees will reduce the quantity of water that can be gathered in a reservoir system because of the evaporation from them so that other forms of vegetation, e.g. heather and grasses, which are not eroded by rapid run-off and which use less water may be preferred to trees by some water authorities. Similarly coppice may be preferred to high forest.

The beneficial role of the forest will continue to be an important factor in water gathering and water control regimes in future.

Substantial areas of abandoned grazing and other farm land will probably be planted – especially in Southern (Mediterranean) Europe – as part of water gathering schemes to improve water quality, control flooding and reduce the role of erosion and silting up of reservoirs.

4.3.2.5 NATURE CONSERVATION

Man has regularly cultivated and fertilized his fields, and time and time again has felled and replanted large tracts of forest. But for one reason or another particular areas of woodland have escaped his attention so that some of the oldest, least changed ecosystems in Europe are to be found in the forest. Even where the forest is regularly exploited, man's intervention often comes at long intervals of time – in contrast to the annual cropping ritual on the farm.

The popular image of forestry is thus often rather different from that of modern agriculture in that the forest is regarded as belonging to nature. Clear evidence of the ambivalent public attitude to the land may be seen in the use of herbicides, insecticides and chemicals in general. Their use in the forest is virtually banned or actively discouraged in a number of countries, e.g. Sweden and Switzerland, and the adverse economic consequences accepted. Yet, in the same countries, large quantities are used on the farmlands – even those totally within forests.

The popular desire to keep forests "natural" can be, at least partly, met by trying to strike a balance between maximising wood production and allowing the naturally occurring trees, flora and fauna to take over the area.

Different types of forest in different areas will merit and will get different management priorities. At one end of the scale the national nature reserves, sites of special scientific interest or heritage woodlands enjoy almost total protection against any form of timber exploitation or interference by man; at the other end of the scale full exploitation of the forest is constrained only by the requirement to safeguard the existence of one or two species of animal or plant. The extent of the national reserves and their importance is discussed in chapter 3. It is sufficient here to note that many ECE countries now have in their forest laws or forest policies safeguards for the protection of nature in the forest.

The growing public awareness of the importance of the forest as a wildlife habitat brings with it a demand by ordinary people to observe and enjoy that wildlife. And so we have the not unusual problem of how to allow people to see the flora and fauna of the forest without damaging or destroying them or their habitat.

Another dilemma arises where the wildlife is unduly influenced by factors that stem from man's activities outside the forest. Forest ecosystems that depend on a high water table with regular inundation of the soil may suffer from drainage schemes that reduce the flow of flood water that is so essential in maintaining their special character; or the flood water may have a high nitrogen content due to run-off from heavily fertilized farm fields and damage delicately balanced plant life. A particular species of plant or animal may be favoured by practices or happenings in the surrounding countryside and may come to dominate other rarer, more valuable species in the forest.

It is not always the case that the objectives of nature conservation are clearly thought out and spelled out; the same also applies to the methods of achieving those objectives. In part this is due to a lack of knowledge as to how best to manage certain types of resource in order to conserve them.

By the turn of the century there will have been a great increase in knowledge about the forest and all that lives in it, above and below the forest floor. Much more will be known about the various balances and checks in nature − of the dependence of one form of life on another; the relationship between trees and other plants and animals and birds and insects and a host of living organisms that are seldom seen. The public will be much more aware, through the medium of television and other popular means of portraying nature, of all the issues involved in forestry. The town dweller no less than the contryman will have a new awareness of the importance of nature conservation and, as he is in a numerical majority, his views will have considerable influence on governments and their perception of the role of the forest. Certainly the benefits of nature conservation will be regarded more highly than they are today, and the considerable costs will be accepted more readily.

At the same time advances in methods of establishment, fertilizing techniques, tree breeding and other such measures to increase the wood benefits of the forest, will mean that it will be possible to grow more usable wood or wood fibre per unit area than is now the case.

One can envisage that understandings will be reached between the producers of wood and what will be a large, well-informed nature conservation movement. Areas where intensive wood production may be undertaken will be agreed. Other forests will have objectives entirely directed to conservation. Some fragile and scarce forest types will be allowed to produce timber so long as the objects and methods of management first guarantee the continued existence of the forest as an ecosystem, e.g. some of the coppice and coppice with standards systems that have persisted since medieval times are worked even today on a profitable basis. Nevertheless, it is not possible to rule out continued, even intensified conflict between different interest groups, with differing, sometimes irreconcilable priorities for the same woodland area.

4.3.3 Recreational functions

4.3.3.1 LEISURE PURSUITS AND TOURISM

There is everywhere a demand for some form of recreation in the forest and the provision of recreation in the forest is often an important part of the forest policy at national or regional level. Recreation in the forest includes:

− Recreation by urban dwellers in their local forest;
− Recreation by the urban and rural population in forests distant from towns;
− International tourism.

Recreational facilities may be provided at varying degrees of intensity. At one end of the scale large tracts of forest may be open to the public for fresh air and exercise and perhaps simple camping, without the provision of any infrastructure other than the normal forest roads and tracks used primarily for forest management and harvesting. At the other end of the scale, the object may be to attract relatively large numbers of people into small areas by providing visitor information centres or picnic areas, or tracks for walking, jogging or skiing: or a combination of these and other facilities. The provision of intensive recreation may help to channel people away from parts of the forest which have a low carrying capacity (e.g. because of unsuitable terrain or which have a flora or fauna sensitive to disturbance by people).

Recreational facilities − even of the extensive kind − nearly always involve some cost, either direct or indirect, through damage caused by people to the soil or to trees or to wildlife generally. Litter can be a nuisance that is costly to control.

There are conflicts between different interest groups using the forest for recreation: walkers and horse-riders can come into conflict because riding often damages paths. The rider perched up on his horse may unwittingly intimidate pedestrians. Wildlife may be disturbed by intensive orienteering. To the extent that it is possible and economically feasible, different groups of activities are separated − walkers and joggers using different routes from horse-riders and both perhaps having different routes from cross-country skiers (*langlauf*).

In the Mediterranean region the recreational demands may differ from those of the rest of Europe. For one thing the region is much less industrialised than the rest of Europe, so that a smaller proportion of the population lives in towns; and country dwellers make less demand for recreational activities than urban dwellers, for they have the countryside everywhere around them.

There is a high demand for tourism in Europe and it makes a major contribution to the economy of countries with quite different climates, geographies and forms of land use. The more northerly or more mountainous forest-rich countries make a tourist feature of their well-forested landscapes as well as suggesting that the forests offer a pleasant ambience in which to walk, picnic or camp; or to ski and walk in the winter. Forests are also a feature in Mediterranean tourism publicity as offering shade and coolness as change from the summer sunshine on the beaches and as making a major contribution to the enjoyment of landscape.

Tourism can have harmful effects on the forest (notably forest fires) and, as mentioned earlier, damage from skiing can be extremely serious in the mountains where it is mainly concentrated.

In future, facilities for urban dwellers will commonly include the sorts of facilities that are currently provided in the vicinity of towns in only a few European countries. There will be separate tracks for walkers and joggers; for horse riders; for skiers where there is winter snow; after dark these will be illuminated so that they can be used at the end of the day's work or before going to work or to school. It will not be uncommon to find changing facilities with showers, information centres, picnic facilities and restaurants.

Public transport will link these forests to the town centres. In forests distant from urban centres, the emphasis will be on leaving them in as natural a state as possible. But there will be overnight accommodation facilities including provision for tents and caravans. These will be on the fringes of the forest or in nearby villages. There will be more hotel and other facilities for tourists catering for forest-based holidays.

The quest for natural surroundings without too many artificial trappings will be evident in the more industrialised parts of central and northern Europe where there is already a large concentration of the population in towns and major industrial conurbations. But throughout Europe much of the forest will be regarded by a large section of the population as a place of relief from the tensions of daily life.

4.3.3.2 SECOND HOMES

Second homes have considerable importance in several parts of Europe. Second home ownership – often of the original family farm and forest – is common in the Nordic countries. It is also important in Eastern Europe and the USSR where the opportunity to leave the city for the week-end is highly prized. Camping – for quite long periods in the summer months – is also popular in some countries (e.g. the Netherlands) as an escape from urban living, and the forest can provide pleasant camp sites screened from the surrounding countryside. The growing demand for second homes can be met in part by surplus farm houses and other buildings adapted for living in. Regular long week-ends will be spent in cultivating plots of land attached to second homes and processing the produce (making wine, preserving by bottling or deep freezing). Garden plots without houses will have tented or caravan-type accommodation for summer use.

It is difficult to envisage a situation in the Nordic countries where increased demand for recreation, tourism and second homes will put the forests under pressure; there is adequate supply to meet demand.

As one moves south from the Nordic countries future supply of forest land for second homes may become more limited in relation to increasing demand. In some regions the concentration of second homes into smaller areas such as coastal strips or lakesides can have a negative effect on the very environment that the home owners wish to enjoy. In the Alps the supply of recreation will be constrained to reduce the risks of further unacceptable damage caused by tourist and local recreational pressures – mainly

from skiing but also from general visitor pressure at other seasons. Some new, relatively small, forests may be created on abandoned agricultural land but these will not make more than local contributions to meeting demand for recreation and tourism.

It is in the southern part of Europe and especially in countries bordering the Mediterranean that demand for forest land for second homes will show the greatest increase and where changes in agriculture and changes to better land use patterns will give the greatest potential for creating new forests to meet this demand. Impetus towards forest-based recreation will be given by the overcrowding of the coastal strip generally and the beaches in particular. Efforts are however being made to resolve the problems of pollution of the Mediterranean.

4.3.3.3 HUNTING AND FISHING

Hunting

Hunting excludes trapping for furs which is dealt with below.

Deer are the animals that are most commonly hunted because they are the animals that are met with throughout European forests. Other animals that are hunted include wild pig, elk, fox, pine marten, lynx, bear, wolf and the humble rabbit. Game birds include pheasants; the much rarer capercaillie; and several species that live more on the fringe of rather than in the forest proper such as the partridge and the pigeon. In some countries birds other than game birds are hunted either for sport or to eat.

Hunting is governed by law in all countries. Some laws are comprehensive and enforced as in the Federal Republic of Germany. In some countries only the hunting of certain animals and birds is covered by law whilst other species may be regarded more as locally important food (e.g. rabbits in Portugal) or as vermin to be hunted at will (e.g. rabbits in Great Britain).

Where there is intensive management of the game including for example, the rearing of large numbers of pheasants, the sporting rental may be far greater than the financial benefit of wood production. This is a reflection of the large sums of money that wealthy individuals are prepared to pay for the privilege of regular shooting where they are guaranteed that there will be ample birds to shoot.

Damage to forests and surrounding agricultural crops from hunting activities stems partly from human interference in the favouring or breeding of species that are prized by hunters and locally in the short term from the general disturbance that may result from intensive hunting (e.g., shooting pheasants by large numbers of guns). In some countries overpopulation of game has caused serious damage to forests. The maintenance of a proper balance between game (particularly deer) and the forest is seen as an economic and ecological necessity (although opinions differ as to a "balanced" level of game).

The desire to protect species from hunters (e.g. moose in Sweden) may reduce hunting to the point where there is a big population increase in the protected species and high damage to the forest and agriculture. Drastic changes in the population of particular species are also caused by quite other forms of human interference e.g. the use of chemicals on bird populations.

Hunting in public and private forests will be more rigidly controlled from the safety aspect because of the numbers of people using the forest for fresh air and exercise and for recreation generally. In some countries where public access to private forests may be currently restricted because of the priority given to hunting, in future, at selected periods hunting will have priority and during those periods access by the general public will be restricted. But public pressure will be such that outside these hunting seasons the forests will be open to the populace generally.

Although the breeding of birds for hunting and the feeding of deer to improve the hunt will always be practised and no significant change is seen concerning hunting by owners themselves in their own woods or by local people with the right to hunt locally, observing and photographing wild animals will become more generally popular with more people than hunting to kill, and will in some cases become revenue-earning for forest owners.

The cost of rented hunting is almost certain to continue to increase in real terms so that the cost of any environmental afforestation which has only marginal timber production interest can be offset against the large sporting revenues that will be obtained – obtained, moreover, early on in the life of the plantation.

Fishing

There is some doubt about the appropriateness of including fishing as a non-wood benefit, since the sport is water based, rather than forest based. Fishing in forest streams and natural and artificially created lakes inside the forest is not much different from fishing outside the forest, except that the forest may provide what some fisherman consider as a more enjoyable ambience. On the other hand, forest services may be the responsible government agency for the development as well as the control of fishing within publicly owned forests, and the regulation of streams and water generally may be a function of the forest. It has therefore been decided to include fishing in this section of the study.

Demand for fishing continues to grow even though the cost of fishing is tending to rise in real terms. Except for salmon and sea-trout fishing and some famous trout streams, the sport is still cheap enough to be afforded by many millions of people in Europe. The revenue from fishing can be substantial. Some local employment is generated by the need to manage the fishing, be it in streams, lakes or specially created ponds.

Afforestation with conifers is said adversely to affect fish in streams if the trees are planted close to them.

It is likely that the demand for fishing for sport will continue to increase. Fishing rents will rise in real terms to compensate for additional expenditure on the management of fishing and the small loss of timber production through leaving areas unplanted or planted with unproductive species along streams and lakes.

4.3.3.4 LANDSCAPE

The landscape of Europe is the direct result of man's activities in agriculture and forestry and in the creation of towns, industries and communication networks.

It is difficult to put a money value on landscape quality, yet the tremendous outcry that often accompanies change in the landscape shows that it is often highly valued, by both town dwellers and country people.

Trees and forests are, along with many types of farming activity, generally thought to be beneficial to the landscape. Many people deplore the destruction of small woods and hedgerows by modern grain farmers for whom large fields are more economic to work with large modern machines. An exception is when open moorlands and rolling hills are afforested on a large scale, since many people prefer the landform and the vegetation of open spaces.

There are moves towards integrated countryside planning and forestry operations will probably be increasingly required to conform to regional landscape plans. Already in some countries landscaping of forests has been considered important enough to make it mandatory for foresters to discuss their plans for reforestation and afforestation with the planning authorities.

As time goes by, public pressure and government reaction to that pressure will mean that plantations will have to be better integrated into the landscape. Society will be willing to accept the reduced efficiency of wood production that is sometimes the cost of good landscaping.

Where the traditional appearance of landscape is threatened because Alpine and other pastures have been abandoned by farmers, measures may be taken to prevent the pastures being invaded by tree growth and shrubs. The frequency of such operations is likely to increase, especially in tourist areas.

4.3.4 Non-wood products

4.3.4.1 FOOD

Early, primitive man certainly relied on the forest for much of his food and most probably his simple medicines too. The fruits of certain broadleaves, e.g. sweet chestnut (*Castanea sativa*) and the seed of some of the conifers (e.g. *Pinus* spp.) undoubtedly formed part of the diet, together with mushrooms and berries in the autumn.

Herbs from the forest floor, the bark of willow (*Salix*) eaten or chewed raw would have provided some relief from some forms of illness.

The wild animals and birds that were killed for food or were eventually domesticated, depended in part on the foliage of trees and in part on the vegetation of the forest floor.

There are several forest foods which still have importance today, including tree fruits and seeds, berries, mushrooms and honey. Most are, of course, also cultivated outside the forest in intensive units which supply by far the greatest part of demand; the exceptions include fruits and nuts such as sweet chestnut (*Castanea sativa*), hazel nuts (*Corylus avellana*) and pine seed (*Pinus pinea*). The forest foods may have considerable local importance — for example in a few parts of Turkey the collection of pine seeds (*Pinus pinea*) is estimated to contribute the equivalent of $US 300 per capita per annum to the local economy. Meat from domestic animals grazed in the forest and from wild animals and birds shot for sport or as part of a control programme is also still an important source of protein in some countries or in others a luxury food. The yield of berries is of economic importance to some countries, e.g. Scandinavia and Poland.

The production of honey from forest-based and forest fed bees is of economic importance in many parts of the world. In the temperate forests of Europe, the forest plants from which honeybees derive much of their nectar can be divided into three groups:

- Trees which supply floral nectar;
- Trees harbouring infestations of Hemiptera which provide honeydew;
- Herbs and shrubs which often cover large areas of the forest floor where there is a break in the tree canopy.

Plants in all the above groups are sources of honey, often in commercially important quantities. Many beekeepers migrate their hives over large distances during one flowering season in order to work a particularly profuse honey flow e.g. from pines in Greece. The lime trees (*Tilia* spp.) of Europe and sweet chestnut (*Castanea sativa*) are also of great importance to beekeepers.

Agricultural research work has shown that it is rarely profitable to base beekeeping solely on field and meadow crops. More predictable honey harvests can be obtained by migrating hives into the forests for the honeydew flows. The indirect benefit of apiculture to fruit and seed production outside the forest through pollination is often greater than the value of honey produced.

Edible forest mushrooms are gathered by families all over Europe. In some countries where people may not know which are the edible and which are the poisonous species, identification facilities are made available during the main gathering season. Forest mushrooms and truffles in some regions may be of considerable importance as a source of additional cash income to rural populations when sold to the restaurant trade.

There is considerable international trade in Europe in chestnuts (*Castanea*) and hazel nuts (*Coryllus avellana*) which may be said to come from the forest proper — as distinct from being cultivated in orchards. The seeds of *Pinus pinea* (umbrella pine) are also traded internationally.

4.3.4.2 FOOD FROM GRAZING ANIMALS

Grazing in the forest is taken to mean grazing by goats, sheep and cattle for the production of milk and its products such as cheese, and meat; and in northern Scandinavia and north-east USSR, grazing by reindeer. It has

also to be noted that wild animals and birds, including those like pheasant that are reared for sport, provide substantial quantities of meat which is often highly prized and relatively expensive if bought on the market.

Throughout the Mediterranean region, domestic animals are grazed in the forest. Cattle damage the forest soil by the compacting action of their feet; regeneration is hindered or stopped altogether. The goat relies on foliage and small branchlets for a substantial part of his diet while the sheep rely more on the grasses and other vegetation of the forest floor rather than foliage. The life style of the goatherds and the shepherds is essentially nomadic with sometimes long traditions of access to large tracts of country irrespective of its ownership. Illegal grazing of the forest is a problem that is difficult to overcome. Even where regeneration is almost totally prevented by grazing and the forest is in danger of extinction it may be difficult in practice to stop nomadic grazing by goats. Shepherds set fire to the forest to clear the forest floor and to get fresh grass.

In other regions there is now very little grazing in the forest except in a few specialised situations. Thus poplar plantations with their wide spacing may also grow sufficient grass to make grazing with cattle feasible, e.g. in northern Italy.

Reindeer husbandry is a traditional but still vital industry, which is practised in the northern part of Scandinavia — Lapland. Reindeer (*Rangifer tarandus*) are semi-domestic animals, which graze in northern forests and fell (tundra) areas. About three-fourths of the total number of reindeer graze in the coniferous zone which is also used for timber production. Reindeer husbandry has traditionally been based on the natural forage supplied by northern forests, peatlands and fell areas. The natural forage is still of vital importance, even though increasingly in the most severe part of the winter supplementary food is carried into the forests and some of the reindeer are tended in corrals. The utilization of forage plants by reindeer and the reindeer management work such as herding, marking, gathering, slaughtering and artificial feeding constitute the output of reindeer husbandry, consisting of meat and some by-products such as horns.

Looking to the future, the food that forests produce will be more highly valued by society that will become more aware of the distinction between "convenience", packaged foods and the more natural foods such as berries and mushrooms and honey. Gathering food from the forest rises with increased leisure time and increased unemployment.

Edible nut-bearing trees, e.g. chestnut, hazel and walnut, and berry bearing shrubs will be planted or encouraged. Because of the large numbers of people wishing to collect these foods in countries with high densities of population, the local authorities may ration the quantities that an individual may gather.

Beekeeping in the forest will increase. Even dense coniferous forests can support bees through the activities of aphids which produce nectar on which the bees can feed.

Grazing animals such as sheep, goats and cattle will reduce in numbers and their importance as a source of food will diminish. In part this will be brought about by the impossibility of getting flockmasters to live a lonely nomadic life and in part because of the competition from animals intensively raised on farmland.

Reindeer husbandry will, however, continue well into the next century, even if some of the current management practices change.

4.3.4.3 MEAT FROM WILD ANIMALS

The meat from wild animals and birds (including those birds reared for sport) shot by sportsmen or in the course of culling by game managers is of considerable economic importance to the restaurant and hotel trade in some countries. It may be traded internationally — the Federal Republic of Germany imports much of the venison (deer meat) arising from the shooting of deer in Great Britain.

4.3.4.4 FODDER

The forest at one time played an important role in providing fodder for domestic animals. Trees along roadsides, ditches and fields were pollarded to provide animal fodder and firewood. Coppices provided young tender shoots and leaves which were harvested for winter consumption by domestic animals. Birch fulfilled a similar role in the Nordic countries. Fodder from the forest is no longer important except in the countries bordering the Mediterranean.

A new interest in utilizing conifer needles as a source of food for hens and other domestic animals has been aroused in the USSR and in Eastern Europe over the past two decades. This utilisation of part of the forest biomass, collected at the time of harvesting the tree for the forest industries, may well increase in the future so that the forest will once again become a source of fodder in some regions.

4.3.4.5 WOOL AND SKINS FROM DOMESTIC
AND WILD ANIMALS

Products other than meat — sheep wool, goat skins and the hides from cattle — are of considerable economic importance in some parts of Europe and their sale helps to maintain the pressure on the forests by graziers, particularly in the Mediterranean region. In Lapland, reindeer hides are locally important in the reindeer economy. The same considerations apply to the outlook for these products as for meat from animals grazed in the forest.

Within the ECE region the trapping of wild animals for their fur is confined mainly to the cold northern regions of the USSR and Canada. The Arctic fox and the mink are the main species trapped for their skins; there is also a market for bearskins. In spite of the growth of animal farming for furs (mostly fox and mink), there is a strong demand for the skins of the wild animals that are regarded as superior in quality to those from farmed animals.

Over the next 30 years, natural furs and skins from animals trapped in the far north of the USSR and Canada will meet increasing competition from farmed animals; and the reduction in skins and wool from goats and sheep grazed in the forest that will follow a reduction in the number of animals grazed will not have any impact on the rural economy.

4.3.4.6 NAVAL STORES

The naval stores that are referred to under non-wood products are the turpentines and resins that are distilled from the gum (resin) tapped from living trees. Other similar products that are obtained by extraction from tree stumps (usually pine) and as by-products of pulping are excluded.

Gum turpentine and gum resin are reported as being produced in substantial quantities by the USA, the USSR, Portugal, Poland and France ("Chemical processing of wood: special lectures to the thirty-ninth session of the Timber Committee", supplement 13 to volume XXXIV of the *Timber Bulletin for Europe*).

Resin tapping is practised in a number of European countries including France (the Landes district), the USSR, the German Democratic Republic and the Mediterranean region. Apart from the use of resins as a base for chemicals for the paint and solvent industries, in Greece resins are used in retsina wines. The resin tapping industry is labour intensive and may be subsidized by governments wishing to maintain rural populations in areas where there is little if any alternative employment.

The importance of naval stores derived from resin tapping will almost certainly diminish over the next two or three decades, except perhaps locally.

Traditional sources of tanning materials (chestnut wood, oak and spruce bark and Valonia, the acorns of the Turkish oak *Quercus macrolepis*) still find a market in spite of the competition from synthetic tanning materials and of the competition that natural leathers encounter from plastics and other synthetic materials.

It seems that this demand will continue since the products made from naturally tanned leathers are sought as luxury goods or are associated with recreation (e.g. the saddlery and other leather used in horse riding) where traditional materials can command a premium.

4.3.4.7 CHRISTMAS TREES AND FOLIAGE; DECORATIVE
LICHENS; MOSSES

The demand for natural Christmas trees is increasing after a fall due to the introduction of various kinds of artificial trees. The traditional species are being supplemented by more decorative species with good needle retention in the warm atmosphere of the house. They are also better prepared after cutting and before sale. In some countries (e.g. Denmark) the trade is an international one yielding an important supplementary forest income; there is also a considerable increase in the growing of Christmas trees by farmers and Christmas tree specialists who are more nurserymen than foresters.

Sales of twigs and foliage to florists can also yield a substantial supplementary income where forests are located near to population centres and where there is a tradition of using such foliage for decoration in the house.

Reindeer lichens are not only important as the winter forage of reindeer but traditionally they have been collected for decorative purposes too. Commercial lichen gathering is an industry practised only in geographically restricted areas. The main areas of the commercial collecting of decorative lichens are located just south of the reindeer management area, where lichens are abundant on dry and barren soils under pure pine stands. The species collected is "ornamental" reindeer lichen (*Cladonia alpestris*).

Mosses used in floral decorations – and at one time for medicinal wound dressings – are gathered in some types of forest, but have relatively little commercial significance.

The growing of Christmas trees, and the cutting of decorative foliage may increase in relative importance, but these markets have little effect on the overall management of forests and indeed Christmas trees may be grown on farm fields rather than on forest land.

Collecting decorative material, e.g. certain types of cones, may increase as a hobby but will not have commercial importance, except locally.

4.3.4.8   CORK

According to an estimation of the FAO (1959), the cork oak once occupied 8.4 million ha in the Mediterranean area whereas there are at present only 2.5 million ha, of which nearly 59% in Europe. The reasons for this decrease in area are connected with constant demographic pressure, often leading to overgrazing, major clear-cuts, thinnings, not always justified, and an increased fire risk.

As the cork oak has the property of renewing its bark after this has been removed, its harvesting is done by successive stages. As it is a slow growing forest tree, the first stripping is usually only done after 20-25 years. The product – virgin cork, is granulated and used for a number of agglomerated cork products.

The trees are stripped again after a minimum of nine years (sometimes ten or more), to produce reproduction cork, which from the second stripping is used to make cork stoppers or other products in natural cork. The residues of this processing are also granulated.

In addition to cork from regular stripping operations, two cultural operations give other types of raw material: pruning and thinning (or clear cutting in some cases).

At present more than three quarters of world cork production is in the Iberian Peninsula, with Portugal accounting for more than half the total.

Other producing countries are in North Africa (Morocco, Algeria, Tunisia) and Europe (France, Italy).

Around 1960 most of the cork industry moved to the producing countries, notably Portugal, and, to a lesser extent, Spain. Before that date it had been concentrated in certain industrialized countries, notably the USA.

As regards the outlook for the future, everything appears to indicate that cork will continue to be a scarce raw material on the world level, given the geographical concentration of the cork oak and its slow growth.

For reproduction cork, major cyclical fluctuations are to be expected, although the broad trend in supply appears quite stable, at least for Portugal and probably for Spain.

Forecasts for world production of virgin cork are much more hazardous.

4.4   SOME POLICY ASPECTS

4.4.1   Introduction

The discussion in section 4.3 has shown that important changes, notably in attitudes have taken place, and more changes can be expected.

Forest policies will have to take note of this changed and changing perception of the role of the forest.

The following section discusses some of the major questions which forest policy will have to address in the field of non-wood benefits.

4.4.1.1   SOCIAL TRENDS

Among the broad social and economic factors discussed in chapter 1, a few may have a rather strong influence on future change in the demand for and supply of non-wood benefits.

The *drift of population* from the countryside to the town (and to large villages near to towns), coupled with higher incomes, implies increased demand for most non-wood goods and services, especially local environmental forests and green belts round large towns, recreation and nature conservation.

To some extent the demand will be modified by the *ageing* of the population in the Nordic countries and the northern half of Europe. The migration of the population from north to south within many countries and even within Europe as a whole will perhaps further dampen demand in the Nordic countries and northern Europe. Greater availability of electronic entertainments in the home may also reduce demand for recreation in the forest.

On the other hand relatively *higher incomes and pensions* will increase older people's ability to travel to the forest and to enjoy its many different aspects.

For the working population, *leisure time* will increase over Europe as a whole as machines and robots take over more of the physical work at present carried out by man. Robots and electronic control systems will also greatly reduce the manpower requirements in many fields potentially creating permanently high unemployment. Job sharing might make a short working week possible.

Educational programmes for adults as well as for children are now much more widely based and cover the whole biological field in a comprehensive and interesting way. There will be a growth in public awareness of and interest in the real issues involved in the management of forests. One effect will be a greater incentive to visit the forest to study nature. Another will be to give informed opinion – and even pressure groups – a more scientifically oriented basis for any particular cause that may be advocated.

Better education will not be a universal panacea: there will be legitimate conflicts of view between individuals and pressure groups, however well informed they may be.

Any *increase in national wealth and personal incomes* in real terms is likely to give governments the opportunity to invest more heavily in creating forests for non-wood benefits and the ability, if they so wish, to accept lower wood production goals in return for increased non-wood benefits. An increase in incomes however also increases the demand for wood and wood products (although these, unlike most non-wood benefits, may be imported).

4.4.1.2 Use of abandoned agricultural land

The tendency throughout Europe has been for the richer farming areas to produce relatively more food per unit of investment and per unit area than the poorer areas where subsidies may have as much of a social purpose (to keep rural communities viable) as a productive function.

There is nothing to indicate a general reversal of this trend although in particular regions the pace of change may slow down. Thus the use of chemical fertilisers may be reduced or abandoned for economical and ecological reasons or a combination of both and less intensive farming methods may be re-adopted, but overall the likelihood is that productive agriculture will be concentrated into a smaller area, which will be more productive.

Grazing in particular is likely to be more concentrated, not only because it is technically more efficient, but because of the difficulty of getting flockmasters (shepherds) and their families to continue to live and work in isolated conditions in remote areas.

Reduced pressure of grazing – especially illegal grazing – will give the opportunity to create more forests with a non-wood production function as well as more wood-producing forests.

In terms of food production it is far less efficient to feed cereals to animals and then to eat the meat rather than eat the cereals direct. Any continuing reduction in the demand for beef and sheep meat will reduce the total area required for food production.

From the above, it could be concluded that abandoned farmland will be afforested, particularly in the Mediter-

ranean countries, or allowed to be colonised naturally by trees and shrubs. In the Nordic countries, already rich in forests, any such changes in land use will be on a very modest scale and could be opposed (desire to maintain open landscape). There will be other regional exceptions too, but since agriculture has such a large share in land use at present (44% of land in Europe), quite small changes from agriculture to forestry could give significant increases in the forest area of individual countries or regions.

The actual farm buildings – houses and other – if no longer needed for farming will often be used as second homes and the land in their immediate vicinity cultivated for hobby farming, market gardening, Christmas tree growing and the like. If the land is totally abandoned, it is likely to revert to woodland sooner or later, although the resulting "forest" may not be very productive.

4.4.1.3 Forests in the rural economy

A major social and economic concern in many countries over recent years has been the future of rural societies. Agricultural efficiency has increased leading to reduced manpower requirements and the economic marginalization of less productive agricultural lands. Urban areas have continued to exert a strong social and economic attraction. The migration from rural to urban areas, which started in the eighteenth and nineteenth centuries, has resulted, in some countries, in near depopulation of some rural areas and great difficulties in maintaining the rural infrastructure. Governments affected by these problems have in many cases instituted programmes to reduce unemployment, prevent depopulation and maintain the infrastructure.

The two major employers in rural areas have been agriculture and forestry and their associated processing and servicing industries; they have been joined by a third – tourism and recreation – in some areas; they are usually therefore the focus of policies in favour of rural areas, alongside direct subsidies e.g. for infrastructure. Agriculture is the major recipient of funds but increasingly policies to aid forestry are justified, at least in part, on the grounds of its contribution to the rural economy.

The provision of rural employment should probably not, strictly speaking, be classified as a "non-wood benefit of the forest", but there is no doubt that this aspect should and will be taken into account when drawing up forest policy.

4.4.1.4 Effect of damage to forests on non-wood benefits

Damage to forests is considered here because there is an interaction between forest damage (e.g. by airborne pollution or forest fires) and non-wood benefits. The damage caused by air pollution, even if it falls well short of killing the forest, will make some forest areas much less able to fulfil their environmental and recreational functions and the production of non-wood products may be severely affected. It seems inevitable that, as demand for forest-based recreation grows in the Mediterranean region and Southern Europe generally, rather expensive measures will need to be taken to reduce the incidence of forest fires, whether caused by tourism, by local town dwelling populations using the forest for leisure pursuits, or by local people. Fire damaged areas remain unattractive for long periods and there can be an adverse effect for tourism. Additional recreational or tourism pressures may be imposed on remaining undamaged forests in the area.

4.4.1.5 Use of chemicals

In those countries where their use is not already severely limited, the use of chemical fertilizers, herbicides and pesticides is likely to be limited on account of public reaction against them, especially in forest where the non-wood benefits are actually (or regarded by the public as) of a higher priority than the production of wood. Wood production may suffer from a reduction in the use of chemicals, but not that of other goods and services.

4.4.2 Policy implications

Most forests can and do fulfil more than one function. Yet a given forest cannot fulfil all the functions at one and the same time in one and the same place. The desire to "optimize the productive, protective and accessory values of the forest for the greatest number, in perpetuity"[1] must therefore be tempered with the reality of what can be achieved in practice on a given site in light of the priorities for that site. There must be trade-offs between the different functions. Even within a function like recreation, the demands and needs of different groups of people more and more require the supply of the required services to be separated physically.

If the demand for the service functions of the forest rises as has been suggested and there is also continuing demand for wood, some form of financial control may be imposed in some situations to limit supply of the service functions to that which the community or region is prepared to pay for, so as to avoid an open-ended commitment to match demand with free supply.

The option to maximize wood production in selected forests to compensate for loss of wood production in forests with high recreational or environmental functions may not be as open as it has been until now. Thus for example, allowances will have to be made in silvicultural practices to leave groups or edges of broadleaved trees where these occur in conifer stands; to avoid straight lines and generally to landscape felling coupes and new plantings. In some cases clear felling may even be prohibited – selection forestry will be the required silvicultural system.

When considering the supply side of the equation a distinction has first to be made between:

- The forest rich Nordic countries with a relatively large area of forest per inhabitant (3.47 ha);
- The industrialized countries of the EEC(9) which have large populations and a relatively small forest area per capita (0.13 ha);
- And the central eastern and southern European countries which have large populations but considerably greater area of forest per capita (0.35 ha, 0.26 ha and 0.41 ha respectively) than the EEC countries.

Given that the *per caput* demand for non-wood benefits, especially in the recreational field, will grow in the Nordic countries, there is *prima facie* an adequate supply to meet that demand. Locally, of course (e.g. near large towns) there may be supply problems and there may be competing demands on the same forest area by different interest groups. Demand for environmental services and benefits is perhaps more likely to constrain management for maximum wood production than demand for recreational services and non-wood products. But the heavy dependence of communities on the forest and forest industries for their livelihood will tend to strengthen the hand of forest managers and politicians who have to consider forestry as a creator of wealth.

In the EEC countries and Central Europe where the forest resources are more limited a rather different future may be envisaged:

- Near large centres of population, forests will be required to provide a wide range of environmental and recreational services; the demand for non-wood products (and for wood) will be high but supply possibilities will be severely limited by the practical constraints imposed when large numbers of people regularly visit and use a forest area. The creation of new local forests designed to provide various environmental services will be a continuing process;
- The forests distant from main centres of population will experience two types of demand for non-wood benefits:
  (a) For intensive use (for winter sports and tourism);
  (b) For remoteness, solitude, "naturalness", special or unique features, or a combination of several of these factors.

The forest resource is limited and the establishment of methodologies for decision-making and policy formulation will be important.

In some countries damage from air pollution is likely to aggravate the supply situation and to give extra environmental and recreational value to those that remain.

One of the main issues to be resolved is the extent to which different types of forest on different sites will be able to fulfil the "multiple use" concept. The natural and man-made forests of the regions cover every forest type from pine on pure dry sand to water-loving species such as alder (*Alnus* spp.) and willow (*Salix* spp.) on permanently wet sites; from steep mountainous Alpine terrain to extensive flat plains.

---

[1] World Forestry Congress, Seattle, 1960.

The resolution of issues not directly connected with the forest *per se* will also have an important bearing on the non-wood role of the forest. These include the use of the micro-light aircraft, already banned in some countries; the use of cross-country, four-wheel-drive cars and motor cycles as well as snowmobiles to reach second homes and to traverse "wilderness" areas.

In Eastern Europe the demands that will arise for non-wood benefits will be similar to those in the EEC and Central Europe but the ability of the forests to supply those benefits will be greater. Similar questions of policy will arise, but the larger area of forest relative to the population should make it easier for governments to meet the aspirations of the different interest groups.

It is in the Southern Europe and the Mediterranean region that there is the greatest potential for change. Reduction in grazing pressures on the forest and the abandonment of farming on very poor land will increase the potential supply of forest through the rehabilitation of degraded forests and new afforestation schemes. On the other hand, in addition to increased demand on the forest from the indigenous population the potential for increased tourism, based partly or wholly on the forest, is very great. In addition the harmonization of fiscal and social policies within the enlarged EEC (i.e. including Greece, Portugal and Spain) could mean that the drift of retired people to the warmer Mediterranean seaboard will increase demands on that seaboard and the forests of its immediate hinterland. Already such permanent north to south movements of elderly populations is evident in Britain and France; temporary winter migration by retired people from northern Europe to the Mediterranean coast is increasing rapidly as the tourist industry seeks for year-round occupation of hotels and self-catering accommodation. Add to this the need for additional protection forests on a substantial scale and there is every incentive to afforest or reforest abandoned agricultural land and areas of forest degraded by grazing, with the primary object of supplying environmental and recreational non-wood benefits.

Common to all the regions will be the need to choose between different objectives and priorities of management. Even in multiple-use forests priorities have to be set between the different functions of the forest. It may be pertinent to ask to what extent is an objective, mathematical and scientific approach appropriate as opposed to a more subjective political approach and to answer the question by suggesting that decisions will be taken "politically" in light of such objective data as can be obtained, whilst recognizing that trying to put a value on the non-market functions such as nature conservation and landscape improvement raises issues that have not yet been satisfactorily solved by economic analyses. This is not the place to debate these issues but simply to draw attention to the added complexity of data gathering, analysis and the use of the information for decision making, when non-wood benefits are given due and full consideration.

One might also ask the question, who will make the choices? The traditional role of the national Forestry Department or Forest Service as policy adviser and administrator is now more and more being questioned, as it is sometimes accused of in-built prejudice in favour of maximizing the production of wood and of managing the forest using commercial management criteria. Other departments of government may claim a much greater stake in policy formulation and implementation because of their interest in the non-wood benefits of the forest.

CHAPTER 5

Outlook for the European forest resource and wood availability

5.1 INTRODUCTION

In 1979-81, wood removals from the European forest amounted to about 341 million m³ underbark. This was by far the largest component amongst the sources of supply to meet the region's requirements for forest products.

Over the past 30 to 40 years, and notably between 1970 and 1980, the relative importance of European wood removals as a source of supply has tended to diminish gradually. Especially during the 1970s, greater use was made of industrial residues and waste paper, which by the end of the decade had, in aggregate, become as important a source of supply as imports from other regions, even though those had also expanded considerably. Most of the industrial residues and waste paper, however, originated as roundwood from European forests; their growing relative importance has been due to more efficient use of the same basic source – European removals. It can be said, therefore, that an assessment of the prospects for wood availability from the European forest is the essential key to the global assessment of forest products supply to the region's market.

The purpose of this chapter is twofold. First, it will present forecasts of wood removals up to the year 2000 and beyond. Second, it will examine these removals in the broader context of their relationship with the forest resource and the outlook for the capacity of that resource to meet demand in the long term. As will be explained in more detail later, the outlook presented in this chapter takes as a starting point the results of the latest forest resource assessment, which was published by ECE/FAO in 1985.[1]

Nevertheless, it has to be pointed out that some differences, mostly relatively unimportant, do exist between the latest forest resource inventory data including those for removals, which were summarized in chapter 3, and those used as the "base period" for the forecasts. There are a number of reasons for these differences: the figures relate to different periods or years; the two sets of data were taken from different sources (in some cases they were prepared by different correspondents); revisions occurred between the dates of submission of one set of data and the other.

The importance of the differences should not be exaggerated. Annex table 5.1 which presents these differences is intended to serve as a reminder to take care in choosing the historical data with which to compare the forecasts presented in this chapter. The ones that should be used come under the heading of "base period", which for most countries is a year or period of years around 1980. With certain reservations, therefore, the base period can be taken as 1980.

---

[1] *The Forest Resources of the ECE Region (Europe, the USSR, North America)*, ECE/TIM/27, ECE/FAO, Geneva, 1985.

TABLE 5.1

Europe: sources of supply of forest products in 1969-71 and 1979-81

	Volume (million m³ EQ)		Change over 10 years (million m³ EQ)	Percentage of total supply	
	1969-71	1979-81		1969-71	1979-81
European removals	338	341	+ 3 (+ 1%)	74	67
Industrial wood residues	29	44	+15 (+52%)	6	9
Waste paper[a]	23	40	+17 (+74%)	5	8
Imports of forest products from other regions	65*	84	+19* (+29%)	14	16
Total supply	455	509	+53 (+12%)	100	100
of which:					
To European market	435	483	+48 (+11%)	96	95
Exports of forest products to other regions	20*	26	+ 6* (+30%)	4	5

[a] Converted from metric tons with the factor 2.5 as used in ETTS III.

The forecasts given below are shown in physical terms (volume, area). They implicitly take into account the factors affecting supply, which were discussed in chapter 2 and, for some countries they also incorporate assumptions concerning the impact of air pollution and other forms of damage on forest productivity and potential wood supply, which are discussed further in chapter 6. These matters will therefore not be raised again in this chapter more than is necessary.

5.2 THE BASIS AND METHODOLOGY OF THE FORECASTS

As for the preparation of ETTS III, countries were invited to provide their own forestry forecasts, and the secretariat is deeply indebted to the correspondents in the majority of countries who took immense trouble to prepare sets of detailed and consistent figures, based on an intimate knowledge of their countries' forestry situation and plans. Full recognition should also be given to the invaluable help given by Professor Kullervo Kuusela (Finland) in helping to devise the enquiry used to collect country forecasts and to analyse the replies.

The framework for the forecasts was originally laid out at the session of the Joint FAO/ECE Working Party on Forest Economics and Statistics in June 1983. Having reviewed the preliminary results of the latest forest resource assessment, the Working Party considered how these could be used as a basis for preparing the forecasts for ETTS IV. The approach suggested was that used in ETTS III for appraising the forecasts offered by countries for inclusion in that study (see section 6.3 of ETTS III and appendix 1:6). This method is based on the fact that there is a relationship between changes over time in the volumes of growing stock, increment and drain (fellings),[2] which may be expressed as follows:

Growing stock at the beginning of a given period *plus* increment during the period *minus* the drain during the period *equals* the growing stock at the end of the period.

Countries were invited to provide forecasts for 1990, 2000, 2010 and 2020. Since the bulk of cuttings takes place on *exploitable closed forest* and the quality of data available is generally the best for this part of the forest area, the forecasts relate to this category. In a supplementary question in the enquiry, countries were also asked for removals other than from exploitable closed forest, that is to say, from other wooded land, trees outside the forest and, under certain circumstances, from unexploitable closed forest. This allowed forecasts to be built up of *total* wood removals, which are presented in section 5.3.

---

[2] For definitions of forestry terms, *ibid.*, appendix I.

In addition to forecasts of growing stock, net annual increment and fellings on exploitable closed forest, countries also provided figures for area, unrecovered felling losses, removals (overbark and underbark) and the proportion of sawlogs and veneer logs in total removals.

Finally, countries were asked to provide sets of forecasts according to "low" and "high" assumptions. The assumptions to be used were elaborated at the 1983 session of the Working Party, mentioned above, and were as follows:

(i) The *low* estimate should be based on modest, but realistic assumptions of biological developments, e.g. for the intensification of silviculture, stand density, species mixture, age-class distribution, length of rotation, and of technical developments, including trends in harvesting or transport technology, labour productivity and methods of road building. For economic development, this estimate should assume low growth in the world economy (as in recent years) with only a small increase in the demand for forest products at best, very competitive conditions in international markets, and no upward pressure on prices of the sort that may be needed to "commercialize" areas at present too costly to exploit;

(ii) The *high* estimate should be based on more expansive (but still realistic) assumptions on developments affecting the biological potential, and more favourable (but again realistic) trends in technical factors which influence the costs of exploitation. For the development of the global economy, this estimate should assume growth rates similar to those achieved in the third quarter of the century, with growing demand for forest products and some upward pressure on the prices of wood and wood products traded internationally. Where increases in relative timber prices (i.e. increases relative to the general level of prices) are important influences on the estimates of removals, countries should state their price assumptions.

Both scenarios should take into account developments in the use of wood for energy and any likely new demands on the forest resource.

The forecasts presented below are for 27 European countries. Those not covered are Iceland and Malta, where forest production is nil or negligible; and Albania, on which adequate information does not exist and which in any case is not dealt with elsewhere in the study. The country data are given in detail in the annex tables and for reasons of space can only be summarized in this chapter.

5.3 FORECASTS OF EUROPEAN WOOD REMOVALS

5.3.1 Total removals

Annex table 5.1, which shows differences between some countries' figures in the latest forest resource enquiry and those used as the base period for the forecasts in this chapter, also gave for underbark removals the figures for

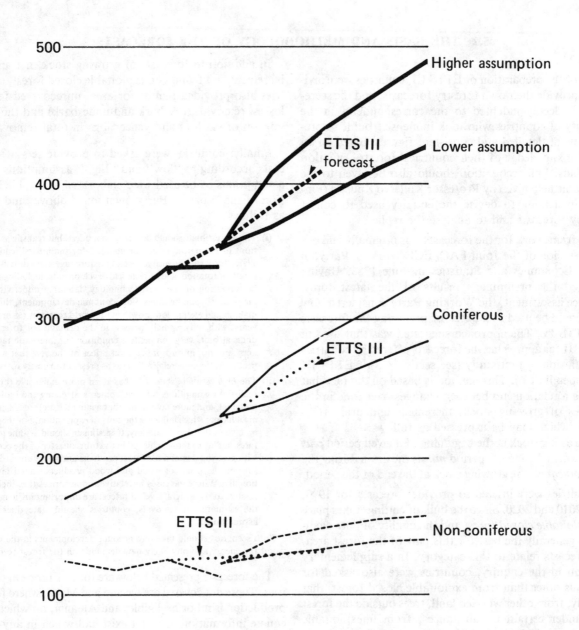

FIGURE 5.1

Removals forecasts for Europe
(*Million m³ underbark*)

TABLE 5.2

**Europe: summary of trends in total removals,
1950 to 1980 and forecasts 1990 to 2020**

Period	Total		Coniferous		Non-coniferous	
VOLUME OF REMOVALS (million m³ underbark)						
Actual						
1949-51	296.6		171.3		125.3	
1959-61	307.5		188.1		119.4	
1969-71	337.5		212.4		125.1	
1979-81	340.7		227.9		112.8	
Forecasts	*Low*	*High*	*Low*	*High*	*Low*	*High*
Base period[a]	350.5		228.7		121.8	
1990	367.0	405.4	239.6	265.8	127.4	139.6
2000	390.8	438.1	256.4	289.2	134.4	148.9
2010	411.7	464.0	272.5	308.7	139.2	155.3
2020	431.2	490.2	287.9	331.2	143.3	159.0
VOLUME CHANGE OVER TEN YEARS (million m³ underbark)						
Actual						
1949-51 to 1959-61	+ 10.9		+ 16.8		− 5.9	
1959-61 to 1969-71	+ 30.0		+ 24.3		+ 5.7	
1969-71 to 1979-81	+ 3.2		+ 15.5		− 12.3	
Forecasts	*Low*	*High*	*Low*	*High*	*Low*	*High*
Base period to 1990	+ 16.5	+ 54.9	+ 10.9	+ 37.1	+ 5.6	+ 17.8
1990 to 2000	+ 23.8	+ 32.7	+ 16.8	+ 23.4	+ 7.0	+ 9.3
2000 to 2010	+ 20.9	+ 25.9	+ 16.1	+ 19.5	+ 4.8	+ 6.4
2010 to 2020	+ 19.5	+ 26.2	+ 15.4	+ 22.5	+ 4.1	+ 3.7
PERCENTAGE CHANGE OVER TEN YEARS						
Actual						
1949-51 to 1959-61	+ 3.7		+ 10.2		− 4.7	
1959-61 to 1969-71	+ 9.8		+ 12.9		+ 4.8	
1969-71 to 1979-81	+ 0.9		+ 7.3		− 9.8	
Forecasts	*Low*	*High*	*Low*	*High*	*Low*	*High*
Base period to 1990	+ 4.7	+ 15.7	+ 4.8	+ 16.2	+ 4.6	+ 14.6
1990 to 2000	+ 6.5	+ 8.1	+ 7.0	+ 8.8	+ 5.5	+ 6.7
2000 to 2010	+ 5.3	+ 5.9	+ 6.3	+ 6.7	+ 3.6	+ 4.3
2010 to 2020	+ 4.7	+ 5.6	+ 5.7	+ 7.3	+ 2.9	+ 2.4

[a] Around 1980, but varies from country to country according to period covered by the latest inventory. See text for explanation of difference between "base period" figures and those for 1979-81.

1979-81 reported in the annual FAO/ECE questionnaires. As said before, the importance of such differences should not be exaggerated, but it is still necessary to draw attention to them to avoid possible misinterpretation of the forecasts. Where differences occur, the figures for the "base period" (shown under the heading "ETTS IV") are almost always higher than the 1979-81 average figures. One explanation to emerge from an examination of the discrepancies is that the data reported in the FAO/ECE questionnaires are not in all instances complete. For example, they may not include removals for auto-consumption.

For Europe as a whole, the difference between reported removals in 1979-81 (average) and those used as a base period for the forecasts is about 10 million m³ underbark or 3%; one country, France, accounts for 7 million m³ u.b. of the total discrepancy. The difference at the European level is not huge, but enough to require a warning to take care in comparing the forecasts for 1990 onwards with the historical figures for 1949-51 to 1979-81. On the other hand, changes over time should be comparable between past and future periods.

The aggregates of European countries' forecasts of removals in the year 2000 are 391 million m³ u.b. (low) and 438 million m³ u.b. (high) (table 5.2 and figure 5.1). Compared with the base period around 1980, these forecasts represent increases over 20 years of 40 million m³ u.b. (12%) and 88 million m³ u.b. (25%) respectively. Reviewing the forecasts in the context of long-term trends since 1950, a number of interesting features emerge:

1. Even under the low forecast, the estimate of the rate of expansion between 1980 and 2000 is more reminiscent of the 1950s and 1960s than of the 1970s. Removals in 1979-81 were only a little higher than in 1969-71. In fact the trend during the decade was in the shape of a shallow "U" with a much deeper "U" in the Nordic countries. Recovery occurred in the second half of the decade from the low point of 1975.

2. There is a contrast between the trends and forecasts for coniferous and non-coniferous removals. Those for coniferous are for a continuation of the growth between 1949-51 and 1979-81 at rates between the base

period and 2000 which fall above and below the average rate of increase between 1949-51 and 1979-81. Even during the 1970s, coniferous removals were on an upward trend (ignoring cyclical fluctuations). For non-coniferous removals, however, the trend between 1949-51 and 1979-81 was generally downwards, mainly due to the fall in the demand for fuelwood. From about 1978, there was a reversal of this trend, as sharply higher prices for fossil fuels brought renewed, even if limited, interest in the use of wood for energy. For the period 1980 to 2000, non-coniferous removals are forecast to continue to recover, at rates only slightly below the growth rates for coniferous. In terms of share of total removals, the trends and expectations are as follows:

TABLE 5.3

Shares of coniferous and non-coniferous species in European removals

| | Percentage share of total removals | | | |
	Coniferous		Non-coniferous	
Actual				
1949-51	57.8		42.2	
1959-61	61.4		38.6	
1969-71	62.9		37.1	
1979-81	66.9		33.1	
Forecasts	Low	High	Low	High
Base period	65.2		34.8	
1990	65.3	65.6	34.7	34.4
2000	65.6	66.0	34.4	34.0
2010	66.2	66.5	33.8	33.5
2020	66.8	67.6	33.2	32.4

3. The estimates in ETTS III, prepared in 1973-75, gave an expansion in European removals between 1980 and 2000 of 55 million m^3 u.b. or 16%. The new forecasts give increases which fall on either side of that prediction prepared a decade earlier. It can be said, therefore, that the outlook for a steadily expanding supply potential of wood from the European forest as a whole has not changed since ETTS III was prepared. The fact that removals rose less strongly during the 1970s than foreseen in ETTS III can be mainly attributed to the weakness of the growth in demand, not to any supply constraints. Indeed the "shortfall" was concentrated in the Nordic countries, which suffered most of the negative consequences of the slowdown in demand.

4. Compared with the ETTS III forecasts, however, uncertainties have increased as to the likely rate of expansion of removals. On the one hand, levels of cutting during the 1970s, which were below those predicted, have left a backlog, particularly of thinnings, which could under favourable market conditions be added to future removals (this has probably been allowed for in the higher forecasts, notably so in the case of Sweden). Furthermore, the present and future volumes of growing stock and increment in the European forest are higher than those foreseen in ETTS III. Against this, negative influences on forest produc-

tivity, of which air pollution is only one, affecting only parts of the region, could have an impact on supply in the long term.

5. The differences between countries' low and high forecasts varies considerably, which presumably reflects differences in their perceptions of the impact of external factors on wood supply, different forecasting methods, different policy objectives, the degrees of uncertainty they attach to their forecasts and the possibilities they see for flexibility in removals to adjust to demand. The high forecast for 2000 for Europe in aggregate is 12% above the low one. As can be seen from table 5.6, countries with big differences between high and low are Sweden (27%), the United Kingdom (25%), Hungary (23%) and Switzerland (2%), while those for Ireland and Yugoslavia are nil and for Denmark 4%.

In replies to the descriptive part of the enquiry on the long-term outlook for forestry, countries explained the assumptions underlying their forecasts. In general, these followed the guidelines suggested: for the "low forecast", a principal factor affecting the rate of removals for many countries would be the continuation of slack demand for forest products, with price levels offering little incentive to raise the cut. For the "high forecast", there would be some growth in demand, but none of the respondents expected prolonged "boom" conditions in the forest products markets. Use of wood for energy could be a positive factor in several countries, including France and Sweden.

In Sweden's "high" forecast, there is a big jump in removals between the base period and 1990 (of 17.5 million m^3 o.b. or 36%), which is deemed realistic under conditions of strong market conditions, and which would enable the sizeable backlog in thinnings to be reduced.

Several countries (Poland and Switzerland, amongst others) have explicitly included assumptions regarding the impact of forest damage in their forecasts, but generally the impact on productivity (increment) is greater than that on the volume of removals over the forecast period (see chapter 6).

Trends and forecasts of removals in volume terms are shown in table 5.4 for the five country groups used in this study. The changes expected between 1980 (base period) and 2000 are given in table 5.5.

In volume terms, the country groups expecting the largest growth in removals between 1980 and 2000, under both the low and high assumptions, are the Nordic countries, the EEC(9) and Southern Europe. In terms of growth rate, however, Central Europe joins those three groups in predicting expansion at a rate above the European average for the high forecast. Under the low forecast, the growth rate in the Nordic countries is well below the European average, as is that for Eastern Europe under both the low and high assumptions.

For Europe as a whole, coniferous removals account for two-thirds of the total increase over the 20-year period, but the share of coniferous in the total increase varies

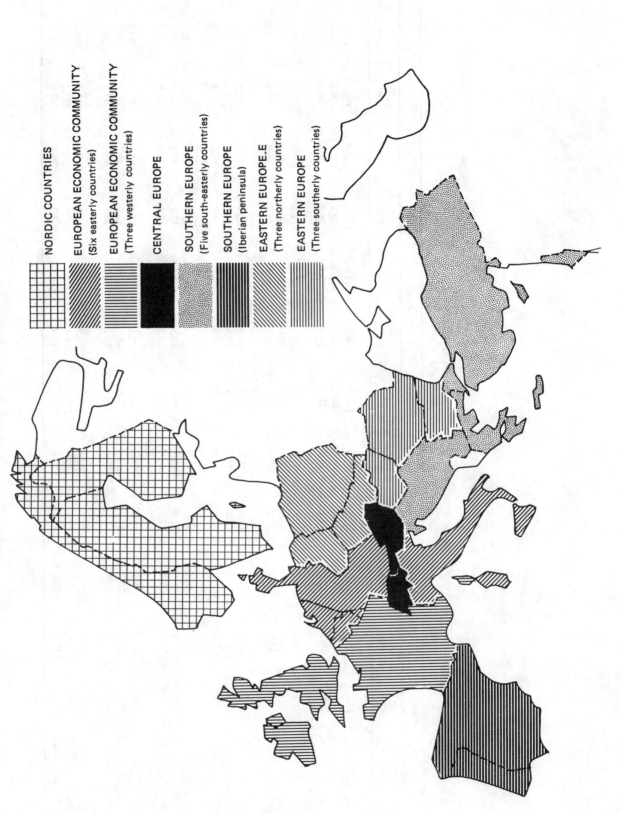

FIGURE 5.2

Sub-division of Europe into eight country groups and sub-groups

NORDIC COUNTRIES

EUROPEAN ECONOMIC COMMUNITY
(Six easterly countries)

EUROPEAN ECONOMIC COMMUNITY
(Three westerly countries)

CENTRAL EUROPE

SOUTHERN EUROPE
(Five south-easterly countries)

SOUTHERN EUROPE
(Iberian peninsula)

EASTERN EUROPE.E
(Three northerly countries)

EASTERN EUROPE
(Three southerly countries)

TABLE 5.4

Europe: forecasts of total wood removals by country group
(Million m³ underbark)

Country and species group	Actual				Base period	Low forecasts				High forecasts			
	1949-51	1959-61	1969-71	1979-81		1990	2000	2010	2020	1990	2000	2010	2020
Nordic countries													
Total	87.4	95.6	110.7	102.9	103.8	104.3	109.8	115.2	119.4	126.9	130.1	135.5	140.4
Coniferous	70.5	79.3	91.2	88.8	89.3	89.2	93.0	97.5	100.9	105.3	109.3	114.2	119.9
Non-coniferous	16.9	16.3	19.5	14.1	14.5	15.1	16.8	17.7	18.5	21.6	20.8	21.3	20.5
EEC(9)													
Total	84.1	83.0	80.2	81.5	91.1	95.3	101.1	106.6	111.1	102.9	114.7	120.1	129.4
Coniferous	36.5	40.2	40.4	44.3	45.8	50.5	56.0	61.7	66.9	56.0	64.4	69.5	79.0
Non-coniferous	47.6	42.8	39.8	37.2	45.2	44.7	45.1	44.9	44.2	46.9	50.3	50.6	50.4
Central Europe													
Total	12.5	15.5	16.1	18.8	16.6	19.2	19.9	20.0	20.0	20.4	21.8	21.6	21.3
Coniferous	10.6	12.7	13.1	15.0	13.0	15.3	15.9	16.0	16.0	16.2	17.4	17.3	17.1
Non-coniferous	1.9	2.8	3.0	3.8	3.6	3.9	4.0	4.0	4.0	4.2	4.4	4.3	4.2
Southern Europe													
Total	50.7	48.6	57.3	59.6	59.6	68.3	77.1	84.6	92.2	71.5	82.3	92.1	98.8
Coniferous	14.7	17.9	27.0	32.5	32.5	37.6	42.8	47.3	52.1	39.8	46.3	52.5	56.7
Non-coniferous	36.0	30.7	30.3	27.1	27.0	30.7	34.3	37.3	40.1	31.7	35.9	39.6	42.1
Eastern Europe													
Total	61.9	64.8	73.2	77.9	79.5	80.0	82.8	85.3	88.5	83.6	89.2	94.7	100.3
Coniferous	39.0	38.0	40.7	47.2	48.1	47.0	48.7	50.0	52.0	48.5	51.7	55.7	58.5
Non-coniferous	22.9	26.8	32.5	30.7	31.4	33.1	34.1	35.3	36.5	35.1	37.5	39.5	41.8
TOTAL EUROPE													
Total	296.6	307.5	337.5	340.7	350.5	367.0	390.8	411.7	431.2	405.4	438.1	464.0	490.2
Coniferous	171.3	188.1	212.4	227.9	228.7	239.6	256.4	272.5	287.9	265.8	289.2	308.7	331.2
Non-coniferous	125.3	119.4	125.1	112.8	121.8	127.4	134.4	139.2	143.3	139.6	148.9	155.3	159.0

TABLE 5.5

**Europe: forecasts of changes in removals
between base period (around 1980) and 2000 by country group**

	Total		Coniferous		Non-coniferous	
	Low	High	Low	High	Low	High
Change in volume (million m³ underbark)						
Nordic countries	+ 6.0	+ 26.3	+ 3.7	+ 20.0	+ 2.3	+ 6.3
EEC(9)	+ 10.0	+ 23.6	+ 10.1	+ 18.6	− 0.1	+ 5.0
Central Europe	+ 3.3	+ 5.2	+ 2.9	+ 4.4	+ 0.4	+ 0.8
Southern Europe	+ 17.6	+ 22.7	+ 10.2	+ 13.8	+ 7.3	+ 8.9
Eastern Europe	+ 3.3	+ 6.1	+ 9.7	+ 0.6	+ 3.6	+ 2.7
Total Europe	+ 40.3	+ 87.5	+ 27.7	+ 60.4	+ 12.6	+ 27.1
Change in percentage						
Nordic countries	+ 5.8	+ 25.3	+ 4.1	+ 22.4	+ 15.9	+ 43.4
EEC(9)	+ 11.0	+ 25.9	+ 22.1	+ 40.5	− 0.2	+ 11.1
Central Europe	+ 19.9	+ 31.3	+ 22.3	+ 33.8	+ 11.1	+ 22.2
Southern Europe	+ 29.5	+ 38.1	+ 31.5	+ 42.4	+ 27.1	+ 33.0
Eastern Europe	+ 4.2	+ 12.2	+ 1.2	+ 7.5	+ 8.7	+ 19.4
Total Europe	+ 11.5	+ 25.0	+ 12.1	+ 26.4	+ 10.3	+ 22.3

varies considerably from group to group. It is well above the European average in Central Europe and the EEC(9); and less than the average in Southern and Eastern Europe.

A grouping of countries is necessary in a multi-country study such as this in order to keep it within reasonable bounds. Nevertheless, any grouping is unsatisfactory from one point of view or another, and inevitably conceals interesting developments at the country or sub-group level. At the risk of overburdening the reader with detail, country forecasts of removals are given for 1990 and 2000 in table 5.6. There is no need for further discussion of the country forecasts here, except to note that, while the Nordic countries and Central Europe are reasonably homogeneous groups in terms of their forestry and wood removals, the same is not the case in the other groups. It has been found worthwhile, therefore, to subdivide the latter, as shown in figure 5.2, which makes it easier to identify where the main developments are expected between the base period and 2000. The results are also shown in table 5.7.

The sub-grouping in table 5.7 reveals some interesting contrasts, particularly within the EEC(9) (figure 5.3). While the six easterly countries of the EEC forecast removals in 2000 slightly down on the base period under the low assumption and only a modest increase under the high assumption, the three westerly EEC countries predict strong growth in removals over the 20-year period, especially for coniferous removals. Reforestation in France and afforestation in Ireland and the United Kingdom over past decades are now beginning to bear fruit in terms of steep rises in potential cut.

Expansion is foreseen throughout Southern Europe, but the sharpest increases are expected in the Iberian Penin-

sula, also reflecting the impact on future wood supply of past and still continuing plantation programmes, especially in Spain (figure 5.4).

The sub-groups in Eastern Europe highlight the predominance of broadleaved species in the forests and in the expected removals (70% or more) of the three southerly countries of the group, in contrast to the three-quarter share of coniferous in removals of the three northerly countries. Under the low assumption, the latter countries' removals of coniferous wood in 2000 are forecast to be slightly below those of the base period around 1980. As for some of the countries of the easterly EEC group, notably the Federal Republic of Germany, where removals could also be down slightly in 2000, the forecasts probably reflect caution in the face of uncertainties about future availabilities, stemming from the damage being inflicted on the forests in these areas by air pollution and other causes.

The result of countries' removals forecasts to 2000 and 2020 in terms of the shares of country groups and subgroups of total European removals is shown in table 5.8 and figure 5.5, which also show historical data in order to view the long-term trend over a period of 70 years. As would be expected the most marked changes in share occur in coniferous removals. Between 1950 and 2020, three groups should double their shares of the total: westerly EEC, Iberian Peninsula and south-easterly Southern Europe, in aggregate from 15% in 1950 to 31% in 2020 (low forecast). The groups whose share may decline noticeably are: Nordic countries, easterly EEC and northerly Eastern Europe, in aggregate from 75% in 1950 to 59% in 2020 (low forecast). Changes in share are far less marked for non-coniferous removals.

TABLE 5.6

Europe: forecasts of total removals by country and country group, 1990 and 2000

(Million m³ underbark)

	Total					Coniferous					Non-coniferous				
	Base period	1990 Low	1990 High	2000 Low	2000 High	Base period	1990 Low	1990 High	2000 Low	2000 High	Base period	1990 Low	1990 High	2000 Low	2000 High
Finland	45.78	45.93	50.21	48.60	53.08	37.99	37.89	39.12	39.00	41.93	7.79	8.04	11.09	9.60	11.15
Norway	9.52	9.20	10.70	9.50	11.30	8.67	8.38	9.78	8.68	10.38	0.85	0.82	0.92	0.82	1.02
Sweden	48.52	49.20	66.00	51.70	65.70	42.61	42.95	56.40	45.35	57.10	5.91	6.25	9.60	6.35	8.60
Nordic countries	103.82	104.33	126.91	109.80	130.08	89.27	89.22	105.30	93.03	109.31	14.55	15.11	21.61	16.77	20.77
Belgium[a]	2.44	2.92	3.08	3.14	3.48	1.47	1.56	1.68	1.63	1.89	0.97	1.36	1.40	1.51	1.59
Denmark	1.89	2.25	2.34	2.34	2.43	1.16	1.43	1.52	1.52	1.61	0.73	0.82	0.82	0.82	0.82
France	38.48	41.85	45.44	43.99	50.88	14.70	17.68	18.79	19.57	21.58	23.78	24.17	26.65	24.42	29.30
Germany, Fed. Rep. of	33.03	30.15	33.15	30.15	33.65	22.59	21.02	24.77	21.02	25.15	10.44	9.13	8.38	9.13	8.50
Ireland	0.53	1.60	1.60	3.04	3.04	0.49	1.55	1.55	2.98	2.98	0.04	0.05	0.05	0.06	0.06
Italy[b]	8.96	9.54	9.80	9.77	10.57	1.71	2.09	2.23	2.40	2.80	7.25	7.45	7.57	7.37	7.77
Luxembourg[b]	0.29	0.28	0.29	0.27	0.29	0.11	0.10	0.11	0.09	0.12	0.18	0.18	0.18	0.18	0.17
Netherlands	1.12	1.02	1.06	1.13	1.26	0.57	0.56	0.56	0.59	0.59	0.55	0.46	0.50	0.54	0.67
United Kingdom	4.32	5.65	6.15	7.25	9.05	3.02	4.55	4.75	6.15	7.65	1.30	1.10	1.40	1.10	1.40
European Economic Community (nine)	91.06	95.26	102.91	101.08	114.65	45.82	50.54	55.96	55.95	64.37	45.24	44.72	46.95	45.13	50.28
Austria	12.17	14.90	15.40	15.35	16.35	10.01	12.44	12.85	12.89	13.75	2.16	2.46	2.55	2.46	2.60
Switzerland	4.39	4.29	4.99	4.59	5.49	2.96	2.86	3.36	3.06	3.66	1.43	1.43	1.63	1.53	1.83
Central Europe	16.56	19.19	20.39	19.94	21.84	12.97	15.30	16.21	15.95	17.41	3.59	3.89	4.18	3.99	4.43
Cyprus	0.07	0.05	0.06	0.05	0.06	0.07	0.05	0.06	0.05	0.06	–	–	–	–	–
Greece[a]	2.59	3.01	3.14	3.35	3.66	0.80	1.04	1.12	1.27	1.43	1.79	1.97	2.02	2.08	2.23
Israel	0.14	0.16	0.20	0.19	0.20	0.06	0.07	0.09	0.10	0.09	0.08	0.09	0.11	0.09	0.11
Portugal	8.42	8.76	9.36	9.98	10.74	5.54	6.10	6.40	6.78	7.16	2.88	2.66	2.96	3.20	3.58
Spain	12.17	16.14	17.64	20.53	22.57	8.13	10.66	11.67	13.61	14.96	4.04	5.48	5.97	6.92	7.61
Turkey[b]	22.38	22.84	23.85	23.73	25.74	14.02	14.68	15.50	15.37	17.04	8.36	8.16	8.35	8.36	8.70
Yugoslavia[b]	13.79	17.30	17.30	19.30	19.30	3.92	5.00	5.00	5.60	5.60	9.87	12.30	12.30	13.70	13.70
Southern Europe	59.56	68.26	71.55	77.13	82.27	32.54	37.60	39.84	42.78	46.34	27.02	30.66	31.71	34.35	35.93
Bulgaria[b]	4.44	4.40	4.74	4.51	5.14	1.20	1.10	1.22	1.14	1.39	3.24	3.30	3.52	3.37	3.75
Czechoslovakia	19.32	17.55	18.45	17.55	18.90	14.82	12.87	13.77	12.87	13.95	4.50	4.68	4.68	4.68	4.95
German Dem. Rep.[b]	9.99	11.05	10.60	11.00	11.18	7.93	8.85	8.34	8.69	8.84	2.06	2.20	2.26	2.31	2.34
Hungary	6.16	6.14	7.35	6.10	7.48	0.38	0.42	0.56	0.57	0.70	5.78	5.72	6.79	5.53	6.78
Poland	21.20	21.45	22.45	23.20	24.96	16.87	16.51	16.87	17.90	18.47	4.33	4.94	5.58	5.30	6.49
Romania[b]	18.42	19.42	20.05	20.49	21.56	6.94	7.22	7.76	7.57	8.38	11.48	12.22	12.29	12.92	13.18
Eastern Europe	79.53	80.01	83.64	82.85	89.22	48.14	46.97	48.52	48.74	51.73	31.39	33.06	35.12	34.11	37.49
TOTAL EUROPE	350.53	367.05	405.40	390.80	438.06	228.74	239.63	265.83	256.45	289.16	121.79	127.44	139.57	134.35	148.90

a Secretariat forecasts prepared on basis of partial forecasts by countries.   b Secretariat forecasts.

FIGURE 5.3

Forecasts of removals by country group or sub-group
(*Million m³ underbark*)

NORDIC COUNTRIES

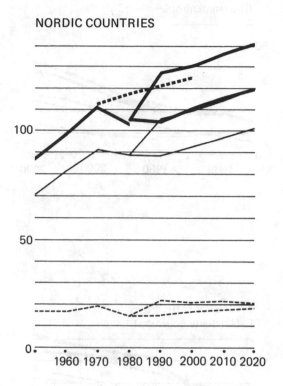

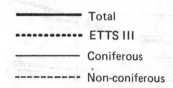

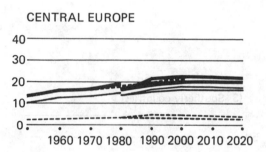

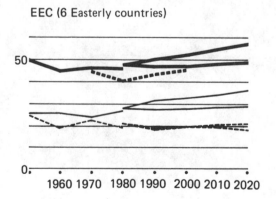

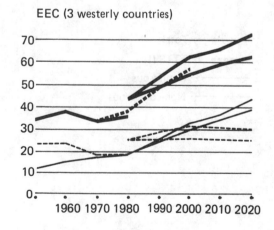

FIGURE 5.3 (*continued*)

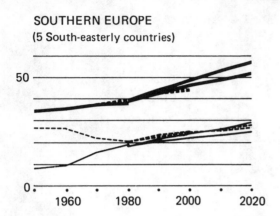

SOUTHERN EUROPE
(5 South-easterly countries)

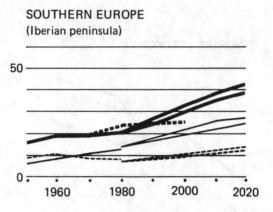

SOUTHERN EUROPE
(Iberian peninsula)

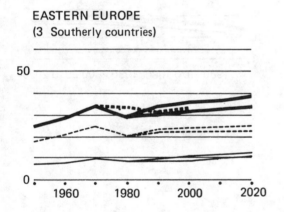

EASTERN EUROPE
(3 Southerly countries)

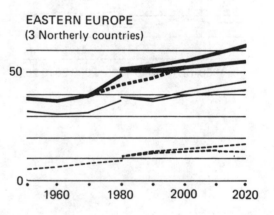

EASTERN EUROPE
(3 Northerly countries)

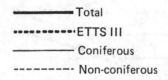

————— Total
••••••••• ETTS III
——— Coniferous
- - - - - - - Non-coniferous

TABLE 5.7

**EEC(9), Southern and Eastern Europe: forecasts of removals in 2000
by sub-group and change on base period (around 1980)[a]**

(*Million m³ underbark*)

Country sub-groups [a]	Total		Coniferous		Non-coniferous	
	Low	High	Low	High	Low	High
1. EEC(9)						
Six easterly countries						
Removals in 2000	46.8	51.7	27.3	32.2	19.5	19.5
Change on base period ..	− 1.0	+ 3.9	− 0.3	+ 4.6	− 0.7	− 0.7
Percentage change.......	− 2.1%	+ 8.2%	− 1.1%	+16.7%	− 3.5%	− 3.5%
Three westerly countries						
Removals in 2000	54.3	63.0	28.7	32.2	25.6	30.8
Change on base period ..	+11.0	+19.7	+10.5	+14.0	+ 0.5	+ 5.7
Percentage change.......	+25.4%	+45.5%	+57.7%	+76.9%	+ 2.0%	+22.7%
2. SOUTHERN EUROPE						
Iberian Peninsula						
Removals in 2000	30.5	33.3	20.4	22.1	10.1	11.2
Change on base period ..	+ 9.9	+12.7	+ 6.7	+ 8.4	+ 3.2	+ 4.3
Percentage change.......	+48.1%	+61.7%	+48.9%	+61.3%	+46.4%	+62.3%
Five south-easterly countries						
Removals in 2000	46.6	49.0	22.4	24.3	24.2	24.7
Change on base period ..	+ 7.6	+10.0	+ 3.5	+ 5.4	+ 4.1	+ 4.6
Percentage change.......	+19.5%	+25.6%	+18.5%	+28.6%	+20.4%	+22.9%
3. EASTERN EUROPE						
Three southerly countries						
Removals in 2000	31.0	34.2	9.2	10.4	21.8	23.8
Change on base period ..	+ 2.0	+ 5.2	+ 0.7	+ 1.9	+ 1.3	+ 3.3
Percentage change.......	+ 6.9%	+17.9%	+ 8.2%	+22.4%	+ 6.3%	+16.1%
Three northerly countries						
Removals in 2000	51.8	55.0	39.5	41.3	12.3	13.7
Change on base period ..	+ 1.3	+ 4.5	− 0.1	+ 1.7	+ 1.4	+ 2.8
Percentage change.......	+ 2.3%	+ 8.9%	− 0.3%	+ 4.3%	+12.8%	+25.7%

[a] *EEC(9)* − *Six easterly countries:* Belgium, Denmark, Federal Republic of Germany, Italy, Luxembourg, the Netherlands.
  − *Three westerly countries:* France, Ireland, United Kingdom.
Southern Europe − *Iberian Peninsula:* Portugal, Spain.
  − *Five south-easterly countries:* Cyprus, Greece, Israel, Turkey, Yugoslavia.
Eastern Europe − *Three southerly countries:* Bulgaria, Hungary, Romania.
  − *Three northerly countries:* Czechoslovakia, German Democratic Republic, Poland.

5.3.2 Removals forecasts by assortment

As part of the enquiry on the outlook for forestry and wood supply, countries were asked to estimate the proportion of their future removals that would be of sawlog and veneer log size and quality. Such estimates are likely to be rough, notably because of the interchangeability between sawlogs and pulpwood of roundwood in the general diameter range of 12 to 25 cm. Conditions in the markets for sawnwood and woodpulp and relative roundwood price levels determine the allocation of raw material in this size range, and the situation can change quite noticeably over short periods of time.

None the less, the results are set out by country and country group in annex tables 5.2 and 5.3, and are summarized in tables 5.9 and 5.10. They are useful in providing the starting point on the raw material side for the consistency analysis in chapter 20.

Removals of sawlogs and veneer logs in Europe are estimated to reach between 184 and 200 million m³ u.b. in the year 2000, which represents increases of 24 million m³ u.b. (15%) and 40 million m³ u.b. (25%) respectively over the base period. The corresponding figures for removals of all other wood (termed "smallwood" in the tables) in 2000 are 207 and 238 million m³ u.b., and increases of 17 million m³ u.b. (8%) and 47 million m³ u.b. (25%) respectively.

Under the low assumption, the share of sawlogs and veneer logs in total removals in 2000 of 47.0% is slightly above that in the base period; while under the high assumption there is virtually no change in the share over the 20-year period. Looked at from an historical perspective, the forecasts suggest that, after climbing steadily from 33% in 1949-51 to 45% in 1979-81, the share of sawlogs and veneer logs in European removals has reached a plateau. ETTS III had assumed that the share would decline between 1970 and 2000, as a result of much stronger growth in consumption of round pulpwood than of sawlogs. That did not happen in the 1970s because of slower than expected growth in consumption of pulpwood and the strong increase in the use of residues and waste paper.

TABLE 5.8

**Share of country groups and sub-groups in total European removals,
1949-51 to 2020 (forecasts)**

(Percentage of European total)

Country groups and sub-groups [a]	1949-51	1969-71	Base period (around 1980)	Low forecasts		High forecasts	
				2000	2020	2000	2020
Coniferous							
Nordic countries	41.2	42.9	39.1	36.3	35.08	37.8	36.2
EEC(9)	21.3	19.0	20.0	21.8	23.2	22.3	23.8
Six easterly countries	14.9	11.3	12.0	10.6	9.8	11.2	10.8
Three westerly countries....	6.4	7.7	8.0	11.2	13.4	11.1	13.0
Central Europe	6.2	6.2	5.7	6.2	5.6	6.0	5.2
Southern Europe	8.5	12.7	14.2	16.7	18.1	16.0	17.1
Iberian Peninsula..........	3.9	5.2	6.0	8.0	9.3	7.6	8.3
Five south easterly countries	4.6	7.5	8.2	8.7	8.8	8.4	8.8
Eastern Europe	22.8	19.2	21.0	19.0	18.1	17.9	17.7
Three southerly countries...	4.1	4.4	3.7	3.6	3.6	3.6	3.9
Three northerly countries...	18.7	14.8	17.3	15.4	14.5	14.3	13.8
Europe.................	100.0	100.0	100.0	100.0	100.0	100.0	100.0
Non-coniferous							
Nordic countries	13.5	15.6	11.9	12.5	12.9	13.9	12.9
EEC(9)	38.0	31.8	37.1	33.6	30.9	33.8	31.7
Six easterly countries	19.6	18.0	16.5	14.6	13.8	13.1	13.0
Three westerly countries....	18.4	13.8	20.6	19.0	17.1	20.7	18.7
Central Europe	1.5	2.4	3.0	3.0	2.8	3.0	2.6
Southern Europe	28.7	24.2	22.2	25.5	28.1	24.1	26.4
Iberian Peninsula..........	7.2	6.7	5.7	7.5	9.1	7.5	8.9
Five south-easterly countries	21.5	17.5	16.5	18.0	19.0	16.6	17.5
Eastern Europe	18.3	26.0	25.8	25.4	25.3	25.2	26.4
Three southerly countries...	13.9	19.5	16.8	16.2	16.2	15.9	16.1
Three northerly countries...	4.4	6.5	9.0	9.2	9.1	9.3	10.3
Europe.................	100.0	100.0	100.0	100.0	100.0	100.0	100.0

[a] For country sub-groups, see footnote *a* to table 5.7.

Nor, according to the new removals forecasts, will it occur between 1980 and 2000. Final judgement on this outlook must be reserved, however, until the material balance is presented and dicussed in chapter 20.

Coniferous species are expected to account for the major part of the increase in removals of both sawlogs and veneer logs and of smallwood between 1980 and 2000 — about three-quarters of the total increase in sawlogs and veneer logs and about three-fifths for smallwood. On the other hand, the *rate* of growth in removals of non-coniferous sawlogs and veneer logs is, somewhat surprisingly, greater than that of coniferous in all country groups except the EEC(9). In that group, the strong increase in supplies from coniferous plantations in the three westerly countries, mentioned above, is reflected in above-average growth forecast for coniferous removals, both of sawlogs and veneer logs and of smallwood.

Removals of coniferous smallwood, at the all-European level, are estimated to grow at a faster rate than those of non-coniferous smallwood; this is also the case in three of the country groups: the Nordic countries and Eastern Europe are the exceptions. One possible explanation for the relatively strong growth in non-coniferous smallwood removals, at least for the Nordic countries, could be the expectation of increasing commercial use of wood for energy generation, with birch which is the principal hardwood species in the area a suitable species for this purpose. In other countries of central and western Europe, the forecasts of appreciable increases in coniferous smallwood removals probably reflect expectations that the backlog in thinning can be reduced over the coming 10 to 20 years (at least under the high forecast) and that there will be increased supplies coming to the market from conifer plantations, notably in westerly EEC countries and Southern Europe. A further factor could be that in several countries with a more or less normal age-class structure, coniferous removals are already closer to the potential cut

FIGURE 5.4

**Share of European removals by country group
and sub-group,[a] 1950 to 2000**

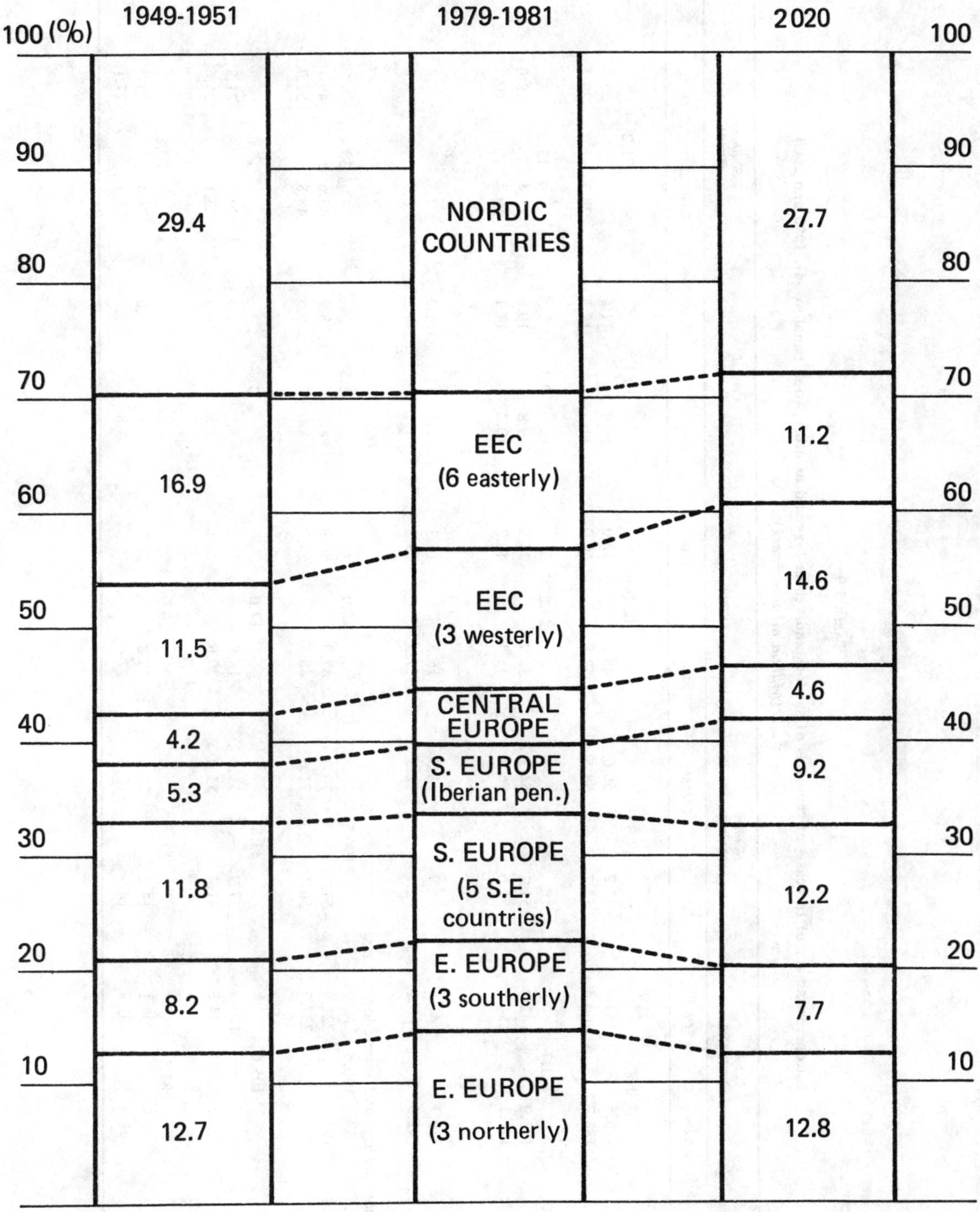

100 (%)	1949-1951		1979-1981		2 020	100
90			NORDIC COUNTRIES		27.7	90
80	29.4					80
70						70
60	16.9		EEC (6 easterly)		11.2	60
50	11.5		EEC (3 westerly)		14.6	50
40	4.2		CENTRAL EUROPE		4.6	40
	5.3		S. EUROPE (Iberian pen.)		9.2	
30	11.8		S. EUROPE (5 S.E. countries)		12.2	30
20	8.2		E. EUROPE (3 southerly)		7.7	20
10	12.7		E. EUROPE (3 northerly)		12.8	10

[a] For country sub-groups, see footnote *a* to table 5.7.

TABLE 5.9

Europe: summary of forecasts of removals of sawlogs and veneer logs and of other wood (smallwood), 1990 and 2000

(Million m³ underbark)

	Total Europe		Nordic countries		EEC(9)		Central Europe		Southern Europe		Eastern Europe	
	Low	*High*	*Low*	*High*	*Low*	*High*	*Low*	*High*	*Low*	*High*	*Low*	*High*
SAWLOGS AND VENEER LOGS, TOTAL												
Base period	160.0		47.3		44.1		10.1		21.2		37.3	
1990	173.6	183.9	53.2	56.6	48.2	52.0	10.9	11.5	24.7	25.5	36.6	38.2
2000	183.7	200.4	53.7	58.5	52.8	60.0	11.3	12.3	28.3	29.5	37.6	40.1
Coniferous												
Base period	123.7		45.5		28.8		9.3		15.0		25.2	
1990	133.9	142.4	51.3	54.3	31.7	34.8	9.8	10.3	17.4	18.0	23.7	24.9
2000	141.1	154.3	51.6	56.2	35.3	40.2	10.2	11.1	19.7	20.7	24.3	26.1
Non-coniferous												
Base period	36.3		1.8		15.3		0.8		6.2		12.1	
1990	39.7	41.5	1.9	2.3	16.5	17.3	1.1	1.2	7.3	7.4	12.9	13.3
2000	42.6	46.1	2.1	2.3	17.5	19.8	1.1	1.2	8.6	8.8	13.2	14.0
SMALLWOOD (wood other than sawlogs and veneer logs), TOTAL												
Base period	190.5		56.5		46.9		6.5		38.4		42.2	
1990	193.5	221.5	51.1	70.3	47.0	50.9	8.3	8.9	43.6	46.1	43.4	45.4
2000	207.1	237.7	56.1	71.6	48.3	54.7	8.6	9.5	48.8	52.7	45.3	49.1
Coniferous												
Base period	105.0		43.8		17.0		3.7		17.6		22.9	
1990	105.7	123.5	37.9	51.0	18.8	21.2	5.5	5.9	20.2	21.8	23.2	23.6
2000	115.3	134.8	41.4	53.1	20.6	24.1	5.8	6.3	23.1	25.6	24.4	25.6
Non-coniferous												
Base period	85.5		12.7		29.9		2.8		20.8		19.3	
1990	87.8	98.1	13.2	19.3	28.2	29.7	2.8	3.0	23.4	24.3	20.2	21.8
2000	91.8	102.8	14.7	18.5	27.6	30.5	2.8	3.2	25.7	27.1	20.9	23.5

TABLE 5.10

Europe: summary of forecasts of changes in removals of sawlogs and veneer logs and of other wood (smallwood), between the base period [a] and 2000 and of share of sawlogs in total removals

	Total Europe		Nordic countries		EEC(9)		Central Europe		Southern Europe		Eastern Europe	
	Low	High	Low	High	Low	High	Low	High	Low	High	Low	High
VOLUME CHANGE, BASE PERIOD TO 2000 (million m³ underbark)												
Sawlogs and veneer logs												
Total	+23.7	+40.4	+6.4	+11.2	+8.7	+15.9	+1.2	+2.2	+7.1	+8.3	+0.3	+2.8
Coniferous	+17.4	+30.6	+6.1	+10.7	+6.5	+11.4	+0.9	+1.8	+4.7	+5.7	−0.9	+0.9
Non-coniferous	+6.3	+9.8	+0.3	+0.5	+2.2	+4.5	+0.3	+0.4	+2.4	+2.6	+1.1	+1.9
Smallwood												
Total	+16.6	+47.2	−0.4	+15.1	+1.4	+7.8	+2.1	+3.0	+10.4	+14.3	+3.1	+6.9
Coniferous	+10.3	+29.8	−2.4	+9.3	+3.6	+7.1	+2.1	+2.6	+5.5	+8.0	+1.5	+2.7
Non-coniferous	+6.3	+17.3	+2.0	+5.8	−2.3	+0.6	−	+0.4	+4.9	+6.3	+1.6	+4.2
PERCENTAGE CHANGE, BASE PERIOD TO 2000												
Sawlogs and veneer logs												
Total	+14.8	+25.3	+13.5	+23.7	+19.7	+36.1	+11.9	+21.8	+33.5	+39.2	+0.8	+7.5
Coniferous	+14.1	+24.7	+13.4	+23.5	+22.6	+39.6	+9.7	+19.4	+31.3	+38.0	−3.6	+3.6
Non-coniferous	+17.4	+27.0	+16.7	+27.8	+14.4	+29.4	+37.5	+50.0	+38.7	+41.9	+9.1	+15.8
Smallwood												
Total	+8.7	+24.8	−0.7	+26.7	+3.0	+16.6	+32.2	+46.2	+27.1	+37.2	+7.3	+16.4
Coniferous	+9.8	+28.4	−5.5	+21.2	+21.2	+41.8	+56.8	+70.3	+31.2	+45.5	+6.6	+11.8
Non-coniferous	+7.4	+20.2	+15.7	+45.7	−7.7	+2.0	−	+14.3	+23.6	+30.3	+8.3	+21.8
PERCENTAGE SHARE OF SAWLOGS AND VENEER LOGS IN TOTAL REMOVALS												
Total												
Base period	45.6	45.6	45.5	45.5	48.4	48.4	60.9	60.9	35.6	35.6	46.9	46.9
2000	47.0	45.7	48.9	45.0	52.2	52.3	56.8	56.5	36.7	35.9	45.4	45.0
Coniferous												
Base period	54.1	54.1	50.9	50.9	62.9	62.9	71.4	71.4	46.0	46.0	52.4	52.4
2000	55.0	53.4	55.5	51.4	63.1	62.5	63.8	63.6	46.0	44.7	49.9	50.5
Non-coniferous												
Base period	29.8	29.8	12.6	12.6	33.8	33.8	22.8	22.8	23.1	23.1	38.5	38.5
2000	31.7	31.0	12.5	10.8	38.7	39.3	28.6	28.2	25.2	24.5	38.7	37.3

[a] See footnote *a* to table 5.3.

than is the case for non-coniferous removals.

5.3.3 Forecasts of removals from exploitable closed forest and other sources

The forecasts presented above were of total removals, that is to say removals from exploitable closed forest together with those from all other sources, namely other wooded land, trees outside the forest and, under certain circumstances, even from unexploitable closed forest. Table 5.11 shows the breakdown by country group of total removals between exploitable closed forest and other sources. For the base period around 1980 removals from exploitable closed forest accounted for 94% of the European total. The percentage was lower for the EEC(9) and Southern Europe. Of removals from other sources, three-

fifths were of non-coniferous species (nearly 70% if the Nordic countries are excluded).

By 2000, removals from other sources may have declined slightly in relative importance – from 6% to 5% of total European removals. It should be noted in passing, however, that a number of countries are not able to identify in their statistics the origin of removals. It appears likely that some of these countries' removals from other sources are either included in the statistics under exploitable closed forests or are not recorded at all. Nevertheless, it is clear that removals from exploitable closed forest account for all but a small fraction of total removals. This is important to bear in mind when considering prospects for the forest inventory on exploitable closed forest, which are discussed below.

TABLE 5.11

Europe: removals from exploitable closed forest and from other sources[a] in base period (around 1980) and 2000 (forecasts)

	From exploitable closed forest						From other sources, volume (million m³ u.b.)		
	Volume (million m³ u.b.)			Share of total removals (%)					
		2000			2000			2000	
	Base period	Low	High	Base period	Low	High	Base period	Low	High
Nordic countries									
Total	100.0	106.6	124.9	96	97	96	3.8	3.2	5.2
Coniferous	86.0	90.3	104.8	96	97	96	3.2	2.7	4.5
Non-coniferous	14.0	16.3	20.1	96	97	97	0.6	0.5	0.7
EEC(9)									
Total	82.2	93.8	107.4	90	93	94	8.9	7.3	7.3
Coniferous	45.6	55.7	64.3	100	100	100	0.3	0.1	0.1
Non-coniferous	36.6	38.1	43.1	81	84	86	8.6	7.2	7.2
Central Europe									
Total	16.2	19.6	21.5	98	98	98	0.4	0.4	0.4
Coniferous	12.8	15.8	17.3	99	99	99	0.2	0.2	0.2
Non-coniferous	3.4	3.8	4.2	94	94	95	0.2	0.2	0.2
Southern Europe									
Total	52.4	69.5	74.5	88	90	91	7.2	7.6	7.8
Coniferous	27.9	37.9	41.3	86	89	89	4.7	4.9	5.0
Non-coniferous	24.5	31.6	33.1	91	92	92	2.5	2.7	2.8
Eastern Europe									
Total	79.0	82.3	87.0	99	99	98	0.5	0.5	2.2
Coniferous	47.9	48.5	51.0	100	100	99	0.2	0.2	0.7
Non-coniferous	31.1	33.8	36.0	99	99	96	0.3	0.3	1.5
TOTAL EUROPE									
Total	329.8	371.9	415.2	94	95	95	20.7	18.9	22.9
Coniferous	220.2	248.5	278.7	96	97	96	8.5	8.0	10.5
Non-coniferous	109.6	123.4	136.5	90	92	92	12.2	10.9	12.4

[a] Other wooded land, trees outside the forest and unexploitable closed forest.

5.4 OUTLOOK FOR THE FOREST RESOURCE

5.4.1 The methodology of the forestry forecasts

The forecasts of removals in the European region, which were presented in section 5.3, showed increases between

the base period and the year 2000 of 40 to 88 million m³ underbark (12% to 25%), with further growth of a comparable order between 2000 and 2020. What are the likely developments in the forest resource that could make an

TABLE 5.12

Europe: forecasts for area and growing stock relating to exploitable closed forest

| | Area (million ha) | | Growing stock | | | | | | | | Volume per ha (m³ o.b./ha) | |
| | | | Total (million m³ o.b.) | | Coniferous (million m³ o.b.) | | Non-coniferous (million m³ o.b.) | | | | | |
	Low	High	Low	High	Low	High	Low	High			Low	High
Nordic countries												
Base period	48.3		4 353		3 635		718				90	
1990	48.8	48.8	4 765	4 625	3 931	3 838	834	787			98	95
2000	49.1	49.1	5 087	4 701	4 146	3 892	941	809			104	96
2010	49.2	49.2	5 383	4 778	4 321	3 936	1 062	842			109	97
2020	49.3	49.3	5 655	4 853	4 458	3 986	1 197	867			115	98
European Economic Community (nine countries)												
Base period	27.9		4 004		2 099		1 905				144	
1990	28.2	28.6	4 267	4 244	2 271	2 260	1 996	1 984			151	149
2000	28.6	29.3	4 520	4 437	2 438	2 402	2 082	2 035			158	151
2010	28.9	30.1	4 755	4 624	2 586	2 557	2 169	2 067			165	153
2020	29.2	30.9	4 977	4 800	2 724	2 707	2 253	2 093			170	155
Central Europe												
Base period	4.1		1 115		889		226				272	
1990	4.1	4.1	1 178	1 194	946	961	231	233			286	289
2000	4.1	4.2	1 234	1 230	1 000	1 001	234	229			299	296
2010	4.1	4.2	1 230	1 206	996	985	234	221			298	290
2020	4.1	4.2	1 227	1 186	994	972	233	214			297	285
Southern Europe												
Base period	26.7		2 580		1 210		1 370				97	
1990	28.0	29.1	2 805	2 792	1 330	1 325	1 475	1 467			100	96
2000	29.4	31.6	2 986	2 956	1 425	1 413	1 560	1 543			102	94
2010	30.9	34.0	3 142	3 090	1 501	1 474	1 641	1 616			102	91
2020	32.3	36.3	3 272	3 211	1 564	1 533	1 708	1 678			101	89
Eastern Europe												
Base period	25.8		4 278		2 530		1 748				166	
1990	26.1	26.4	4 488	4 547	2 644	2 699	1 844	1 848			172	172
2000	26.3	26.7	4 682	4 785	2 730	2 849	1 952	1 936			178	179
2010	26.5	27.0	4 822	4 991	2 778	2 978	2 044	2 013			182	185
2020	26.7	27.3	4 925	5 163	2 808	3 083	2 117	2 080			185	189
TOTAL EUROPE												
Base period	132.8		16 330		10 363		5 967				123	
1990	135.2	137.0	17 503	17 402	11 122	11 083	6 380	6 319			129	127
2000	137.5	140.9	18 509	18 109	11 739	11 557	6 769	6 552			135	129
2010	139.6	144.5	19 332	18 689	12 182	11 930	7 152	6 759			138	129
2020	141.6	148.0	20 056	19 213	12 548	12 281	7 508	6 932			142	130

TABLE 5.13

Europe: forecasts for net annual increment and fellings on exploitable closed forest

	Net annual increment (NAI)								Fellings							
	Total (million m³ o.b.)		Coniferous (million m³ o.b.)		Non-coniferous (million m³ o.b.)		Volume per ha (m³ o.b./ha)		Total (million m³ o.b.)		Coniferous (million m³ o.b.)		Non-coniferous (million m³ o.b.)		Felling/NAI Ratio (%)	
	Low	High	Low	High	Low	High	Low	High	Low	High	Low	High	Low	High	Low	High
Nordic countries																
Base period	146.2		117.1		29.0		3.0		123.8		105.2		18.6		84	
1990	157.0	158.6	124.3	128.2	32.7	30.4	3.2	3.2	124.8	151.2	104.6	122.3	20.2	28.9	79	95
2000	161.3	162.5	126.1	131.9	35.2	30.7	3.3	3.3	131.7	154.8	109.3	127.1	22.4	27.7	81	94
2010	166.1	168.8	127.8	137.8	38.3	31.1	3.4	3.4	138.4	161.5	114.7	133.0	23.7	28.5	83	95
2020	169.2	175.2	128.0	143.0	41.2	32.2	3.4	3.6	143.7	167.6	118.8	139.9	24.9	27.7	84	95
European Economic Community (nine countries)																
Base period	128.1		74.4		53.7		4.6		98.2		54.1		44.1		77	
1990	130.1	134.6	76.8	80.5	53.3	54.2	4.6	4.7	104.8	114.0	60.3	66.7	44.5	47.3	81	85
2000	138.4	146.8	83.8	90.0	54.6	55.9	4.8	5.0	113.0	129.3	67.1	76.9	45.9	52.4	82	88
2010	141.5	156.5	86.5	99.9	55.0	56.6	4.9	5.2	120.5	136.7	73.9	83.1	46.6	53.6	85	87
2020	145.7	164.9	90.1	107.7	55.6	57.2	5.0	5.3	127.1	149.0	80.5	94.8	46.7	54.2	87	90
Central Europe																
Base period	24.8		20.3		4.5		6.0		20.0		15.9		4.1		81	
1990	26.0	26.3	21.4	21.7	4.6	4.6	6.3	6.3	23.4	25.0	18.9	20.1	4.5	4.9	90	95
2000	27.3	27.2	22.5	22.6	4.8	4.6	6.6	6.5	24.3	26.6	19.7	21.5	4.6	5.1	89	98
2010	27.4	27.0	22.6	22.4	4.8	4.6	6.6	6.5	24.4	26.4	19.8	21.4	4.6	5.0	89	98
2020	27.5	27.1	22.6	22.5	4.9	4.6	6.7	6.5	24.4	26.1	19.8	21.2	4.6	4.9	89	97
Southern Europe																
Base period	91.7		47.8		43.9		3.4		66.7		35.0		31.6		72	
1990	95.8	98.3	50.6	52.3	45.2	46.0	3.4	3.4	76.7	80.8	41.1	43.7	35.4	37.0	80	82
2000	101.3	105.2	54.3	57.6	47.0	48.2	3.4	3.3	87.5	93.5	47.2	51.3	40.3	42.1	86	89
2010	106.0	112.4	57.4	61.9	48.6	50.5	3.4	3.3	96.7	105.6	52.8	58.9	43.9	46.7	91	94
2020	110.7	119.0	60.7	66.6	50.0	52.4	3.4	3.3	105.8	113.4	58.6	63.9	47.2	49.5	96	95
Eastern Europe																
Base period	113.3		66.8		46.5		4.4		93.1		56.4		36.7		82	
1990	115.1	122.4	66.4	72.7	48.7	49.7	4.4	4.6	93.8	96.8	55.4	56.6	38.4	40.2	81	79
2000	112.2	124.5	63.3	74.1	48.9	50.4	4.3	4.7	96.8	102.2	57.4	60.2	39.4	42.0	86	82
2010	112.0	126.7	63.0	75.5	49.0	51.2	4.0	4.5	99.7	109.9	58.4	64.8	41.3	43.7	91	88
2020	111.5	128.3	62.7	76.4	48.8	51.9	4.2	4.7	103.4	113.1	61.4	67.5	42.0	45.6	93	88
TOTAL EUROPE																
Base period	504.1		326.4		177.6		3.8		401.8		266.6		135.2		80	
1990	524.0	540.2	339.5	355.4	184.5	184.9	3.9	3.9	423.5	467.8	280.3	309.4	143.0	158.3	81	87
2000	540.5	566.2	350.0	376.5	190.5	189.8	3.9	4.0	453.3	506.4	300.7	337.0	152.6	169.3	84	89
2010	553.0	591.4	357.3	397.5	195.7	194.0	4.0	4.1	479.7	537.8	320.2	360.3	159.5	177.5	87	91
2020	564.6	614.5	364.1	416.2	200.5	198.3	4.0	4.2	504.4	569.2	339.1	387.3	165.4	181.9	89	93

FIGURE 5.5

**Europe: net annual increment on exploitable closed forest,
by country group or sub-group, base period to 2020 (forecasts)**

(*Million m³ overbark*)

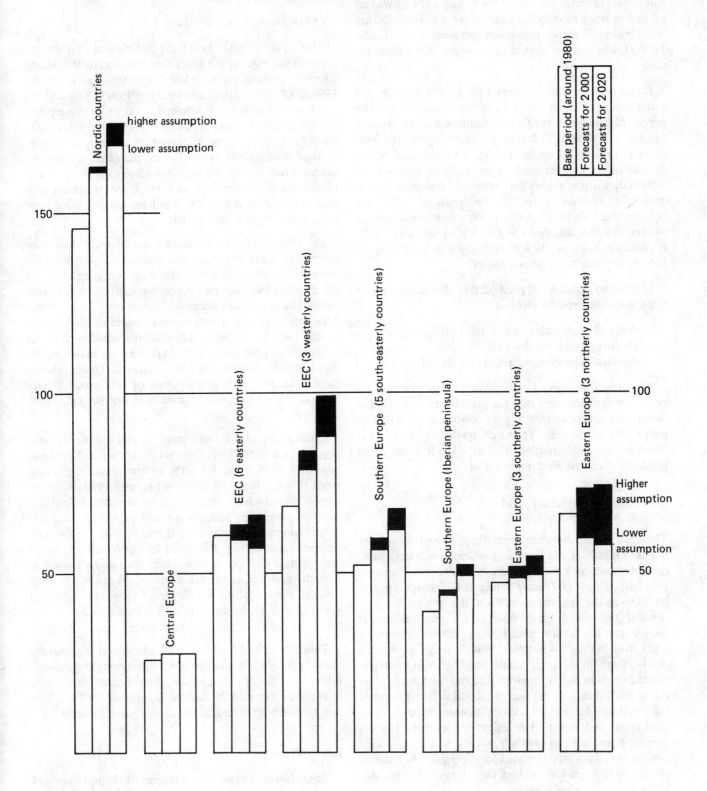

expansion of this order possible? After all, it would be quite substantial in volume terms: the average of the lower and higher forecasts is an increase of 110 million m³ u.b. in removals over 40 years, or 2.75 million m³ u.b. a year. The actual increase between 1949-51 and 1979-81 was an annual average of only 1.5 million m³ u.b. Even during the period of active expansion between 1949-51 and 1969-71, the average annual increase was 2.0 million m³ u.b.

Based on countries' forestry forecasts, a drain of the order indicated above would be fully compatible with the predicted changes in the forest production potential on a sustained yield basis. That is to say, assuming that the forecasts of the area of exploitable closed forest and of net annual increment per hectare are realistic and will be achieved, Europe's growing stock will continue to rise gradually, at least up to 2020, in terms of both total volume and volume per hectare. This conclusion applies not only to the region as a whole, but to virtually all the countries in it, as may be seen from the individual country forecasts given in the annex tables.

In order to assess wood production potential, countries were asked to provide forecasts of:

- Area of exploitable closed forest (ECF);
- Growing stock on the ECF;
- Net annual increment (NAI) on the ECF.

As noted above, exploitable closed forest provides all but a small percentage of total wood removals, and it seemed rational to concentrate the outlook enquiry on this part of the forest area. To match against the wood production potential, countries were asked to forecast wood production on the ECF in terms of:

- Fellings overbark;
- Removals overbark; and
- Removals underbark.

The relationship between these three is as follows: removals underbark *plus* bark on the wood removed *equals* removals overbark (o.b.); removals o.b. *plus* unrecovered harvesting losses *equals* fellings o.b. Fellings o.b. can be taken as the drain of wood from the forest. The relationship between drain or fellings o.b. and net annual increment o.b. will determine how the volume of growing stock will change over time. Models may be used to demonstrate the dynamic relationship between these elements, which in their more sophisticated form may incorporate such details as the age- or size-class distribution of the forest, site productivity classes, thinning regimes, rotation ages, and so on. Not all countries have the resources or the data to use such models, and a much simpler methodology was suggested for application by country correspondents in formulating their forecasts. This was to use the following equation:

$$GS_1 + [(NAI_1 + NAI_2) \times t/2] - [(F_1 + F_2) \times t/2] = GS_2.$$

where:

GS_1 and GS_2 are the growing stock volumes at the beginning and end respectively of the period of t years;

NAI_1 and NAI_2 are the net annual increment volumes at the beginning and end of the period of t years;

F_1 and F_2 are felling volumes at the beginning and end of the period of t years.

(All volumes in m³ overbark.)

This equation takes as a basic assumption that the average of the NAI at the beginning and end of the period covered (forecasts were asked for at ten-year intervals: 1990, 2000, 2010, 2020) accurately reflects the annual average for the period as a whole; and the same assumption for fellings. This assumption should be quite reasonable for NAI, but less so for fellings which are subject to quite marked fluctuations. None the less, provided this qualification is recognized, the equation allows the compilation of a series of forestry forecasts which are consistent with each other. Most of the correspondents who provided forecasts worked with this model.

For those countries which did not provide their own forecasts, the secretariat was obliged to prepare estimates, also using the above equation. Lacking much of the detailed knowledge of the forestry situation available within the countries concerned, it had to follow a somewhat mechanical approach which involved making certain assumptions as to how two key elements would evolve over time: (1) NAI per hectare; and (2) the ratio between fellings and NAI. By also applying certain other assumptions, such as changes in area and share of coniferous species in total NAI and fellings, complete sets of forecasts were built up.

Correspondents of those countries that did supply forecasts, which were 70% of the total number of European countries accounting for 74% of the region's growing stock on exploitable closed forest around 1980, will have applied their own hypotheses and assumptions in building up their forecasts, which were probably more sophisticated than those employed by the secretariat for the non-responding countries. Forecasts for the Nordic countries and Central Europe were provided entirely by the countries themselves. At the other end of the scale, just over half of the countries of Eastern Europe (in terms of growing stock) provided their own forecasts and over two-thirds of the southern European countries.

Tables 5.12 and 5.13 summarize by country group the forestry forecasts which are given by country in the annex tables. For Europe as a whole, the forecasts of changes in aggregate between the base period around 1980 and 2000 and between 2000 and 2020 are as shown in table 5.14.

5.4.2 Fellings

The difference between the figures of fellings overbark presented in this section and removals underbark presented in section 5.3 is due to the exclusion in the former of fellings other than on exploitable closed forest; and the inclusion of (i) bark, and (ii) unrecovered harvesting losses. As was seen in table 5.11, removals from "other sources" (than exploitable closed forest) amounted to about

TABLE 5.14

Europe: forecasts of changes from base period (around 1980) to 2000 and from 2000 to 2020
of forest inventory data relating to exploitable closed forest

| | | Base period to 2000 | | | | 2000 to 2020 | | | |
| | | Area or volume | | Percentage | | Area or volume | | Percentage | |
	Unit	Low	High	Low	High	Low	High	Low	High
Area...............	Million ha.	+ 4.7	+ 8.1	+ 3.5	+ 6.1	+ 4.1	+ 7.1	+ 3.0	+ 5.0
Growing stock	Million m³ o.b.	+2 179	+1 779	+ 13.3	+ 10.9	+1 547	+1 104	+ 8.4	+ 6.1
Net annual increment ..	Million m³ o.b.	+ 36.4	+ 62.1	+ 7.2	+ 12.3	+ 24.1	+ 48.3	+ 4.5	+ 8.5
Fellings	Million m³ o.b.	+ 51.5	+104.6	+ 12.8	+ 26.0	+ 51.1	+ 62.8	+ 11.3	+ 12.4

5% of total removals around 1980 and it could be assumed that the same percentage − or possibly a slightly smaller one − could be applied to fellings (overbark and including unrecovered harvesting losses). Leaving those relatively unimportant volumes aside, table 5.15 shows the relationship between fellings and removals overbark and removals underbark on exploitable closed forest.

The difference between fellings overbark and removals underbark on exploitable closed forest for Europe as a whole around 1980 amounted to 72 million m³, of which 27 million m³ were unrecovered harvesting losses and 45 million m³ bark. The approximate nature of the figures for quite a number of countries needs to be stressed, especially for unrecovered harvesting losses, the true volume of which is probably much higher for Europe as a whole than that shown. The point of setting down these figures is, however, to show that if it became technically and economically possible to achieve fuller use of the available forest biomass, the volumes of harvesting losses and bark that might be recovered and used are quite substantial. In addition, there are also the parts of the felled tree which are left *in situ,* the stumps and roots.[3] Even under the most optimistic of the "realistic" assumptions regarding the fuller utilization of total forest biomass, however, it is likely that only a part of the "residuals" could be harvested and used, how large a part depending on various economic, technical and environmental factors. A dramatic rise in the real price of wood would be needed to make it economic to recover the more expensive harvesting residuals. This does not appear probable in the foreseeable future, even though in some countries there is a clear movement towards harvesting a larger part of the forest biomass. Wood from branches and tops is finding increasing outlets as energy wood. Amongst the factors weighing against whole tree harvesting there is the question of loss of soil fertility.

With regard to fellings on exploitable closed forest, the outlook for 2000 is for the total in Europe to reach between 453 million m³ o.b. (low forecast) and 506 million m³ o.b. (high forecast), increases of 12% and 26% respectively compared with the base period around 1980.

Fellings are expected to expand further in the first two decades of the twenty-first century.

5.4.3 Net annual increment (NAI) (figure 5.6)

Europe's net annual increment (NAI) on exploitable closed forest around 1980 (the base period) of 504 million m³ overbark is forecast to rise to between 540 and 566 million m³ o.b. by 2000. The expected rates of increase over the 20-year period of 7% to 12% are lower than those of fellings, with the result that the ratio of fellings to NAI will rise from 80% in the base period to 84% (low forecast) or 89% (high forecast) by 2000. In other words, there will be some closing of the gap between NAI and fellings on exploitable closed forest (figure 5.7), but increment will still be in excess of drain, with the result that growing stock will continue to expand in Europe. One expects to find in countries with large planting programmes in recent decades that the felling/NAI ratio is lower than average. In Ireland, for example, the ratio is only 26% in the base period but rises to 77-79% by 2000 as more plantations reach maturity. The corresponding ratios for the United Kingdom are 41% for the base period and 54-65% for 2000. Even in countries with a more "normal" forest in terms of age-class distribution, however, there seems to be a tendency for the ratio to rise, as in Austria, from 78% for the base period to 85% to 88% in 2000.

One exception to the generally expected trend in felling/NAI ratios seems to be Hungary, where the ratio for non-coniferous species is expected to fall from 83% in the base period to 68% to 80% by 2020, as a result of the new forest policy which tends to lengthen the rotation ages for the more valuable non-coniferous species, such as oak and beech, and the stands of high productivity.

Despite the further narrowing of the gap between NAI and fellings in Europe generally between 2000 and 2020, increment is expected still to be more than drain at the end of the period with average felling/NAI ratios in Europe of 90% (low forecast) and 93% (high forecast). The trend towards a higher felling ratio is another example, along with the rise in utilization of residues and of waste paper, of the more intensive use of the forest resource.

---

[3] Further information about volume and mass of tree and other woody biomass may be found in Part II of *The Forest Resources of the ECE Region (Europe, the USSR, North America).* See also section 3.8.

FIGURE 5.6

**Europe: forecasts of net annual increment (NAI)
and fellings on exploitable closed forest**

(*Million m³ overbark*)

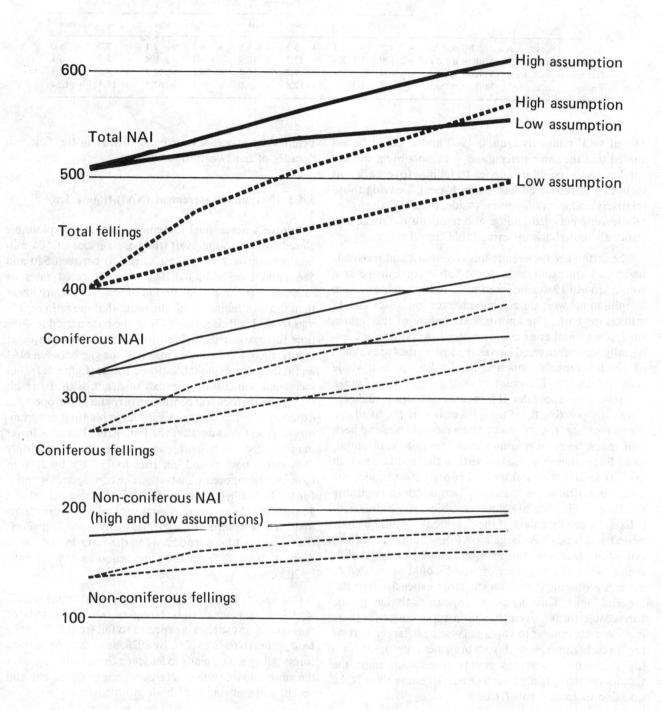

TABLE 5.15

Europe: relationship between fellings and removals (overbark) and removals (underbark) in exploitable closed forest, by country group around 1980

		Unrecovered harvesting losses			Volume of bark on removals		
	Fellings overbark (million m³ o.b.)	Volume (million m³ o.b.)	Percentage of fellings	Removals overbark (million m³ o.b.)	Volume (million m³ o.b.)	Percentage of overbark removals	Removals underbark (million m³ u.b.)
Nordic countries	123.8	9.6	7.8	114.2	14.2	12.4	100.0
EEC(9)	98.2	5.8	5.9	92.4	10.2	11.0	82.2
Central Europe	20.0	1.0	5.0	19.0	2.8	14.7	16.2
Southern Europe	66.7	6.5	9.7	60.2	7.8	13.0	52.4
Eastern Europe	93.1	4.3	4.6	88.8	9.8	11.0	79.0
TOTAL EUROPE	401.8	27.2	6.8	374.6	44.8	12.0	329.8
of which:							
Coniferous	266.6	14.8	5.6	251.8	31.6	12.5	220.2
Non-coniferous	135.2	12.4	9.2	122.8	13.2	10.7	109.6

5.4.4 Growing stock (figure 5.7)

As explained earlier, the assumptions underlying the low and high forestry forecasts relate essentially to the expected level of demand for wood from the forest. At the same time it could be reasonably expected that under the high assumption, the greater level of demand would also stimulate to some extent activity in forestry, which would be reflected in more intensive silviculture leading to improved increment rates and, in many countries, a higher rate of conversion of non-forest land or low productivity woodland into productive, i.e. exploitable, forest. This does appear to be the case in quite a number of countries' forecasts.

Nevertheless, such measures would take time to bring results. It is no surprise, therefore, that in the majority of countries, and for Europe as a whole, the forecasts of growing stock give a slower rate of increase under the higher forecast than under the lower. Thus, Europe's growing stock on exploitable closed forest is expected to reach 18.5 thousand million m³ o.b. by 2000 under the low forecast but 18.1 thousand million m³ o.b. under the high, increases of 13% and 11% respectively over the 20-year period from the base period. Further growth is foreseen up to 2020, but at lower rates than in the 20 years to 2000.

Another feature to note concerns the proportion of coniferous and non-coniferous species in total growing stock. Many countries have been placing emphasis on the use of coniferous species for afforestation and reforestation. This is reflected, especially under the high forecast, in a gradually increasing share of coniferous species in total increment. At the same time, the felling/NAI ratio for coniferous is expected to remain several percentage points above that of non-coniferous species. The net result is that the proportion of coniferous in total growing stock may remain more or less constant at between 60% and 65% between the base period and 2020, at least at the all-European level. There will be exceptions at the country level, notably in the westerly EEC and Iberian Peninsula sub-groups and Hungary, where coniferous growing stock will take a gradually larger share of the total.

5.4.5 Area of exploitable closed forest (figure 5.8)

With regard to area, countries were asked to forecast changes only to exploitable closed forest. Consequently, no precise figures for the trend in the total area of forest and other wooded land can be presented, although some indications will be given later.

The area of exploitable closed forest (ECF) is expected to reach between 138 and 141 million hectares by 2000, an increase of 4.5 to 8 million ha compared with the base period. The largest increases in area of ECF are forecast to occur in the EEC(9), notably the westerly EEC countries, and in Southern Europe, notably Spain and Turkey.

Whilst the general direction of changes in forest area can be seen, since they will be a reflection of national forest policy, it is often more difficult to assess the rate at which they will occur, given that forest policy is only one part of overall land use policy. Many imponderables exist, for example with regard to the intensity and extent of agriculture, policies towards rural population and employment, and the demand for land for building, industry and infrastructure. These questions were already raised in chapter 1.

In the corresponding chapter of ETTS III (chapter 6), two tables were presented, one on extension and losses of forest land between 1950 and 1973 and forecasts for 1973 to 2000 (ETTS III table 3/6); the other on afforestation, reforestation and plantations, covering the same periods (ETTS III table 4/6). In a supplementary enquiry[4] to that on forestry forecasts, ETTS IV correspondents were asked to update those tables, and the information received is shown in tables 5.16 and 5.17. There are not enough figures to allow European totals to be estimated, but some general comments may be made:

1. It may be roughly estimated that the average area of plantations primarily intended for wood production established in Europe yearly between 1980 and 2000

---

[4] The help of Mr. H. Marchand (France) with the preparation of this enquiry is gratefully acknowledged.

FIGURE 5.7

**Europe: growing stock on exploitable closed forest,
by country group or sub-group, base period to 2020 (forecasts)**

(*Million m³ overbark*)

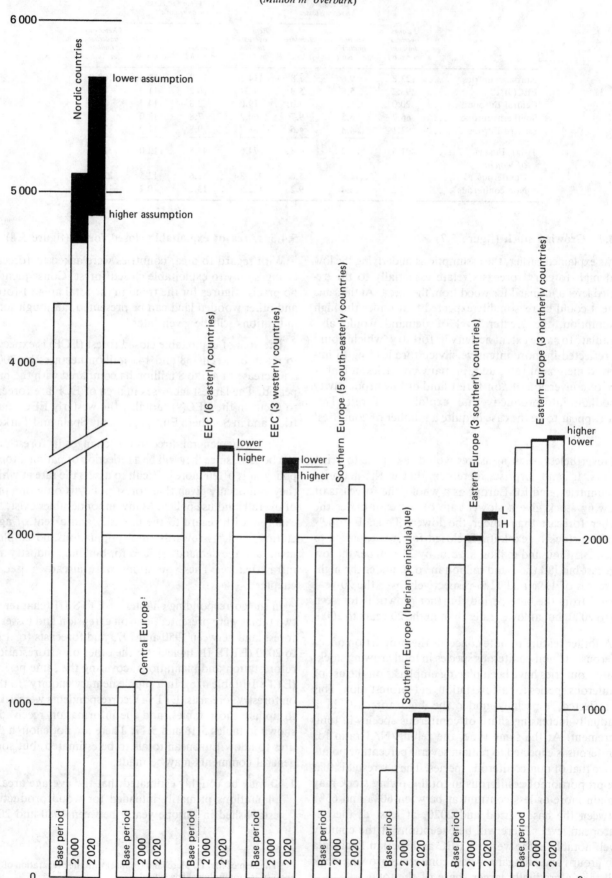

FIGURE 5.8

Europe: area of exploitable closed forest, by country group or sub-group, base period to 2020 (forecasts)
(Million hectares)

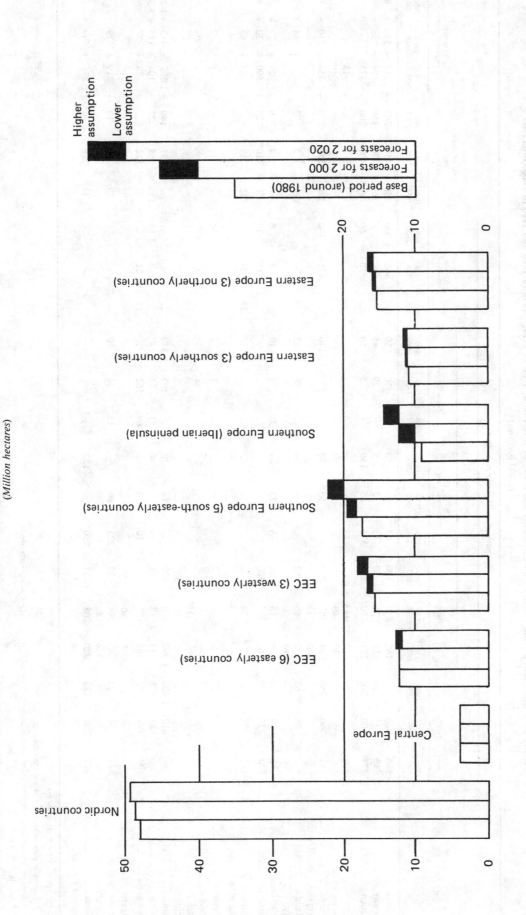

TABLE 5.16

Europe: available information from countries on extensions and losses of forest land, 1950 to 1973,[a] 1970 to 1980 and 1980 to 2000 (forecasts)

(1 000 hectares)

	Extension of forest on non-forest land															Loss of forest land to other uses			Net change in forest area		
	Total			**Afforestation – Total**			**Afforestation – of which: on agricultural land**			**Afforestation – On other land**			**Natural extension**								
	1950-1973	1970-1980	1980-2000	1950-1973	1970-1980	1980-2000	1950-1973	1970-1980	1980-2000	1950-1973	1970-1980	1980-2000	1950-1973	1970-1980	1980-2000	1950-1973	1970-1980	1980-2000	1950-1973	1970-1980	1980-2000
Finland	565	1 020	1 100	165	158	100	67	88	50	98	70	50	400	862	1 000	450	450	200	+ 115	+ 570	+ 900
Norway	15	31	50	15	31	50	15	2	5	—	29	45	—	—	—	169	70	150	− 154	− 39	− 105
Sweden	375	100	200	375	70	140	375	65	130	—	5	10	—	30	60	−*	100	200	+ 375*	—	—
Belgium	20	..	..	20	(13 ....)	..	15	..	..	5	..	..	—	..	..	7	..	..	+ 13	..	..
Denmark	..	40	40[c]	..	40	40[c]	..	..	..	..	..	..	..	..	..	..	..	..	..	..	..
France	634	360	350	634	185	210	..	55	70	..	130	140	..	175	140	180	150	250	+ 454	+ 210	+ 100
Germany, Fed. Rep. of	200	..	..	200	16[c]	..	..	..	..	..	..	..	..	..	..	..	..	..	..	..	..
Ireland	..	86	62	..	86	62	..	—	—	..	86	62	..	—	—	..	2	2	..	+ 84	+ 59
Italy	530	..	..	530	..	..	..	..	..	..	..	..	..	..	..	..	..	..	..	..	..
Netherlands	35	26	40	26	18	24	3	13	24	23	5	..	..	8	6	6	4	6	+ 29	+ 22	+ 34
United Kingdom	640	325	500	640	325	500	..	..	..	..	..	..	50	15*	10	50	10*	20	+ 590	+ 330	+ 510
Austria	135[d]	78	..	57[d]	52	..	..	..	..	..	..	..	78	26	..	12	15	..	+ 123	+ 63	..
Switzerland	..	3[c]	..	..	3[c]	..	..	..	..	..	..	..	..	..	..	..	..	..	..	..	..
Cyprus[e]	..	2*	2	..	1	2	..	—	..	..	1	..	..	1	—	..	—	—	..	+ 2	+ 2
Greece	..	..	5	..	..	5	..	..	5	..	..	—	..	..	—	..	..	..	..	..	..
Israel	..	15[c]	..	..	15[c]	..	..	..	..	..	..	..	..	..	..	..	..	..	..	..	..
Portugal	190	60	300	190	60	300	..	..	..	..	..	..	..	..	..	—	..	..	−	..	..
Spain	1 892	610	1 400	1 892	610	1 400	..	18	70	..	592	1 330	..	..	..	..	197	140	+ 1 892	+ 413	+ 1 260
Turkey	..	183	635	11	13	41	11	13	41	—	—	—	..	170	594	115	68	27	− 104	+ 115	+ 1 373
Yugoslavia	100	199	..	85	178	..	16	16	..	69	162	..	15	21	..	42	122	..	+ 58	+ 77	..
Bulgaria	5	80[c]	..	5	80[c]	..	..	..	..	..	..	..	..	..	..	..	..	..	..	..	..
Czechoslovakia	248	94	30	248	94	30	223	80	30	15	14	..	..	..	..	29	15	..	+ 219	+ 79	..
Hungary	354	86[c]	..	354	86[c]	..	..	..	..	..	..	..	..	..	..	..	..	..	..	..	..
Poland	920	208	122	920	208	122	920*	208	122	—	..	..	—	..	..	44	18	30*	+ 876	+ 190	+ 92*
Romania	54	..	..	54	..	..	..	..	..	..	..	..	..	..	..	..	..	..	..	..	..

[a] Taken from ETTS III, table 3/6.

[b] Finland: mainly on treeless peatland after drainage.

[c] Based on average annual data for recent years in document TIM/EFC/WP.1/R.57 - Country Facts Sheets.

[d] Austria: 1965 to 1973.

[e] Cyprus: data for 1970-1983 and 1981-2000 (forecasts) respectively.

TABLE 5.17

Europe: available information from countries on afforestation, reforestation and plantations, 1950 to 1973,[a] 1970 to 1980 and 1980 to 2000 (forecasts)

(1,000 hectares)

Country	Afforestation 1950-1973	Afforestation 1970-1980	Afforestation 1980-2000	Reforestation 1950-1973	Reforestation 1970-1980	Reforestation 1980-2000	Production plantations — Total 1950-1973	Total 1970-1980	Total 1980-2000	of which: Coniferous 1950-1973	Coniferous 1970-1980	Coniferous 1980-2000	Non-coniferous 1950-1973	Non-coniferous 1970-1980	Non-coniferous 1980-2000	Other plantations 1950-1973	Other 1970-1980	Other 1980-2000
Finland	165	158	100	2 105	1 114	2 400	2 270	1 292	2 500	2 239	1 223	2 360	31	49	140	-	-	-
Norway	15	31	50	579	370	800	594	401	850	594	390	830	-	11	20	-	-	-
Sweden*	375	70	140	2 275	2 000	4 000	2 650	2 070	4 140	2 650	2 070	4 140	-	-	-	-	-	-
Belgium	20	13 →		133	179 →		152	192 →		115	..		37	..		1	..	
Denmark	40	40[b] →		-	45[b] →		35	..	..	30	..	..	5	..	..	5	-	-
France	634	185	210	724	325	500	1 358	510	710	1 175	430	450	183	80	160	-	-	-
Germany, Fed. Rep. of	200	16[b]	62	77	600[b]	64	277	92	126	194	88	115	83	4	11	-	-	-
Ireland	530	86	-	..	6	-	175*	..	..	..	..	..	..	..	..	..	..	..
Italy	..	-	-	227	..	..	79	..	..	10	..	..	69	..	..	678	..	..
Luxembourg	1	3 →		3	5 →		4	8 →		4	8 →		-	-		-	10	-
Netherlands	26	18	34	25	75	40	19	34	64	8	28	48	11	6	16	32	4	-
United Kingdom	640	325	500	238	75	300	878	400	800	826	385	750	52	15	50	-	-	-
Austria	57	52	-	150	188	-	212	240	-	..	..	..	..	..	..	-	-	-
Switzerland	..	3[b]	-	..	2[b]	-	..	..	..	..	..	..	..	..	..	..	..	..
Cyprus[c]	..	1	2	..	20	15	..	21	17	..	20	16	..	1	1	..	..	..
Greece	..	..	..	..	43	300	120	40	250	15	36	200	105	4	50	..	..	..
Israel	..	15[b]	-	..	5[b]	..	..	..	..	..	..	..	..	..	..	..	..	..
Portugal	190	60	300	10	35	[f]	190	95	300	90	75[e]	210[e]	100	20	90[g]	1 437*	591*	1 140*
Spain[d]	1 892	610	1 400	505	211	500	960*	330*	760*	840*	240*	550*	120*	90*	210*	..	..	..
Turkey	11	13	41	297	265	2 441	320	278	2 482	289	253	1 812	31	25	670	..	..	..
Yugoslavia	85	178	-	28	10	..	34	17	..	4	3	..	30	14	..	..	..	..
Bulgaria	5	80[b]	30	..	370[b]	..	884	..	..	600	..	..	284	..	..	..	..	..
Czechoslovakia	248	94	-	1 658	413	1 000	..	..	..	..	..	..	..	..	..	..	..	..
German Dem. Rep.	..	86[b]	-	1 240	195[b]	-	680	..	..	65	..	..	615	..	..	170	..	..
Hungary	354	-	122	496	850	1 610*	..	1 058	1 732	..	740*	1 130*	..	318*	602*	..	..	..
Poland	920	208	-	2 517	..	..	3 437	..	..	..	..	..	..	..	..	-	-	-
Romania	54	-	-	1 434	..	..	1 322	..	..	822	..	..	500	..	..	166	..	..

[a] Taken from ETTS III, table 4/6.
[b] Based on average annual data for recent years in document TIM/EFC/WP.1/R.57 - Country Fact Sheets.
[c] Cyprus: the 1970-80 column shows data for 1970-83; the 1980-2000 column for 1984-2000.
[d] Spain: estimated that 40% of plantations are primarily for wood production and 60% for soil protection.
[e] Portugal: mainly Maritime pine.
[f] Portugal: will depend on the forest fire situation.
[g] Portugal: mainly eucalyptus.

will be one million ha, with a further 50,000 ha a year of plantations for other purposes, mainly soil protection. Of the total plantation area, approximately three-quarters (700,000 to 800,000 ha) will be reforestation, that is planting on previously forested land, including conversion of low productivity woodland to exploitable forest. The remaining 25%, or 200,000 to 250,000 ha, will be afforestation on previously non-forested land;

2. Adding natural extension, which is expected to occur in Finland and a number of other countries, to afforestation, the annual extension of forest on non-forest land in Europe could occur at a rate of 300,000 to 350,000 ha between 1980 and 2000, giving a total of 6 to 7 million ha over the 20-year period;

3. Loss of forest land to other uses – agriculture, building, transport infrastructure, sports areas, etc. – could be at the rate of 50,000 to 100,000 ha a year in Europe as a whole. Deducting this estimate from that for the annual rate of forest extension, the net increase in forest area in Europe works out at 200,000 to 300,000 ha yearly, or 4 to 6 million ha over the 20-year period 1980 to 2000;

4. As noted earlier, the area of exploitable closed forest in Europe is forecast to rise by between 4.5 and 8 million ha over the same 20-year period. Since a good part of this will be the conversion of other wooded land to productive forest, the estimates of changes in forest area shown in tables 5.15 and 5.16 appear quite consistent at the regional level with those of the area of exploitable closed forest in table 5.11. That is to say, the increase in the area of exploitable closed forest in Europe will be somewhat greater than that of the total area of forest and other wooded land; that of unexploitable closed forest may rise relatively faster as more forest is reserved for parks and nature reserves; and the area of other wooded land will, as a result of conversion to productive forest, diminish. What is difficult to foresee is the rate at which these changes will occur.

While still on the outlook for the area of exploitable closed forest (ECF), the question may be asked whether the country forecasts adequately take into account the possibilities for changes in land use over the coming decades. For Europe as a whole the forecasts show an increase in area of ECF of 9 to 15 million ha over the 40-year period to 2020, some of which would be the result of reclassification of other wooded land as a result of reforestation or improvements in silvicultural treatment. Of the total European increase in the area of ECF over 40 years, around three-fifths is expected in Southern Europe, and one-fifth or less in the EEC(9).

Around 1980, exploitable closed forest accounted for 24% of Europe's total land area: a transfer of 5 million ha from non-forest land to ECF would raise this by one percentage point and of 10 million ha by two percentage points to 25% and 26% respectively. Even at the higher rate this does not result in a really significant shift in land use.

How realistic would such a shift in land use be, if it took place over, say, four decades, when account is taken of developments in other land use sectors, notably agriculture? Productivity in agriculture is continuing to rise steadily and the result of many countries' agricultural policies is already serious over-production of certain agricultural products. This has led to a growing debate, not only on whether some farming land should be converted to other uses, but how much and to what. Unofficially, it has been suggested that this might apply to 500,000 to one million ha in Sweden and around ten million ha in the EEC countries. Forestry is, of course, one of the obvious alternatives, not only in wood deficit countries such as those of the EEC, but many questions have to be answered before it can be recommended as a viable alternative to agriculture on a large scale. Some of the questions, notably those concerned with the likely level of long-term demand for wood from new plantations should be answered by this study. Others concerned with the selection of land for transfer, choice of species, closer integration of forestry with agriculture, use of new forest land for purposes other than wood production, the scope for energy plantations, and so on, require much more study.

The national forecasts in this chapter do not appear to allow for a significant shift in land use policies. This is especially the case for the EEC(9), where the problems of agricultural production and deficiency in wood are most acute. It should not be forgotten, however, that wood supplies from plantations established after the mid-1980s will not come onto the market until well into the twenty-first century and beyond the time scale of ETTS IV.

5.5 APPRAISAL OF THE FORECASTS

It was seen in section 3.7, that for the fifteen countries that provided comparable change data between 1970 and 1980 in response to a special supplementary enquiry, the aggregate changes over the 10-year period were for:

- Area of exploitable closed forest (ECF)............ + 2.5%
- Growing stock on ECF .... +12.3%
- Net annual increment on ECF .... + 8.9%

If these countries, which accounted for two-thirds of the European forest resource, were representative of the region as a whole (which, looking at their coverage, it is not unreasonable to assume), the equivalent change over 10 years for Europe as a whole would have been:

- Area of exploitable closed forest (ECF) .... +3 million ha
- Growing stock on ECF .... +1 790 million m³ o.b.
- Net annual increment on ECF........... +41 million m³ o.b.

Comparing these with the forecast changes over 20 years in table 5.12, it appears that countries are, in general, expecting a slower rate of expansion of growing stock and NAI after 1980, but not of area of ECF. For growing stock this could be explained if the felling/NAI ratio were to be higher than before 1980, as expected. For NAI, there could be several explanations: that the forecasts are unduly conservative; or that negative factors such as damage, such as that attributed to air pollution, or an increasing proportion of over-mature stands, are coming into play. It is not possible here to make a judgement on this question.

An analysis was attempted in ETTS III[5] of the potential impacts, positive and negative, on future wood supply from the European forest. The nine factors considered were: land use policy, ownership structure, investment and fiscal incentives, labour and mechanization, silvicultural techniques, environmental factors and non-wood benefits, overseas trade and development policies, fuller use of the available biomass, and assortment composition of supply and demand. It was concluded on a largely subjective basis that, taking the most positive hypothesis, supply could be raised by 100 million m³ u.b. over and above the removals forecast for Europe as a whole (408 million m³ u.b. in 2000). Taking the most negative hypothesis, it would be 45 million m³ u.b. less than the forecast, but nevertheless 26 million m³ u.b. higher than removals in 1970.

Whatever the limitations from a methodological point of view of the assessment of the impact of various factors on wood supply in ETTS III, it did serve a valuable purpose in identifying the factors involved and the sort of impact they might have. It led to a range of forecasts, the higher of which was 40% above the lower, which after further review was narrowed to a range of 17% (386 to 453 million m³ u.b.) in Europe's removals in 2000, with a "net effect" forecast of 430 million m³ u.b.

In ETTS III, the predicted increases in removals over the 20-year period from 1980 to the year 2000 were as follows:

"Realistic low" forecast . . . . +30 million m³ u.b.
"Realistic high" forecast . . . . +80 million m³ u.b.
"Net effect" forecast . . . . +65 million m³ u.b.

These may be compared with the forecasts of change over the same 20-year period presented in this chapter of

"Low" forecast . . . . +40 million m³ u.b.
"High" forecast . . . . +88 million m³ u.b.
Average . . . . +64 million m³ u.b.

The conclusion may therefore be drawn that, compared with the outlook for European wood supply as seen in the early 1970s for ETTS III, the latest outlook has not changed significantly, at least at the aggregate European level. At the country level, there are some interesting changes. In Sweden, for example, the increase in removals between 1980 and 2000 in ETTS III was forecast to be only 2 million m³ u.b., and even that small increase would result in a decline in growing stock; in other words there

would be over-cutting. Sweden's latest forecast for the same 20-year period is for growth of from 3 to 17 million m³ u.b., and even at the higher rate, growing stock in 2000 would still be somewhat greater than in the base period around 1980.

The volume of growing stock in Europe has risen appreciably over the past 40 years, continuing a trend of recovery that began probably in the early part of the nineteenth century (although interrupted during wartime). According to the forecasts, this development will continue over the 40 years to 2020, even if it may begin to slow down after 2000. This is not the place to discuss the merits or otherwise of further expansion of growing stock in Europe: there are many factors involved, some specific to individual countries. It may be said, however, that the policy to build up or to rebuild the forest resource, which was a central plank of many countries' policies in the past, certainly justified in most cases, must reach a culmination point sooner or later. In other words, once the forest has reached a reasonably "normal" state in terms of age-class distribution, fellings may be planned to be about level with increment. This may be too simplistic a view, since there are other factors than age-class distribution determining the volume of potential cut and, even more so, of actual cut. Nevertheless, if it could be supposed that a state of equilibrium were reached by 2020, it would mean, according to countries' forecasts, that fellings in Europe could be in the order of 560 to 600 million m³ o.b. on exploitable closed forest, compared with the forecast of 504 to 569 million m³ o.b. in table 5.12. And this level of fellings, equivalent to removals underbark of 460 to 500 million m³ u.b. from exploitable closed forest alone, could be achieved without drawing on the forest capital, that is to say, without reducing the volume of growing stock in the region.

This discussion is, of course, theoretical, but is intended to demonstrate that policy towards wood supply in Europe still has a certain degree of flexibility, if demand should grow sufficiently strongly. The reverse of the coin is that, under conditions of persistently weak demand for wood, fellings will not take place, growing stock will expand to the point where policy would have, sooner or later, to be directed towards stimulating consumption or taking measures to discourage further growth in supply. Alternatively, in the absence of such measures, natural losses would increase and net annual increment would decline, as would the vitality of the forest resource.

The forecasts presented in this chapter will be weighed against those of demand later in the study, at which point it will be possible to pursue this discussion on the basis of a scenario or scenarios of the long-term wood supply/demand balance. Here, it is necessary just to underline that the supply forecasts which have been presented do appear to have an element of flexibility, which should be allowed for in drawing up the prospective wood balance.

Are there factors which might affect this flexibility? There could be, the most obvious being anything that

---

[5] Section 6.4.

TABLE 5.18

Europe: forecasts of area of exploitable closed forest and growing stock on exploitable closed forest for the year 2000, by country and country group

| | Area of exploitable closed forest (ECF) (1000 ha) | | | Growing stock on exploitable closed forest | | | | | | | | | | | |
| | | | | Total (million m³ o.b.) | | | Coniferous (million m³ o.b.) | | | Non-coniferous (million m³ o.b.) | | | Growing stock per ha (m³ o.b./ha) | | |
	Base period	2000 Low	2000 High	Base period	2000 Low	2000 High	Base period	2000 Low	2000 High	Base period	2000 Low	2000 High	Base period	2000 Low	2000 High
Finland	19 445	20 300	20 300	1 568	1 741	1 650	1 290	1 424	1 370	278	317	280	81	86	81
Norway	6 600	6 600	6 600	575	714	695	459	541	523	116	173	172	87	108	105
Sweden	22 230	22 230	22 230	2 210	2 632	2 356	1 886	2 181	1 999	324	451	357	99	118	106
Nordic Countries	48 275	49 130	49 130	4 353	5 087	4 701	3 635	4 146	3 892	718	941	809	90	104	96
Belgium[a]	600	600	635	73	95	96	40	49	51	33	46	45	122	158	151
Denmark	365	371	383	47	53	55	29	37	39	18	16	16	129	143	144
France	13 340	13 490	13 590	1 550	1 732	1 688	605	657	663	945	1 075	1 025	116	128	124
Germany, Fed. Rep. of[b]	6 960	6 960	7 040	1 500	1 582	1 549	1 065	1 170	1 129	435	412	420	215	227	220
Ireland	347	423	469	32	48	49	25	40	40	7	8	8	92	114	104
Italy[c]	3 868	3 925	3 975	557	623	627	201	239	243	356	384	384	144	159	158
Luxembourg[c]	80	80	81	13	13	13	2	2	2	11	11	11	161	159	160
Netherlands[c]	294	320	335	29	33	33	21	24	24	8	9	9	99	103	99
United Kingdom	2 017	2 400	2 800	203	341	327	111	220	211	92	121	116	101	142	117
EEC(9)	27 871	28 569	29 308	4 004	4 520	4 437	2 099	2 438	2 402	1 905	2 082	2 035	144	158	151
Austria	3 165	3 175	3 210	803	925	955	679	791	817	124	134	138	254	291	298
Switzerland	935	950	950	312	309	275	210	209	184	102	100	91	334	325	289
Central Europe	4 100	4 125	4 160	1 115	1 234	1 230	889	1 000	1 001	226	234	229	272	299	296
Cyprus	100	100	100	4	4	4	4	4	4	–	–	–	35	37	37
Greece[a]	2 300	2 400	2 500	159	180	180	85	98	100	74	82	80	69	75	72
Israel	66	80	80	3	5	5	2	3	2	1	2	2	44	56	58
Portugal[c]	2 590	2 890	3 690	189	218	209	116	146	144	73	72	65	73	75	57
Spain	6 506	7 800	8 300	453	660	638	280	408	394	173	252	244	70	85	77
Turkey[c]	6 642	7 440	8 240	637	639	640	412	417	418	225	222	222	96	86	78
Yugoslavia	8 500	8 700	8 700	1 135	1 280	1 280	311	350	350	824	930	930	134	147	147
Southern Europe	26 704	29 410	31 610	2 580	2 986	2 956	1 210	1 425	1 413	1 370	1 560	1 543	97	102	94
Bulgaria[c]	3 300	3 350	3 400	298	309	308	101	115	115	197	194	193	90	92	91
Czechoslovakia[c]	4 185	4 185	4 200	923	934	933	686	690	687	237	244	246	221	223	222
German Dem. Rep.[c]	2 590	2 590	2 650	440	475	489	335	353	365	105	122	124	170	183	184
Hungary	1 596	1 726	1 736	253	303	283	29	44	43	224	259	240	159	176	163
Poland	8 410	8 755	8 880	1 162	1 266	1 373	897	989	1096	265	277	277	138	145	155
Romania[c]	5 723	5 720	5 820	1 202	1 395	1 399	482	539	543	720	856	856	210	244	240
Eastern Europe	25 804	26 326	26 686	4 278	4 682	4 785	2 530	2 730	2 849	1 748	1 952	1 936	166	178	179
TOTAL EUROPE	132 754	137 560	140 894	16 330	18 509	18 109	10 363	11 739	11 557	5 967	6 769	6 552	123	135	129

[a] Forecasts prepared by the secretariat, based on partial forecasts by the country.   [b] Unofficial forecast.   [c] Forecast prepared by the secretariat.

TABLE 5.19

Europe: forecasts of net annual increment on exploitable closed forest for the year 2000, by country and country group

	Total (million m³ o.b.)			Coniferous (million m³ o.b.)			Non-coniferous (Million m³ o.b.)			NAI per hectare (m³ o.b./ha)		
	Base period	2000 Low	2000 High	Base period	2000 Low	2000 High	Base period	2000 Low	2000 High	Base period	2000 Low	2000 High
Finland	61.93	67.90	68.00	48.12	52.44	53.71	13.81	15.46	14.29	3.2	3.3	3.4
Norway	17.31	18.81	18.04	13.71	14.16	13.46	3.60	4.65	4.58	2.6	2.9	2.7
Sweden	66.94	74.60	76.50	55.31	59.50	64.70	11.63	15.10	11.80	3.0	3.3	3.4
Nordic Countries	146.18	161.31	162.54	117.14	126.10	131.87	29.04	35.21	30.67	3.0	3.3	3.3
Belgium[a]	4.50	4.38	4.89	2.44	2.37	2.74	2.06	2.01	2.15	7.5	7.3	7.7
Denmark	2.80	2.80	3.10	2.00	2.00	2.20	0.80	0.80	0.90	7.7	7.5	8.1
France	54.00	59.73	63.44	23.50	28.42	31.35	30.50	31.31	32.09	4.0	4.4	4.7
Germany, Fed. Rep. of[b]	39.67	38.28	40.13	30.15	28.33	30.50	9.52	9.95	9.63	5.7	5.5	5.7
Ireland	2.53	4.40	4.50	2.45	4.20	4.30	0.08	0.20	0.20	7.9	9.1	9.2
Italy[a]	11.88	11.78	13.00	4.28	4.24	4.94	7.60	7.54	8.06	3.1	3.0	3.3
Luxembourg[a]	0.33	0.31	0.35	0.12	0.11	0.14	0.21	0.20	0.21	4.1	3.9	4.3
Netherlands	1.24	1.38	1.45	0.91	0.92	0.92	0.33	0.46	0.53	3.7	3.9	3.9
United Kingdom	11.20	15.30	15.90	8.60	13.20	13.80	2.60	2.10	2.10	5.6	6.4	5.7
EEC(9)	128.15	138.36	146.76	74.45	83.79	90.89	53.70	54.57	55.87	4.6	4.8	5.0
Austria	19.58	22.59	23.35	16.66	19.22	19.87	2.92	3.37	3.48	6.2	7.1	7.3
Switzerland	5.20	4.70	3.80	3.60	3.30	2.70	1.60	1.40	1.10	5.6	4.9	4.0
Central Europe	24.78	27.29	27.15	20.26	22.52	22.57	4.52	4.77	4.58	6.0	6.6	6.5
Cyprus	0.09	0.09	0.09	0.09	0.09	0.09	–	–	–	0.9	0.9	0.9
Greece[a]	4.10	4.56	5.00	1.90	2.15	2.40	2.20	2.41	2.60	1.8	1.9	2.0
Israel	0.20	0.21	0.20	0.09	0.10	0.09	0.11	0.11	0.11	3.0	2.6	2.5
Portugal[c]	11.45	14.97	15.69	8.33	10.77	10.99	3.12	4.20	4.70	4.4	5.2	4.3
Spain	27.83	29.35	29.82	19.53	20.60	20.93	8.30	8.75	8.89	4.3	3.8	3.6
Turkey[c]	19.21	21.58	23.90	11.70	13.59	15.53	7.51	7.99	8.37	2.9	2.9	2.9
Yugoslavia	28.85	30.55	30.55	6.15	7.00	7.00	22.70	23.55	23.55	3.5	3.5	3.5
Southern Europe	91.73	101.31	105.25	47.79	54.30	57.03	43.94	47.01	48.22	3.4	3.4	3.3
Bulgaria[c]	6.00	6.70	7.48	2.14	2.21	2.62	4.39	4.49	4.86	1.8	2.0	2.2
Czechoslovakia[c]	22.50	19.50	21.00	17.05	14.00	15.10	5.45	5.50	5.90	5.4	4.7	5.0
German Dem. Rep.[c]	15.00	14.51	15.37	11.50	11.17	11.99	3.50	3.34	3.38	5.8	5.6	5.8
Hungary	9.71	10.04	10.28	1.24	1.45	1.51	8.47	8.59	8.77	6.1	5.8	5.9
Poland[c]	28.45	30.00	37.16	23.43	23.50	30.60	5.02	6.50	6.56	3.4	3.4	4.2
Romania[c]	31.59	31.46	33.17	10.99	11.01	12.27	20.60	20.45	20.90	5.5	5.5	5.7
Eastern Europe	113.25	112.21	124.46	66.81	63.34	74.09	46.44	48.87	50.37	4.4	4.3	4.7
TOTAL EUROPE	504.09	540.48	566.16	326.45	350.05	376.45	177.64	190.43	189.71	3.8	3.9	4.0

[a] Forecasts prepared by the secretariat, based on partial forecasts by the country.   [b] Unofficial forecast.   [c] Forecast prepared by the secretariat.

would significantly disturb the productivity of the forest. It has been suggested that the damage inflicted by air pollution, in combination with other agents, could be one such factor, at least in the central parts of Europe. It is not immediately clear, however, whether damage of this type would lift or lower the quantities of wood put onto the market in the period up to 2000, although it could be assumed that the longer-term impact on potential supply would be negative. Some countries have attempted to build this factor into the forecasts they have prepared for ETTS IV. Others feel that the information base is still inadequate on which to assess the potential impact. It cannot be ignored, however, and consequently some further consideration is given to this question in the following chapter.

By way of concluding this chapter, the detailed forestry forecasts given country by country in the annex tables are summarized for the base period and the year 2000 in tables 5.18 and 5.19.

CHAPTER 6

Consequences of forest damage, including that attributed to air pollution

6.1 INTRODUCTION

In presenting the outlook for wood supply in the previous chapter, it was pointed out that there appeared to be an element of flexibility; this could apply to the forecasts of quantities felled and the assortment composition, as well as to the timing. The factors involved could be anything that significantly disturbed the productivity of the forest. One such factor could be air pollution and it was the specific request of the European Forestry Commission that this question be addressed in ETTS IV.

Damage to forests attributed to air pollution is, however, neither an entirely new phenomenon, nor is it the only one that may affect the health and productivity of the forest. The forest ecosystem exists in a state of equilibrium, the fragility or stability of which depends on a large number of internal and external factors. The objective of silviculture is to maintain that equilibrium, while at the same time seeking to adapt it in order to obtain from the forest an optimum mix of the goods and services needed by society. Man is also capable of disturbing the equilibrium, deliberately or unwittingly, often with adverse consequences, of which forest fire is one example. It may also be affected by natural causes, such as storms and avalanches.

The purpose of this chapter is to discuss in a general way the causes of damage to forests and their impact on forest health and productivity. Inevitably, for lack of adequate information, this discussion has to be more in qualitative than in quantitative terms. In theory, a distinction should be made between two classes of damage:

1. Damage to trees that may affect the timing of the availability of the wood for the market and perhaps its quality, but does not seriously affect the long-term productivity of the site. For example, a stand may be attacked by foliage-eating insects. The productivity of the trees may be reduced temporarily until the epidemic is over (or permanently, if they are killed), but the nutrients will be recycled through the forest floor and soil and the site will maintain its growing capacity;

2. Damage to the forest ecosystem as a whole that, mainly through its impact on the soil, may affect the productivity of the site, the increment rates of the tree stand and hence the volume of wood supply. The quality may also be affected, as well as the timing. Site degradation may result from the removal of topsoil, for example by erosion by heavy rain following a forest fire that has destroyed the vegetation and humus. Alternatively, the soil can be poisoned or have its pH altered by chemicals, for example, as a result of contamination by air or water pollutants. The practice of removing litter for bedding for farm animals, which used to occur in parts of Europe, also had a negative influence on site productivity which is still sometimes apparent today.

In practice, these two categories of damage cannot be so easily separated. Often, there is a chain-effect: a weakening of the tree's or forest's vitality caused by one type of stress leads to reduced resistance to other stresses. Nevertheless, for the purposes of a long-term study such as the present one, it is the second category of damage which is of concern.

6.2 TYPES OF FOREST DAMAGE

As a contribution to the work of the Joint FAO/ECE Working Party on Forest Economics and Statistics concerned with forest health and forest damage, Polish experts prepared a list of types of damage, which is reproduced in table 6.1. While the classification shown in the table may need refinement — further work on forest health and forest damage is planned by the Working Party — it is useful in grouping damage according to the main classes of factors and in listing the main types of damage. The Polish paper suggests that the most important of these, so far as the threat to European forests is concerned, are: air pollution and the resultant acid precipitation reactions, other factors resulting from economic activities, primary damaging insects, secondary invaders, forest fires and

TABLE 6.1

Main factors causing forest damage

Anthropogenic factors	Biotic factors	Natural abiotic factors
1. Unfavourable impact of industrial activity – energy use – mining – other industrial processes	1. Insects – primary insects – secondary insects 2. Parasitic fungi – root fungi – other 3. Bacteria and viruses	1. Unfavourable atmospheric factors – low temperatures (frost) – high temperatures – precipitation (especially snow) – wind – lightning – drought
2. Unfavourable impact of agricultural activity – overgrazing on forest land – other	4. Birds and animals – game animals – other 5. Other biotic factors (e.g. plants, small animals)	2. Unfavourable property of soil – insufficient humidity – excessive humidity – excessive changes of water level – deficit or excess of nutrients in the soil
3. Unfavourable impact of traffic		
4. Unfavourable impact of community activity		
5. Unfavourable impact of tourism and recreation		
6. Forest fires		
7. Damage resulting from methods and techniques of forest management and use		

Source: Joint FAO/ECE Working Party on Forest Economics and Statistics, TIM/EFC/WP.2/R.85, slightly adapted by secretariat.

weather phenomena. Other factors causing local, but large-scale damage include parasitic fungi and game animals.

Of the natural abiotic factors, high winds or tornadoes pose a threat to forests virtually everywhere, but especially in the more exposed regions, for example on the Atlantic seaboard, and where rooting conditions are unstable. Every year, some damage occurs somewhere, but major instances of widescale damage are not frequent. One of the worst occurrences in living memory was in February/March 1967, when a volume of standing timber estimated in excess of 30 million m³ was blown down, mainly in central Europe. The Federal Republic of Germany, Czechoslovakia and France were the worst affected countries, with losses of 10.7, 6.1 and 3.6 million m³ respectively, but Austria, Denmark, the German Democratic Republic and Switzerland were also seriously touched. The damage in Switzerland was equivalent to two-thirds of normal annual fellings, and in Czechoslovakia, the Federal Republic of Germany and Denmark to 48%, 42% and 38% respectively.

A more recent case was the storm-felling of 10 million m³ of trees in central France in November 1982, with some damage also in neighbouring countries.

Storm-fellings are normally followed by clearance – with as little delay as possible to avoid the build-up of bark beetle infestations – and replanting. They can be considered as being one of the categories of damage which do not significantly affect the long-term productivity of the forest site, even if they seriously disturb wood supply schedules, locally and even nationally, over lengthy periods of time.

Insect epidemics generally fall into the same category, even if sometimes in areas that have not been brought under systematic management they may reach ecologically catastrophic proportions, for example the spruce budworm outbreaks in North America. In Europe, the most serious case in recent times was the nun moth infestation in Poland, which at its peak in the early 1980s covered an area of about 3 million ha. Sanitation fellings over several years accounted for around three-quarters of the total annual fellings of over 20 million m³ and the damage also resulted in a marked decline in the average increment rate for the country as a whole.

It is quite possible that the nun moth attack in Poland was associated with other factors causing weakened resistance on the part of the forest stand. Forest fires, on the other hand, can be said mostly not to be associated with other damaging agents, although even in this case, increased litter on the forest floor from windthrow or other causes, adds to the fire risk. Forest fires are an annual occurrence, the extent of which varies considerably from year to year, the principal factor influencing the fluctuations being the summer weather patterns in southern Europe. As table 6.2 shows, those countries account for nearly all the forest fires each year in Europe.

As visible damage extends, a political decision is taken to reduce depositions of pollutants on forest areas (for this decision to be effective, it must take into account transboundary aspects and be designed to reduce those pollutant depositions which are in fact causing the damage).

It is almost certain, however, that there will be two types of time lag between the policy decision and a cessation of the increase in forest damage:

(a) The time necessary to carry out the measures decided and thus to reduce depositions. Some measures (e.g. vehicle speed limits) can be enacted quickly, others (e.g. catalysers on all cars, scrubbers on polluting industries) require a certain time to be implemented;

(b) The time necessary for the forest to react favourably to the reduction in depositions.

During this period between the policy decision and an improvement in the health of the forest, visible damage would continue to spread and increment to drop.

Variant 1

In this variant, before the onset of the pollution damage, increment and drain are in balance, and the growing stock is therefore constant. As a result of the spread of visible damage, and the appearance of areas classified as dying or dead, sanitation fellings are undertaken. After a time, it becomes impossible to counterbalance these necessary sanitation fellings by adjusting existing felling plans (i.e. by not carrying out planned removals in undamaged forests), so that total removals and, thereby, drain on the forest increase. Simultaneously, increment drops, due to the effects of pollution. As a result, drain is higher than increment, and the growing stock decreases. This continues until the area of forest classified as dying or dead ceases to expand. At this stage, sanitation fellings are no longer necessary and the level of fellings is adjusted to the new level of increment. This is lower than at the beginning of the period because the reduction in growing stock has reduced the increment. Increment is now high, relative to growing stock, but is low in absolute terms and relative to earlier levels.

Future developments from this point will depend on the decisions taken about future forestry objectives. If the new, lower level of increment and growing stock is considered acceptable, then the new rate of drain can be equivalent to that of increment. If however, it is decided to recover the "pre-pollution" level of growing stock, then drain must be kept below increment until earlier levels are recovered. (The latter course is shown in figure 6.1.)

Variant 2

As in variant 1, at the beginning of the period, increment and drain are in balance. In variant 2 however, it is possible to maintain the level of fellings constant, in order to avoid disruption of the roundwood markets even with a high proportion of sanitation fellings, by strongly limiting all fellings from undamaged stands. Nevertheless, increment is reduced by air pollution, so that the volume of growing stock declines (but less than in variant 1). As in variant 1, when the damaged area ceases to expand, fellings must be adjusted to the new level of increment which is lower than before (but higher than in variant 1, as the level of total fellings was lower than in variant 1). A decision must also be taken about the desirable future level of growing stock: figure 6.2 shows a new balance of increment and drain at a lower level than before i.e. drain is not kept below increment to increase growing stock.

A significant difference between variants 1 and 2 is that in the latter, less felling takes place on undamaged areas. There is a danger that on these areas thinning may not take place, or that stands become overmature.

Variant 3

In this variant, before pollution damage becomes apparent, increment is above drain and the growing stock is increasing. As pollution damage spreads, sanitation fellings are undertaken, and increment drops. However, the existing situation makes it possible to absorb this increase in drain, without a drop in growing stock, at least in a first phase. In variant 3, sanitation fellings do ultimately reach a level where they cause a decline in the growing stock, but it is conceivable, if drain is at the beginning of the period very much less than increment, that increased sanitation fellings could be absorbed without a decrease in growing stock.

As in the other two variants, when the damage ceases, drain must be adjusted to increment and decisions taken as to the future level of growing stock.

Needless to say, these three variants are not the only possible ones and, in the present state of knowledge it is not possible to assign a degree of probability to any one of them. They are intended as examples of the type of interactions which might take place if the forest damage attributed to air pollution does follow a negative course.

It appears from the discussion above that there are three major areas for policy decision:

(a) Decision to reduce pollutant emissions. This is the most important and necessary decision of all (although it lies outside the area of direct influence of the forest administration or forest owners);

(b) Decision on whether to undertake sanitation fellings, and on what basis (compulsory or voluntary, subsidized or not). For forest hygiene reasons, however, some type of sanitation felling appears inevitable;

(c) Decision on future forestry objectives. Although final decisions are probably only possible after forest damage has ceased to expand, provisional decisions, e.g. on species for regeneration, need to be made at an early stage.

After each sanitation felling, it will be necessary to decide on the future use of the land. If it is decided to regenerate the forest cover (probable in the great majority of cases), further decisions will have to be taken on species and provenance, and silvicultural regime. In evitably, there will be a significant element of experimentation in these choices, as the reaction of "new" species to the prevailing conditions would not be well known. It will be necessary to monitor carefully the success of the different types of regeneration. In the light of these results, it would be possible to build up gradually a more general forest policy for the affected areas, which would take into account all relevant factors. One possibility would be the reconstruction of the "pre-pollution" forest, but the decision might well be taken to create a slightly or radically different type of forest. The debate on the type of forest which is to be established on the area of sanitation fellings and what its major functions should be will no doubt be at the centre of forest policy discussions in affected countries for many years.

Satisfactory decision-taking requires adequate information, in this case on extent and type of forest damage, pollutant depositions, volume, location and assortment of sanitation fellings and changes in increment rates. It is a prerequisite for the establishment of new forest policies for changed circumstances that monitoring systems are set up, if this has not already been done, to provide accurate and up-to-date information on these matters.

6.5.2 Roundwood markets

As mentioned above, trees which are considered dying or dead should be removed for silvicultural and phytosanitary reasons, notably to avoid insect infestations. Clearly, if there are large volumes of sanitation fellings and if normal felling levels are not sufficiently adjusted, there is a risk of market disruption due to oversupply. Up to the mid-1980s, however, it appears that in most cases it has been possible to avoid disruption at the national level by reducing planned fellings. There have however been effects at the local level. This procedure is easier for large-scale forest owners, public or private, who have more flexibility in their felling plans than for small-scale owners, e.g. those who may only harvest occasionally. The latter, when faced with the prospect of silviculturally necessary sanitation fellings, may not have the possibility to reduce fellings elsewhere on their land. Furthermore, those small-scale owners who are *not* faced with sanitation fellings may be unwilling or unable to reduce their own fellings to preserve the balance of the roundwood market.

The market situation is unclear in many cases, as no separate statistics are kept in most countries on sanitation fellings and no precise definition of them is agreed in practice. Furthermore, forest owners are understandably unwilling to weaken their negotiating position by admitting publicly that they are obliged to fell a certain volume of timber.

A very rough estimate was prepared of the volume of sanitation fellings of trees damaged by air pollution in 1984. It was assumed that the whole growing stock in dying/dead areas would be removed in three to five years and that 25% of average removals on "moderate damage" areas and 5% on "minor damage" areas would be sanitation fellings: from this it was estimated that there were 12-16 million m³ of sanitation fellings which could be

wholly or partly attributed to pollution damage in Europe around 1984 − 3-4% of total fellings. In some of the countries with heavy damage, this proportion may have amounted to 10-16%. There has been no significant disruption of the roundwood markets at the national or European level which could be attributed to the emergence of additional supplies from sanitation fellings in pollution affected areas. Other special factors, notably the clearance of storm-damaged timber, seem to have played a more important role in causing some price weakening in some parts of the region.

It is, of course, impossible to predict the future level of sanitation felling which will be determined by the progression of forest damage, itself probably determined by pollution levels and the trees' capacity to resist. If the damage spread rapidly, sanitation fellings could reach a level where market disruption occurred. It is not possible at the present stage of knowledge to estimate the probability of market disruption or the volumes which might be concerned. In the long run, the determining factor would probably be the rate at which damaged forest areas passed from the "minor" or "moderate" damage class to the "dying/dead" class. It should however be borne in mind that markets may be extremely flexible (see for instance the events after the "oil shock" of the mid 1970s) and may well adapt satisfactorily to changed circumstances.

Another factor is the type of harvesting method chosen: if selection cutting is used to remove only damaged trees, the volume of sanitation fellings will equal the volume of damaged trees; if clear cutting is used on areas with a certain proportion of damaged trees, the volumes removed will be considerably greater. Harvesting costs would also be affected.

If market disruption were to occur, some or all of the following could take place:

Market reactions

(*a*) A fall in roundwood prices;

(*b*) A surplus of some assortments (e.g. logs) or species (e.g. spruce/fir);

(*c*) Increased transport costs as certain areas were saturated with roundwood and the surplus had to be transported elsewhere.

Measures to minimize market disturbance

(*d*) Storage of harvested wood (e.g. with sprinkler systems or in ponds); this can, however, only be a temporary measure;

(*e*) Exports of surplus wood in unprocessed or processed form;

(*f*) Increased co-ordination of fellings, notably the reduction of planned fellings in favour of sanitation fellings. This co-ordination could be undertaken by associations of forest owners or by official bodies, and might therefore implicate unaffected areas in the management of the consequences of forest damage;

(*g*) Rapid build up of processing capacity and efforts to increase markets (see next sections).

Apart from the actual progression of the damage, the level of sanitation fellings and the capacity of the roundwood markets to absorb them are the most uncertain aspects of the pollution damage question. Most of the negative consequences for industry, trade or consumption result more or less directly from a high level of sanitation fellings (possibly followed by a low level of fellings as suggested above). It should be borne in mind, however, that changes, such as a fall in roundwood prices may be negative for some participants in the market, but positive for others.

6.5.3 The forest industries

Before large volumes of sanitation fellings can be used by the forest industries, it is necessary to ascertain whether this material is technically suitable for the uses for which it is intended. A number of research projects are under way, examining the suitability of damaged wood as raw material for different industries. Although these projects are not yet completed, it appears that, if the wood is not allowed to dry out in the forest or in storage after felling, it can be used as raw material for most processes, although there may be increases in processing costs or reduction in yield (the situation varies according to the process concerned). If it is allowed to dry out, however, its value is much reduced − so that it is sometimes only suitable for fuelwood. The period during which the tree has been affected before being felled also appears to play a role.

It is possible that wood prices would diminish because of oversupply, although this hypothetical benefit for the industries could be partly counteracted by higher harvesting and transport costs and poorer yields. It is also possible that, in the affected areas, official bodies might take measures to safeguard the income of forest owners, which might or might not affect prices of wood raw material, according to the way the measures were defined.

The first consequence for the forest industries would probably be a more intense use of existing capacity in affected areas, in order to use the extra volumes of wood becoming available. In some of the affected countries, capacity utilization rates are at present quite low, especially for the sawmilling industry.

If the price of damaged wood did fall and the quantities available were greater than the needs of local industries, it is likely that buyers in other parts of the country or in other countries would seize the opportunity to secure a source of cheap raw material. In this way, the surplus of damaged wood could depress wood raw material prices far beyond the area suffering from forest damage. This might apply particularly to the more homogeneous assortments, such as pulpwood (roundwood or residues).

If it became apparent that significant volumes of cheap wood raw material were likely to become available in the areas of damaged forests over a period of several years, it is possible that forest industries would seek to instal new capacity to process this raw material locally. As the availability of the raw material would be limited in time, it would be desirable to choose a type of industry for which the capital invested could be recovered fairly rapidly (e.g. sawmills, particle board mills, MDF, CTMP, wood energy installations).

In the longer run, however, it is most unlikely that forest industries would rely on raw material transported long distances or resulting from sanitation fellings. The size and location of the European forest industries would essentially be limited by the annual increment of the forests within a reasonable transport radius. If forest increment, and therefore allowable cut were to be lower in the long term than at present (see for instance variants 1 and 2 above), the forest industries in affected areas would eventually have to reduce capacity, leading to mill closures and loss of employment, after the period of high capacity utilization and possible installation of new capacity.

Some adjustments might also be necessary where mills are dependent on raw material of a particular type. If this raw material were no longer available in sufficient quality and quantity, technical modifications to the installations could prove necessary, or even a change in the type of product produced.

6.5.4 Demand for forest products

How would demand for forest products be affected by large volumes of sanitation fellings and the corresponding industrial adjustments? It is possible that any fall in wood costs could lead to cheaper products and thereby to stronger demand, which would, of course, contribute to managing the situation. It is not clear however to what extent markets for forest products are price sensitive, and whether a fall in prices would significantly increase consumption (see chapters 10-12).

It is also likely that forest industries, faced with the prospect of marketing the products of sanitation fellings, would mount major promotion campaigns to encourage the use of forest products, preferably of domestic origin. It has been suggested that such campaigns could be supported by public funds made available to alleviate the consequences of forest damage.

In the new atmosphere of awareness of forest problems − notably forest damage attributed to air pollution − psychological aspects cannot be ignored. Cases have already been reported of fears that "infected" wood could prove harmful to human health as well as of fears, apparently unjustified, that wood from damaged trees is not suitable for structural purposes. It is also conceivable that some consumers would not understand how "the forest can be helped" by cutting down trees and making products from them. Such attitudes could have a negative effect on consumer acceptance of forest products, but should be overcome through public information measures.

6.5.5 International trade in roundwood and forest products

The developments in trade would be largely determined by the developments in the forest industries which have been briefly examined above. The consequences for particular areas would, of course, vary according to the underlying trade position. If a major wood surplus due to sanitation fellings did occur in northern continental Europe, and prices did fall, one consequence would probably be the export of wood raw material to industries elsewhere in Europe, for instance, in the Nordic countries. This has occurred in the past, for example after cases of severe storm damage.

Further developments would be determined by decisions on industrial capacity in the damaged areas and on the attitudes of the governments of these countries towards the need to protect domestic industries. Objections might well be made to the export of raw material which was subsequently reimported in the form of processed products. If new industrial capacity, based on sanitation fellings, were installed in net importing countries, there might be pressure for both new and old capacity to be protected by tariffs or quotas, if they were not competitive with imports without such protection. Exporting countries would wish to increase their exports to dispose of their surplus. With large volumes of sanitation fellings, they would be under strong pressure to increase the volume of exports, possibly by lowering prices.

The arrival on the market of large volumes of sanitation fellings could cause major changes in the pattern of trade flows. Whether such changes occur will depend partly on relative costs and effectiveness of marketing, but also on whether governments remain loyal to the principles of free trade in a situation where strong protectionist pressures are likely to arise.

6.6   OVERVIEW OF COSTS OF FOREST DAMAGE

It is of interest to evaluate the costs of forest damage as these costs, along with the cost of other damage due to air pollution, must be set against the costs of reducing pollution. Such an evaluation could also have a legal significance: some cases have already been brought in the Federal Republic of Germany claiming large sums from polluting enterprises as compensation for damage to forests.

While many of the costs of reducing pollution are concentrated and quantifiable, this is not the case for the cost of forest damage which raises a number of difficult methodological questions which the secretariat is not competent to discuss.[2]

The costs might, however, be classified under the following headings:

A.   COSTS TO THE FOREST OWNERS (AND OTHERS DIRECTLY CONCERNED WITH THE FOREST)

1.   Loss in value of forests and forest land, as well as of other property in the affected area.

2.   Reduced income from harvests due to lower roundwood prices and higher harvesting costs.

3.   Losses to fauna and flora, resulting in reduced earnings from hunting, mushroom- and berry-gathering, etc.

4.   Loss in income from recreation and value of recreational facilities, including hotels, restaurants, tourist facilities etc. in the affected areas.

B.   COST TO THE COMMUNITY

5.   Cost of subsidies (if governments decided to pay them):
    – to forest owners;
    – to industries.

6.   Employment: at first there would be no drop, or even an increase due to the felling and processing of damaged stands, but later there could be a scaling down of the workforce and capacity in industry, partly counterbalanced by increases in reforestation and silviculture.

7.   Losses from damage to the environment through increased danger of soil loss, avalanches, loss of water retention capabilities, danger of flooding, etc.

8.   Landscape and spiritual values.

In addition, there would be a number of cases where the changed circumstances benefit some but harm others, for instance if a local processor of damaged wood were to take the markets traditionally occupied by suppliers in another region or country.

All concerned with the forestry and forest industries in Europe are certain to examine carefully the possible economic consequences for themselves of the forest damage noted in northern continental Europe.

---

[2] Work on costs and benefits is being carried out under the auspices of the Executive Body for the Convention on Long-range Transboundary Air Pollution.

6.7   REVIEW OF FORESTRY FORECASTS IN THE LIGHT OF FOREST DAMAGE

Should the forestry forecasts in chapter 5 be modified in the light of the analysis above? It is first necessary to ascertain to what extent countries themselves have taken forest damage into account when preparing their scenarios, concentrating on nine countries which appear to be most affected,[3] divided into three subgroups: four easterly EEC countries, Central Europe and three northerly eastern European countries.

Around 1980, these nine countries accounted for 20.5% of the area of European exploitable closed forest, but 32.2% of the growing stock, 27.1% of the net annual increment and 29.9% of the fellings. Growing stock and net annual increment for the group were well over the European average (m³ o.b./ha):

	Nine most affected countries	Rest of Europe	European average
Growing stock per hectare	193	105	123
Net annual increment per hectare	5.0	3.5	3.8

The national forecasts in chapter 5 show that the nine countries expect their growing stock, increment and fellings to grow more slowly than those in the rest of Europe, and their net annual increment to diminish (see table 6.5). The damage attributed to air pollution is only one of the possible factors which could explain this. Furthermore, circumstances vary widely between countries, and it is necessary to refer to the comments of national correspondents on this subject, which are reproduced below.

Hungary

"Air pollution, which is affecting the region more and more and has the effect of lowering increment, as well as considerable damage by game ... (is) having a negative impact on the development of growing stock. The increase in harvesting will be limited or influenced by, amongst other things, the increasing damage due to air pollution and to game damage."

Poland

"The GAI (gross annual increment) in 1978 was less than in 1970 and also than (the level forecast for) 1990. From 1976 to 1984 we experienced the nun moth infestation. Due to control measures the loss of timber was relatively small, because the trees regained their assimilation systems. But a strong decrease in increment occurred.

"The lower forecast of increment is the result of negative impacts on forest, of which the most important is air pollution."

Sweden

"We agree it would be valuable to make estimates of probable effects. So far, however, we have no knowledge of the size of damage and of the consequences of such damage. In my opinion, it will take several years before we can give some quantitative estimates of effects."

Switzerland

"The effect of noxious emissions has led, over the last 10-15 years, to a diminution in growth rates in our forests.

---

[3] Austria, Belgium, Czechoslovakia, the German Democratic Republic, the Federal Republic of Germany, Luxembourg, the Netherlands, Poland, and Switzerland.

A study of stumps of 3 800 samples confirms a sharp loss of increment in sick trees. We estimate that this loss in increment up to now is about 5-10% on average and we fear that the situation can only deteriorate further. (Moderate damage forecast: 10% drop in increment, with maximum effect between 1990 and 2000; serious damage forecast: 30% drop, with maximum effect between 2000 and 2010.)

"The damage to forests from air pollution has become a reality in Switzerland, so that the forecasts in Part A (of the forestry forecasts enquiry) follow the following senarios:
  "low": moderate damage;
  "high": serious damage.

"These forecasts can only be intuitive, because the assessment of damage is not yet complete and solid information is lacking."

German Democratic Republic, Luxembourg

Forestry forecasts are secretariat estimates. For the lower forecast, it is assumed that there will be a gradual decline of NAI under the impact of air pollution and other causes of damage.

On the other hand, some of the forecasts did not incorporate assumptions regarding air pollution damage.

Austria

The effects of air pollution were not included in the calculation.

Denmark

The correspondent was "not afraid of 'forest death' ", (although some reduction in increment and fellings due to this cause was included in the forecasts).

France

"The effects of pollution or insect attacks, as well as those of storms, remain negligible for the moment, while fires only concern those stands generally with very low wood production."

It may be concluded on the basis of the above remarks and analysis that:

(a) Some at least of the removals forecasts presented in chapter 5 do take into account the possible effects of air pollution damage on increment and fellings; and

(b) Estimates of future effects of pollution damage on increment or fellings are in the present state of knowledge speculative. It is not possible to assign a degree of probability to any particular scenario.

In the circumstances, it was decided not to modify the quantitative forecasts presented in chapter 5, but to bear in mind, when analysing the outlook for the sector as a whole, that future developments for the countries with significant forest damage at present or for other countries could be quite radically changed. In particular, an increase in removals followed by a fall to levels below those of earlier years, cannot be ruled out.

TABLE 6.5

**Forecasts (low assumption) of changes in net annual increment and fellings
on exploitable closed forest in countries most affected by air pollution
and in the rest of Europe, base period[a] to 2000 and 2000 to 2020**

	Base period to 2000		2000 to 2020	
	Volume (million m³ o.b.)	Percentage	Volume (million m³ o.b.)	Percentage
Net annual increment				
Four easterly EEC countries[b]	− 1.39	− 3.0	− 1.35	− 3.0
Central Europe[c]	+ 2.51	+10.1	+ 0.20	+ 0.7
Three northerly eastern Europe[d]	+ 1.94	− 2.9	− 1.22	− 1.9
Sub-total (most affected countries)..	− 0.82	− 0.6	− 2.37	− 1.7
Rest of Europe	+37.21	+10.1	+26.49	+ 6.5
Total Europe..................	+36.39	+ 7.2	+24.12	+ 4.5
Fellings				
Four easterly EEC countries[b]	− 2.39	− 5.8	+ 0.75	+ 1.9
Central Europe[c]	+ 4.28	+21.4	+ 0.10	+ 0.4
Three northerly eastern Europe[d]	+ 1.70	+ 2.9	+ 4.12	+ 6.8
Sub-total (most affected countries)..	+ 3.59	+ 3.0	+ 4.97	+ 4.0
Rest of Europe	+47.91	+17.0	+46.13	+14.0
Total Europe..................	+51.50	+12.8	+51.10	+11.3

Source: Chapter 5.

[a] Around 1980 (see text of chapter 5 for explanation).

[b] Belgium, the Federal Republic of Germany, Luxembourg, the Netherlands.

[c] Austria, Switzerland.

[d] Czechoslovakia, the German Democratic Republic, Poland.

In any case, it will be essential for forest policy-makers to monitor carefully the development of forest damage and its consequences for the rest of the forest and forest industries sector, and to adapt their policies accordingly.

Analysts, at the national and international level, will be expected to improve the data base and the scientific construction of scenarios of the impact of forest damage on the forest sector.

CHAPTER 7

Trends in consumption of forest products

7.1 INTRODUCTION

The objective of this chapter is to present the broad outline of developments in consumption of forest products, purely in quantitative terms. This chapter will therefore provide the quantitative background to the chapters analysing the outlook for consumption.

An exhaustive presentation of the trends for each product and each region is not possible, for reasons of space. For more detailed analysis, readers are referred to the annual reviews of forest products markets and the medium-term surveys of markets for particular product groups regularly published by the Timber Committee.

Detailed series of data on apparent consumption of forest products are being published separately. All data

are for apparent consumption i.e. production plus imports minus exports, as real final consumption is not usually measured, at least on an annual basis. The data are therefore subject to distortion, notably by stock changes. In this chapter, however, three-, four- or five-year averages are used, minimizing the danger of distortion due to stock movements.

The four- and five-year averages for earlier years are those presented in the original publication. The use of averages tends to conceal strong movements in real consumption as well as artificial movements due to stock changes. This is particularly the case for the 1970s where the 1973 boom and the subsequent depression do not appear on the graphs.

7.2 TRENDS FOR EUROPE AS A WHOLE, 1913 TO 1980

It appeared interesting and relevant to present these very long-term trends (nearly 70 years) in order to provide a context for developments in later years. It is also a principle of analysis and projection of time series that the longer the historical series, the more reliable are the inferences which may be drawn from it.

The source for the data from 1913 to 1947-49 is the data base collected for the first timber trends study.[1] These data may not be entirely comparable with those from the Timber Section's data base, which are constantly revised, but it is unlikely that there are any serious comparability problems, except for the breakdown of paper and paperboard consumption, for which the classification used in the 1953 publication is different from that used subsequently. The total "paper and paperboard" is comparable over the whole period.

Figure 7.1 and table 7.1 present the broad outline of developments in the consumption of forest products in Europe in the twentieth century up to 1979-81. The main points to be noted are the following:

– Up to about 1950 there were few significant changes in the pattern of consumption. In volume terms, more fuelwood was consumed than any other product, although sawnwood accounted for a roughly equivalent volume of raw material. Paper, paperboard and wood-based panels accounted for insignificant volumes

– After 1950, during the great post-war economic expansion, consumption of all the products of industrial wood grew steadily and strongly while that of fuelwood declined even more rapidly. Fuelwood was widely supplanted, even in rural areas, by other fuels which were much more convenient and, in many cases, cheaper. In any case, Europe's demand for energy grew so strongly over this period that fuelwood soon became a marginal energy source. To a certain extent the rise in the consumption of the products of pulpwood was made possible by the decline in fuelwood consumption, as this "liberated" supplies of small-size wood. However, the decline in fuelwood consumption slowed down after the first energy crisis (1973/74) and was reversed in 1978. (This development is not visible in figure 7.l as averages are used.) The implications of this are discussed in chapters 18 and 19;

---

[1] *European Timber Statistics 1913-1950*, Geneva 1953.

FIGURE 7.1

Europe: consumption of forest products

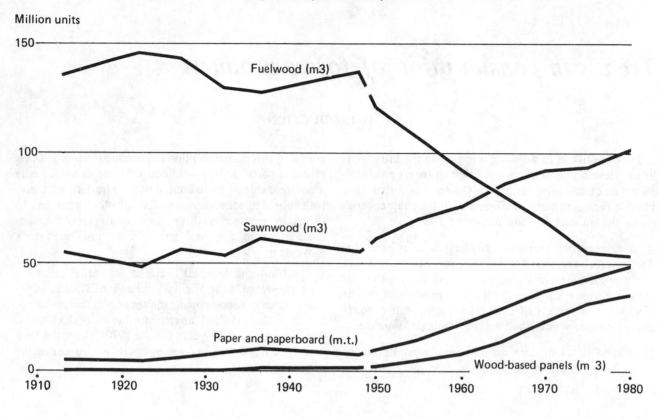

FIGURE 7.2

Europe: consumption of sawnwood and panels

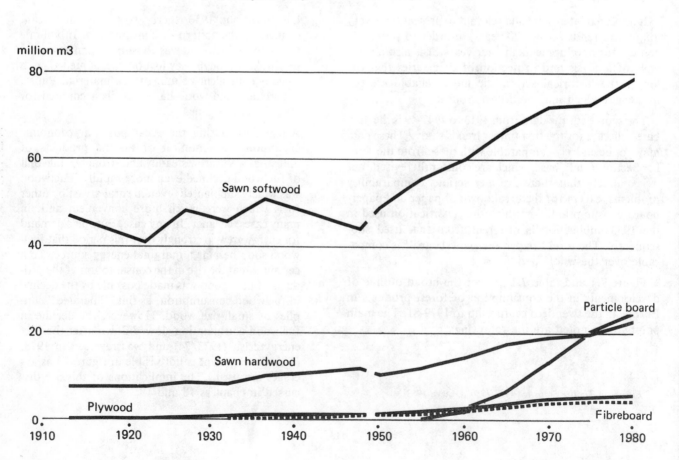

FIGURE 7.3

Europe: consumption of paper and paperboard

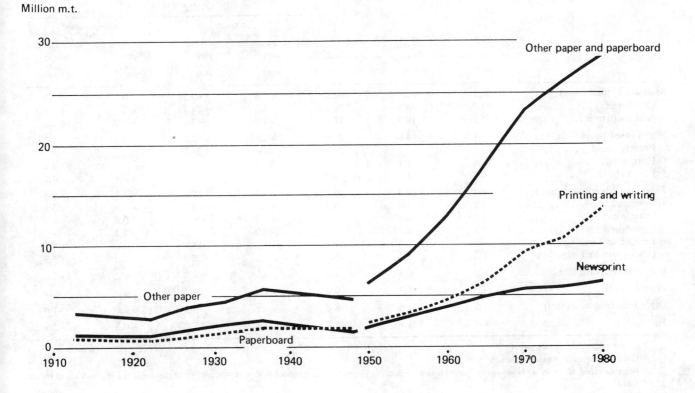

FIGURE 7.4

OECD Europe: consumption/GDP ratios

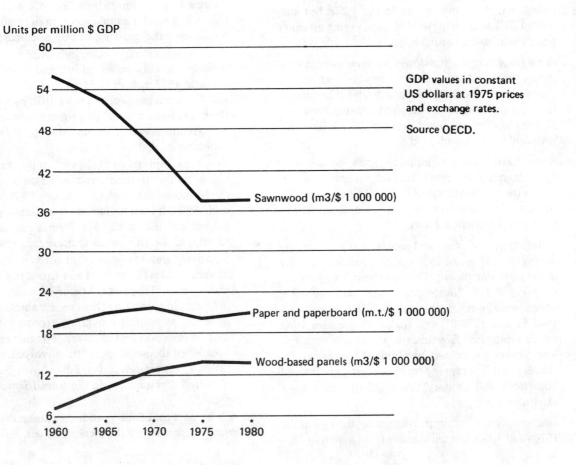

TABLE 7.1

Europe: apparent consumption of forest products 1913 to 1979-81

	Apparent consumption (million units)			Change			
				Volume (million units)		Average annual (%)	
	1913	1949-51	1979-81	1913 to 1949-51	1949-51 to 1979-81	1913 to 1949-51	1949-51 to 1979-81
Sawnwood, total[a] (m³)........	54.5	62.1	102.3	+7.6	+40.2	+0.4	+1.7
Coniferous................	47.0	48.5	78.2	+1.5	+29.7	+0.1	+1.6
Non-coniferous............	7.5	10.3	22.4	+2.8	+12.1	+0.9	+2.6
Wood-based panels[b] (m³)	0.1	3.0	35.6	+2.9	+32.6	+9.6	+8.7
Particle board..............	–	–	23.8	–	+23.8	–	c
Fibreboard................	–	1.1	4.4	+1.1	+3.3	c	+4.7
Plywood..................	0.1	1.4	5.4	+1.3	+4.0	+7.4	+4.6
Paper and paperboard (m.t.)...	5.2	10.5	49.2	+5.3	+38.6	+1.9	+5.3
Newsprint	1.1	1.9	6.5	+0.8	+4.6	+1.5	+4.1
Other paper	3.3	4.6[d]	..	+1.3	..	+0.9	..
Paperboard	0.8	1.7[d]	..	+0.9	..	+2.1	..
Printing and writing	..	2.3	13.7	..	+11.4	..	+6.1
Other paper and paperboard .	..	6.3	29.0	..	+22.7	..	+5.2
Dissolving pulp (m.t.)	..	0.8	1.6	..	+0.9	..	+2.6
Pitprops (m³)................	14.0	15.7	6.5	+1.7	−9.2	+0.3	−2.9
Other industrial wood (m³)	11.0	22.5	16.8	+11.5	−5.7	+2.0	-1.0
Fuelwood (m³)	136.6	121.8	53.8	−14.8	−68.0	−0.3	−2.7

[a] Includes sleepers, not shown separately.

[b] Includes veneer sheets, not shown separately.

[c] Percentage change of infinity due to non-existent quantity in earlier period. For particle board in the period between 1954-56 and 1979-81 the average rate of growth was 17.8% p.a.

[d] 1947-49.

– By far the fastest post-war growth was for particle board, which hardly existed in the 1950s but surpassed 24.1 million m³ in 1980, overtaking all other panels and sawn hardwood;

– Sawn softwood, despite slower post-war growth than most other products, remained pre-eminent among the products of industrial wood. Indeed, in volume terms, growth in sawn softwood consumption was higher than for any other product (29.7 million m³ between 1949-51 and 1979-81);

– Sawn hardwood consumption grew consistently faster than that of sawn softwood, more than doubling in the last 30 years, but in 1979-81 was only 29% of that of sawn softwood, and 1.4 million m³ less than that of particle board;

– Consumption of paper and paperboard grew slightly during the 1920s and 1930s, and expanded strongly in the post-war period. The expansion was particularly rapid for "other paper and paperboard", which multiplied three-and-a-half times between 1949-51 and 1979-81, and for printing and writing which multiplied five times over the same period. The "other paper and paperboard" category includes notably wrapping and packaging grades and household and sanitary tissue, all of which have expanded strongly;

– Consumption of pitprops fluctuated strongly in the 1920s and 1930s but fell steadily from a high of 16.4

million m³ in 1964-66 to less than 40% of that volume (6.5 million m³) in 1979-81. The causes of this were a fundamental change in deep mining techniques and the growing use of open-cast mining;

– The consumption of "other industrial wood", which includes notably poles, piling and posts, doubled between 1913 and 1949-51, was stable in the 1950s and 1960s but has declined since 1970 by about 15%. (It should be mentioned that some doubts exist about real trends for this group, due to classification problems);

– The consumption of dissolving pulp, used in the manufacture of rayon and other non-paper products, more than doubled between 1949-51 and the mid-1960s. It then remained on a plateau of around 1.8 million m.t. until 1977 when a gradual decline to around 1.6 million m.t. occurred;

– Relatively small volumes, but often of high value, of veneer sheets have been consumed in Europe for many years. There are, however, many statistical problems in separating these veneer sheets (used, for instance, as furniture overlay) from veneer sheets used in plywood manufacture. For this reason and because of the small quantities involved, consumption of veneer sheets is not shown separately, although it is included in the total for wood-based panels.

It will be seen that forest products can be divided into two groups according to the trends in their consumption

TABLE 7.2

Europe: per caput consumption of forest products, 1920-24 to 1979-81

					Average annual percentage change	
	Unit	1920-24	1949-51	1979-81	1920-24 to 1949-51	1949-51 to 1979-81
Sawnwood	m³/1000 cap.	141	150	193	+0.2	+0.8
Wood-based panels ..	m³/1000 cap.	1	7	67	+8.7	+7.7
Paper and paperboard	m.t./1000 cap.	13	25	3	+2.3	+4.4
Fuelwood	m³/1000 cap.	425	295	102	−1.3	−3.5

since 1950: those whose consumption has increased more or less strongly (sawnwood, paper and paperboard, wood-based panels) and those whose consumption has been stable or declined (fuelwood, pitprops, other industrial wood).

If *per caput* consumption is considered (table 7.2), very much the same pattern is seen, as population has grown relatively slowly in most European countries this century – at least when compared to economic growth. It should be borne in mind, however, that the averages for Europe as a whole presented here include wide country-to-country differences, which are analysed below (section 7.3).

Changes in consumption are due to a combination of technical, economic and social reasons. A new material may be developed which performs the same function as the previous one more effectively or more economically, or whole new processes may be developed which use greater or lesser amounts of forest products. Such processes have affected, in a positive or negative direction, a number of different forest products. Pitprops and, to some extent, sawnwood may be cited as forest products whose consumption has been negatively affected by techno-economic trends, and particle board and many grades of paper and paperboard as examples of positive developments.

It is instructive to examine briefly the context in which these developments took place. From the 1950s to 1973 Europe went through an unparalleled period of rapid and sustained economic growth. Growth has even continued in most years since 1973, although at a slower rate. In such circumstances, one would expect the growth of a product which was holding its ground on the techno-economic level

to be roughly parallel with that of Gross Domestic Product (GDP). Figure 7.4 and table 7.3 present the consumption/GDP relationship for the three main product groups.[2] Paper and paperboard consumption has expanded roughly in line with the economy as a whole, but wood-based panels (particle board) considerably faster: the consumption/GDP ratio nearly doubled over 15 years (i.e. consumption of panels grew twice as fast as GDP). It should be noted however that after 1975, consumption of wood-based panels grew slightly more slowly than GDP.

Sawnwood, however, lost ground relative to GDP between 1960 and 1975. Between 1975 and 1980, on the other hand, there was only a marginal decline. Interestingly enough, between 1975 and 1980, for the first time, the consumption of all three product groups expanded at about the same speed as GDP. The question of future development of the relationship between consumption and macroeconomic indicators such as GDP is of central importance for projections of consumption and is discussed in more detail in chapters 11 and 12.

Another underlying factor which should be mentioned even in this brief overview is the decline of the rural population, traditionally large consumers of forest products in a wide variety of uses about the house and farm (fuelwood, construction and repairs of buildings, farm tools, fencing, furniture etc.).

---

[2] For reasons of availability and comparability of data, only data from 1960 for European OECD member countries are presented. In 1979-81 these countries accounted for 75% of total European sawnwood consumption, 77% for panels and 86% for paper and paperboard.

TABLE 7.3

OECD Europe: ratio of consumption of forest products to GDP 1959-61 to 1979-81

		Ratio of consumption to GDP (units/$ million)					Average annual percentage change 1959-61 to 1979-81
	Unit	1959-61	1964-66	1969-71	1974-76	1979-81	
Sawnwood	m³	56.1	52.7	46.0	37.6	37.5	−1.3
Wood-based panels	m³	7.1	9.7	12.0	13.7	13.5	+2.2
Paper and paperboard........	m.t.	19.1	20.9	21.8	20.1	20.6	+0.2

Source for GDP data: OECD. Dollars are constant US dollars at 1975 prices and exchange rates.

7.3 TRENDS BY COUNTRY GROUP, 1964-66 TO 1979-81

The previous section presented, for the sake of simplicity, trends for Europe as a whole, but there are often marked differences between the trends in different country groups and countries. (Indeed, there are frequently differences between trends for different parts of the same country, but these are clearly beyond the scope of a study at the international level.) Tables 7.4 to 7.7, and figures 7.5 to 7.7 present the trends for product groups with detail by country group. The data for the four largest consumer countries (France, the Federal Republic of Germany, Italy and the United Kingdom) are also presented in the text as each of these countries has consumption equivalent or superior to that of some of the smaller country groups. In the figures of this section, annual data are shown, in order to show fluctuations and recent changes in trend which are not always apparent if three-year averages are used. Annual apparent consumption figures are affected by stock changes (e.g. in the mid-1970s) and this should be borne in mind when examining the graphs.

Annual time series, by country, for each product are being published separately.

It should be borne in mind that trends for Europe as a whole are dominated by those for the EEC(9) which accounted in 1979-81 for 49% of European sawnwood consumption, 59% for wood-based panels and no less than 67% for paper and paperboard. The Federal Republic of Germany alone accounted for 14-22% of European consumption. Thus the European total reflects principally developments in the EEC(9) and may mask developments elsewhere which are not in the same direction. After the EEC(9), in order of volume of consumption, come Eastern Europe, Southern Europe, the Nordic countries and Central Europe. The shares of these country groups have, of course, changed between 1965 and 1980, as is clear from table 7.7. It is interesting to note the few cases where the share of a country group in the European total for one product is significantly different to its share for another product. Thus Eastern Europe's share of European consumption of paper and paperboard (12%) is significantly lower than its share for sawnwood (21%) or for panels (19%). The share of Southern Europe and the Nordic countries in the European total for sawnwood is higher than their share of the totals for panels and paper.

Before summarizing the developments for the different product groups, it is necessary to make two general remarks:

– Year-to-year fluctuations are especially marked for the EEC(9), notably after 1973. These are in large measure due to stock movements in that period, linked in part to various speculative actions and to governmental counter-cyclical measures in some

countries. These fluctuations were so large that they caused serious perturbations in the sector and obscure the trend in the late 1970s. The mid-1970s cycle was the most violent for all products, but fluctuations around 1980 were also marked. Some other country groups, notably the Nordic countries and Southern Europe show similar swings, although not as large;

– An examination of figures 7.5 to 7.7 shows that, apart from these year-to-year fluctuations, a fundamental change in trend appears to have occurred in the mid-1970s, notably for wood-based panels. The period after the mid-1970s not only shows larger swings, but also a slower underlying rate of growth, notably in the EEC(9) and Eastern Europe but also in the other country groups. This is partly due to macro-economic factors, notably the slowdown in growth of nearly all economies, and particularly of the housing sector, and partly due to factors specifically linked to the forest products sector such as relative prices and technical developments. It is hoped that the analysis of chapter 8 will be able to shed some light on the relative importance of these factors. This question is, of course, crucial for establishing realistic projections of future consumption.

Tables 7.4 to 7.6 show developments between 1964-66 and 1979-81 for the three major product groups. For the sake of simplicity, only broad trends over the fifteen-year period are shown i.e. with no account taken of any change in trend in the mid-1970s, which is apparent from the graphs.

SAWNWOOD

For sawnwood (table 7.4), growth in consumption was slower than for both the other major product groups (1.2% p.a. for Europe as a whole), but the volumes concerned remain very large – 102.3 million m^3 in 1979-81. The fastest growth over the period (4.1% p.a.) was for Southern Europe whose consumption rose by 6.5 million m^3 to reach 14.3 million m^3. Central Europe also registered quite rapid growth (+ 2.4% p.a.). Within the EEC(9), there was a wide variety of growth rates in member countries. While consumption in the group as a whole rose at 0.7% p.a., growth rates were significantly higher in the southern EEC countries, Italy (+ 3.0% p.a.) and France (+ 1.5% p.a.). The United Kingdom on the other hand showed a drop in absolute terms of no less than 2.4 million m^3 to 8.6 million m^3. The United Kingdom's share of total European sawnwood consumption thus fell from 12.8% to 8.4%.

FIGURE 7.5

Consumption of sawnwood

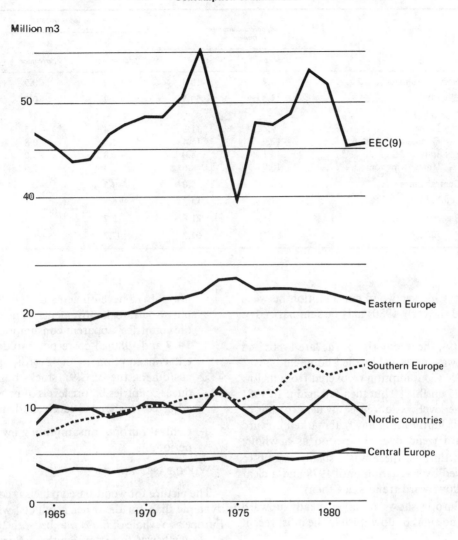

FIGURE 7.6

Consumption of wood-based panels

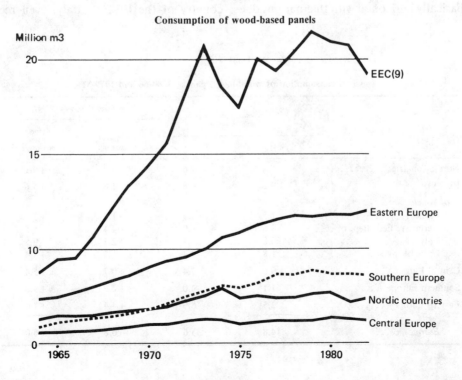

TABLE 7.4

Apparent consumption of sawnwood, 1964-66 and 1979-81

	Apparent consumption (million m³)		Change 1964-66 to 1979-81	
	1964-66	1979-81	Volume (million m³)	Average annual percentage
Nordic countries	9.4	10.8	+ 1.3	+ 0.9
EEC(9)	45.3	50.5	+ 5.2	+ 0.7
of which:				
France	9.4	11.8	+ 2.4	+ 1.5
Germany, Fed. Rep. of	12.9	14.4	+ 1.5	+ 0.8
Italy	5.1	8.0	+ 2.8	+ 3.0
United Kingdom	10.9	8.6	− 2.4	− 1.6
Central Europe	3.5	5.0	+ 1.5	+ 2.4
Southern Europe	7.8	14.3	+ 6.5	+ 4.1
Eastern Europe	19.1	21.8	+ 2.7	+ 0.9
Europe	85.1	102.3	+ 17.2	+ 1.2

The development in sawnwood consumption between the mid-1960s and the early 1980s may be summarized as follows:

- In the EEC(9), the steady rise of the late 1960s and early 1970s was succeeded by the "boom and bust" of 1973-1975. Consumption recovered from its low point in 1975 until 1980 but then fell again. In 1981 and 1982 the level was about equivalent to that of 1967-68. It is hard to find any real marked upward or downward trend over the period as a whole;

- For Eastern Europe, there was a steady rise up to 1975 followed by stagnation until 1978 and a clear if gradual downward trend since then;

- Southern Europe shows a rather steady upward trend, with no signs of downturn in the most recent years;

- The trend for the Nordic countries may be characterized as basically horizontal with fluctuations due not only to developments on the domestic markets but also to stock movements. (The peaks in the Nordic countries' apparent consumption in 1965, 1974, 1977 and 1980 all correspond to declines in apparent consumption in the Nordic countries' major customers, the EEC(9): stocks built up in the producer countries before levels of production could be adjusted to changed levels of demand);

- Central Europe consumption grew slightly over the period.

WOOD-BASED PANELS

The picture for wood-based panels is considerably more dynamic than for the other products, with a growth for Europe as a whole of 6.0% p.a. between 1964-66 and 1979-81. Growth was fastest in Southern Europe (9.0% p.a.), followed by Eastern Europe (7.0% p.a.). The southernmost country of the EEC(9), Italy, even recorded a rate of

TABLE 7.5

Apparent consumption of wood-based panels, 1964-66 and 1979-81

	Apparent consumption (million m³)		Change 1964-66 to 1979-81	
	1964-66	1979-81	Volume (million m³)	Average annual percentage
Nordic countries	1.4	2.6	+ 1.2	+ 4.1
EEC(9)	9.2	20.8	+ 11.6	+ 5.6
of which:				
France....................	1.5	3.3	+ 1.8	+ 5.3
Germany, Fed. Rep. of	3.6	8.0	+ 4.4	+ 5.5
Italy	0.8	3.1	+ 2.3	+ 9.5
United Kingdom	1.8	3.4	+ 1.6	+ 4.4
Central Europe	0.6	1.4	+ 0.7	+ 5.4
Southern Europe	1.1	4.0	+ 2.9	+ 9.0
Eastern Europe	2.5	6.9	+ 4.4	+ 7.0
Europe	14.8	35.6	+ 20.8	+ 6.0

TABLE 7.6

Apparent consumption of paper and paperboard, 1964-66 and 1979-81

	Apparent consumption (million m.t.)		Change 1964-66 to 1979-81	
	1964-66	1979-81	Volume (million m.t.)	Percentage
Nordic countries	2.2	3.3	+1.2	+2.9
EEC (9)	20.9	32.8	+11.9	+3.1
of which:				
France	3.6	6.2	+2.6	+3.6
Germany, Fed. Rep. of	5.8	9.6	+3.8	+3.4
Italy	2.4	5.2	+2.8	+5.3
United Kingdom	6.2	7.1	+1.0	+1.0
Central Europe	1.1	1.9	+0.8	+3.8
Southern Europe	1.7	5.4	+3.7	+8.0
Eastern Europe	3.2	5.7	+2.5	+3.8
Europe	29.2	49.2	+20.0	+3.5

growth of 9.5% p.a. over the period (from a low base). Developments in the European market for wood-based panels are dominated by those in one country – the Federal Republic of Germany, the 'home' of particle board, which accounted for no less than 24.3% of the total in 1964-66 and still for 22.5% in 1979-81 (growth in other countries started later and was often faster, while in the Federal Republic many markets were already reaching maturity). All country groups showed quite high rates of growth, the lowest being the Nordic countries, with 4.1% p.a., possibly because of the relatively greater importance in that area of fibreboard, and, in Finland, of plywood, which has grown less fast than particle board.

The trends over time (figure 7.6) in consumption of wood-based panels show steady, almost unbroken growth for Eastern Europe, Southern Europe and, to a lesser extent, Central Europe. The Nordic countries appear to show a levelling off after 1974.

The situation for the *EEC(9)* is rather more difficult to interpret, at least as regards the period after the steep unbroken rise in consumption which ended in 1973. Since then, fluctuations have been stronger, and the upward trend is much less marked. Consumption of panels in the EEC(9) dropped for three consecutive years after 1979. The peaks and troughs in consumption between 1973 and 1980 coincide exactly with those for sawnwood. These factors all seem to indicate that the markets for panels are showing signs of saturation (like the sawnwood markets) and have passed the phase of rapid market penetration and fast growth.

PAPER AND PAPERBOARD

The rate of growth of paper and paperboard consumption between 1964-66 and 1979-81 was between the rapid development of consumption of panels and the rather stagnant situation for sawnwood. Most countries and

TABLE 7.7

Shares of country groups and selected countries in European total of apparent consumption, 1964-66 and 1979-81

(Percentage of European total)

	Sawnwood		Wood-based panels		Paper and paperboard	
	1964-66	1979-81	1964-66	1979-81	1964-66	1979-81
Nordic countries	11.0	10.6	9.5	7.3	7.4	6.7
EEC (9)	53.2	49.4	62.2	58.4	71.5	66.7
of which:						
France	11.0	11.5	10.1	9.3	12.5	12.6
Germany, Fed. Rep. of	15.2	14.1	24.3	22.5	20.0	19.5
Italy	6.0	7.8	5.4	8.7	8.2	10.6
United Kingdom	12.8	8.4	12.2	9.6	21.1	14.4
Central Europe	4.1	4.9	4.1	3.9	3.8	3.9
Southern Europe	9.2	14.0	7.4	11.2	5.8	11.0
Eastern Europe	22.4	21.3	16.9	19.4	11.1	11.6
Europe	100.0	100.0	100.0	100.0	100.0	100.0

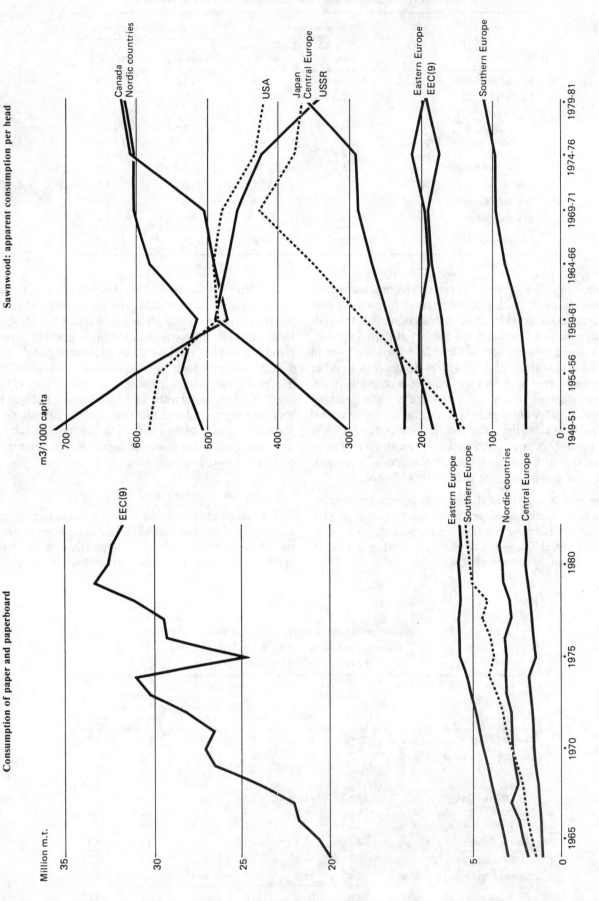

FIGURE 7.8

Sawnwood: apparent consumption per head

FIGURE 7.7

Consumption of paper and paperboard

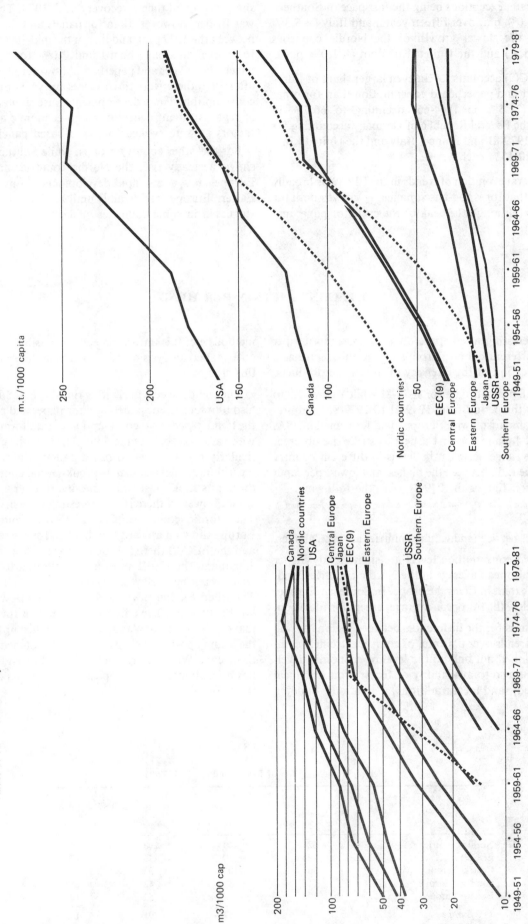

FIGURE 7.10

Paper and paperboard: apparent consumption per head

FIGURE 7.9

Wood-based panels: apparent consumption per head
N.B. Log scale
Values below 10 m³/1000 cap not shown

country groups shown in table 7.6 expanded at roughly the same rate, exceptions being the fast pace of Southern Europe (8.0% p.a. over fifteen years) and Italy (+5.3% p.a.) and the slower growth of the Nordic countries (+2.9% p.a.) and the United Kingdom (+1.0% p.a.).

The EEC(9) accounts for an even larger share of European paper and paperboard consumption than for other products (71.5% in 1964-66, declining to 66.7% in 1979-81), the Federal Republic of Germany accounting for 19.6% in 1979-81 and France, Italy and the United Kingdom for 10-15% each.

The general evolution of trends in the EEC(9) is broadly similar to that for wood-based panels − growth until the mid-1970s (although the peak of the cycle for paper and

paperboard was 1974 rather than 1973), followed by a sharp drop and then a recovery until 1979. The recovery was firmer, however, than for panels and the difference between the 1979 peak and that of the mid-1970s was 7.4% for paper and paperboard and 3.0% for wood-based panels. For paper and paperboard, however, as for panels, after 1979 there were three successive years of decline in consumption. These developments raise similar questions as regards the influence of macro-economic and market factors as were evoked for wood-based panels.

For the other country groups, while Southern Europe shows a steady rise, the Nordic countries and Central Europe show a less rapid development. Consumption in Eastern Europe grew steadily until 1975 but since then has stagnated in some countries or declined in others.

7.4 CONSUMPTION PER HEAD

It is interesting to compare data on consumption per head in different country groups taking this figure as a crude indication of the intensity of use of the products.

Table 7.8 presents data for 1979-81 while figures 7.8 to 7.10 cover the period from 1949-51 to 1979-81, and present, for comparison with Europe, data for Canada, USA, USSR and Japan. The first aspect to strike the observer is the wide range between the levels in different country groups. The ratio between the highest and lowest per caput consumption figures in 1979-81 were the following:

Sawnwood	Wood-based panels	Paper and paperboard
5.4 : 1	5.0 : 1	8.6 : 1

Differences between individual countries are even greater.

Per caput consumption levels are significantly higher than the European average in the Nordic countries, and, to a lesser extent in Central Europe. In Southern Europe, they are below the European average, except for fuelwood.

The reasons for the differences between regions as well as for the trends over time are, of course, extremely complex and linked not only to the level of economic development but also to availability of forest products (local or imported), and local traditions, regulations, building

methods etc. It is unfortunately not possible to undertake an extended analysis of these factors in the framework of this study.

The trends over a fairly long time period (30 years) are also of interest. For panels and for paper and paperboard the trend in *per caput* consumption was uniformly upward, with variations between regions. The growth rate was particularly rapid for wood-based panels (figure 7.9 has a logarithmic scale!). Japan overtook several country groups during its rapid post-war economic recovery. For sawnwood, however, there is an interesting diversity of trends. The Nordic countries, Central Europe and Southern Europe show an upward trend, while for Eastern Europe and the EEC(9) there is little movement over the period. Japan and the USSR show a rise followed by a decline, which was longer (20 years) and more severe for the USSR. The trend for the USA has been steadily down over the last thirty years. The pattern for Canada is the reverse of that of the USSR: down to 1959-61 and then up, to recover the leading position held by that country immediately after the Second World War. Since 1959-61 the trend in Canada has been almost exactly opposite to that of its neighbour, the USA.

TABLE 7.8

Per caput consumption of forest products in 1979-81

	Sawnwood (m³ /1000 cap.)	Wood-based panels (m³ /1000 cap.)	Paper and paperboard (m.t./1000 cap.)	Fuelwoood (m³ /1000 cap.)
Nordic countries	621	149	190	575
EEC(9)	194	80	126	70
Central Europe	367	101	137	158
Southern Europe	111	31	42	233
Eastern Europe	198	63	52	107
Europe	193	67	91	136

7.5 TRENDS SINCE 1980

The first years of the 1980s were marked by further strong cyclical fluctuations in consumption of forest products and the absence of marked growth trends. This was a continuation of the picture of the late 1970s.

In 1982, apparent consumption of the major product groups was below that of 1979-81, sometimes significantly so (e.g. 95.6 million m³ for sawnwood, compared to 102.3 million m3 for 1979-81). 1982 marked the trough of the business cycle and apparent consumption was higher in 1983 and 1984, although the 1979-81 level was not recovered for sawnwood and hardly surpassed for wood-based panels. For paper and paperboard, however, 1984 (and probably 1985) were record years.

The exception to this picture of cyclical movement was fuelwood, which appeared to reach the lowest point of its secular decline in 1978. Since that year recorded European fuelwood consumption has risen by 1-2 million m³ every year (except possibly 1985 for which only preliminary data are available). The rise in consumption may be even larger as it is widely accepted that recorded fuelwood consumption does not account for all consumption.

TABLE 7.9

Europe: apparent consumption of forest products from 1979-81 to 1985

(Million units)

	1979-81	1982	1983	1984	1985 [a]
Sawnwood, total[b] (m³)	102.3	95.6	96.8	98.6	97.2
Coniferous	78.2	73.2	74.7	75.8	74.8
Non-coniferous	22.4	20.9	20.7	21.6	21.2
Wood-based panels[c] (m³)	35.6	33.5	34.2	35.2	34.9
Particle board	23.8	22.6	23.1	23.9	23.7
Fibreboard	4.4	4.2	4.2	4.2	4.2
Plywood	5.4	4.9	4.9	5.0	4.9
Paper and paperboard (m.t.)	49.2	48.2	49.6	53.2	53.3
Newsprint	6.5	6.3	6.3	7.0	6.9
Printing and writing	13.7	13.6	14.5	16.0	16.0
Other paper and paperboard	29.0	28.3	28.8	30.3	30.4
Dissolving pulp (m.t.)	1.6	1.5	1.4	1.3	1.3
Pitprops (m³)	6.5	6.6	6.8	6.2	6.2
Other industrial wood (m³)	16.8	20.9	20.2	21.5	21.0
Fuelwood (m³)	53.8	58.9	59.8	60.1	56.8

[a] Preliminary data.   [b] Includes sleepers.   [c] Includes veneer sheets.

CHAPTER 8

Trends in end-uses of sawnwood and wood-based panels

8.1 INTRODUCTION

The purpose of this chapter is firstly to show the relative importance of the main end-use sectors for sawnwood and wood-based panels and the pattern of utilization within them, in particular as regards the construction sector, and secondly to analyse trends over time in the utilization pattern and intensity in different end-uses and their underlying factors. This qualitative analysis may be compared to the econometric analysis in chapters 11 and 12. End-uses for paper and paperboard are treated in chapter 9.

Sawnwood and wood-based panels are for the most part destined to serve as a material component in investment and consumer goods, notably construction and furniture. They may also be put to various temporary uses, such as concrete shuttering, during the production and distribution of such goods.

Demand for sawnwood and panels is therefore basically a function of the pattern and level of activity in the different economic sectors using forest products and of the intensity of their utilization for these different uses. The level of consumption of forest products is therefore affected both by changes in the level of activity of the user sectors and by changes in patterns of use. The latter are the result of many different technical, economic and social factors which may have relatively short-term effects (e.g. fashions in furniture design) or medium- and long-term effects (e.g. performance requirements and regulations in building or the promotion of building methods such as timber frame construction).

End-use analysis is plagued by severe data problems, which are as yet unresolved. In particular, in most countries only rough estimates are available as to the volumes of sawnwood and panels used in each end-use sector. It follows that analysis of trends over time is almost impossible. In addition, information on the situation and trends for activity in each end-use sector is often incomplete.

Research in these questions is costly, often covers only small sub-sectors or is not publicly available, for commercial reasons.

It is unfortunately the case that the quality of information in this field has not improved but declined. ETTS I and ETTS II treated the subject quite extensively, on the basis of relatively satisfactory information (ETTS I devoted nine chapters of Part I to the various uses for sawnwood and panels, and ETTS II six chapters). Coverage in ETTS III was considerably less satisfactory, and does not provide a basis for analysis of *trends* since the early 1970s, as the only estimates available could not be considered comparable with the estimates made for ETTS IV.

The following analysis is based on the results of an enquiry circulated for this purpose to which 11 countries[1] responded. It was complemented by information and data originating from other sources, such as trade and industry journals, publications, studies, experts, etc.

Even though caution must be used when interpreting and comparing data on end-uses, the available information makes it possible to distinguish a number of trends in the field of forest products utilization and to identify some points of major interest for the future.

---

[1] Czechoslovakia; Finland; Ireland; the Netherlands; Norway; Poland; Portugal; Sweden; Switzerland; the United Kingdom; and Yugoslavia.

8.2 BROAD UTILIZATION PATTERNS OF SAWNWOOD AND WOOD-BASED PANELS

8.2.1 General

Sawnwood and wood-based panels are put to a very wide variety of uses in all spheres of human activity. There are three areas, however, which appear as the most important users in terms of overall volumes:

1. *Construction* (including new building, building repair and maintenance work, and heavy construction);

2. *Furniture* (for domestic purposes, offices, schools, hotels, hospitals etc.); and

3. *Packaging*, the latter term being used here in a broad sense to cover all types of recipients and means for the collection, handling, transport and storage of solid and liquid goods, such as boxes, cases, crates, drums, vats, casks, tanks, containers, pallets etc.

End-use sectors *other* than those mentioned above include a broad diversity of uses, ranging from boats and transport vehicles to household utensils, technical articles, musical instruments, toys, and handicrafts, none of which individually may be a very large user of forest products but which, taken together, constitute a sizeable volume for each main product.

It continues to be difficult to obtain a clear idea of the relative importance of the different end-use sectors in terms of volumes and percentage shares of forest products consumption.

The information on broad end-use patterns received from countries in reply to the enquiry referred to earlier, supplemented by information collected for some other countries from various sources is set out in detail in annex tables 8.1 to 8.5 and summarized in table 8.1 below.

TABLE 8.1

Europe: estimated relative importance of different end-use sectors for sawnwood and wood-based panels around 1980

(Percentage of total consumption of each product)

	Construction	Furniture	Packaging	Other
Sawn softwood	62 – 92	3 – 10	2 – 14	2 – 22
Sawn hardwood....	20 – 73	15*– 40	2 – 30	3 – 22
Plywood	12 – 65	10 – 40*	5 – 25*	7 – 22*
Particle board	8 – 70	20 – 90	1 – 3*	1 – 20*
Hardboard	20 – 84	6 – 41	2 – 10*	1 – 28
Insulating board ...	71 – 100	–	–	2 – 29

Source: Special enquiry (see text).

One noteworthy feature of the data is the rather wide variation in the percentage shares which countries reported. These variations may be due to several factors, such as the composition of a country's forest resource, trade patterns, traditions and preferences in wood use, structure of the wood-using industries, or methodological differences, but it is difficult to determine the extent to which any one factor may be involved.

8.2.2 Sawn softwood

The data indicate that sawn softwood finds its single most important application in construction. In all countries listed in annex table 8.1, construction accounts for more than half of total apparent consumption of sawn softwood.

Packaging appears as the second main end-use sector for sawn softwood, though well behind construction. It is followed by furniture, with under 10% of the total, but the difference between packaging and furniture does not appear to be as important as estimated in ETTS III.

It is not easy to explain the differences in the construction share at the country level. In the eastern European countries usage of materials is guided by overall economic policies, and these have generally been in the past to limit the use of wood materials in favour of mineral and metal-based materials. The share of construction in sawn softwood consumption tends therefore to be lower than in western Europe.

The importance of the coniferous forest resource (in absolute terms and relative to the non-coniferous forest), would appear to be one of the major factors determining the level of sawn softwood utilization in construction, and, possibly even more important, in other uses. Indeed, it might be concluded from the available data, that the more important hardwoods are in a country's forest resource, the smaller the relative share of sawn softwood utilization in end-uses other than construction. At the same time, the intensity of sawn softwood use in construction (i.e. sawn softwood consumption per unit of construction), particularly in building, tends to be higher in countries with extensive coniferous forests.

8.2.3 Sawn hardwood

The utilization of sawn hardwood in Europe is less markedly centred on the construction sector than that of sawn softwood, reflecting the greater diversity of this product group, which between temperate-zone and tropical species comprises a very wide range of properties making it suitable for a wide variety of end-uses.

Leaving aside the apparently special cases of Poland and Portugal, the pattern suggested by the 1980 enquiry shows very roughly a third of the apparent consumption of sawn hardwood in western Europe being accounted for by the furniture sector, somewhat less than a third by packaging, and about a third for construction.

8.2.4 Plywood

The data in table 8.1 indicate that in western Europe construction is also the single most important utilization field for plywood, accounting in most of the countries listed for just over 50% of total consumption. Packaging, including transport, appears as the sector second in importance, followed by furniture. In Eastern Europe, on the other hand, construction appears as a relatively minor use of plywood though the trend has been rising.

The share of furniture was considerably higher (35%) in ETTS III than it appears from the present data. There has no doubt been a trend, observed already well before 1970, for plywood in furniture to be substituted by other panels, in particular by particle board.

Among the wood-based panels, plywood is probably the one most sharply affected by a variety of substitution processes, although it has also found new, and wider application in a number of fields, for instance in packaging and transport, and for agricultural buildings; thus plywood has been increasingly replaced in furniture by particle board (and to some extent hardboard) and in building it has met with sharp competition notably from particle board, waferboard, oriented structural boards, and structural particle board. Plywood cores have been substituted by particle board. In some countries, core plywood has been increasingly replaced by oriented structural boards in certain end-uses.

8.2.5 Particle board

The data in table 8.1 and annex table 8.4 confirm the predominant role of construction and furniture for particle board consumption. They also confirm the marked differences between the relative shares of these two end-use sectors between countries. These differences are also evident in the data published by the European Federation of Associations of Particle Board Manufacturers (FESYP), which confirm the overall picture. Other end-use sectors than construction and furniture, such as packaging, transport, ship building, and Do-it-yourself (DIY) are of minor importance, accounting generally for rather less than 5% of total consumption.

Some interesting trends in the utilization of particle board occurred in some countries during the 1970s, according to the FESYP data. Thus construction was the main use in the early 1970s in the Nordic countries with a share of 65% to 75%; since then the share of this utilization sector has decreased while the share of furniture has increased correspondingly. In the countries of Southern

Europe, furniture remained the main end-use; France was the only major EEC country to show a rise in the share of construction as a market for particle board in the 1970s; it was also, in the early 1980s with Finland, the only one where the share of construction was above 50%. On the other hand, the share of construction as an outlet increased in some of the smaller countries.

8.2.6 Fibreboard

The data in table 8.1 and annex table 8.5 show that considerable variations exist between countries, in particular for hardboard. While construction appears generally to be the main end-use sector for hardboard, accounting in some countries for three-quarters of consumption and more, furniture continues to be an important outlet in a number of countries.

For insulating board, building construction is the predominant utilization sector, with generally only minor volumes used for other purposes.

For medium-density fibreboard (MDF), as a newer material with a still relatively small consumption in Europe, the main use is at present in furniture, both as a surface material (partly with a veneer overlay) and as profiles. Its utilization extends, however, also to the construction sector (door frames, mouldings) and other uses (toys, picture frames, etc.), substituting notably sawnwood and particle board.

8.3 UTILIZATION BY MAIN END-USE SECTOR

8.3.1 Construction

8.3.1.1 BROAD TRENDS IN THE USE OF SAWNWOOD AND PANELS IN CONSTRUCTION

From the utilization patterns outlined in the preceding section, the importance was evident of construction as the single most important outlet for sawnwood and panels. Construction also seems to be the end-use for forest products which has been the object of greater attention than most others and for which relatively more information is available. An example is a study carried out in Finland in the early 1980s.[2]

For an overall review of trends in construction in Europe since the early 1960s and the outlook to 2000, reference is made to chapter 1, section 1.4.

Statistical data on the use of sawnwood and panels in construction are presented in annex table 8.6. Despite the fragmentary nature of the data, it is possible to draw some general conclusions. Among the main types of construc-

tion activity, new building construction is the single most important outlet for these products in western European countries, accounting for between one- and two-thirds of total consumption of sawnwood and panels in construction, or even more in some cases. Repairs and maintenance, and other construction work, taken together, account generally for less than half of the volume used in the construction sector as a whole. The relative importance of repairs and maintenance consumption of sawnwood and panels has shown a clearly rising trend in recent years, with particular significance for some types of panels.

Within the new construction sector, residential building appears as the main end-use, accounting generally for over half of the volume of sawnwood and panels used in new buildings, although residential building accounts for a smaller proportion of building volume. This is indicative of the more intensive use of forest products in new residential buildings and, in particular, in new, low-rise family houses, than in commercial, industrial and other types of buildings.

The utilization of sawnwood and panels in construction has been influenced by technical, economic and environmental factors.

---

[2] Valtonen, K.: *Use of sawnwood and wood-based panels in new building construction in the 1970s*, Folia Forestalia 529: 1-42 Helsinki, Finland, 1982 (in Finnish, with English summary). This study is summarized in annex 8.7.

As far as the *technical aspects* are concerned, notable features were:

- The industrialization of the manufacture of building joinery (in particular doors, windows, wall elements, wooden stairs, and built-in furniture). The positive effects on cost and consistent quality of the products were an important factor in maintaining the position of wood in these uses;

- The prefabrication of wood-based building components, such as roof structures (which led to a certain decline in traditional roofs), wall and floor panels, housing sections and whole housing units;

- The development of glue-laminating, edge-jointing and finger-jointing techniques, which opened up a wide field of application and use of wood for engineered structures that had long been the domain of metal and reinforced concrete structures. Jointing techniques also made it possible to upgrade lower wood qualities.

With regard to *economic factors*, the rapid rise in the cost of many raw materials, together with that of other factors including labour, contributed to a general sharpening of cost consciousness in the building field, as in other sectors. One immediate effect of the increase in energy costs was the greater attention paid to insulation as one means of energy conservation. This led in some countries to a reduction in the size of windows, while window frame dimensions were made larger to accommodate double or triple glazing. In many cases, building regulations were amended to provide for minimum standards of insulation. On the other hand, the dimensions of interior joinery (skirtings, architraves and the like) were reduced.

A more general effect, however, was that owners and specifiers of buildings paid particular attention to the quality and performance of various building elements and components, not least in so far as they affect subsequent durability in use and maintenance costs. Increasingly, performance criteria were prescribed to meet precise end-use requirements. This was particularly important, as there were considerable changes in the sources of supply of wood and wood products which sometimes led to uncertainty about product properties. The stress grading of sawnwood (mainly coniferous) became more frequently applied. In turn, competition between alternative materials (wood, aluminium, plastics) used for building purposes increased, particularly for exterior joinery and related applications. Especially in the case of windows, aluminium, often combined with wood, found increasing application, notably in non-residential building. In the Federal Republic of Germany, windows made of plastic materials, sometimes with an imitation wood textured frame, had a market share of about 40% in the early 1980s. Aluminium and plastic materials were also substituted for wood to a significant extent in the manufacture of blinds.

With regard to *environmental* aspects, which in general have grown significantly in importance since the early 1970s, concern arose over potential health hazards of certain chemicals associated with the preservation of wood products, and with glues, an example being formaldehyde emissions from glues used in panels.

At the same time, wood enjoyed a high degree of acceptance as a decorative natural material and, as such, since the late 1960s, has found growing application as cladding for walls and ceilings, primarily in residential buildings and buildings serving social and cultural activities, and sports.

Looking now at the individual products, the following observations can be made about trends in utilization since the late 1960s.

8.3.1.2 SAWN SOFTWOOD

It has been estimated that, of the total consumption of sawn softwood accounted for by construction, just over one-third each is used for structural framing and building joinery (including flooring, etc.), and roughly one-fifth for heavy (engineering) construction and formwork, the remainder being taken up for other, mainly temporary uses.

With regard to *structural uses*, competition and substitution seem to have worked not so much between sawn softwood and other wood products and materials, as between one construction method and another. Thus, the design and building techniques used for high multi-storey housing, with flat concrete roofs, were increasingly applied also in housing with a limited number of storeys (up to about four or five), and with flat roofs, using no wood. This applied equally to many of the industrial, commercial and other buildings. Also, for low-rise, one- or two-family houses, flat roofs were not exceptional. There have been indications lately, however, that the problems sometimes associated with flat roofs, as well as changing architectural concepts, may be leading to a return to pitched roofs for low- and medium-rise housing and other buildings in a number of countries. In this connexion the wider use of stress-graded sawn softwood may help to increase its competitive position.

It appears also that in several countries in Europe, if not all, the substitution of sawn softwood (and sawn hardwood) by wood-based panels in building uses may not yet have come to an end. One example appears to be that of floors in low-rise housing in the Netherlands and the United Kingdom. Wooden floors (i.e. with sawnwood structural members) are, however, often associated with insufficient sound insulation, although this does not seem to be a problem in the United Kingdom.

Sawn softwood also continued to be substituted by wood-based panels and mineral/plastic insulating materials as roof underlay components. On the other hand, expanding fields of application of sawn softwood in a number of countries were timber-frame construction, glue-laminated and engineered wooden structures for different types of commercial, industrial, sports and agricultural buildings. In this field, however, competition between steel and wood was increasingly marked in the latter half of the 1970s and in the early 1980s, especially as far as standard types of such buildings were concerned. This was partly due to the depressed level of prices of steel.

From Norway, it was reported that as a result of new building regulations requiring greater insulation thicknesses in outer walls of timber-frame houses, the dimensions of the studs had to be made larger, increasing timber consumption per house.

In contrast to the situation relating to structural uses, sawn softwood used for *joinery* has encountered increasing competition both from other wood products, especially sawn hardwood, as well as from other materials, such as plastics, aluminium, steel and concrete.

In several countries, including France, the Federal Republic of Germany and the Netherlands, a significant proportion of building joinery, particularly windows and doors, but also trim, is made of tropical hardwoods. In the Netherlands this percentage is estimated at 60% for doors and windows taken together. The reasons for this development are said to be the better quality control and more efficient production processes that are possible when using tropical hardwoods as well as the latter's decorative value. Recently, however, cost differentials between certain tropical hardwoods largely used for joinery, and sawn softwood, seem to have led to a certain reversal of this trend.

In Italy, on the other hand, exterior joinery is reportedly made in most cases from imported sawn softwood, while tropical sawn hardwood is used predominantly for interior joinery. In the United Kingdom also, sawn softwood continues to be the main timber used for windows, sawn hardwood being used only for large windows and prestige buildings.

There was also a general decrease in the use of sawn softwood for stairs (except in timber-frame houses), as it was partly substituted by precast concrete and by hardwoods; for joinery trim (by plastic materials); for blinds (by aluminium or plastic materials); and also for door frames (by metal frames, especially in non-residential buildings).

As a material for internal panelling, sawn softwood maintained its position generally and even showed an increase in some western and southern European countries. It is more difficult to assess trends in the use of sawn softwood as an external cladding material. In several countries in western Europe, for instance in parts of the Federal Republic of Germany, the Netherlands and Switzerland, its use for this purpose seems to have decreased; the reason is said to be its tendency to deteriorate in exposed situations, although unsuitable design and construction may also have played a role.

The use of sawn softwood increased markedly in the field of "Do-It-Yourself" (DIY) work. While part of this increase may concern such uses as cladding, interior and exterior, there are undoubtedly many other uses, indoors (for instance, shelving, built-in furniture) and outdoors (garden sheds and furniture); if sawn softwood were not used, these objects would probably either not be realized at all or would be bought as objects made of other materials.

Finally, an interesting development could be observed in the United Kingdom where, for fencing, sawn softwood (treated with preservatives) substituted increasingly sawn hardwood – the latter, however, maintained its position where prestige was a factor.

As far as temporary site uses are concerned, sawn softwood maintained its position in the case of walking boards used in scaffolding, but has been replaced to a considerable extent by panels in concrete formwork.

8.3.1.3 SAWN HARDWOOD

Compared to that of sawn softwood, the use of sawn hardwood in the construction sector is limited, although in some end-uses, particularly joinery, hardwoods play an important role.

As far as *structural* uses are concerned, domestic, temperate-zone, sawn hardwood is used to a certain extent in load-bearing applications in building, primarily in countries with a large hardwood forest resource. A number of tropical hardwoods are used extensively for marine construction and other works demanding special strength and resistance against decay. Furthermore, some tropical hardwood is used, for instance in the United Kingdom, for structural purposes in agricultural buildings, and, in the Netherlands, for general structural elements.

The main field of utilization for sawn hardwood in the construction sector is, however, that of building *joinery*. In this area, the expanding use of tropical hardwoods, which replaced softwoods, was the most significant development since the 1960s. In several of the main European importing countries of tropical hardwood, the latter accounts for a substantial share of all sawn hardwood used in construction, an example already mentioned being that of France, where tropical species account for about half of all sawn hardwood used in construction and are among the main timbers used in certain end-uses such as windows and doors. In the Federal Republic of Germany and in the Netherlands, significant volumes of tropical hardwood are also consumed for these as well as for other joinery uses such as stairs, interior wall cladding, balcony railings and joinery trim.

In the United Kingdom, the use of tropical hardwoods for building joinery, notably industrially-produced joinery components for residential buildings, has been of lesser importance although in recent years their use in this field, in particular for doors and doorsets, including exterior doors, skirting, and shopfittings, has tended to increase quite strongly. A major use of tropical hardwoods is said to be the interior wood surrounds of double-glazed aluminium windows. The reasons for this are said, at least partly, to be in the appearance and inherent status enjoyed by tropical hardwoods.

Relatively minor volumes of tropical hardwoods are used for flooring (parquet), this field having largely, and partly for reasons of cost, remained a domain for (primarily lower quality) temperate-zone hardwoods and, in some countries, for softwoods. But for both, the trend in the majority of western European countries since the 1960s has been for wood utilization to decline as textile wall-to-wall carpeting in residential as well as in office buildings has become increasingly popular. In recent years, however, there have been indications that wooden flooring may again be gaining in popularity.

8.3.1.4 PLYWOOD

The single most important end-use for plywood in construction in Europe, according to available data, is in concrete formwork (shuttering). It is generally estimated that this end-use now accounts for about half of all plywood used in construction. Other important uses for plywood are external cladding, roof sheathing, wall sheathing, and interior walls. An important utilization sector for these uses of plywood is timber-frame construction and prefabricated houses in several countries of the EEC(9).

Plywood has also found increasing application for warehouses and for agricultural buildings, including silos, particularly in the United Kingdom. The renovation and modernization of farms, which has been extensive since the late 1960s, has provided an expanding market for this product.

Another field where plywood has been used increasingly is that of engineered structures – roof elements, beams and related applications – where its strength properties are of particular importance and can be put to use very effectively.

Further improvement in the properties of particle board and the development of special types and grades (moisture-resistant, insect/fungus resistant, flame/fire resistant, and special grades for farm buildings, flooring etc.), has led to increasing utilization of particle board in uses in construction where plywood was previously found. This is true of exterior uses, as well as of interior joinery uses, for instance built-in cupboards and kitchens. Another joinery end-use where the importance of plywood has declined, but in favour of hardboard, is that of flush doors.

8.3.1.5 PARTICLE BOARD

In the majority of European countries, particle board was originally used in the furniture industry. In the construction sector it was first used for built-in furniture, primarily cupboards, and later for built-in kitchens, now quite an important end-use.

With the development of special types of particle board, its use in construction expanded rapidly into such areas as roof underlay, partition walls, wall cladding, doors, and flooring, the latter end-use being said to be the single most important outlet for particle board in the United Kingdom. Indeed, particle board appeared in most end-uses of plywood, even engineered structures and formwork, although the volumes in such end-uses may at present be relatively small.

Apart from the standard particle board grades, the particle board industry has developed a wide variety of special boards with specific properties, such as exterior and structural grades, to meet the requirements of particular end-uses. The development of thin particle board has added further to the competition in many end-uses between plywood, hardboard and particle board.

An important utilization field in a number of countries in northern as well as in continental Europe, especially France and the Federal Republic of Germany, is that of prefabricated houses. In the United Kingdom, on the other hand, the use of particle board in timber-frame construction is relatively limited at present, though growing.

In the countries of Eastern Europe, building is generally a minor utilization sector for particle board.

The appearance of thin particle board brought an entirely new element into the wood-based panels markets. This new type of panel complemented the existing range of thin panels enabling them to compete with plywood and hardboard in many of their uses.

8.3.1.6 FIBREBOARD

For both hardboard and insulating board, construction is by far the most important utilization sector and accounts for the bulk of consumption. The main end-uses are for walls, roof underlay, floors, ceilings, doors, and built-in furniture. Also not negligible are the temporary uses of both hardboard and insulating board on building sites.

In the Nordic countries, the main uses of *insulating board* are for internal facings of walls and ceilings, and as windbreak boards (often bitumen-impregnated) in external walls, mainly of timber-frame houses. In this connection, the use of insulating board for insulating purposes in DIY work has increased considerably. The same can be said for repair and maintenance work, where better insulation is often one of the main objectives.

Apart from thermal insulation, insulating board – mainly in the form of tiles and planks – is also used for acoustic insulation, particularly in non-residential buildings. Some insulating board is also used in shop and office fittings.

The *hardboard* sector has been characterized by the relatively large variety of so-called improved boards, to meet specific end-use requirements, such as tempered, decorative, and perforated boards. Since the mid-1970s the share of such boards in total production has apparently decreased, mainly for reasons of cost.

The main end-uses of hardboard in construction are for flush doors, interior and exterior wall claddings, wall sheathing in timber-frame construction, roof underlay, floors (sub-flooring and decks) in new buildings and, increasingly, in repair and modernization work, as well as for built-in furniture – the latter use having declined as hardboard has been substituted by particle board.

Structural grades of hardboard are used in certain types of engineered elements and structures, especially where board flexibility is important.

It seems that the utilization of hardboard in east European countries may be more limited and confined mainly to doors and built-in furniture.

As regards *medium-density fibreboard* (MDF), despite its application potential for joinery in the construction sector, it seems that its use in this field is still very limited, the furniture industry being by far the largest user.

8.3.2 Furniture

To a higher degree perhaps than most other material goods, furniture is destined to combine functional and aesthetic qualities in order to serve man in his varied needs and activities, both in his home and in other places, and to help create a congenial environment.

For office furniture and some other types of furniture serving professional purposes, technical and economic criteria may play a more prominent role, but for domestic furniture (by far the most important among the different categories of furniture), two main groups of factors determine the pattern of utilization of sawnwood and panels for furniture. These are:

(a) *Technological factors* influencing the construction of furniture and manufacturing techniques, including basic materials; and

(b) *Social factors* influencing the functional aspects of furniture as well as design, and the choice of decorative surface materials.

With regard to *technological* developments, the close of the 1960s saw the introduction of a process of rationalization of the furniture industry aimed at increasing operating efficiency, rather than continuing to expand production capacity. Important aspects of this rationalization trend were the following:

– Simplification and shortening of the manufacturing process;

– Marked reduction in the amount of waste in the preparation, cutting and working of pieces;

– Increasingly, certain manufacturing stages, for instance the cutting to dimension and prefinishing of sawnwood, panels and veneers, or the manufacture of certain parts, were no longer carried out by the furniture manufacturer but by specialized suppliers to the industry;

– Most recently, the introduction of computer numerical control (CNC) lines and other computer-based techniques into the manufacture of industrial furniture is bringing a degree of flexibility not previously known.

Besides these developments affecting furniture manufacturing techniques, important developments also occurred in the field of materials used by the industry. These concerned notably glues (for instance, improving lipping); surface finishing materials (plastic laminates, foils); plastic materials used for framing and for drawers; connecting materials and devices used in assembly technique construction; and, among wood products, the development of thin particle board, laminated boards, pre-finished panels, and medium-density fibreboard.

All these developments had a marked impact on the utilization of both sawnwood and wood-based panels for furniture. Particle board became the wood-based panel used most widely in this field while the application of plywood increasingly concentrated on special end-uses, such as formply or moulded plywood. The main use of hardboard was increasingly oriented towards situations where cost effectiveness was more important than technical factors, such as the backs of cupboards and shelving units, simple (low price) furniture, drawer bottoms, and the like.

Among the *social* factors which were of importance for the furniture sector as an end-use for sawnwood and wood-based panels were the formation of households by young, single, divorced or separated adults (leading to increased demand for furniture), and greater mobility and more frequent changes of dwellings (resulting in increased replacement of furniture). A second important factor tending to increase the demand for furniture was the increase in living space per dwelling in the majority of European countries (although this trend may have been reversed in the early 1980s). At the same time, certain former notions, for instance, that bedrooms should be separated from the other living areas, tended to give way to new lifestyles, resulting in their closer integration with the general activities.

An important exception to this trend, however, was kitchen furniture, one of the main growth areas in the furniture sector since the 1960s. Kitchens in new residential buildings were being equipped increasingly with a complete, built-in range of elements for all main functions. The same trend could be observed in renovation and modernization of residential properties. The aesthetic aspect and the sophistication of the equipment became increasingly important in the 1970s, when kitchens were transformed into living areas using furniture with real wood fronts, etc.

The social trends noted above were also reflected in the design of furniture, its dimensions, form and components, as well as in the materials selected for its exterior appearance: until the latter part of the 1960s, the preferred wood species was teak followed by walnut (especially from North America), mahogany, and oak. These, in turn, were followed in the early 1980s by lighter coloured species, such as ash, cherry, and birch. There was not, of course, a strict succession and separation of species, but more a change in the emphasis or intensity of use of any one species, most of the ten to fifteen main species of wood used in furniture, including a few tropical ones, being used concurrently to a greater or lesser extent and with fairly distinct regional and national preferences.

There has been a transition since the late 1960s from traditional, craftsmanship-oriented, furniture manufacturing techniques to an engineering-oriented approach. With the steady penetration of particle board in furniture manufacturing, the traditional hardwood-frame construction with infilled, veneered plywood panels, gave way increasingly to "solid" particle board construction for cabinet furniture. The new approach also affected the selection of materials; instead of the previously accepted traditional criteria, which presupposed the choice of particular wood products, species and qualities for a given purpose, the selection of materials was based increasingly on an overall value analysis of potentially useable materials – whether wood or not – to determine the most useful material for any given purpose, and the one which gave the best value for money.

The industrialization of the furniture production process, growing specialization, and the availability of new materials led to new furniture construction techniques. Among these, special mention should be made of the introduction of "knock-down", unassembled units which permitted large quantities of furniture in unassembled form to be transported long distances at greatly reduced costs.

In the eastern European countries also, the furniture industry went through a substantial technological transformation, characterized by specialization, the development of material-supplying industries (particle board, veneer imitation foils, etc.), new design and construction techniques, and the expanding manufacture of unitized products, especially kitchen furniture pieces and wall units. A major means for rationalizing furniture production was the use of plastics, laminates and foils.

The single most important trend for forest products in furniture appears to have been the increasing penetration of particle board into the market as the dominant panel material, accounting for well over half of all panels used for furniture.

The use of solid wood (primarily sawn hardwood) both for visible parts, especially in edgings and lippings to match veneered surfaces, and in visible ones for framing purposes, has decreased.

On the other hand, wood as such, as a natural material, seems to have maintained its position well. The age of plastic furniture, foreseen at one stage, especially in the late 1960s/early 1970s, has not materialized. Plastic materials found extensive application in end-uses where surface-resistance is important (laminates), such as tops in kitchens and outdoor furniture, where resistance to bio-degrading agents is a factor, and, partly, where furniture is much exposed to wear such as in public places (but wood was said to be preferred where image is important). Foils as a surfacing material also achieved a certain, albeit limited, importance in western Europe, but a much greater one generally in the countries of Eastern Europe.

It was characteristic of the 1970s that consumer preference for real wood increased. A special trend in this connexion was the growing use of solid wood furniture, a typical example being solid pine furniture which became increasingly popular for use indoors, including kitchens, as well as for use outdoors. Furthermore, there was a marked revival for beech bentwood furniture, for which some countries in Central and Eastern Europe have a long tradition.

Sawnwood, notably of spruce and beech, is also said to have maintained its position at least in some countries as a framing material for upholstered furniture.

8.3.3 Packaging

Packaging as understood in the broad sense mentioned above may well be the end-use sector for sawnwood and wood-based panels that has seen the greatest changes, not only because new packaging materials and methods have appeared, but also because methods for handling, transporting, storing, and distributing goods have changed tremendously over the last few decades.

For nearly all traditional types of packaging once made of wood there are now available other materials (in particular plastics, metals, paperboard) and other methods of packaging which are used at the same time and for the same purposes as sawnwood and wood-based panels without having completely substituted the latter in any one end-use.

At the same time, new handling and transport techniques have led to the development of new types of packaging, such as pallets and containers, both of which are now important uses for sawnwood and panels and which have to some extent compensated for the decline in other fields. Pallets are an important outlet, especially for low quality sawnwood, but are subject to cyclical fluctuations now that most of the initial palletization process is complete.

Wood, notably sawn softwood, and softwood plywood, continues to be used extensively for the packaging for shipment of products of the engineering and metal industries.

But the use of wood in packaging, in particular sawn softwood and hardwood as well as veneer sheets has, on the whole, shown a declining trend, with its share falling in such end-uses as cases, crates, and boxes (for instance for cheese and fruit). Concurrently, the volume of wood used per unit of packaging tended to decline, due to smaller wood sections and new packaging construction methods. The reasons for these developments were generally economic, packaging being in many cases an important cost factor.

On the other hand, the use of plywood, not only for containers, but also for certain types of boxes was reported to have increased in several countries, for instance the Nordic countries, while in Yugoslavia the use of veneer sheets for boxes rose quite markedly.

Attention has also been drawn to the raw material aspect for the use of wood for packaging. Thus the availability of low cost, small-sized softwood suitable for processing into wooden packaging was the reason for the development of this industry in Ireland, with a strong export market in the United Kingdom. Portugal has also specialized in exports of pallet stock.

8.3.4 Other end-uses

A vast array of very diverse uses comes under this heading, ranging from boats to household utensils, brushes, pencils, musical instruments, toys and so forth. Individually, any one utilization segment may consume from a few hundred m^3 (certain musical instruments) to several thousand or even tens of thousand m^3 (boat industry) in any one country. Too few data are generally available for most of the utilization segments to describe developments and trends in more than the broadest terms. Some of these industries are important in size, such as the railway carriage industry, and boat-building which since the late 1950s has expanded sharply, although the consumption of sawnwood and wood-based panels by the industry has not increased correspondingly, due to the now widespread use of plastic materials, especially for smaller boats.

The wooden brush-manufacturing industry has generally seen a decline in activity, especially as small wooden brushes were increasingly replaced by plastic brushes. In addition, this industry has been affected, particularly in western Europe, by growing imports of such brushes from developing countries which are producers of bristles.

It seems to be a general phenomenon that in many of these varied end-uses, the utilization of plastic materials has grown markedly, one example being office equipment. Sometimes these are used in combination with wood, for instance in the case of many sports goods.

On the other hand, no major changes seem to have occurred in the manufacture of traditional musical instruments.

8.4 ESTIMATION OF CONSUMPTION PATTERNS FOR MODELLING PURPOSES

One of the econometric models for projecting future levels of consumption of sawnwood and panels (the "end-use elasticities" model, which will be presented in chapter 11) required estimates of the importance of major end-use sectors for each major assortment of sawnwood and panels. This information is needed to construct indices of activity in end-use sectors for these assortments (sawn softwood, sawn hardwood, particle board, plywood, fibreboard). The three broad end-use sectors are construction, furniture and other uses.

The estimates prepared by the secretariat, on the basis of the information in this chapter are presented in table 8.2. It is stressed that these are secretariat estimates, based on an unsatisfactory data base and intended only as an input to the modelling exercise. They should not be taken as authoritative in any way.

TABLE 8.2

Secretariat estimates of importance of major end-use sectors for sawnwood and wood-based panels

(*Percentage of consumption of each product*)

	Construction	Furniture	Other	Construction	Furniture	Other	Construction	Furniture	Other
	Sawn softwood			*Sawn hardwood*			*Plywood*		
Austria	70.0	5.0	25.0	30.0	23.0	47.0	57.0	25.2	17.8
Finland	70.0	5.0	25.0	30.0	23.0	47.0	69.0	13.9	17.1
France	71.5	5.5	23.0	30.0	23.0	47.0	62.7	12.8	24.5
Germany, Fed. Rep. of	66.5	4.4	29.1	30.0	23.0	47.0	60.4	17.0	22.6
Italy	65.3	6.9	27.8	30.0	23.0	47.0	39.9	34.3	25.8
Netherlands	70.0	5.0	25.0	30.0	23.0	47.0	93.8	3.1	3.1
Norway	87.0	5.0	8.0	30.0	23.0	47.0	57.0	8.0	35.0
Portugal	70.0	5.0	25.0	30.0	23.0	47.0	50.0	35.0	15.0
Spain	70.0	5.0	25.0	30.0	23.0	47.0	50.0	35.0	15.0
Sweden	70.0	5.0	25.0	30.0	23.0	47.0	69.4	5.3	25.3
Switzerland	70.0	5.0	25.0	30.0	23.0	47.0	75.4	10.1	14.5
United Kingdom	71.0	3.0	26.0	30.0	23.0	47.0	54.3	12.0	33.7
	Particle board			*Fibreboard*					
Austria	31.7	60.0	8.3	64.0	17.0	19.0			
Finland	71.7	12.3	16.0	64.0	17.0	19.0			
France	55.0	36.3	8.7	64.0	17.0	19.0			
Germany, Fed. Rep. of	37.7	52.3	10.0	64.0	17.0	19.0			
Italy	5.0	90.0	5.0	64.0	17.0	19.0			
Netherlands	43.0	44.0	13.0	64.0	17.0	19.0			
Norway	66.0	21.0	13.0	85.0	12.0	3.0			
Portugal	15.0	84.0	1.0	64.0	17.0	19.0			
Spain	30.7	57.3	12.0	64.0	17.0	19.0			
Sweden	49.0	46.6	4.4	64.0	17.0	19.0			
Switzerland	39.7	49.6	10.7	64.0	17.0	19.0			
United Kingdom	33.8	38.7	27.5	64.0	17.0	19.0			

Note: Estimates for use in "end-use elasticities" model. For reliability of data, see text.

8.5 PROSPECTS FOR THE UTILIZATION OF SAWNWOOD AND PANELS IN THE MAIN END-USES

This section briefly presents the points made by countries regarding possible future trends in connection with the information provided on the utilization of forest production in the main end-use sectors. It should be noted that in this section the views presented are those of correspondents, which do not necessarily coincide with those of the secretariat. The hypotheses on these subjects used as a basis for the projections are presented in chapter 1.

8.5.1 Construction

From a technological point of view, quality requirements regarding buildings will generally be higher and energy conservation will continue to be a major concern.

Competition between materials will increase, as the overall level of demand for building materials would be lower than in some past years. On the other hand the physical and technical properties of wood are in general not yet used to their full potential and there is scope therefore further to improve the effectiveness of wood utilization. Examples are load-bearing walls and ceilings (end-uses, where, for instance, special high quality plywoods could be used), as well as exterior uses. A precondition for the latter is that products would be developed with a high degree of durability and, therefore, a long life span. In some countries, however, doubts have been expressed as to the degree of technological innovation which the wood processing and using industries could muster.

Another aspect of great importance is the future share of timber-frame houses, which use more sawnwood and panels per unit built than other types, in residential construction. This share has increased in a number of EEC countries since the mid 1970s, partly because of promotional efforts and partly because of the advantages of the method (transfer of work from building site to factory, faster erection time). This trend was, however, sharply reversed in one country, the United Kingdom, because of adverse publicity and problems with quality of workmanship. It appears possible that the share of timber-frame housing could continue to increase, particularly if great care is taken to maintain standards of workmanship and to continue promotional efforts. This would have a positive effect on the consumption of both sawnwood and panels in construction.

One factor affecting the use of all forest products in consumption is their behaviour in fire. This behaviour is now well known and safe practice laid down in building regulations. Nevertheless, the fear, sometimes irrational, of fire, can be a constraint on expanding the use of sawnwood and panels in construction. A prerequisite for such an expansion is the education of the public, including builders and architects, to accept practices in the use of wood products in construction which have been proved to be safe.

A wider range of composite products would be developed and used, for instance glued and laminated products. Furthermore, a wider range of tropical hardwood species would be used. This is not to say that a greater volume will be used: in the United Kingdom, it was felt that these would not substitute softwoods for traditional building purposes to any greater extent than at present.

Another point made was that the use of preservative treatments for greater durability would increase strongly. The need to conserve energy would result in thicker building joinery, especially doors and windows. At the same time aesthetic appearance and fashion trends and influences would become increasingly important factors for building joinery products.

As far as the countries of Eastern Europe are concerned, a main future objective would be to develop further the rational use of wood, for instance by means of stress grading, preservative treatment, knot plugging and finger-jointing. Changes would be made in the standard classification of sawnwood in order to take better account of changing raw material quality. Strong growth is expected for the use of particle board in building construction.

8.5.2 Furniture

No predictions were made regarding the priority accorded by consumers to furniture in their total expenditure. It seems doubtful, however, that it would increase significantly from its present level without some major influence, for instance much improved design and versatility, as well as marketing by manufacturers, or much reduced durability which would raise the replacement rate. On the other hand, increased leisure time might lead to increased demand for furniture to suit different indoor and outdoor leisure activities.

There seemed to be general agreement amongst the correspondents that due to its appearance, wood will continue to be the preferred material for furniture. Demand for real wood, especially as far as surfaces are concerned, will continue to rise in the Eastern European countries where the strongest growth within the furniture sector is expected for veneered furniture.

On the other hand, as the higher qualities become scarcer, their price will inevitably rise. Edging and other solid hardwood parts will increasingly be veneered. Sought-after species may be substituted or imitated by others made to resemble them by special treatments.

Few countries in western Europe expect a further relative increase in the use of sawnwood for furniture, one exception being Finland.

In Eastern Europe, Poland foresees further expansion in the use of hardboard in furniture, especially of the further processed (e.g. enamelled) type. Significant progress in rationalization and technological innovation in manufacturing techniques and materials (such as the wider use of medium-density fibreboard) was mentioned as a general objective of the industry. At the same time, the role of the specialized supplier of pre-finished furniture parts and components is expected to increase markedly, a trend that has already been observed in western Europe.

8.5.3 Packaging

Sawnwood and panels are expected to maintain their relative share and position in packaging, although there may be some further changes in the shares of the individual products and wood consumption per unit of packaging may decrease as new manufacturing techniques for wooden packaging are developed.

8.6 CONCLUSION

From the foregoing analysis one very general trend seems to emerge, namely that sawnwood and wood-based panels have maintained their position, and even improved it, in uses where higher product quality in a broad sense (higher wood quality, higher technical content, better performance in use) is important, and where products have to meet more complex requirements. They have lost ground, however, in uses where quality requirements, particularly in the case of solid wood, are of lesser importance and where the wood material used is generally of lower quality. In such utilization fields, if the end-use continued to exist, wood has been increasingly replaced by other materials offering scope for more rational manufacturing techniques, but in several cases the end-use itself has declined as new technologies emerged, for instance, in the fields of packaging.

For all those concerned with sawnwood and wood-based panels, from the forester to the user, the future will bring big challenges to which they will need to respond actively. On the one hand, these products will have to meet increasingly complex and demanding end-use requirements effectively, taking into account economic, technical and other, not least environmental, factors. On the other hand, constraints will be felt increasingly as regards the raw material base, as the share of high quality (clear surface, large dimension, slow grown) logs in the forest will diminish (and their prices rise so as to make the gap between higher and lower qualities ever wider); secondary species (both tropical and temperate-zone) will increasingly have to be used and greater technical and promotional efforts made to encourage the use of lower qualities.

The wood-processing industries will thus have to produce higher quality products with lower quality materials. Will their innovative spirit succeed in maintaining the position of forest products in uses where high quality (technical and/or decorative) wood is required, possibly in combination with other materials, or will part of these outlets be lost for wood?

The technical input in processed forest products of all types will have to be increased further. The most effective use will have to be made of the available wood raw material, notably by:

(a) Ensuring the highest possible yield;

(b) Making the best use of the physical and other properties of the wood and bringing out to the maximum the advantages of wood;

(c) Minimizing the negative effects of potentially less positive properties, such as biodegradability, flammability, dimensional instability;

(d) Improving surface finishing, both for durability and appearance.

It will not be sufficient, however, to produce a better product. It has also to be brought efficiently to the market. To this end, the product will in many cases need to be promoted as part of a solution to a problem, for instance as part of a building component, wall unit etc. i.e. as part of a system. Research, development and technical innovation, will take on added importance in the wood-processing and wood-using industries. In this connection, prefabrication may well gain further in importance as new techniques (computer numerical control, CNC, and computer aided design, CAD) permit mass production in combination with considerable flexibility.

A further important point is that with technically more sophisticated products, great emphasis needs to be put on their correct application, still frequently a source of problems for sawnwood and panels, especially in construction.

To meet the challenge of other materials – which will no doubt be strong – the wood processing industries will need to look beyond their own sector to anticipate changing conditions in other sectors, especially as regards emerging new technologies, in order to perceive areas where the use of forest products might be affected, as well as new opportunities and outlets for forest products that might arise. Sawnwood and panels can be substituted by other materials and products in virtually all their applications.

The future level of consumption of sawnwood and panels will depend to a large extent on the success of the forest industries in meeting these challenges.

Finally, neither Governments nor individual companies can plan for the future without a sound, quantified, knowledge of the present situation of markets, past trends and outlook for the future. At present, even the broad framework for this type of analysis − knowledge of how much sawnwood, or panels, is used in each end-use sector, in a given country − is not available. In these conditions, analysis of the outlook at the international level can only be speculative in nature. A pre-requisite for the drawing up of long-term strategies, by industry or government, is the carrying out at the national level, in a systematic way, of basic, quantitative research into the use patterns of sawnwood and wood-based panels, on which analysis of the outlook can be based. These studies need to be repeated at regular intervals, so that trends can be detected. A few countries, including Finland, the Netherlands and Switzerland, have carried out this work. It is essential for the future competitivity of sawnwood and panels that other countries follow their example. Not only will this be vital to improve the situation at the national level, but in the long term it is the only way to improve the quality of end-use analysis in future timber trends studies.

CHAPTER 9

*End-uses of paper and paperboard**

9.1 INTRODUCTION

This chapter deals with the qualitative factors which will underpin the demand for paper and paperboard in western Europe for the rest of this century.

Exogenous factors such as political, economic and social trends as well as those driven by technological developments in the fields of television, computers and plastics will be examined. The main paper grades will be cross-indexed by end-use and those end-use trends will be further detailed.

This chapter draws heavily on work carried out during the preparation of the study *Outlook for Pulp and Paper to 1995*, which will be published by FAO, Rome in 1986. This study was prepared by the FAO secretariat in close co-operation with an industry working party. One of the major contributions of the latter group to the study has been their intimate knowledge of the industry structure and markets. This chapter presents the use structure of the main grades of paper and paperboard, general trends and prospects and draws on the opinion of the industry working party on the outlook for specific grades.

The market based approach used cannot produce quantitative projections of consumption: these have been prepared, for major product groups, using econometric methods and are presented in chapters 11 and 12. The discussion in this chapter is intended as a complement to this econometric analysis. In particular, it is possible to evaluate the reasonableness of the quantitative analysis and to draw some conclusions for particular paper grades, which have not been separately analysed in the econometric work.

Since this qualitative summary derives from a global and macro-economic study, specific investment decisions to satisfy demand in Europe must be supported by detailed quantitative analysis.

---

* This chapter was prepared with the help of Mr. A. Whitman (United States of America), consultant to the secretariat.

9.2 CLASSIFICATION OF PAPER GRADES AND END-USES

9.2.1 End-uses for paper grades

GRADE CATEGORY	END-USE
Coated wood free[1]	Very high quality print required in general advertising, special interest magazines and books
Uncoated wood free	Photocopying and duplicating, continuous stationery, educational writing paper, books and in-plant printing
Coated wood containing[2] (includes lightweight coated papers – LWC)	Special interest magazines, direct mail and mail order catalogues, inserts
Uncoated wood containing (includes supercalendered paper – SC – and upgraded newsprint)	Mass circulation magazines, newspaper supplements, directories, paperback books
Newsprint	National, local and free newspapers

Linerboard and fluting medium	Both virgin fibre and waste based grades used for corrugated shipping cases and display packs
Folding carton boards	Both virgin fibre and waste based boards for folding cartons used mainly for fast moving consumer goods including those requiring liquid tight packaging
Other packaging papers and boards	Includes sack papers, bag papers and a wide range of other papers and boards
Cellulose wadding, soft and crepe tissue	Includes personal and industrial wiping and cleaning products for in-home and away-from-home use

9.2.2 Paper grades listed by end-use

END-USES	PAPER GRADES USED
Graphic papers	
Newspapers	Body of newspaper-newsprint, inserts – LWC, supplements – SC and upgraded news

---

[1] With less than 10% mechanical or thermo-mechanical pulp.
[2] With 10% or more of mechanical or thermo-mechanical pulp.

		Packaging papers	
Magazines	Mass circulation – SC, special interest – LWC, some coated wood free	Shipping cases and display packs	Linerboard and fluting medium
General advertising	Coated wood free is largest single grade – most grades used	Consumer goods packaging	Folding boxboard (includes liquid packaging board)
Direct mail /mail order	LWC	Sack and bag paper, others	Kraft wrapping and other packaging papers and boards
Office papers	Uncoated wood free		
Books	Uncoated mechanical and all wood free	Household and sanitary papers	
Others (writing and scholastic, telephone directory, stationery, labels)	Mainly uncoated mechanical and wood free	Toilet tissue, towels, facial, serviettes in home and away-from-home use, including hospital and medical products	Cellulose wadding, soft tissue, towel and crepe paper

9.3 ECONOMIC AND SOCIAL FACTORS

The future economic development in Europe will continue to have great influence on paper demand. Projections of consumption to the year 2000, based on analysis of past relationships between economic development and paper consumption will be presented in chapter 12. In preparing the FAO study, the industry working party commented that while growth is expected to continue, it might not be as robust as the projections indicate. The qualitative reasons for this opinion are spread throughout this chapter and are best summarized by saying that the negative trends will exert a downward pressure on demand while the positive trends will be slightly delayed.

Different *life-styles* are being adopted as society changes in the post-industrial era and this is also affecting paper demand. The decline of traditional family life, increasing divorce rates, higher numbers of smaller family units, more working women, increasing concerns about health and ecology, continuing education, more leisure time and higher unemployment are all examples of changing life-styles. These changes, amongst others, will have differing impacts on the consumption of paper as well as on the grades of paper consumed.

For example, changing family structure has left less time to shop. So buying from catalogues has become a practical necessity. There is a growing emphasis on labour-saving products – notably convenience foods, for which new pack types have been launched (oven-proof trays, retortable pouches, aseptic cartons for sauces, etc.) and disposable paper products – diapers, kitchen towels and facial tissues.

Reduced shopping time has also helped the development of "one-stop" shopping, stimulating the demand for packaging and packaging systems which increase product shelf life (aseptic systems) and contributing to the power of the food chain-stores.

As a consequence of these changes, it now takes less time to shop and since work hours have also been reduced, there is more leisure time. This generates demand for travel brochures and specialist hobby magazines. It also stimulates demand for packaging for leisure oriented products – for gardening, do-it-yourself, sports products, hifi, video and home computers.

These changes will of course occur at different speeds and to a different extent in different countries, so that uniform effects on consumption are not expected.

9.4 TECHNOLOGICAL FACTORS

The industrial base in Europe is in the process of transition from manufacturing towards information and service. Technology is assisting this trend. The long term impact on paper demand is significant and somewhat in favour of paper.

Paper has sometimes been characterized as the sure loser in the competition between television and print media, between electronic machines and paper in the office and between plastic and paper packaging but the real picture is more complex. The main factors determining the results of this competition are costs and demand for more and better information, communicated in more timely ways.

Paper has been losing ground in the fields of education and entertainment, and demand for books, national newspapers and general interest magazines is expected to continue downward. Paper has also lost a lot of ground to plastics in packaging, though this trend is largely over. At the same time, paper demand is actually increasing in the larger areas of communication and information. This can be seen by examining trends in advertising and print technology, packaging and office communications. Therefore the overall demand for printed matter is not likely to decline although there will be significant shifts between grades.

9.4.1   Advertising

The advertising sector is one of those service industries popular in a society based on information. In the past, the advertising industry has grown at well above the average rate for industry. With the expanding importance of information, it will remain a high growth industry. Attitudes to advertising have changed in Europe. Industry has become more aware that advertising is a key tool with which to defend market share, and invaluable when attacking new markets.

Television advertising has been severely restricted by Governments in western Europe, particularly when the situation there is compared with that in the USA and Japan. These restrictions may be liberalized over the forecast period due to more liberal consumer opinion. This will have a major impact on the media used for advertising and thus on the demand for paper. There are two trends at work, however, and the net impact is a positive one for overall paper demand.

On the negative side will be the trend to more TV advertising (including cable and satellite TV), which will improve efficiency in reaching mass audiences. This will result in a reduction of newsprint used in national newspapers and printing papers used in wide-circulation magazines. The presence of high quality colour pictures on television in most homes is generating demand from advertisers for higher quality colour print – in magazines and now in newspapers. There is also a growing demand for packaging with improved visual appeal on the supermarket shelf: the growth in pre-printed liners for corrugated board, and in gravure printing for liquid packaging cartons are prime examples. In the intense competition for attention in this increasingly colourful world, design consultants and advertising agents are becoming more involved in pack design.

At the same time, there is an increasing demand to improve the cost-effectiveness of advertising by reaching specific, target markets. Print is clearly the medium of choice, supported by computer technology (e.g. electronic mailing lists). The growth of specialized magazines, direct mail campaigns, and local and "free" newspapers will have a major positive impact on paper demand. The improved quality demands on print advertising may well follow the trend observed in the US to high quality paper inserts in newspapers, thus creating a shift in the grade used.

The net impact of advertising on paper demand may well be a positive one. Television's importance will grow strongly in support of mass appeals. At the same time, the need to pinpoint advertising to the specific person who is likely to buy the product will increase the use of paper in a significant way.

While paper demand may increase overall as a result of this trend, the impact on each grade will be different and could be dramatic. Newsprint and supercalendared uncoated mechanical paper will be the most likely losers. Coated grades for specialized advertising and all the mechanical grades for local newspapers and inserts will gain the most.

9.4.2   Printing and packaging technology

The dramatic improvement in the ability to deliver messages by the printed word will continue to help paper demand. Computers and electronics have been introduced into almost every aspect of printing, from text storage and image analysis through process and waste control to binding and finishing. As a result, the printing industry has assisted in increasing the use of paper by improving quality and speed of response to demand while reducing the relative cost of the printed message.

The printing industry is expected to continue using new technology further to increase productivity and thus the cost-effectiveness of print. The growth of satellite printing plants closer to the ultimate market will further improve response time.

Similar developments in packaging technology – in particular quicker setting up and making ready of corrugated boxes and folding cartons, have helped suppliers of board-based packaging to keep pace with the increasing fragmentation of their markets. This has helped to offset the higher costs of shorter runs and to cut downtime on the press.

9.4.3   Competition from plastics

In 1984, paper and board accounted for 42% of packaging demand (by weight) in most west European countries, compared with over 45% for the United States. The share held by plastics was about 13% in western Europe, compared with only around 9% in the United States. By the year 1995 some market researchers suggest that the share of paper and board in western Europe may drop to 41%, while plastics increases to 18%.

The major gains by plastics will be at the expense of glass and tinplate, not paper and board. This is because the major opportunities for the replacement of paper and board by plastics monowebs have already been developed – shrinkwrapping, grocery bags, refuse and fertilizer sacks, and food wrappings are good examples. The major emphasis for plastics is now in high barrier resins and co-extruded materials capable of replacing glass and tinplate.

The development of plastics technology is actually assisting paper and board sales in a number of areas where plastics are used to enhance the properties of paper and board. Examples include liquid tight cartons, bag-in-box, oven-proof boards, and the paper can.

9.4.4   The office

Perhaps nowhere has the struggle with what many refer to as "that blizzard of paper" been less understood than in the office. Information is vital for all economic and administrative activities and new electronic techniques generate ever more information. The crucial question, however, is how this information is generated, stored and distributed. Theoretically, it is already possible to do this entirely electronically. However sociological and psychological factors have clearly overriden technological factors and it is at present extremely difficult to foresee how much paper will be needed in the office of the future.

In fact another technology in this area has actually increased the use of paper. Developments in processing data and presenting the results (colour, graphs, etc.) have made it convenient and cheap to create hard copy (i.e. on paper) of information. This has secured a bright future for the grades of paper used in these applications.

9.4.5 Cost reduction in packaging

The pressure on manufacturing costs has meant a growing emphasis on labour saving through the use of automatic and semi-automatic pack erection, filling, sealing and handling arrangements:

- Shrink wrapped trays and other automated systems are gaining ground from regular slotted cases in the corrugated board market;
- Virgin fibre-based folding cartonboards are gaining ground because they run better on automatic equipment than waste-containing boards;
- The development of the liquid tight carton is a prime example of a successful systems approach;
- Paper sacks are keeping their hold in a number of important markets because they can be more readily filled on automatic equipment than can plastic sacks.

9.5 OUTLOOK BY END-USE SECTOR

This section sets out some subjective judgements on the outlook, arranged by end-use sector.

9.5.1 Graphic papers

NEWSPAPERS (INCLUDING INSERTS AND SUPPLEMENTS)

- The total demand for paper for newspapers will probably continue to grow, albeit more slowly than in the past;
- Newspapers will continue to be a major and growing medium for the advertising industry, but they will lose share to television;
- National newspaper circulations have been in decline for many years and are expected to continue this trend;
- The shift in the newspaper industry will continue to be away from national to local and freely distributed newspapers;
- Colour inserts and supplements are likely to be more popular than run-of-the-press advertising spreads. Growth is expected for the former sectors;
- Newspapers printed in four colours will grow in popularity;
- Basis weights of newsprint have declined from 49 g/m^2 in 1975 to 47 g/m^2 in 1984. By 1995 the average weight is expected to be 44 g/m^2 or less;
- Recycled fibre and thermo-mechanical or chemi-thermo-mechanical pulps will largely replace the traditional sheet blended from mechanical and chemical pulps.

MAGAZINES

- Wide circulation magazines are forecast to decline, because of competition with television, for both advertising and entertainment;
- The growth of specialized magazines will more than compensate for this decline, as far as demand for paper is concerned;
- Increased leisure time and more hobbies will support the demand for specialized magazines;
- The shift to specialized titles will benefit web-offset printing, whereas rotogravure will suffer the consequences of the decline of mass circulation magazines;
- The trend is for a shift from supercalendered uncoated mechanical paper (SC) to light-weight coated (LWC) and other offset mechanical papers.

GENERAL ADVERTISING

- This sector covers a wide range of end-uses, from travel brochures to company calendars. These end-uses will generate overall at least average growth for paper demand;
- The central trend in the type of paper used will be towards higher quality, particularly coated wood-free grades.

DIRECT MAIL/MAIL ORDER

- The direct mail sector has excellent prospects for paper demand growth, as direct mail is a fast growing advertising medium, popular for specialized, targeted advertising;
- The appeal to the reader of colour catalogues has been proved;
- This sector is aided by the wider availability of computerized mailing lists;
- The longevity of printed advertisements (compared to those on television) favours direct mail;
- The future for mail-order catalogues depends on their competitivity compared to retailing in shops;
- The possibility of home shopping through computer terminals could be realized in the next fifteen years, but widespread use in Europe is unlikely before the next century. Even then, catalogues will play a major role in this development;

- Light-weight coated (LWC) papers will be favoured, because of the high proportion of postage costs in total costs.

OFFICE PAPERS

- The use of papers in the office may well grow significantly, even beyond 1995;

- The driving force is the greater demand for information, coupled with new technologies;

- Widespread use of photocopiers, word-processors and computers is currently creating fast growth for paper demand;

- The technology to eliminate paper entirely from the office has existed for some time, but experience has shown that new technologies have instead generated even greater demand for paper. Enormous investment costs are a major deterrent to the installation of radically new office systems;

- The social acceptability of paper, its low costs compared to the value of information and its effectiveness in conveying the message will continue to support strong paper demand in this sector;

- The practice of in-plant printing is likely to grow substantially, particularly when type-setting and copying become more directly linked to computers and word-processors, as has become possible in recent years;

- Some paper grades are expected to show rapid growth, for example laser copier papers and other grades used by the new technologies. Others will decline, for example duplicating, stencil, filing and transactional papers;

- The benefits of economies of scale will lead to further standardization of paper sizes in the office towards A4.

BOOKS

- The book market may further stagnate or decline in the coming years;

- Leisure habits are changing to the detriment of books, particularly due to television viewing;

- Paperbacks are expected to be the sector with the best performance, but educational paperback books will decline. This shift will increase the share of uncoated mechanical paper at the expense of wood-free grades.

OTHER END-USES FOR GRAPHIC PAPERS

- These cover a wide range of markets (writing and scholastic paper, traditional stationery, telephone directories, labels, filters, tickets, etc.). Individual markets are small in size compared with the major markets listed above;

- Markets are generally mature, and paper demand is forecast to stagnate or decline. For example, French national policy to introduce subsidized electronic terminals in the home will have the effect of replacing traditional telephone directories.

9.5.2 Household and sanitary papers

- The European market shows signs of maturity and saturation similar to that which the US market has already experienced;

- Substitution (for example fluff pulp for cellulose wadding in diapers) has largely taken place already.

9.5.3 Packaging and wrapping papers

SHIPPING CASES AND DISPLAY PACKS

- Factors limiting growth include:

  (a) Lower population and GDP growth;

  (b) Design economies favour automated packaging systems and therefore wraparound and shrink-wrap trays rather than the conventional corrugated case;

  (c) Market saturation in the major applications for corrugated board;

- Factors favouring continued growth are:

  (a) Steady replacement of wooden trays by corrugated boards in agricultural markets (especially France, Italy and Spain);

  (b) Continued modest development of heavy duty corrugated boards replacing wood for export packing;

- A drive by European packaging manufacturers to reduce stocks and thereby working capital has increased the share taken by waste-based local suppliers at the expense of virgin liner;

- Specification changes, stressing crush resistance and print surface quality, rather than mullen or fibre content, have supported this trend;

- The share of fluting in corrugated sheet is increasing;

- The need for improved print quality at the point of purchase is causing a steady increase in the share of unbleached liner with a white top. This trend will probably continue and while mottled liner will hold its own, fully bleached liner is losing share due to higher cost, as well as improved performance of the waste-based sheet.

CONSUMER GOODS PACKAGING

- Generally a static market for traditional folding carton end-uses is expected, as market saturation and end-use stagnation (e.g. cigarettes, detergents, sweets) limit growth;

- Further grammage reduction and design economies (reduced flap sizes) will continue to reduce the amount of board per unit;

- Cost pressures and the performance requirements of fast running automatic filling equipment support the following trends:

  (a) Marked increase in share held by virgin fibre grades including CTMP (in contrast to other packaging materials);

(b) Increased emphasis on surface characteristics with a wide variety of coating methods;

(c) More stringent quality control and more waste paper in the top layer of waste based boards.

LIQUID PACKAGING BOARDS

- Growth will probably slow as market saturation nears;

- Increased use of duplex (bleached/unbleached) board which provides a cost saving compared with bleached and semi-bleached, and also gives a better barrier to ultra-violet light;

- Increased use of coated boards (unbleached duplex, or triplex) for gravure and offset printed cartons for fruit juice and similar products;

- Average board weights were reduced by around 10% between 1978 and 1983 and there is scope for some further reduction because of the increased stiffness/weight ratio of modern boards.

SACK PAPER

- Though the rate of decline has slowed, demand for sack paper has been falling steadily for many years, due to intense competition from:

(a) Plastic sacks, which have taken the refuse and fertilizer markets;

(b) Bulk transportation, which has been particularly significant in the cement and chemicals market;

(c) Intermediate bulk containers (plastic bags or lined boxes holding 500 kg to 1 m.t. of product), which have been particularly important in the chemicals market.

- In response to this competition, sack makers have made substantial reductions in the number and weight of paper plies, further reducing demand by weight;

- Technical developments to combat this market decline include:

(a) Growth in bleached grades in support of point-of-sale advertising;

(b) Extensible and twin wire grades to improve strength/weight ratios.

9.6 OUTLOOK AS SEEN BY THE INDUSTRY WORKING PARTY

This section summarizes the opinion of the industry working party. Most of these opinions are the result of a global assessment and many need slight adaptation for European conditions.

A number of perceptions of the possible pattern of development of *newsprint* consumption emerged. First it was considered reasonable to expect that at high levels of wealth and high relative levels of consumption, the marginal propensity to increase consumption would diminish, implying a diminution in income elasticity as the level of income increased.

Newsprint consumption, based on newspaper circulation, was considered to be strongly related to population dynamics. In countries where the population is static, circulation may diminish. In countries with growing literacy rates and expanding population, one might also expect increasing newspaper circulation. Several developing countries mentioned special policies to promote newspaper distribution.

It was not possible to identify clear cut indications of the impact of technological change in communications and the movements in advertising to new media. In general, the idea of somewhat slower growth in the developed market economies in future periods was considered probable.

As regards the outlook for *printing and writing papers*, alternative (lower) forecasts for European countries were formulated. The reasons for expecting lower growth in printing and writing papers included the fact that rapid growth in special interest magazines and inserts may already have taken place and come to an end. In the Federal Republic of Germany, restrictions on advertising on television were being relaxed and changes were taking place in other European countries which would lead to an expansion of television advertising. Rotogravure printing for magazines was stagnant. Book and telephone directories showed no growth. Areas of opportunity included small catalogues and new areas of use. Paper use in offices could be expected to continue its expansion up to 1990. Advertising in electronic media frequently required print advertising in support. Taken altogether it was considered that there would be increasing inroads of electronic communications in areas previously provided for by print media.

For household and sanitary papers it was considered that the market showed signs of maturity and was characterized by heavy substitution. Maturity meant that the market for a specific product had reached its maximum of consumption. Substitution was a two-sided phenomenon: on the one hand, paper substituted for other materials, mostly textile, but, on the other, was substituted for by other materials, mostly cellulose fluff and non-woven fabrics. In all western European countries, for example, baby napkins made from paper had replaced textile napkins while paper napkins, themselves, were being replaced by napkins consisting mainly of fluff and non-woven fabrics. As the substitution by fluff pulp and non-woven material started at a much lower (*per caput*) level of paper consumption, western Europe would never reach the *per caput* consumption of household and sanitary papers of the United States; low income countries would never experience the *per caput* consumption of high income countries. Apart from the type of substitution mentioned above, it was expected that the use pattern would change very little in future.

For *containerboard* the following qualitative factors were singled out for particular attention:

- Attempts by end users to reduce their overall packaging requirements to control packaging costs;

- Streamlining of distribution systems as well as inventory control procedures that reduced packaging needs;

- Government regulations that encouraged the use of returnable forms of packaging in some countries;

- Saturation of traditional markets for containerboard in some countries counterbalanced newly "opened" markets (i.e. fruits and vegetables) in others;

- Loss of some markets to competing packaging materials, particularly plastic film.

In Europe the traditional markets for corrugated board were either approaching or had reached saturation, and only a few "new" markets could be penetrated (i.e. fruits and vegetables). Some markets had been lost to competitive packaging materials, and there were trends to lighter basis weights, and towards stricter cost control in the use of packaging materials. Government legislation encouraging the use of returnable packaging materials was also cited.

For *folding boxboard* it was reported that plastic materials had made inroads into the traditional uses of cartons. This tendency could very well continue, especially if cost relations between raw materials were changing in favour of plastics. At the same time, packing based on a combination of board and plastic was increasing. The growing demand for packaging for convenience food and for oven board should give carton board new opportunities.

The trend towards lower grammage of board had already resulted in a situation where little further development could be expected. Too much rigidity would then be lost.

In western Europe there seemed to have been a relatively balanced situation between virgin fibre and waste-based board over the last few years. The existing food packaging regulations in western Europe were not expected to lead to any changes in the use of different types of board.

There was, however, for reasons of hygiene, a trend to use virgin fibre or pre-consumer waste paper for some specific papers and boards, which come into long term direct contact with wet or greasy food-stuffs. Another trend is towards the combination of materials such as board and aluminium foil or board and plastic foil.

Other packaging grades cover a wide variety of materials subject to similar pressures, including:

(a) Intense competition from increasingly sophisticated plastic materials based on declining feedstock prices;

(b) Increased use of waste paper because of cost pressures.

What of the outlook in quantitative terms? The FAO prepared a reference scenario using econometric methods, which will be used for the ETTS IV consumption scenarios. The methodology and results of this projection will be presented in chapter 11 and 12. The industry working party, taking the FAO reference scenario as a basis for discussion, prepared an alternative scenario for several products, which took into account qualitative factors of the type discussed in this chapter. For western European countries (except Cyprus, Israel, Malta, Turkey, Yugoslavia) the two scenarios were as follows (in million m.t.):

	1984	FAO reference scenario	Working party alternative	Difference
Scenarios for 1995:				
Newsprint	6.2	7.7	6.2	− 1.5 (− 19.5%)
Printing and writing paper	14.5	20.9	16.8	− 4.1 (− 19.6%)
Other paper and paperboard	25.6	27.7	32.9[a]	+ 5.2 (+ 18.8%)
Total	46.3	56.3	55.9	− 0.4 (− 0.7%)

[a] Midpoint of range of 30.1 -35.7 million m.t.

In total, the working party's alternative scenario was very similar to the FAO reference scenario, although higher for other paper and paperboard and lower for newsprint and printing and writing paper. This judgement should be borne in mind when considering the overall supply/demand balance.

9.7 CONCLUSION

Both exogenous and technological factors will continue to slow the growth in the demand for paper and board in western Europe.

Paper is largely an intermediate product and thus tends not to initiate movements but to react to external trends: however its recyclability and renewability as a resource

make it adaptable. This makes it hard to predict future demands as second and third-effect trends often favour the use of paper or moderate theoretical rates of decline.

The exogenous factors demonstrate as never before the interdependence of western Europe and the rest of the world.

While technology will cause paper to lose share in most major market segments, total demand for paper will continue to grow.

The most significant shifts will be between paper grades within particular end-use demands.

Although many of these factors apply equally to western and eastern Europe, others apply essentially to eastern Europe only. Section 11.4 discusses some of the specific factors affecting eastern European consumption patterns.

CHAPTER 10

Trends in prices for forest products

10.1 INTRODUCTION

Prices play a role in determining demand for forest products, and have therefore been included in the demand models used in this study (see chapter 11). Price levels also help to determine the economic health of various industrial sectors. Furthermore, many users of outlook studies, notably those considering investment decisions, wish to have indications on the outlook for prices.

The present study therefore examines, in greater detail than was possible in earlier studies, the trends and outlook for the prices of forest products and of wood in the rough. The data base was originally compiled for the purposes of modelling demand, using the end-use elasticities model to be described in chapter 11. National correspondents were requested to supply uniform long-term price series; series available in international publications were also used. Although this data base concerned primarily sawnwood and wood-based panels, it was extended by the addition of readily available series to cover wood in the rough and paper and paperboard. The secretariat wishes to repeat its thanks to the correspondents who made it possible to collect the data base on prices.

This chapter presents trends in prices between 1964 and 1981. One major objective is to determine whether there exist strong underlying price trends which should be projected into the future when modelling demand.

For reasons of space, it is not possible in this chapter to do more than present briefly the series, with very little analysis of underlying causes, especially where these are of a local nature.

For this chapter all series have been converted to indices (1965 = 100) and put on a constant price basis. The deflator chosen was the producer price index of the country concerned, in order to remove the effects of inflation. The original price series and the corresponding deflated indices (and combined indices) are available on request from the secretariat. The figures in the text of the chapter are graphs of the trends in the deflated indices over time (1964 to 1981). They are on a uniform scale (with a very few exceptions which are mentioned in the text).

As a result of this choice of methodology, it is possible:

- To see whether prices of forest products have progressed faster or slower than those of manufactured products in general;

- To see whether there have been marked fluctuations in price levels;

- To compare, within one country, price trends of forest products competing against each other, or of raw material and product (e.g. sawlogs and sawnwood).

It is doubtful whether valid conclusions may be drawn from comparisons of trends in different countries, partly because of distortions due to exchange rates and partly because many of the original series were also in index form. (The indices were mostly derived from series in local currency as the method was originally conceived in the context of modelling demand on national markets.)

Over 150 price series were collected, whose direction and degree of fluctuation vary widely between countries and between products. As this is too many to enable a clear overall picture to be presented, the series have been combined to produce one series for each product or product group. (In this process the national price series were weighted by the importance of consumption of the product in question in that country.) As the countries for which series were collected account for a large part of European consumption, these combined series may be taken as being representative of trends in real prices of forest products in Europe (excluding Eastern Europe). It should be borne in mind, however, that:

- Trends in individual countries sometimes differ very markedly from the overall trend;

- For products other than sawnwood and panels, especially paper and paperboard, only a relatively small number of series have been included, sometimes with a limited coverage of different grades. In particular, the series for "other paper and paperboard" is in fact constructed almost exclusively from series for packaging grades and is therefore *not* representative of trends in other parts of this very heterogeneous product group.

The series for sawnwood and wood-based panels were originally constructed for the demand analysis in chapter 9. For this analysis, for technical reasons, it was necessary that there be a price series for each product and that all series be complete (i.e. from 1964 to 1981). For that reason, estimates have sometimes been made (e.g. for the first or last year of the series), even on the basis of fairly tenuous information. For the other products, however (e.g. wood in the rough, pulp and paper), no such estimates were made.

It should be noted that this chapter aims only at presenting the trends, rather than explaining them. Market conditions have varied widely between countries, as do the processes of price formation which are often strongly affected by factors peculiar to a given country, such as the presence or absence of associations of buyers or sellers, the holding of formal price negotiations (e.g. on roundwood), government regulations (e.g. price freezes in the context of anti-inflation policies), openness or otherwise of domestic markets to trends on the international markets.

The role of prices is different in the centrally planned economies from other countries in the region. They are therefore treated separately, in section 10.4.

10.2   TRENDS IN PRICES OF FOREST PRODUCTS IN WESTERN EUROPE

Figures 10.1 to 10.5 present in graphical form[1] the combined series for the different products between 1964 and 1981, prepared by the methods described above. Although not every country is covered, the combined series may be considered representative of trends in western Europe.

Figure 10.1 shows quite clearly the price trends for the three major product groups and the clear differences between them. Sawnwood prices were significantly higher (by 20-30%) at the end of the period than at the beginning. The weighted series does not however show a steady upward trend but rather two periods of little or no movement, separated by a violent spasm in 1973-1976, which left the price level in the second half of the 1970s above that before 1973. Paper and paperboard on the other hand, were only slightly more expensive, in real terms, in 1981 than in 1964. Real prices for wood-based panels in aggregate, however, fell steadily over the whole period. For sawnwood and paper and paperboard, the mid-1970s saw a violent price rise (nearly 50 points for sawnwood, about 25 for paper and paperboard) followed by a steep fall. The movement was earlier and more violent for sawnwood, while wood-based panels were hardly affected at all. At the same time, many commodities other than forest products showed up-and-down movements, which were as strong as that for sawnwood, if not more so.

Prices of *sawn softwood* (figure 10.2) remained roughly stable, in real terms, in most countries until the early 1970s. Thereafter, almost all countries showed a rise in real prices in 1973 and 1974, followed by a drop. In some countries this rise was very sharp indeed (over 80 points in the United Kingdom, over 60 in France) an indication of the extraordinary excitement, not to say hysteria, of that period on the markets of many commodities, including sawn softwood. In the Federal Republic of Germany, however, the movement in prices was quite modest. In Portugal, no price boom took place, possibly because at that time there were considerable political upheavals in that country.

After 1974, trends diverge between countries. In some, the "gains" of the early 1970s were entirely lost and the level around 1980 was the same as or only slightly higher than that of the mid-1960s (Austria, Finland, the Federal Republic of Germany, the Netherlands, United Kingdom). In Portugal, the real price of sawn softwood dropped sharply between 1974 and 1976 and remained roughly stable thereafter. In other countries, although prices did fall after 1974 or 1975, either the fall was relatively minor or the recovery in the late 1970s was strong, so that by 1980, real prices were significantly higher than in the 1960s (France, Italy, Spain). The weighted series shows price levels in the late 1970s between those of the late 1960s/early 1970s and the 1974 peak.

The trends for *sawn hardwood* until 1977 were broadly similar to those for sawn softwood in most countries. The combined series for sawn hardwood diverged from that for sawn softwood in the late 1970s to reach a second peak in 1979/80, at almost the same levels as 1974. In 1981, the series for sawn softwood and sawn hardwood again coincided. In the Federal Republic of Germany, the Netherlands and Portugal, sawn hardwood prices were at a higher level at the end of the period than those of sawn softwood, while the opposite is true of Italy and Spain.

Prices for all three wood-based panels fell over the period, but the real price of *particle board* (figure 10.3) declined more than those of plywood and fibreboard. In almost all countries, the real price around 1980 was significantly lower than in the mid-1960s. Only in the Netherlands and the United Kingdom did particle board prices participate in the mid-1970s "commodity boom". This steady fall in prices, made possible by a reduction in costs (new processes, economies of scale, cheaper raw material from sawmill residues), was certainly a contributing factor in the dynamic expansion of particle board consumption, although the gain in convenience from using particle board and the new utilization possibilities it offered may well have played an even more important role.

For *plywood*, the general trend is not so clear, although only one of the countries shown (Finland) had a significant upward trend over the period. There was little change in the Federal Republic of Germany and Spain, while France, Italy, the Netherlands and the United Kingdom showed significant downward movement, comparable to that of particle board in some cases. In several countries,

---

[1] The numerical values for these and the constituent series (original series and deflated indices) are available on request from the secretariat.

FIGURE 10.1

Western Europe: price indices of forest products
(*1965 = 100*)

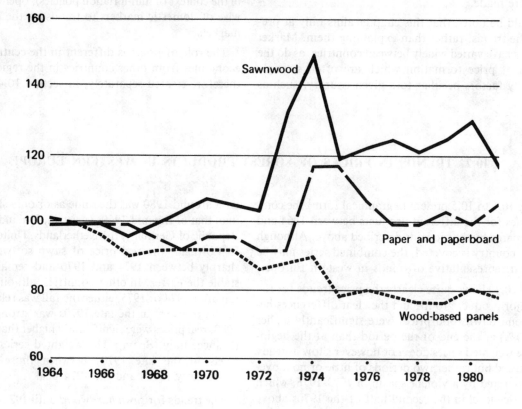

FIGURE 10.2

Western Europe: price indices of sawnwood
(*1965 = 100*)

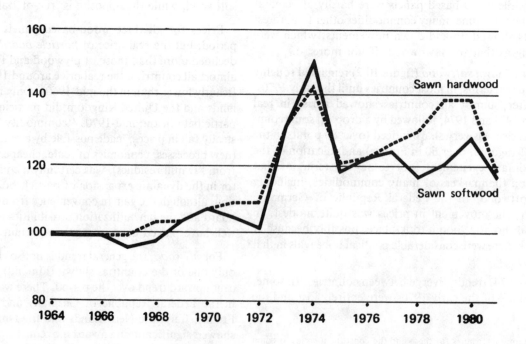

plywood prices fluctuated strongly. The combined series shows a fairly marked downward trend after 1974. It is likely that several of these changes are due to changes in product quality as the series are import or export unit values (higher value added for Finnish plywood exports, lower cost imports from south-east Asia for France, Italy, the Netherlands and the United Kingdom).

Real prices of *fibreboard* followed overall a similar trend to those of plywood. They tended to fall in the Federal Republic of Germany and in Switzerland, but in Sweden, a major fibreboard producer and consumer, the price remained stable in real terms until the end of the 1970s when it rose quite sharply. This situation may be linked to the fact that in the 1970s there was serious over-capacity in the fibreboard industry, leading to the closure of a number of mills, particularly in Sweden.

Although the three types of panel are not directly comparable from a technical point of view, it appears that the successful market penetration and falling real prices of particle board put the other two panels under pressure to reduce their prices, even if they were unable to reduce their costs as much as particle board. The cost of the high quality raw material necessary for plywood made price reductions particularly difficult for this assortment.

The trends in prices of *paper and paperboard* (figure 10.4), like those of sawnwood and panels, vary widely between countries and between grades: generalizations must be treated with caution. Prices for many grades, however, notably the "bulk grades" (with essentially uniform characteristics, produced in large quantities, such as newsprint or Kraftliner) are mostly determined on highly competitive and transparent international markets. Prices for these grades are often quoted in US dollars which would lead to a uniformity of trend between countries, were it not for the movement of the different currencies relative to the dollar.

Rather fewer series were available for paper and paperboard than for sawnwood and panels, and this should be taken into account, especially with regard to the combined series.[2] Furthermore, the series for two major con-

sumers were incomplete, as those for the Federal Republic of Germany start in 1968 and those for the United Kingdom in 1966, as opposed to 1964 (or before) for the other countries. Combined series have been prepared for the group with and without those two countries (see figure 10.5), which show that their inclusion in fig. 10.4 would not have modified the broad outline.

In the group "other paper and paperboard", the series for all countries except Sweden refer to Kraft or packaging grades only (the Swedish series is a global one, covering "other paper and paperboard" as a whole). The combined series therefore hardly takes into account any of the non-packaging grades, notably of household and sanitary paper, which make up an important part of this group.

Prices of *newsprint* remained fairly stable in real terms until the early 1970s with a slight downward trend in some countries (e.g. Austria, the Federal Republic of Germany, the Netherlands, Portugal, Sweden). In 1973 and 1974 there was a sharp upward movement in all countries, followed in many countries by a decline. After 1975, however, apart from the general slightly downward trend, it is hard to see a common development as local factors appear to predominate, including the strength of demand for newsprint and of the local currency compared to the dollar.

The few series available for *printing and writing* grades show a very similar overall trend to newsprint: stability to 1973, sharp rises in 1974, followed by a downward movement.

The combined series for *"other paper and paperboard"* which, as mentioned above, covers essentially only Kraft and packaging paper and paperboard, shows a slow fall in real terms to 1973, followed by a rise in 1974/75 and then a steep fall to 1978, and a recovery thereafter. Nevertheless, in 1981 the combined series for "other paper and paperboard" was still significantly lower than in 1964. In the international markets for bulk grades in the capital intensive pulp and paper sector, supply factors, notably over- or under-capacity, and the costs of the lowest cost producers play a role as important as that of demand in determining prices. The final price will also take account of movements in currency values.

---

[2] Newsprint: seven countries (Austria; Finland; France; the Federal Republic of Germany; Portugal; Sweden; the United Kingdom). Printing and writing: four countries (Austria; France; the Federal Republic of Germany; Sweden). Other paper and paperboard: five countries (Austria; the Federal Republic of Germany; Portugal; Sweden; the United Kingdom).

10.3 TRENDS IN ROUNDWOOD PRICES AND COMPARISON OF TRENDS BETWEEN RAW MATERIAL AND PRODUCT PRICES

This section presents trends for roundwood prices together with those for the major forest products manufactured from them. It intends, therefore, to show what was the trend not only in the price received by the forest owner but also, in very rough terms, in the raw material costs of the forest industries, when compared to the prices they received for their products.

Here, as elsewhere in this chapter, readers must be aware of a number of factors which make the drawing of conclusions and the making of comparisons extremely difficult. Roundwood markets, with a few major exceptions (e.g. Japanese imports), even more than forest products markets, are extremely local and traditional. The structure of markets varies widely between, and even within,

FIGURE 10.3

Western Europe: price indices of wood-based panels

(1965 = 100)

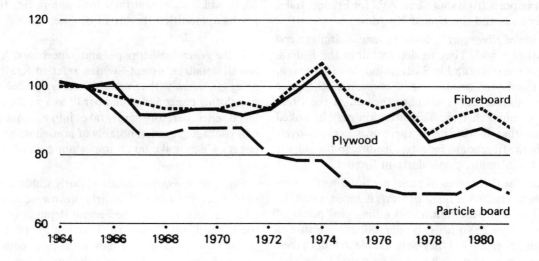

FIGURE 10.4

Western Europe: price indices of paper and paperboard

(1965 = 100)

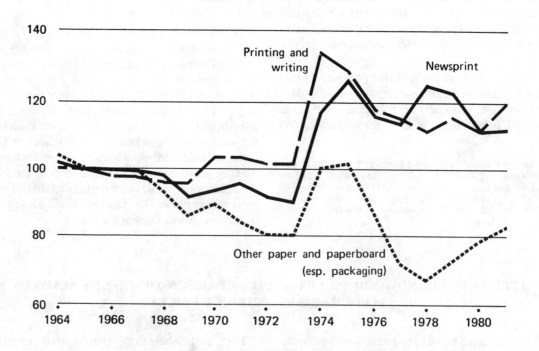

N.B.   Excludes Federal Republic of Germany and United Kingdom
(see text and Fig. 10.5)

FIGURE 10.5

Western Europe: price indices of paper and paperboard

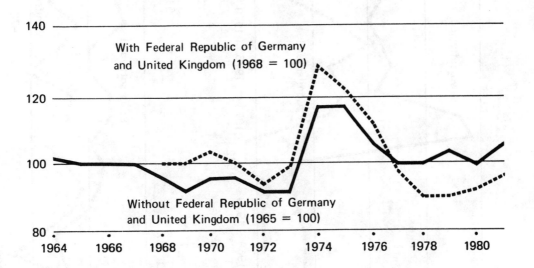

countries. In some cases, prices and/or quantities are agreed on in regional or national negotiations between industries and forest owners. Sometimes the level of the market is set at auctions by major forest owners, usually state forest services; sometimes there is a multitude of individual agreements. Sometimes the system differs according to the quality or species of the roundwood. Wood is sold standing, at roadside or delivered at mill, barked or unbarked. The harvesting is carried out by the forest owner, by the buyer, by contractors working for one or the other or by middlemen who buy the wood to resell it.

In each country price statistics, when available, are collected in the framework of the prevailing local system. It is not therefore possible to collect series which are strictly comparable on the international level, especially if the aim is to compare absolute prices. The series presented here, however, do give an indication of trends, although it should not be forgotten that in some cases price movements may be due more to changes in harvesting and transport costs than to changes in roundwood prices as such. (For further information, readers are referred to the detailed specifications of the series.)

As regards the comparison between raw material and product prices, it should be borne in mind that there are strong influences preventing the two trends from following radically different courses: rises in roundwood prices force the industries to attempt to raise their own prices while if product prices are raised, roundwood producers, aware of these rises, will claim at least a part of the increase for themselves. (Of course, there are frequently lags between

price movements for raw material and for product, due to the imperfections of the market mechanism.) When there are significant differences over the long-term in the two trends, the causes might be sought in the changing supply/demand balance, in technological progress or in cost movements due to events outside the sector (e.g. energy prices) or a combination of these (for instance, a technological improvement in manufacturing which permits the use of lower quality raw material and thereby changes the market balance).

It should also be borne in mind that raw materials represent a major part of total costs for many forest industries, reaching as high as 60-70% in some cases (e.g. sawnwood). Even in the pulp and paper sector, where wood costs account for a smaller share of total costs, notably because of high capital costs, the fact that the technology is broadly similar all over the world means that wood costs are one of the major areas whereby individual companies seek to gain a competitive advantage.

Because of the local nature of most roundwood markets, no combined (international) series have been prepared for wood in the rough, as these could be seriously misleading. Rather, national series for roundwood are compared with national series for the products of that particular assortment.

Prices for *coniferous logs* in most of the countries shown, like sawn softwood, showed a very sharp rise around 1973-74, after a less eventful period. This rise was followed by a more or less abrupt decline which brought prices of coniferous sawlogs around 1980 to approxi-

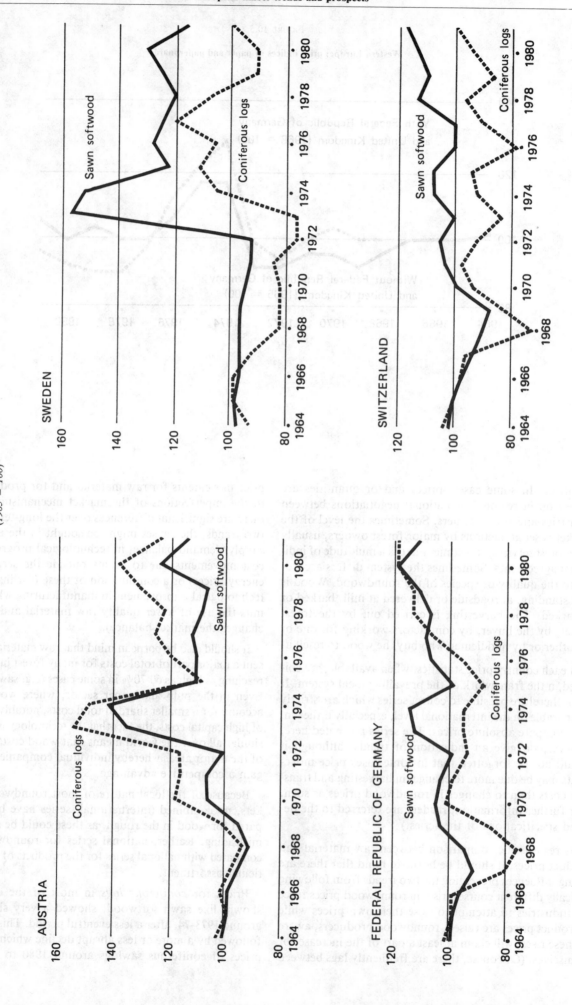

FIGURE 10.6

Price indices for sawn softwood and coniferous logs
(1965 = 100)

FIGURE 10.7

Price indices of broadleaved logs and sawn hardwood
(1965 = 100)

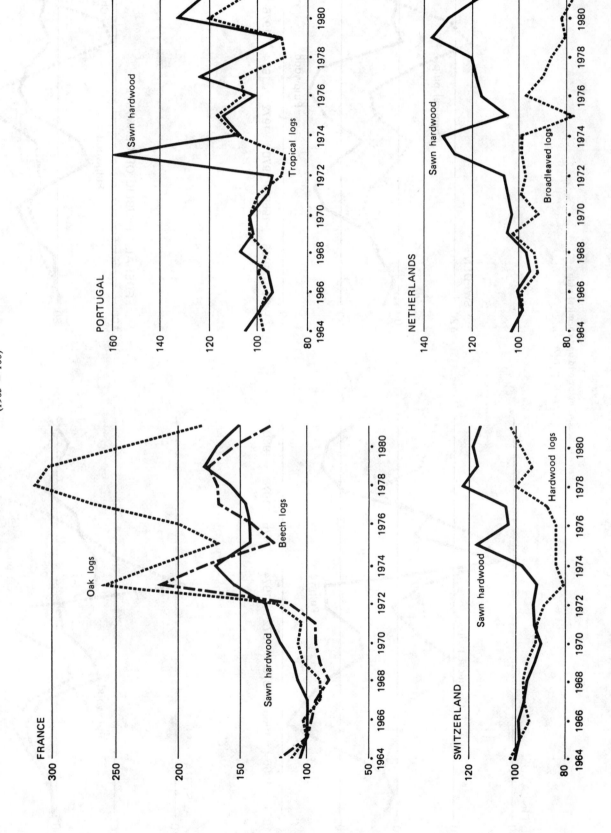

FIGURE 10.8

Price indices of pulpwood and chemical pulp
(1965 = 100)

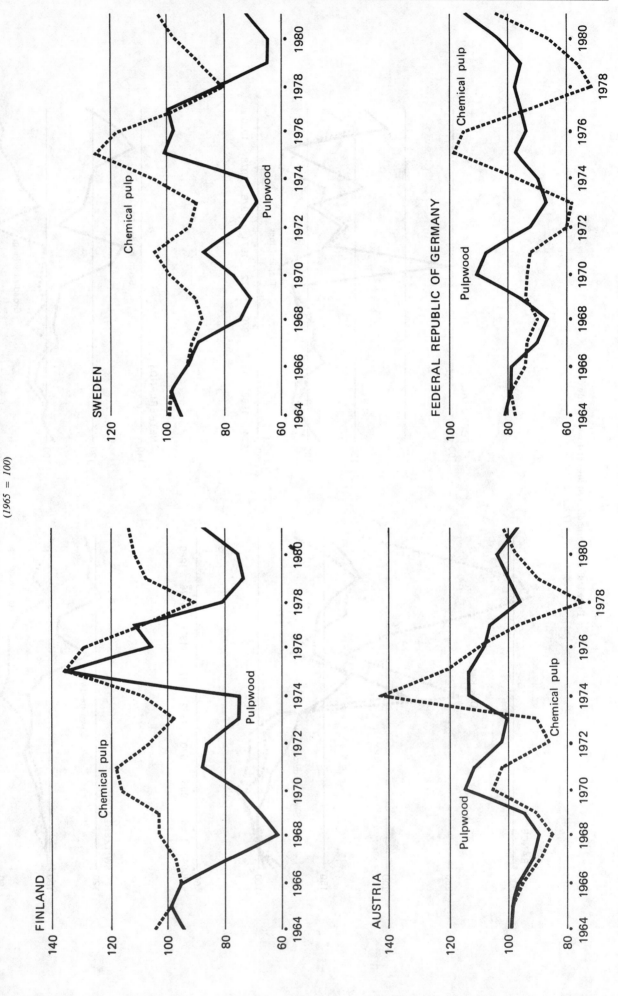

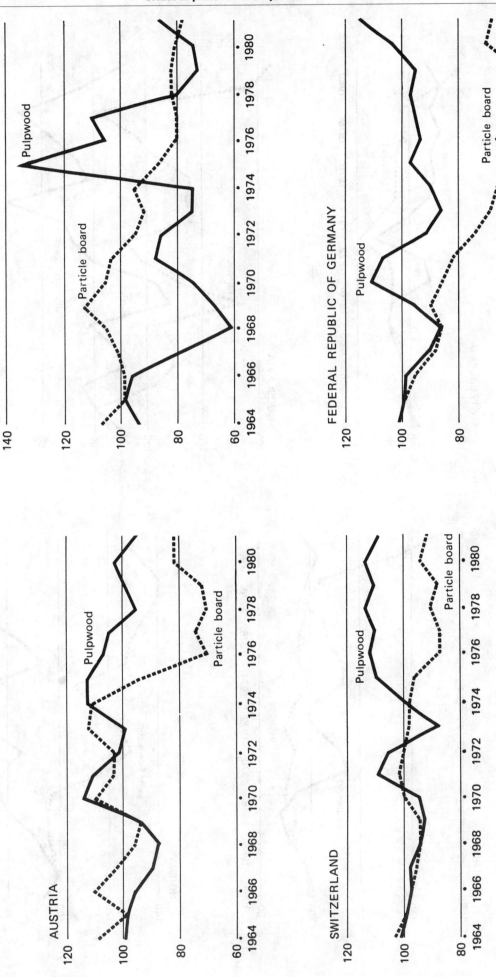

FIGURE 10.9

Price indices of pulpwood and particle board
(1965 = 100)

FIGURE 10.10

Price indices of pulpwood and fuelwood
(1965 = 100)

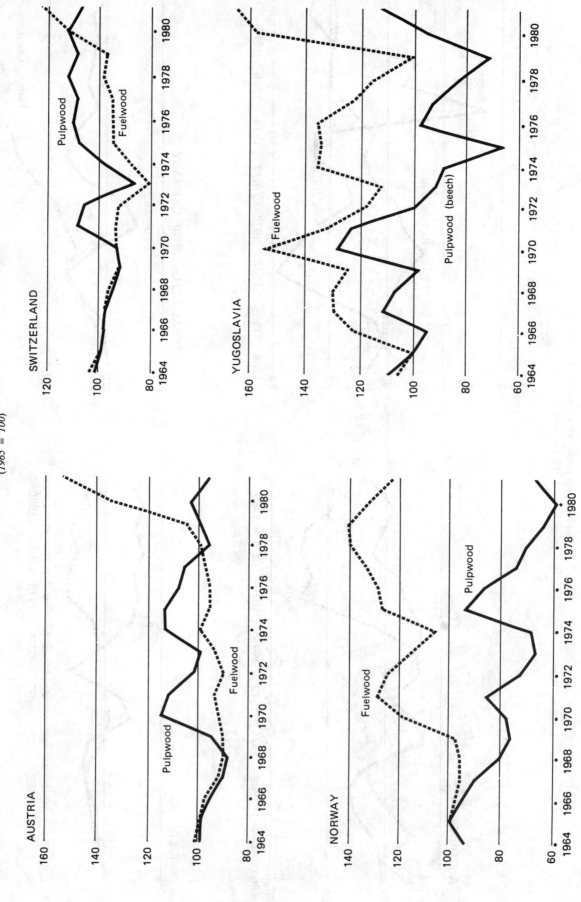

mately the same level, in real terms, as in the mid-1960s. In most cases, too, the broad trend for coniferous sawlogs was similar to that for sawn softwood; although the two series often diverge, sometimes for several years, they usually come together again, as in Sweden where sawnwood prices rose much faster than sawlog prices in 1973, but thereafter fell, while sawlog prices rose more or less steadily from 1974 to 1977. In Switzerland, however, it appears that coniferous sawlog prices have lost some ground, in the long term, relative to sawnwood prices.

The situation is more heterogeneous for *hardwood (nonconiferous) sawlogs* as fewer series were available and price movements in one country (France) were particularly violent. In France,[3] the price of beech and especially of oak logs rose very fast between 1971 and 1973 and again between 1975 and 1978 to reach high points (index 1965 = 100) for beech of over 210 in 1973 and of 315 for oak in 1978. The corresponding peaks for sawn hardwood (all species combined) were 170 and 180 in 1974 and 1979, i.e. one year after the peaks for logs. Both of these peaks in sawlog prices were, however, followed by corresponding declines to or below the level of sawnwood prices. This still represented, however, a strong gain in real terms (about 50%) over the period as a whole. Among the causes for this phenomenon may be mentioned strong demand for French sawn hardwood, growing scarcity of high quality logs (prices rose consistently faster for the better grades) and a feverishness which took hold of the market. It should also be mentioned that these very large price rises were undoubtedly a factor in the increased exports of oak sawnwood from the USA to Europe. In

the Netherlands, there was a steady decline in the price of hardwood logs, in real terms, after 1975, while the price of sawn hardwood was rising. In Portugal and Switzerland, the real price of hardwood logs remained roughly stable over the period, but in Switzerland, sawn hardwood prices rose faster while in Portugal the prices for sawn hardwood and tropical logs moved closely together.

In most countries, the real price of *round pulpwood* was stable, with a quite marked downward trend in the Netherlands and Norway. In Finland and Sweden, the two major European pulp exporters, round pulpwood prices fell in the late 1960s, recovered and then rose quite sharply in 1975 – one year after a very strong year in the pulp market but a year when that market was in severe depression. In the late 1970s in these two countries, pulpwood prices fell again, but recovered in the early 1980s. If pulpwood prices are compared to those of chemical pulp,[4] it will be seen that the two series have approximately the same trend, but that the fluctuations for chemical pulp are more marked. Particle board prices, however, fell compared to those of round pulpwood: this may be misleading, however, as the raw material for particle board is often not round pulpwood, but sawmill residues, whose price trends are not reflected in the series available.

In Austria, the Federal Republic of Germany and Switzerland, *fuelwood* prices remained stable, following a similar trend to pulpwood prices until 1979, when they started moving upward due to a sharp rise in fuelwood demand. The generally upward trend in Norwegian fuelwood prices since the mid-1960s is difficult to explain and may be due to confusion in specifications with birch pulpwood.

[3] Because the maximum levels achieved are so much higher than for other products, the scales of the graphs for French beech and oak logs are different from the uniform scales used for other countries and products.

[4] In the interests of brevity, the trends for other grades of pulp or of fibreboard are not examined in this section.

10.4 TRENDS IN PRICES IN EASTERN EUROPE

Until recent reforms, domestic prices in centrally planned economies, including those for forest products, were not the result of the interplay of supply and demand forces, but were, in most cases, set by central authorities as part of the planning process. They were adjusted from time to time to take account of changes in market conditions.

In the early 1980s, however, a number of eastern European countries introduced reforms which aimed, among other things, at increasing the flexibility and efficiency of the economies and their responsiveness to economic stimuli. One of the main measures in these reforms has been the introduction of new price formation systems whereby more autonomy is given to enterprises to form prices (for some products) by decentralized negotiation.

In Poland, for instance, whereas prices for some products are still centrally fixed, others are formed by nego-

tiation between producers and consumers and vary between enterprises and over time. In the forest products sector, sawlogs, sawnwood and pulp are in the first group, and wood-based panels, pulpwood, fuelwood, sawdust and pitprops in the second.

In Hungary, before 1980 forest products' prices were set by central authorities, similarly to other eastern European countries. At the beginning of 1980, however, as part of the economic reform, domestic prices were raised to the level of foreign market prices and were closely linked with them. In 1984, forest products prices were, with the exception of fuelwood, almost entirely liberalized. At present, producers are authorized to form their prices themselves, but they are obliged to adjust them to the prices on the main markets. Thus, for example sawn softwood and fibreboard prices are guided, on the basis of official price statistics, by the average of the import prices of the Federal Republic of Germany from the USSR, from Scandinavia and Austria, for oak sawnwood by the average of the producer prices in the Federal Republic of Germany and France, for standard particle board by the producer prices in the Federal Republic and so on. In this way domestic prices of forest products are closely connected with foreign markets: in the future, price trends in Hungary are expected to be the same or very similar to those in western European countries.

These reforms are relatively recent and it appears that price levels as such have not affected levels of consumption of forest products in the historical period under review. This is one reason why different methods have been employed to forecast eastern European consumption of sawnwood and panels (see chapter 11). These reforms may have a fundamental impact on patterns of consumption of forest products in eastern Europe in the long term, although it is at present too early to estimate, even in broad terms, what this effect would be.

From the data supplied for Poland (price indices with no correction for changes in general price levels), it is possible to draw some conclusions about price trends for forest products relative to each other. These data refer to centrally fixed prices i.e. before the price reforms mentioned above. Up to 1981, price movements were rather slow, although there were marked differences between products: prices of coniferous sawlogs and sawnwood, co-

niferous pulpwood and plywood rose by 60-100% over 16 years, while those for non-coniferous sawlogs and sawnwood, particle board and fibreboard rose by less than 40% over the same period, with price declines for sawn hardwood and insulating board in the early 1970s. Fuelwood, however, rose by nearly 200%. 1982 saw very sharp price increases in all parts of the economy, including for forest products. Most forest product prices at least doubled between 1981 and 1982, with especially sharp rises for coniferous sawlogs (+200%), sawn hardwood (+236%) and fuelwood (+213%).

In Hungary, too, where data are for annual prices (i.e. with no correction for changes in the general level of prices), there are marked differences between developments for different products. The change between 1959 and 1982 was around 200% for most types of roundwood (from +157% for first-class oak logs to +240% for coniferous logs, class II, poplar pulpwood and oak logs, class II). For products, however, there is a wider range: particle board prices only increased by 12% over the period and hardboard by 75% while sawn softwood prices (class II and class III) rose by 270-340% and sawn hardwood (oak) by no less than 500-640% (i.e. considerably more than for oak logs). It might be said therefore, that relative to raw material price, the prices of particle board and fibreboard fell, while those for sawnwood rose.

10.5   CONCLUSIONS

The following main conclusions may be drawn:
- No strong underlying long-term trends were identified of such a nature that they could be expected to continue into the future;
- Special and local factors play a major role in the formation of forest product prices;
- There have been strong fluctuations in these prices, notably in the mid-1970s;
- A few medium-term trends were identified, notably the decline in particle board prices.

What of the future outlook? There are two, rather different, questions which must be asked:
- What specific assumptions should be made about the future development of the price variable in the econometric modelling?
- Are there any indications, however uncertain, of future price trends for forest products, or of factors which might play a role in determining them?

For the econometric projections, which necessitate conservative assumptions, there does not therefore appear to be any satisfactory basis for making specific assumptions about trends in prices of forest products in future. As a first step, therefore, it will be assumed that forest products prices will stay roughly stable in real terms. It might be possible to draw conclusions about the outlook for forest

products prices at a later stage when the broader outlooks for supply and demand have been brought together.

It is perhaps also worth pointing out that, although no strong trends in prices have been identified, the inclusion of price series in the demand models has significantly improved these models. In particular, the end-use elasticities derived from the regressions are in some cases significantly changed by the inclusion of a price variable.

On a more general level, although it does not appear possible to make price forecasts, it is possible to identify certain major factors:
- Costs must of course be kept under control to avoid price rises which could have a potentially damaging effect on markets. Changes in cost of wood raw material, of processing and of transport, distribution and marketing must all be closely controlled. In a period when markets are expected to grow only slowly and where inflation is low, cost control becomes particularly important if competitivity is to be maintained;
- The experience of recent years has shown that the higher quality assortments in each product group have had less difficulty in maintaining or improving their prices than lower quality assortments. The price differential between lower quality and higher quality grades has widened for several products (e.g.

sawn softwood, plywood, particle board). This trend is expected to continue, particularly as supplies of large old-growth timber become rarer and more expensive;

- The presence or absence of competition in particular market sectors will also affect prices. For "bulk grades" (e.g. chemical pulp, newsprint, Kraftliner, some sawn softwood) differences between products are minimal, and markets are relatively transparent, so competition between producers is chiefly based on price and the lowest cost producers dominate the market. On the other hand when the product is well differentiated, often further processed in some way, so that it suits well the needs of the end-user and is exposed to little direct competition, a price pre-mium can be obtained. Sometimes competition is lacking on a particular market because potential competitors have not established an effective distribution and marketing system for that particular area.

It could be concluded from the above general observations that the forest industries can raise their profitability by concentrating on developing higher quality products intended for specific market sectors and controlling their costs. The markets for bulk grades will continue to be dominated by the lowest cost producers.

Finally the broad supply/demand balance will also play a role. This will be examined in greater detail in the concluding part of the study.

CHAPTER 11

Methods of projecting demand for forest products

11.1 INTRODUCTION

This chapter will examine, in a quantitative way, the relationship in the past between levels of consumption of forest products and a number of independent variables, notably GDP, residential investment, manufacturing production and furniture manufacture. These results, taken with the estimates of possible developments for the independent variables proposed in chapter 1, will provide the basis for the outlook in chapter 12. One section deals with the specific problems of forecasting demand in the centrally planned economies of Eastern Europe.

Although the quantitative, model-based approach is at the centre of the methodology used, it is by no means to be used in isolation. For some products (e.g. fuelwood, dissolving pulp, wood used in the rough) it did not appear feasible to prepare reliable models; for those products for which models were used, the results should be examined in the light of the experience of experts in the field, and the analysis in chapters 8 and 9.

When using the results of this section, it is important to bear in mind the nature of all projections based on models. They are *not* authoritative "forecasts" of the future, based on mysterious, apparently infallible, methodology ("crystal ball gazing"). The projections which will be presented in chapter 12 are the end of a chain of reasoning which depends on the following assumptions:

(a) That the models developed in this chapter accurately reflect the relationship in the past between consumption of forest products and the independent variables;

(b) That the assumptions (scenarios) for the independent variables (chapter 1) are reasonable;

(c) That· the relationships between independent variables and consumption observed in the past do not change in the future.

As ETTS IV is intended for readers from all parts of the forest sector, theoretical discussion and mathematical presentation have been kept to a minimum in the text and much of the supporting argument and data have been relegated to annex 11.1. For a clear discussion of the theoretical background to projecting forest products consumption, readers are referred to *Analysis of the demand for forest products; a preliminary survey of objectives, methods and data* by Anders Baudin and Lars Lundberg, prepared in the framework of the FAO Programme in Outlook Studies for Supply and Demand of Forest Products.

Two different model systems are presented in this chapter:

(a) "GDP elasticities" prepared on a uniform basis for nearly all countries in the world for three major product groups in the context of the above-mentioned FAO Programme in Outlook Studies;

(b) "End-use elasticities" applied to twelve major west European consuming countries for sawnwood and wood-based panels which is a more differentiated, data-intensive approach.

For Eastern Europe, a non-model-based approach is suggested, to be used in conjunction with the models.

These models are presented below. The secretariat considers it useful to compare and contrast the results, where appropriate, as there is no one "correct" model of demand – different models are appropriate for different objectives or different circumstances. However, the basic data used for the different models (on consumption of forest products, GDP, etc.) as well as the assumptions for the future (see chapter 1) are the same.

The work on the "End-use elasticities" model was carried out by Mr. G. Lundbäck, an expert put at the disposal of the secretariat by the Government of Sweden. The secretariat wishes to express its warm appreciation of Mr. Lundbäck's invaluable contribution.

11.2 "GDP ELASTICITIES" MODEL[1]

The model expresses consumption of forest product

---

[1] This section is a summary of the full documentation of the model, which may be found in the documents of the FAO Outlook Programme.

groups (sawnwood and sleepers, wood-based panels, paper and paperboard) as a function of GDP (in $US at 1980 prices and exchange rates) and prices. The series used for

TABLE 11.1

**Summary of results of the GDP elasticities model,
based on data for 60 consuming countries**

	Sawnwood and sleepers	Wood-based panels	News-print	Printing and writing paper	Other paper and paperboard
Income elasticity with country groups according to GDP *per caput* per year:					
$2 000	0.93	1.23	0.83	0.83	1.37
$2 000 - $4 000	0.75	0.68	0.68	0.91	0.70
$4 000 - $9 000	0.48	0.97	0.60	0.70	1.06
$9 000	0.56	1.16	0.59	0.74	1.19
Price elasticity	−0.02	−0.12	−0.09	−0.15	+0.02
Linear trend	−0.008	−0.036	+0.006	+0.018	−0.007
R²	0.96	0.94	0.95	0.97	0.97
Standard error of residuals	0.081	0.178	0.082	0.069	0.043

Note: Two time variable co-efficients were found, with the following values:

	1961-1973	1974-1984
Wood-based panels	+0.051	−0.036
Other paper and paperboard	+0.019	−0.007

the latter are for deflated border unit values, i.e. the unit value of exports or imports (whichever is larger for a particular country) divided by a general price index.

The model used aggregate data (i.e. not *per caput*), and a linear trend variable to account for substitution. It took account of the different levels of consumption in countries with similar incomes by using dummy variables by country. For the income elasticities, countries were divided into four groups according to their level of GDP *per caput* in 1981.

The model was estimated for the period 1961-1984 for the 60 largest consumers of sawnwood and panels in the world. The results of this estimation are presented in table 11.1 above.

The main results of the model estimation are:

– Income elasticities, as well as the price elasticity, were higher for panels than for sawnwood;

– The income elasticities estimated for sawnwood, newsprint and printing and writing paper decrease as the level of *per capita* income rises. For wood-based panels and other paper and paperboard, the income elasticity first fell then rose as *per caput* income increases across the sample of countries.

11.3 "END-USE ELASTICITIES" MODEL

The "GDP elasticities" model, described above, was designed, among other things, to exploit to the maximum an international data base (*FAO Yearbook of Forest Products*) in order to obtain results for a large number of countries all over the world. It is clear, however, that GDP is, at best, a rather general indication of activity in sectors consuming forest products and that border unit values do not always accurately reflect general price trends on domestic markets.

It was therefore decided to undertake a more detailed exercise, for a limited number of countries – those for which satisfactory data were available. These countries were further subdivided into groups according to their type of economy (both general and as concerns forest products). The twelve countries were:

Group I (West-Central Europe)
France
Federal Republic of Germany
Netherlands
Switzerland
United Kingdom

Group II (Southern Europe)
Italy
Portugal
Spain

Group III (Exporters)
Austria
Finland
Norway
Sweden

Between them, these countries account for most of west European consumption of forest products (for Eastern Europe, see section 11.5).

Annex 11.1 describes the model in full in technical terms. This section summarizes the main lines of annex 11.1, concentrating on the results rather than the methodology.

Two major end-use sectors were identified and series selected to represent them:

- *Construction*, which was represented by the series for investment in residential construction (source: OECD);

- *Furniture*, which was represented by the index for industrial production under ISIC heading 332 – Furniture manufacture (source: *UN Yearbook of Industrial Statistics*);

- In addition, *all other user sectors*, such as packaging, transport etc. were represented by the general index of manufacturing production (ISIC Group 3, also from the *UN Yearbook of Industrial Statistics*).

When these series were used together, technical problems arose due to multi-collinearity (similar movements in the different series, due to wider economic movements). They were, therefore, combined[2] for each product and country, into a product-specific index (referred to as ENDR), which represents the *activity in sectors using the forest product in question*. There is therefore one ENDR series for each product in each country (60 in all).

For prices, the series presented in chapter 10 were used (indices of real prices). The price series used had a distributed lag i.e. for 1966, the data used were those for 1964 (30%), 1965 (50%) and 1966 (20%). This appeared reasonable from the theoretical point of view, as it is likely that prices in earlier years affect consumption levels at least as much as prices in the same year. Results obtained using these series were significant in many cases whereas unlagged prices were never significant.

From a theoretical point of view it would have been desirable to compare prices of forest products with those of substitutes, but this proved impossible because of the difficulty of accurately identifying substitutes (often building *systems* substitute for other systems, not materials) and of finding satisfactory series for substitutes, with the limited resources available to the secretariat. The forest products prices were therefore compared, not to prices of substitutes, but to an indicator of general price movements, the producer price index.

As with the "GDP elasticities" model, a trend factor was also used for some products, notably particle board, to take into account the "autonomous growth rate".

Experiments were carried out with a number of different forms of equation. That chosen was the following:

$$AC_t = a \cdot ENDR_t^b \cdot WP_t^c e^{dT},$$

where:

AC_t = index of apparent consumption of product at time t;

$ENDR_t$ = user sector activity index (see text) at time t;

WP_t = real price of product, with weighted lag (see text) at time t;

T = trend variable;

a, b, c, d, e are constants.

The model was applied to the five individual products and to wood-based panels as a group (to minimize the effects of substitution between panels and the arrival of new composite panels). It was applied both to individual countries and to pooled data (time series and cross-sectional data), as the number of observations (16) was rather small if countries were taken individually. Dummy variables, by country, were used to take into account differences between countries.

The detailed results, with measures of the quality of the regression, are given in annex 11.1. The elasticities only are presented in tables 11.2 and 11.3. The elasticities shown for each country and product are those of the form of the basic equation which gave the best results.[3] Where an elasticity was not significant, nothing is shown.

These results appear satisfactory as, of the 84 relationships (between a country or country group and a product) modelled, 70 produced significant results. Where this was not the case, the reason may be sought among the following:

- Fluctuations in apparent consumption due to stock changes (notably in the Group III exporting countries);

- The product concerned is of minor importance, so that small changes assume large relative importance;

- In country by country analysis, there may be insufficient observations to establish a statistically significant relationship;

- The econometric approach chosen is not appropriate for the country or product in question e.g. where supply factors, not covered in this chapter, play a major role;

- Shortcomings in the data quality.

No significant results were found for Portugal, which is therefore omitted from the analysis.

For *sawn softwood*, in Group I there was a striking conformity between countries, both for end-use and price elasticities (around 1.0 and −0.8 respectively) except for France. When data for the Group were pooled, the end-use elasticity fell to 0.7, while the price elasticity was about −0.6. For Group II (South), the price elasticity was not significant, while the end-use elasticity was also around 0.7. In the case of Group III (exporters), no price elasticity and few of the end-use elasticities were significant, probably because of distortions due to stock changes. The pooled data give an end-use elasticity for the exporting

---

[2] Weighted according to the importance of each sector in the end-use pattern, drawing on some of the information presented in chapter 8.

[3] In order to correct problems linked to auto-correlation of residuals, the "Cochran-Orcutt" procedure was tested. In some cases, where this produced a significant improvement, these results were used.

TABLE 11.2

**End-use, price and trend elasticities for sawn softwood,
sawn hardwood and wood-based panels, total**

	Sawn softwood			Sawn hardwood			Wood-based panels (as a group)		
	ENDR	WP	T	ENDR	WP	T	ENDR	WP	T
Group I	0.708	−0.578	−	1.151	−0.463	−	1.024[a]	−	0.021[a]
France	0.513	−0.363	−	0.753	−0.279	−	0.848	−	0.017
Germany, Fed. Rep. of......	1.036	−0.795	−	0.974	−0.342[a]	−	1.055	−	
Netherlands	1.056	−0.836	−0.031	1.508		−	0.735	−	0.019
Switzerland	1.118	−	−	1.481[a]	−1.070[a]	−	1.610	−	0.015
United Kingdom	1.083	−0.613	−	1.211	−0.486	−	1.487	−	0.046
Group II	0.744	−	−	0.797	−0.889	−	0.532	−	0.038
Italy	0.870	−	−	0.949	−1.876	−	−	−	−
Spain	0.670	−	−	0.797	−	−	1.736	−	0.027
Group III	0.352	−	−	−	−	−	0.961[a]	−	0.014[a]
Austria	0.610	−	−	1.503	−	−	1.436	−	−
Finland	−	−	−	−	−	−	1.223	−	−
Norway	0.321	−	−	−	−	−	1.151	−	−
Sweden	−	−	−	−	−	−	1.441	−	0.032

For explanation of terms, see text. ENDR − Activity in user sectors. WP − Price (lagged). T − Trend variable.
[a] With Cochran-Orcutt correction for auto-correlation.

group as a whole of 0.35. Experience in Sweden at least appears to show that sawn softwood consumption is quite stable and inelastic.

In a number of countries, *sawn hardwood* may be considered a marginal product, which would explain the relatively high number of non-significant results. The wide variation in elasticities may be due to this factor as well as to the differences in consumption patterns. For Group I (West-Central), the pooled data give an end-use elasticity of 1.15 and a price elasticity of −0.46, and for Group II (South) 0.80 and −0.89 respectively. The price elasticity for Italy (−1.88), although significant, appears unreasonably high.

Similar problems arise, at the individual country level, and even for some country groups, for *plywood*. For instance, the elasticities both for end-use (2.77) and for price (−1.44) seem unreasonably high for the United Kingdom. It would appear more prudent, here as elsewhere, to use the pooled data, notably for Group I, with an end-use elasticity of 1.17 and a price elasticity of −0.24.

Perhaps the major problems in modelling demand for mechanical forest products arise for *particle board*, because of the very rapid rise in consumption over the estimation period.

What caused the fast growth for particle board? It could not have been the growth of the end-use sectors, for these

TABLE 11.3

End-use, price and trend elasticities for individual wood-based panels

	Plywood			Particle board			Fibreboard		
	ENDR	WP	T	ENDR	WP	T	ENDR	WP	T
Group I	1.170	−0.241	−	1.120[a]	−	0.006[a]	0.939[a]	−	−0.027[a]
France	0.707	−0.491	−0.021	1.137	−	0.041	0.771	−1.478	−0.100
Germany, Fed. Rep. of......	1.125	−0.485	−	0.951[a]	−	−0.009[a]	1.061	−0.906	−0.026
Netherlands	0.572	−0.871	0.066	1.722	−	−	−	−	−
Switzerland	1.667	−	−	1.420[a]	−	0.018[a]	1.595[a]	−	0.096[a]
United Kingdom	2.768	−1.442	−	2.275	−	0.112	1.778	−0.769	−
Group II	−	−	−	0.641[a]	−	0.045[a]	−	−	−
Italy	1.110	−	−	−	−	−	2.517	−	−
Spain	1.956	−0.880	−0.095	2.440	−	0.029	−	−	−
Group III	0.683	−0.952	−	0.876[a]	−	0.030[a]	0.844	−	−0.032
Austria	−	−	−	1.735	−	−	1.045	−	−0.044
Finland	−	−	−	2.041	−	−	0.841	−	−0.028
Norway	1.256	−0.630	−	0.889	−	0.040	1.506	−	−0.063
Sweden	0.955	−0.368	0.021	2.717	−	0.077	1.072	−	−0.028

For explanation of terms, see text. ENDR − Activity in user sectors. WP − Price (lagged). T − Trend variable.
[a] With Cochran-Orcutt correction for auto-correlation.

have grown relatively slowly during the period. Was it decreasing relative prices? This was probably the case to some extent but the most important explanation may have been rising productivity within the end-use sectors made possible by the use of particle board. It became very much simpler thanks to particle board to make a cupboard door, to make a floor etc. Particle board thus produced a remarkable technological change at the secondary or tertiary product stage. Therefore the development of consumption has been like that for a new product.

If, in such a situation, we use the equation

$$AC = a \cdot ENDR^b \cdot WP^c,$$

(i.e. without the time trend variable) we get an unreasonably high end-use elasticity. Furthermore, in many cases the price elasticity has a positive sign. If we add the trend-variable (T) to the equation, the parameter for the price variable turns out to be not significant and/or with a wrong sign. It appears, therefore, that the effect of changes in relative prices is very small compared to that caused by technological change. Therefore, we do not get acceptable results for the price variable. As a consequence, WP was excluded and the equation

$$AC = a \cdot ENDR^b \cdot e^{cT}$$

was tried.

This approach seems to work quite well. When estimating separate countries, however, the end-use elasticities are still too high in some cases. Sometimes this can be explained by a high correlation between $ENDR$ and T, which makes the estimation of the parameters uncertain.

However, low Durbin/Watson values indicated that there was a problem of auto-correlation. The Cochran-Orcutt correction, however, changed the elasticities considerably and gave a much better fit and better Durbin/Watson values. In the original estimation, the fitted values, plotted on a graph, often lay in a straight line and did not pick up variations in the original values. The corrected fitted values picked up these variations much better and therefore seemed to give a more accurate estimate of the systematic influence on the dependent variable. The Cochran-Orcutt values were therefore used in many cases for particle board.

One problem raised by this equation is the modifications which will occur when the period of technological change is over, as the trend variable will then be no longer relevant. It was unfortunately not possible to investigate this question in detail. Many experts, however, consider that the "market penetration" phase is now over for par-

ticle board, an opinion which seems to be confirmed by the behaviour of the trend variable in the "GDP elasticity" model. It would appear to be a reasonable assumption, therefore, that the trend variable should *not* be used in projecting future demand, only the end-use elasticity.

For *fibreboard*, similar problems arose as for plywood and sawn hardwood, which are all marginal products in many countries. For Groups I (West-Central) and III (Exporters), there is also a strong negative co-efficient for the trend variable which might be considered a mirror image of that for particle board. Thus the price elasticity was only significant for three countries (of which one result, for France, appeared unreasonable) and there is a negative trend factor for most countries. The situation is rather different for Group II (South) which saw an upward trend from very low levels: in general, the results for fibreboard in Group II were not satisfactory.

In view of the substitution between panels and the development of new composite panels which are difficult to classify in the accepted categories, consumption of wood-based panels as a group was also analysed. As the group is dominated by particle board, the same method was used as for particle board. The results (table 11.2) appear quite satisfactory, with an end-use elasticity around 1.0 (mostly between 0.7 and 1.4) and a marked positive co-efficient for the trend variable.

Figure 11.1 shows that the "end-use elasticities" model fits recorded trends quite well.

It is clear from this discussion that the situation is complex and that there are wide variations between countries which are partly real and partly arise out of the methods employed. What equations should be used for the projections in chapter 12? It was decided to follow these broad principles:

(a) Although it is theoretically desirable to use elasticities for individual countries, the number of observations is rather small. Therefore, the basic method will be to employ the elasticities arising from the pooled data by group;

(a) When the results for a group are not significant, the elasticities for Group I should be used;

(c) Where possible, projections should also be made on an individual country basis as a check on those derived using country group elasticities;

(d) Consumption of wood-based panels (total) should be calculated as a check on the sum of the projections for the individual panels.

11.4 PROJECTING CONSUMPTION IN EASTERN EUROPE*

11.4.1 Introduction

The preceding sections have presented models based on

* This section has been prepared with the invaluable help of Dr. A. Madas (Hungary).

econometric methods, including notably an assessment of factors affecting demand and a price element. In the centrally planned economies of Eastern Europe, however, supply factors play a major role in determining consumption levels, and the price mechanism has functioned in a

FIGURE 11.1

Fit of the end-use elasticities model with real trends
(1975 (actual value) = 100)
GROUP 1

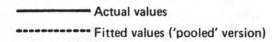

——— Actual values

■■■■■■■■■■ Fitted values ('pooled' version)

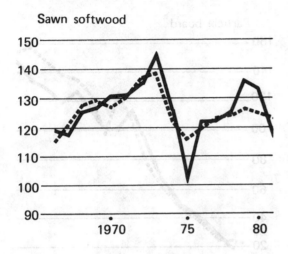

Sawn softwood

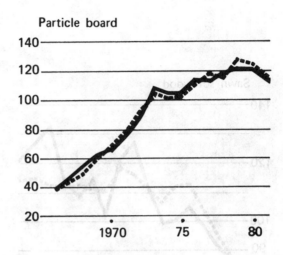

Particle board

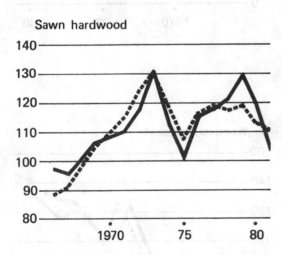

Sawn hardwood

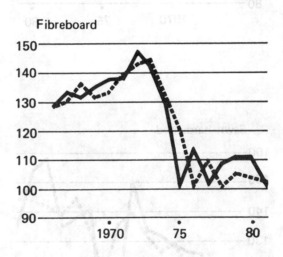

Fibreboard

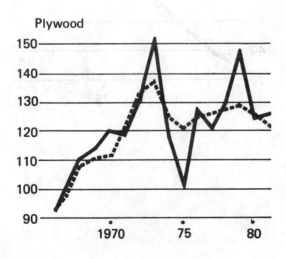

Plywood

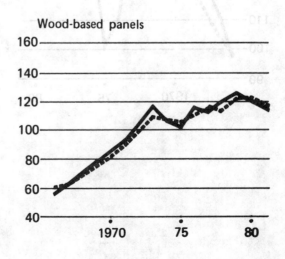

Wood-based panels

FIGURE 11.1 (*continued*)

GROUP 2

—————— Actual value

•••••••••••• Fitted value ('Pooled' version)

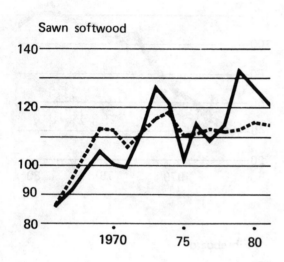

Sawn softwood

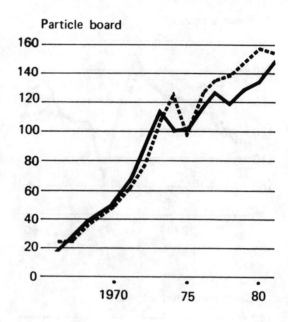

Particle board

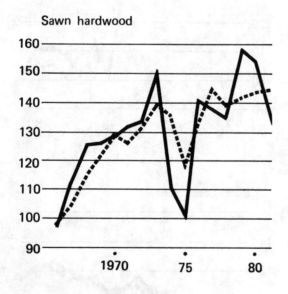

Sawn hardwood

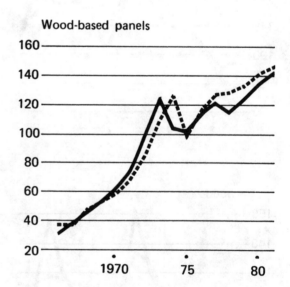

Wood-based panels

FIGURE 11.1 (*concluded*)

GROUP 3

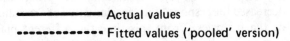
—————— Actual values
••••••••••••••• Fitted values ('pooled' version)

Sawn softwood

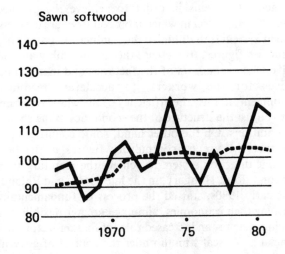

Fibreboard

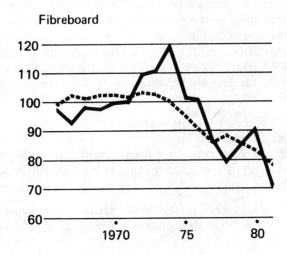

Plywood

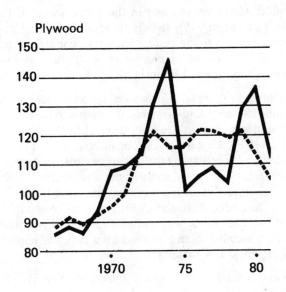

Wood-based panels

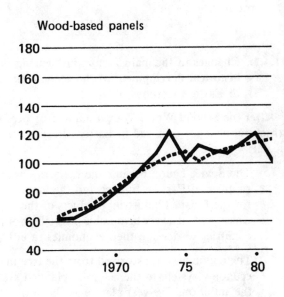

Particle board

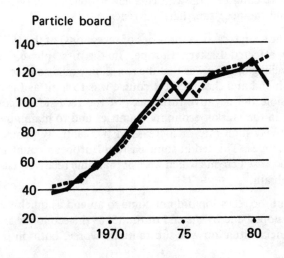

very different way from the systems in market economies over almost all the historical period considered. There is therefore some uncertainty about the theoretical validity of applying this type of econometric forecasting to Eastern Europe. Other econometric models, taking account of the specific conditions of this region might be developed, but this was not possible in the context of ETTS IV.

The problem has been increased by the fact that the five-year plans for the period 1986-1990 were not available at the time of drafting. The eastern European countries have now issued their long-term plans covering the period 1985-2000 (at least in draft form), but these plans could unfortunately not be taken into account when finalizing the study.

For these reasons it was necessary to analyse for a longer period and to a broader extent the factors influencing timber consumption in Eastern Europe. During the course of analysis special attention was paid to the following factors:

- Changes in the main factors influencing the historical development of the economy of Eastern Europe;

- The long-term time trends for the main forest products.

11.4.2 Changes in the main factors influencing the historical development of the economies of Eastern Europe

After the Second World War, a number of profound structural changes occurred in Eastern Europe. In particular:

(a) The Soviet Union became the main raw material supplier of Eastern Europe and the main market for the finished industrial products of the region, which created a very favourable possibility for these countries to develop their economies rapidly;

(b) The region was transformed from the free market economy system to the centrally planned system: the ambitious five-year plans and obligatory planning figures created the necessary basis in the post-war reconstruction period and in the extensive phase of economic development for rapid economic development. Industrial production doubled or even trebled in a decade in the eastern European countries.

At the same time, wood consumption rose; the Soviet Union supplied the two wood importing countries of the region (the German Democratic Republic and Hungary) with the necessary material; the other countries also increased their production and consumption.

In the 1960s, the centrally planned economies approached or attained the phase of transition from the extensive phase of economic development to the intensive phase; on the one hand, some of these countries found

that the methods used with success in the reconstruction period after the war and in the extensive phase had become more and more inefficient. The enormous number of compulsory planning figures led to some extent to an increasing centralization of decision-making processes and decreased the responsibility and initiative of the managers of the factories. From this time the different planned economies developed and improved their economic management systems in order to achieve the planned targets. The direction in which most countries are developing their system is to diminish the number of compulsory production figures, to increase the responsibility and initiative of the managers of the factories and material incentives for the workers, to accelerate technical development, to raise the efficiency of the whole economy and to adjust the structure of the economies to the changing conditions. On the other hand, some countries consider that the use of compulsory figures is the best economic direction system to achieve the plan targets in their conditions. Hungary, in the late 1960s, and Poland, in the early 1980s, started the process of fundamentally reforming their economies, aiming at greater flexibility of planning systems, an increased role for market factors and financial and fiscal stimuli under the control of governmental authorities.

Due to structural changes in the forest sector of the USSR (see chapter 19), the continuous increase of raw material imports to the eastern European countries from the Soviet Union has come to an end; this has also heavily affected wood imports.

According to official information, the Soviet Union will maintain its energy and raw material exports to the eastern European countries at the level reached in the first half of the 1980s; the eastern European countries have the possibility of contributing their own resources by developing different sources of energy and raw material in the USSR and thereby raising their imports from the Soviet Union. In addition, in the 1970s a system of modified world market prices was introduced into the CMEA system, which led to an increase in the prices of imports of energy and raw materials.

In this situation the eastern European countries increased their imports from the developing and developed western countries; this caused a new problem, that of providing the necessary hard currency.

The deep recession (or series of recessions) of the 1970s also affected Eastern Europe. In the first phase, the prevailing opinion was that the recession would soon come to an end and that the best course was to limit as far as possible the negative influences of the recession occurring in the market economy countries and to maintain a relatively high rate of increase in the centrally planned economies. This led in some eastern European countries to increased credits which were at that time relatively easy to obtain.

But the recession did not come to an end as quickly as was hoped and by the end of the 1970s the economic and financial situation was substantially worse, both on the world scale and consequently in the eastern European countries, which were faced with the following problems:

- Production costs increased substantially in the eastern European countries;
- Towards the end of the 1970s, the western financial organizations changed their policy and practically stopped new credits on the previous scale to the developing and eastern European countries. This meant that some of the eastern European countries had financial problems arising from the fact that instead of obtaining additional resources for their development they had to pay back credits and interests;
- The eastern European countries are now faced unavoidably with a basic reconstruction of the structure of their economies aimed at accelerating technical development, raising efficiency and adapting the economies to the new conditions. For this, capital investment is necessary, which is now the bottleneck;
- The increase in population slowed or stopped by the end of the seventies. This also influences the growth of GDP (labour constraints).

All the above-mentioned factors tend to keep down the rate of growth of the economy and of consumption. They thus affect wood consumption which will also increase only very slowly.

Poland's share in eastern European population and NMP is approximately one third: therefore, events in Poland strongly influence the figures of the total eastern European economic development and consumption. The recession in Poland was much deeper (and probably will be longer) than that of the other eastern European countries. For this reason, the situation and the future outlook must be studied separately for Poland and for the other eastern European countries.

It may be expected that the Polish economy will be constrained by the factors mentioned above but, in addition, that it will go through a period of recovery from recession, possibly along the lines suggested by Jánossy for a post-war reconstruction period (figure 11.2), which may also apply to recoveries from other economic disturbances. The main feature of this is a rather rapid recovery from recession until the underlying pre-recession trend line is reached, when growth resumes its "normal" course.

The scenarios for economic growth prepared by FAO, essentially on the basis of past trends, foresee an average annual growth of 3.7%-4.7% for Eastern Europe as a whole between 1980 and 2000, compared to 5.2% between 1960 and 1980. In the light of the considerations set out above, even the lower forecast seems rather too high. A figure of 3% a year for the region as a whole might be considered more realistic.

11.4.3 Long-term trends in consumption of forest products

In addition to the econometric methods used elsewhere in this study, it is also useful to examine long-term trends in consumption and use them as a basis for projection. This method is based on the observation that the movement of consumption over time very often follows a remarkably stable trend line. This line often takes the characteristic form of a logistic curve. For this method it is, of course, important to use very long time series, as the use of shorter series may be seriously misleading.

This method is especially useful when used in conjunction with other methods, notably as a check on their "reasonableness": a deviation from a very long-term trend line needs to be justified.

Graphs have been prepared for the period 1913-1982 for eastern European consumption of sawnwood, wood-based panels and paper and paperboard, as well as of industrial wood as a whole.[4] These graphs, in figure 11.3, show the characteristic logistic curve, with a marked slowing down in growth in the mid-1970s.

The lines have been projected, with two variants, for the upper and lower end of a range, to give the following projections (table 11.4). For comparison, the corresponding projections, obtained by the "GDP elasticities" model, are also shown.

It appears that the results obtained by analysis of the long-term trends are in every case significantly lower than those obtained by the "GDP elasticities" model, which was designed to be applied to all countries in the world.

---

[4] Data before 1948 are estimated on the basis of overall European trends, as most eastern European countries did not exist with their present boundaries at that time.

TABLE 11.4

Eastern Europe: comparisons of projections to 2000

			2000	
	Unit	1979-81	Graphic projection	"GDP [a] elasticities"
Sawnwood	million m³	21.56	23.0 − 25.0	34.3
Wood-based panels	million m³	6.88	10.0 − 11.0	24.4
Paper and paperboard.	million m.t.	5.67	7.0 − 8.0	12.8
Total industrial wood	million m³EQ	81.1	95.0 − 100.0	..

[a] Base scenario.

FIGURE 11.2

Typical scheme of the development of output during a post-war reconstruction period
(*According to Jánossy*)

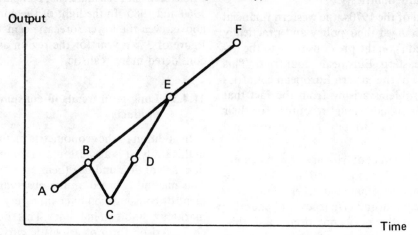

AF = long-term trend of output
AB = prewar development
BC = wartime recession
CE = reconstruction period (prewar level) reached again at D,
    level the level of long-term trend attained at E)
EF = development of output after the reconstruction period

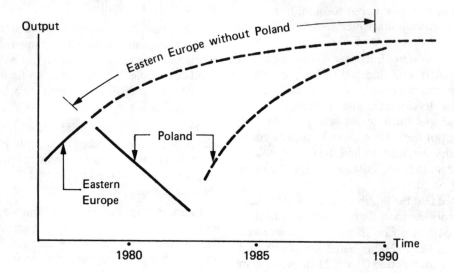

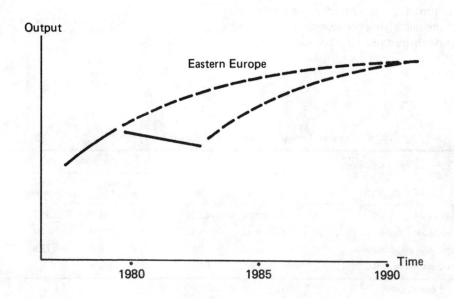

FIGURE 11.3

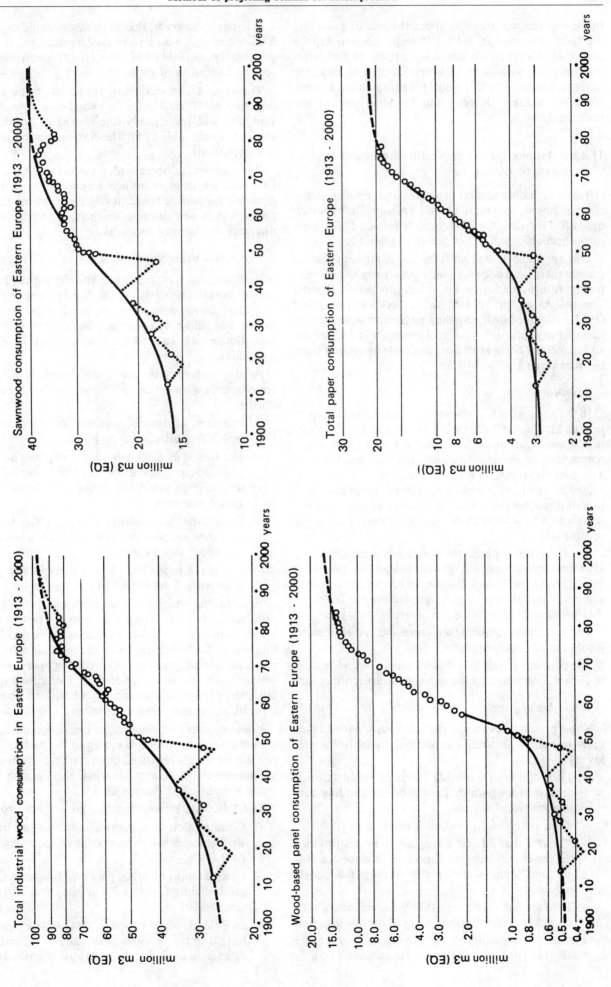

Total industrial wood consumption in Eastern Europe (1913 - 2000)

Sawnwood consumption of Eastern Europe (1913 - 2000)

Wood-based panel consumption of Eastern Europe (1913 - 2000)

Total paper consumption of Eastern Europe (1913 - 2000)

In view of the uncertainties about the role of price and the GDP scenarios in the GDP elasticities model, the results of this approach would not appear to reflect realistically the outlook for Eastern Europe. In fact, the increases projected by this model for consumption in some countries, notably Bulgaria and Romania, appear unrealistically high.

11.4.4 Outlook for the major forest products in eastern Europe

This concluding section presents, on the basis of the analysis above, assumptions and assessments by which quantified forecasts for 2000 could be prepared. The forecasts themselves will be set out in chapter 12.

The approach is not uniform, as it appears that at present there is no one really satisfactory method for estimating future consumption of forest products in Eastern Europe. As an aid to analysis, indices were prepared (1964 = 100) for NMP, dwelling construction, consumption of forest products and total removals in each of the six countries. (These graphs are available on request from the secretariat.)

Sawnwood

In the past, growth in sawnwood consumption has been slow and has borne little relation to developments in NMP. A stagnation, even decline, since the mid-1970s is apparent in some countries (Bulgaria, Hungary and Poland). There seems no reason why consumption should start to grow fast again, as the likely continued rise in construction of dwellings and secondary residences may be counterbalanced by economies in use of sawnwood and supply constraints.

It is therefore assumed that average annual growth in sawnwood consumption is between 0% and 0.75% a year from 1979-81. In Poland, however, where sawnwood consumption has fallen since 1975, growth may be faster, so that previous peak levels are recovered.

The breakdown between coniferous and non-coniferous would remain unchanged.

The estimates obtained in this way are in fact quite close to the total obtained by the graphic estimation method.

Wood-based panels

These products, especially particle board, have advantages in eastern European conditions, notably the following:

- Their ability to substitute for other products, such as sawnwood, which have more demanding raw material requirements;
- Their potential for technical development;
- The fact that all east European countries produce the panels, and most of them also produce machinery and equipment for the wood-based panels industry;
- The fact that they are relatively undemanding with respect to the species and quality of the raw material.

As a result, this consumption has grown at around the rate of NMP (for NMP elasticities see annex table 11.5).

It appears, however, that the phase of rapid expansion is also coming to an end in Eastern Europe. It is also possible that availability of capital may prove a bottleneck for the expansion of capacity.

The estimates set out below are based on the assumption that NMP elasticities for wood-based panels consumption will fall steadily (the elasticities assumed are shown in annex table 11.5). The NMP scenario used is the low hypothesis.

It is assumed, as fibreboard, plywood and veneer sheets have shown little growth since the mid-1970s, that all the growth is for particle board. It is likely, however, that with the arrival of new composite panels, the traditional distinctions will become less relevant.

Paper and paperboard

Both the long-term trend line and the indices for individual countries show that in the mid-1970s consumption of paper and paperboard in Eastern Europe grew more slowly than before, or not at all (Czechoslovakia, the German Democratic Republic, Poland, where it fell, and Romania).

Among factors constraining the growth of paper and paperboard consumption in Eastern Europe are the following:

- Problems of domestic availability of raw material (species, quality, quantity);
- The size of modern pulp mills and paper mills is so great that investment costs and raw material supply requirements sometimes exceed the potential of individual countries;
- Possibilities for increasing imports from western countries are also very limited, for the reasons set out above (hard currency);
- It may not be realistic to expect any further growth in imports from the USSR.

In addition, it may be mentioned that the role of the print media is very different in centrally planned economies from that in market economies. In particular, the link, through advertising, between economic activity and consumption of printing and writing paper (including newsprint) is not nearly so strong. This situation may, however, change to a certain extent with the economic reforms at present being undertaken.

It also appears that whilst paper consumption was fairly closely related to NMP growth up to the mid-1970s, the two growth curves separated quite markedly from then on in some countries (Czechoslovakia, the German Democratic Republic and Romania).

The following assumptions for growth are proposed:

- Consumption of newsprint and other printing and writing paper will remain stable or grow very slowly (+0.5% a year);
- Consumption of other paper and paperboard will grow slightly faster (1-2% a year), but still slower than NMP;
- In Poland, consumption in 1990 will be equivalent to that in the previous peak year, 1977, and thereafter grow at similar rates to those described above.

CHAPTER 12

Outlook for consumption of forest products to the year 2000

12.1 INTRODUCTION

Earlier chapters have presented trends in consumption, end-uses and prices of forest products, as well as different demand models developed in conjunction with the study.

It is probably not wise, in the present state of knowledge, to rely exclusively on one demand model, as each model has different advantages and disadvantages. For instance, the GDP elasticities model may be applied to all countries in the world, but with a rather limited range of independent variables, while the end-use elasticities model is able to analyse influences in greater detail, but can only be applied to a limited number of countries, because of data limitations. In addition, it is necessary to examine the projections in the light of more qualitative analysis, such as that in chapters 8 and 9.

This chapter will present the projections derived from the various models separately, as well as an examination of the particular situation for Eastern Europe, and a brief review of the outlook for energy wood (a recapitulation of chapters 18 and 19) and for other products. It will present a unified set of consumption scenarios (low and high variants) for use in the synthesis section of the study, analyse the sensitivity of some of the projections to changes in assumptions for independent variables, and compare the projections to each other and to those of ETTS III.

The assumptions about independent variables are consistent for all the models and are those presented in chapter 1. It will be recalled that in addition to the scenarios presented for GDP, residential investment and manufacturing production, it is assumed that prices of forest products remain constant in real terms, for the reasons presented in chapter 10.

12.2 PROJECTIONS BY THE GDP ELASTICITIES MODEL

The methodology of this model has been described in section 11.2. The base scenario of the model, assuming low GDP growth, is set out in tables 12.1 and 12.2, with country detail in annex tables 12.1 to 12.6.

TABLE 12.1

Projections, by the GDP elasticities model, for sawnwood and panels

	Volume (million m³)				Average annual percentage change
	1979-81	1990	1995	2000	
Sawnwood, total					
Nordic countries	10.75	11.00	12.11	13.33	+ 1.1
EEC(9)	50.46	49.90	53.19	56.86	+ 0.6
Central Europe	5.04	5.54	5.99	6.48	+ 1.3
Southern Europe	14.32	15.41	17.58	20.12	+ 1.7
Eastern Europe	21.75	26.08	29.69	34.30	+ 2.3
Europe	102.32	107.93	118.56	131.09	+ 1.3
Wood-based panels, total					
Nordic countries	2.57	3.30	4.03	4.92	+ 3.3
EEC(9)	20.78	25.35	30.77	37.45	+ 3.0
Central Europe	1.35	1.72	2.06	2.46	+ 3.0
Southern Europe	4.02	6.07	8.44	11.83	+ 5.5
Eastern Europe	6.88	12.16	17.05	24.41	+ 6.5
Europe	35.60	48.60	62.35	81.07	+ 4.2

Note: Projections by country are presented in annex tables 12.1 and 12.2.

TABLE 12.2

Europe: projections by the GDP elasticities model, for paper and paperboard

	Quantity (million m.t.)				Average annual percentage change	Quantity (million m.t.)				Average annual percentage change
	1979-81	1990	1995	2000		1979-81	1990	1995	2000	
	Newsprint					Printing and writing				
Nordic countries	0.52	0.68	0.79	0.93	+3.0	0.98	1.47	1.91	2.46	+4.7
EEC(9)	4.58	5.25	5.95	6.73	+1.9	10.05	14.09	17.67	22.23	+4.1
Central Europe	0.34	0.43	0.49	0.56	+2.5	0.47	0.61	0.77	0.96	+3.6
Southern Europe	0.50	0.71	0.84	1.03	+3.7	1.31	1.87	2.39	3.09	+4.4
Eastern Europe	0.51	0.67	0.81	0.98	+3.3	0.85	1.31	1.72	2.23	+4.9
Europe	6.47	7.74	8.88	10.23	+2.3	13.66	19.35	24.46	30.97	+4.2
	Other paper and paperboard					Total paper and paperboard				
Nordic countries	1.84	1.92	2.04	2.19	+0.9	3.34	4.07	4.74	5.58	+2.6
EEC(9)	18.11	19.12	20.83	22.75	+1.1	32.75	38.45	44.43	51.71	+2.3
Central Europe	1.12	1.18	1.25	1.32	+0.8	1.94	2.24	2.51	2.84	+1.9
Southern Europe	3.63	4.61	5.68	7.06	+3.4	5.44	7.18	8.93	11.16	+3.7
Eastern Europe	4.35	6.05	7.55	9.58	+4.0	5.71	8.05	10.08	12.78	+4.1
Europe	29.05	32.88	37.35	42.90	+2.0	49.18	59.99	70.69	84.07	+2.7

Note: Details by country are provided in annex tables 12.3, 12.4, 12.5 and 12.6.

The projections were obtained in two steps:
- 5-year average *growth rates* in consumption were calculated for each country from the GDP elasticities, adjusting for the time trend;
- *Levels* of consumption were calculated by applying these growth rates to the historical data. (For the GDP elasticities model, the base period is a 5-year average, 1980-84. Nevertheless, in the tables the projections are compared with data for 1979-81 as this is the base period for the removals forecasts and the material balance.)

These projections show annual average growth rates for Europe as a whole between 1% and over 4%. By product, the slowest growth is foreseen to be for sawnwood (1.3%), with an intermediate rate of growth for other paper and paperboard (2.0%) and newsprint (2.3%), but 4.2% for printing and writing paper and 4.2% for wood-based panels. By region, in general slower growth is foreseen for the EEC(9) and Central Europe, with faster growth for Eastern and Southern Europe, (especially Bulgaria, Romania, Turkey and Yugoslavia). This is due in large part to the GDP growth hypotheses used.

12.3 PROJECTIONS BY THE END-USE ELASTICITIES MODEL

The methodology of this model has been described in section 11.3. It will be recalled that the model related consumption of sawnwood and panels to an index of activity in user sectors (construction, furniture, other), lagged prices and, in some cases, a time trend (for the autonomous growth rate). For the projections in this chapter, the assumptions for the independent variables are as follows:

- The assumptions for construction and manufacturing are those described in chapter 1;

- It is assumed that prices will remain stable in real terms, for the reasons explained in chapter 10;

- It is assumed that the phase of expansion due to technical trends (market penetration) is over.

In essence, therefore, it is assumed that growth in consumption will be due only to increases in activity in user industries.

The model was developed using data for eleven west

European countries.[1] Elasticities with respect to end-use activity, price and a trend variable were obtained, for groups of countries and for individual countries (annex table 11.2). For a number of reasons the elasticities by country group were considered preferable, although the elasticities by country, when available, were considered useful as a check on the results obtained by the former method.

It proved possible to generate hypotheses about the independent variables (residential investments, furniture and total manufacturing production) for west European countries other than those covered by the original research. It was therefore possible, using the country group elasticities, to prepare projections for nearly all west European countries (18 countries in all), broken down into the following groups (which do not coincide with groupings used elsewhere in this study):

---

[1] Austria; Finland; France; Germany, Federal Republic of; Italy; Netherlands; Norway; Spain; Sweden; Switzerland; United Kingdom.

Group I (west-central)	Group II (south)	Group III (exporters)
Belgium-Luxembourg	Greece	Austria
Denmark	Italy	Finland
France	Portugal	Norway
Germany, Fed. Rep. of	Spain	Sweden
Ireland	Turkey	
Netherlands	Yugoslavia	
Switzerland		
United Kingdom		

The projections were derived at the level of individual products: sawn softwood, sawn hardwood, plywood, particle board and fibreboard. In addition, as a check, projections were prepared for consumption of wood-based panels as a whole (except veneer sheets). It is thus possible to compare projections using elasticities for panels as a whole with the sum of projections for individual panels.

In this section, sleepers and veneer sheets are not included because of the difficulties of modelling consumption of small quantities. Estimates for these products will of course be included in the consolidated forecasts (section 12.10).

The base period for any economic projection should satisfy the following criteria:

- Be as recent as possible, to take into account the latest developments;
- To minimize the effects of market cycles (for long-term projections);
- To be within the estimation period for the regressions.

For the end-use elasticity model, the base period chosen was the 5-year average for 1979-83 (the last year of the estimation period was 1981, as data were not available for some of the independent variables after that date). For several products consumption in the base period 1979-83 was lower than for the 3-year average 1979-81, used as a base period for other parts of the study. As a consequence, those products for which slow growth is

foreseen on the 1979-83 base period, will show negative growth when compared to 1979-81 data.

Tables 12.3 and 12.4 present the projections by the end-use elasticities model, with country detail in annex tables 12.7 to 12.12.

The following broad comments may be made:

- All the growth rates for wood-based panels, while still above those for sawnwood, are significantly below the rates foreseen by ETTS III, being mostly below 3% p.a.;
- Projected growth rates are higher for sawn hardwood than for sawn softwood and for particle board than for the other two panels. Sawn softwood and fibreboard show the slowest projected rates of growth;
- By country group, projected growth rates are highest in Group II (south), followed by Group I (west-central). The lowest growth rates for domestic consumption are foreseen in the exporting countries.

When projections are made using national elasticities (for those countries for which it was possible to derive such elasticities), these projections are generally higher than those calculated using country group elasticities. The major exceptions to this are France and, to a lesser extent, the Federal Republic of Germany. The differences are greater for panels, especially plywood. The reasons for this situation are not clear. As the number of observations in the series used to obtain the national elasticities was rather small — sixteen — it was decided to use the projections obtained with the country group elasticities. With a few exceptions however (e.g. particle board in Spain, Italy and Austria, fibreboard in Italy), the differences between the two types of projection are not so great as radically to distort the results.

12.4 COMPARISON OF PROJECTIONS BY THE GDP ELASTICITIES AND END-USE ELASTICITIES MODELS

The two previous sections have presented projections obtained using two different models. This section intends briefly to compare these results (where the same products i.e. sawnwood and panels, are covered) and to discuss briefly the reasons for the differences between them.

12.4.1 Theoretical background

Sawnwood and wood-based panels are used as inputs in a number of production activities — mainly in residential construction but also in furniture production, other joinery production, transport and distribution etc. Therefore, to explain the demand of these products, the following three types of variables were used in both models:

- A variable, which describes the end-use activities;
- The price for the product in question, in relation to substitutes;

- A trend factor, which is meant to represent shifts in demand depending on, e.g. technical changes within the end-use sector.

The main difference between the two models is in the way end-use activities are measured.

It is desirable to use a variable which describes the development within the end-use activities as precisely as possible. The main reason why such a rough variable as GDP was chosen is the fact that it was the only activity measure which was available for all countries and for which projections had been made up to year 2000.

It was possible to get a rough description of the end-use pattern for sawnwood and panels in western Europe (chapter 8). It also proved possible to find statistical measures which describe the activity in these sectors. Therefore in the end-use elasticity model, it has been poss-

TABLE 12.3

Western Europe: projections of consumption of sawnwood and panels, by the end-use elasticity model

(Million cubic metres)

	Actual 1979-83	Low hypothesis				High hypothesis			
		1990	1995	2000	Average annual percentage change	1990	1995	2000	Average annual percentage change
Sawn softwood									
I. West-central	33.73	34.79	36.13	37.45	+0.6	36.66	40.38	44.01	+1.4
II. South	14.93	16.33	17.91	19.43	+1.4	17.91	21.21	24.94	+2.7
III. Exporters	12.07	12.38	12.68	12.96	+0.4	12.87	13.63	14.32	+0.9
Total	60.73	63.50	66.72	69.85	+0.7	67.44	75.23	83.27	+1.7
Sawn hardwood									
I. West-central	9.22	10.32	11.82	13.34	+2.0	10.98	13.57	16.30	+3.0
II. South	6.18	7.39	8.76	10.07	+2.6	7.86	9.86	12.20	+3.6
III. Exporters	0.76	0.91	1.06	1.23	+2.5	1.00	1.28	1.57	+3.9
Total	16.16	18.62	21.64	24.63	+2.2	19.83	24.71	30.07	+3.3
Plywood									
I. West-central	3.26	3.46	3.70	3.94	+1.0	3.75	4.39	5.06	+2.3
II. South	0.72	0.87	1.04	1.23	+2.8	0.95	1.25	1.62	+4.3
III. Exporters	0.35	0.37	0.39	0.41	+0.8	0.39	0.44	0.49	+1.8
Total	4.33	4.69	5.13	5.58	+1.3	5.09	6.09	7.16	+2.7
Particle board									
I. West-central	12.42	13.55	14.99	16.44	+1.5	14.52	17.38	20.35	+2.6
II. South	4.76	5.50	6.32	7.07	+2.1	5.75	6.89	8.10	+2.8
III. Exporters	2.09	2.29	2.52	2.73	+1.4	2.49	2.94	3.37	+2.6
Total	19.26	21.35	23.82	26.24	+1.6	22.76	27.21	31.82	+2.7
Fibreboard									
I. West-central	1.37	1.44	1.52	1.60	+0.8	1.53	1.74	1.95	+1.9
II. South	0.70	0.78	0.88	0.97	+1.8	0.86	1.06	1.27	+3.2
III. Exporters	0.67	0.71	0.76	0.80	+1.0	0.78	0.89	1.01	+2.2
Total	2.74	2.93	3.15	3.37	+1.1	3.17	3.69	4.23	+2.3

TABLE 12.4

Western Europe: projections of consumption of sawnwood and wood-based panels, by the end-use elasticity model

(Million cubic metres)

	Actual 1979-83	Low hypothesis				High hypothesis			
		1990	1995	2000	Average annual percentage change	1990	1995	2000	Average annual percentage change
SAWNWOOD[a]									
I. West-central	42.95	45.12	47.95	50.79	+0.9	47.63	53.95	60.30	+1.8
II. South	21.11	23.72	26.66	29.50	+1.8	25.78	31.08	37.13	+3.0
III. Exporters	12.83	13.28	13.74	14.19	+0.5	13.86	14.91	15.90	+1.1
Total	76.89	82.12	88.35	94.48	+1.1	87.27	99.94	113.34	+2.1
WOOD-BASED PANELS, TOTAL[b]									
A. *Projected as a group*									
I. West-central	17.05	18.32	19.92	21.52	+1.2	19.56	22.90	26.32	+2.3
II. South	6.18	6.91	7.69	8.40	+1.6	7.18	8.30	9.44	+2.3
III. Exporters	3.10	3.41	3.73	4.06	+1.4	3.74	4.45	5.15	+2.7
Total	26.33	28.64	31.34	33.98	+1.4	30.48	35.65	40.91	+2.4
B. *Sum of separate projections*									
I. West-central	17.05	18.44	20.21	21.98	+1.4	19.80	23.51	27.36	+2.5
II. South	6.18	7.15	8.24	9.27	+2.2	7.56	9.20	10.99	+3.1
III. Exporters	3.10	3.37	3.66	3.94	+1.3	3.66	4.27	4.86	+2.4
Total	26.33	28.97	32.10	35.19	+1.5	31.03	36.99	43.21	+2.6

[a] Sum of projections for sawn softwood and sawn hardwood. Excludes sleepers.

[b] Excludes veneer sheets.

ible to use a variable which better describes the end-use pattern.

If it were possible to define and measure the end-use variable perfectly, then, if all other explanatory variables were unchanged, the demand for the product in question would have the same development as the end-use variable. In other words, the elasticity according to end-use activity would be equal to 1, and would not need to be estimated. However, we cannot reach that precision, so it is necessary to estimate the elasticity according to end-use activity.

12.4.2 Past trends in residential investment and in GDP

The residential construction sector is the major end-use sector for sawnwood and panels. So it is important to analyse the development of GDP and residential investment during the last 25 years in western Europe:

TABLE 12.5

OECD Europe: changes in GDP and residential investment

(*Annual average percentage change*)

	1960-68	1968-73	1973-79	1979-83	1960-73	1973-83	1960-83
GDP	+3.6	+4.0	+1.9	+0.3	+3.8	+1.3	+2.7
Residential investment	+6.0	+4.0	−1.1	−1.8	+5.0	−1.4	+2.3

Source: OECD, Historical Statistics 1960-1983

As the two variables have quite different developments, it cannot be desirable to use only GDP as a measure of the activity within the sectors using these products.

In addition, from the demand side, GDP mainly consists of factors which have practically no influence on the demand for sawnwood and panels. Examples of such factors are machinery investment, consumption of cars and consumption of services.

12.4.3 Comparison of projections

The projections by the two models are as follows (for the 18 western European countries in the end-use elasticities model, in million m³):

	1979-81 real	Projections for 2000	
		GDP elasticities	End-use elasticities
Sawnwood (incl. sleepers)	80.2	96.3	95.6[a]
Wood-based panels	28.5	56.1	36.7[b]

[a] Including independent estimates for sleepers.
[b] Including independent estimates for veneer sheets.

For *sawnwood*, the two models give very similar projections. How can this be explained? Answers may be sought in the following:

- The end-use elasticity is higher than the GDP elasticity;

- Residential investment, the main component of the end-use indicator (ENDR) is assumed to be constant over the projection period, leading to rather slow growth in ENDR and markedly lower than that for GDP;

- Thus for the end-use model, we have a higher elasticity and slower growth of the independent variable (end-use activity), while for the GDP model we have a lower elasticity and a faster growth of the independent variable (GDP). By chance, the combination of the two factors leads to a similar result.

For *wood-based panels*, the GDP elasticity model gives a projection for 2000 which is much higher than the projection obtained by the end-use elasticity model. The primary explanation for this lies in the fact that, in the GDP model, the trend variable observed in the past is assumed to continue to effect the projections for the future. In the end-use model, however, the trend effect (i.e. the means by which technical change, leading to market penetration, is taken into account in the model) is assumed to have ended around 1980. The two projections are thus based on different judgements as to whether wood-based panels, especially particle board, will continue to increase their market share in the future as they have in the past.

In addition, while the observed elasticities in the past for GDP and the end-use indicator are similar, GDP is expected to grow faster than the end-use indicator in the future. This also tends to make the GDP elasticity projection higher than that by the end-use elasticity model.

12.5 SENSITIVITY ANALYSIS OF PROJECTIONS

12.5.1 Introduction

The quantitative projections set out above depend on a specific set of assumptions about a number of variables including GDP, residential investment, manufacturing production, the prices of forest products and technical change (represented by the trend variable). If these assumptions are changed, the projections will be different. Although it is impossible to present all the possible permutations between the variables, it is possible to explore some alternative scenarios in order to show in quantitative terms what might be the consequences for the projections if circumstances were different.

This sensitivity analysis can be used in two ways:

- By those who consider that the assumptions of the basic scenarios are not realistic;

- By those who wish to know in quantitative terms what might be the results of a particular policy.

This analysis was carried out for both the GDP elasticities model (all products and countries) and the end-use elasticities projections (sawnwood and panels in western Europe).

The type of sensitivity analysis possible is limited by the structure of the model, in particular by the independent variables used in the analysis of past trends in consumption. These are:

- An indicator of activity in user sectors. In the GDP elasticities model this is GDP and in the end-use elasticities model the specially constructed variable ENDR. The sectors taken into consideration for ENDR are residential investment, furniture manufacture and other user sectors, represented by total manufacturing production;

- The relative prices of forest products (both models);

- A trend variable (both models), representing changes not due to end-use activity or prices – notably technical change (e.g. product development).

12.5.2 GDP elasticities model

This sensitivity analysis covered the possible effects on consumption of forest products of different rates of growth of GDP. The base scenario, presented above, assumed GDP growth for Europe as a whole, of 2.6% per year. (National rates are shown in annex table 1.2). For the sensitivity analysis two other scenarios were constructed:

- A *low GDP* scenario, for which the average annual rate of growth in GDP was 1 percentage point less than in the base scenario i.e. 1.6% per year for Europe as a whole;

- A *high GDP* scenario, for which the average annual rate of growth in GDP was 1 percentage point more than in the base scenario i.e. 3.6% per year for Europe as a whole.

In the present state of uncertainty about long-term economic growth, both these new scenarios for GDP growth are well within the bounds of possibility.

The resulting projections are summarized in table 12.6. As expected, consumption growth is lower than the base scenario in the low GDP scenario, and higher in the high.

TABLE 12.6

Europe: sensitivity analysis of projections by GDP elasticities model

	Quantity consumed (million units)				Average annual percentage change 1979-81 to 2000		
		2000					
	1979-81	Low GDP	Base	High GDP	Low GDP	Base	High GDP
Sawnwood (m³)	102.3	115.5	131.1	149.0	+0.5	+1.3	+1.9
Wood-based panels (m³)	35.6	68.6	81.1	95.9	+3.3	+4.2	+5.1
Paper and paperboard (m.t.) ...	49.2	70.9	84.1	99.9	+1.8	+2.7	+3.6
of which:							
Newsprint	6.5	9.1	10.2	11.5	+1.7	+2.3	+2.9
Printing and writing	13.7	26.6	31.0	36.0	+3.4	+4.2	+5.0
Other	29.0	35.2	42.9	52.4	+1.0	2.0	+3.0

Note: GDP scenarios are as follows (for Europe as a whole):
    Low GDP .... +1.6% p.a.
    Base................................ +2.6% p.a
    High GDP.......................... +3.6% p.a.

For explanation, see text.

12.5.3 End-use elasticities model

A wider range of sensitivity analysis was undertaken for this model. The scenarios chosen are the following:

- The two *base scenarios*, presented above;

- A scenario entitled *recession in user sectors*. As pointed out in chapter 1, it is not inconceivable that the negative factors influencing construction should outweigh the positive factors. These negative factors include high real interest rates, satisfaction with existing housing stocks, and lack of availability of government funds to subsidise housing. Manufacturing production could also grow more slowly than assumed in the base low scenario as service industries gain in relative importance (a larger share of manufacturing could be done outside the ECE region, possibly in areas with lower labour costs);

- There are also alternative scenarios for *product development and marketing*, represented by the technical change (trend) variable. A positive trend in this variable could be attributed to successful marketing efforts or product development: if the forest industries were able to develop and promote new types of wood-based product (or improve existing ones or the services provided with the products), it might be possible to increase consumption even if construction and other user activities were stagnant and forest product prices were stable. On the other hand, if competing industries were successful in product development and marketing, sawnwood and panels could lose market share (represented by a negative trend for this variable);

- Two alternative scenarios for the *relative prices of sawnwood and panels* are also proposed. Chapter 10 indicated that there are not sufficient grounds to assume *a priori* that these prices will change in the future. Yet such a change cannot be ruled out. Control of raw material costs, and technical advances in processing as well as optimization of the whole transport and distribution system might lower the relative prices of sawnwood and panels (as was the case for particle board in the past). On the other hand failure in these fields could have the opposite effect.

The two base scenarios and the five alternative scenarios are described in tabular form in table 12.7. The construction of the scenarios is rather arbitrary (e.g. as regards the rate of change for the trend effect or the prices): the results should be taken rather as an indication of the type of change which might occur in the base scenarios if the broad trend of developments was as suggested in the alternative scenario.

For the price scenarios, the projections only differ from those in the base low scenario for those products for which price elasticities were found over the estimation period:

Sawn softwood − Group I (West-central);

Sawn hardwood − Groups I and II (West-central and South);

Plywood − Groups I and III (West-central and exporters).

No price elasticities were found for particle board and fibreboard. However, the historical data base for particle board (the panel with by far the largest consumption) covers a period of rapid market penetration, essentially due to technical change. This phase is probably over and particle board is now seen essentially as a "mature" product. In these circumstances, it may be that particle board consumption will be more price sensitive than in the past. As this change is relatively recent however, there is no historical data base with which to measure the price elasticity of particle board as a mature product. The possibility cannot therefore be excluded that consumption of particle board in the 1980s and 1990s will be more price sensitive than in the 1960s and 1970s, the base period for these projections.

TABLE 12.7

Western Europe, end-use elasticity model: scenarios for sensitivity analysis

	GDP [a]	Residential investment	Other user sectors [b]	Relative price of sawnwood and panels	Trend effect
Base scenarios					
Low	Low	Constant	Low	Constant	None
High	High	growth at half rate of GDP	High	Constant	None
Alternative scenarios					
Recession in user sectors	Low	Decline 1.5%/year	Constant	Constant	None
Successful marketing and product development	Low	Constant	Low	Constant	+0.5%/year
Loss of competivity	Low	Constant	Low	Constant	−0.5%/year
Price rise	Low	Constant	Low	+1%/year	None
Price fall	Low	Constant	Low	−1%/year	None

[a] FAO scenarios. See chapter 1.

[b] Data for total manufacturing production are used as proxy for both furniture and other user sectors. The scenarios for manufacturing production are set out in table 1.4.

TABLE 12.8

Western Europe, sawnwood and wood-based panels: alternative scenarios

	Real 1979-1983	Base		Scenarios for 2000				
				User sector recession	Alternative			
					Marketing and product development		Prices of sawnwood ands panels	
		Low	High		Success	Failure	Rise	Fall
				Million m³				
Sawn softwood	60.7	69.8	83.3	54.0	76.8	63.5	65.9	74.1
Sawn hardwood	16.2	24.6	30.1	15.0	27.1	22.4	21.8	27.5
Sawnwood, total^a	76.9	94.5	113.3	69.1	103.9	85.9	87.7	101.6
Plywood......................	4.3	5.5	7.2	3.6	6.1	5.1	5.4	5.8
Particle board	19.3	26.2	31.8	17.5	28.8	23.9	26.2^c	26.2^c
Fibreboard....................	2.7	3.4	4.2	2.3	3.7	3.1	3.4^c	3.4^c
Panels, total^b A	26.3	35.2	43.2	23.5	38.7	32.0	35.0	35.4
B	26.3	34.0	40.9	23.7	37.4	30.9	..	..
				Average annual percentage change, 1979-83 to 2000				
Sawn softwood	..	+0.7	+1.7	−0.6	+1.2	+0.2	+0.4	+1.1
Sawn hardwood	..	+2.2	+3.3	−0.4	+2.8	+1.7	+1.6	+2.9
Sawnwood, total^a	..	+1.1	+2.1	−0.6	+1.6	+0.6	+0.7	+1.5
Plywood......................	..	+1.3	+2.7	−0.9	+1.8	+0.8	+1.2	+1.6
Particle board	..	+1.6	+2.7	−0.5	+2.2	+1.1	+1.6^c	+1.6^c
Fibreboard....................	..	+1.1	+2.3	−0.8	+1.6	+0.6	+1.1^c	+1.1^c
Panels, total^b A	..	+1.5	+2.6	−0.6	+2.0	+1.0	+1.5	+1.5
B	..	+1.4	+2.4	−0.6	+1.9	+0.8	..	..

Note: for data by country group, see annex tables 12.27 to 12.29.

^a Excluding sleepers. Sum of projections for sawn softwood and sawn hardwood.

^b Excluding veneer sheets. Total A is the sum of projections for individual panels. Total B is the independently calculated projection for wood-based panels as a group, used as a cross check.

^c Same as in base low scenario, as no price elasticities found for this product.

The resulting projections for western Europe are set out in table 12.8 (with data by country group in annex tables 12.27 to 12.29). The projections for 2000 with the different scenarios may be compared to the base low projection as follows (percentage difference in the year 2000 for the 18 countries):

	Sawnwood	Panels
Base high	+19.9	+22.7
User sector recession	−26.9	−23.2
Successful marketing etc.	+9.9	+9.9
Unsuccessful marketing etc...	−9.1	−9.1
Price rise	−7.2	−0.6
Price fall	+7.5	+0.6

Two major conclusions may be drawn from this analysis:

— If there is a recession in user sectors, notably construction, while other circumstances (price, product development) are unchanged, consumption of sawnwood and wood-based panels could be significantly lower than in the base low scenario;

— The forest and forest industries sector has it within its power to influence the levels of consumption of its products. Success in marketing and product development, or in reducing the relative prices of forest products would be expected to raise levels of consumption. On the other hand failure in this enterprise, or success by competing industries, could bring consumption below projected levels.

12.6 CONSUMPTION OF SAWNWOOD, PANELS, PAPER AND PAPERBOARD IN EASTERN EUROPE

Models based on econometric methods are often considered inappropriate for the special situation of a centrally planned economy. For this reason it was decided to analyse trends and outlook in a more pragmatic way, drawing on experience of conditions in these economies and of the policy outlook of the central planners. This analysis was presented in section 11.4, which described broad trends in consumption of the major forest products in Eastern Europe and proposed methods for estimating future consumption. In particular, a number of hypotheses for future developments in this respect were proposed. The results of these estimates are summarized in table 12.9 (for Eastern Europe as a whole, country data are presented in annex tables 12.13 to 12.20).

Wood-based panels are expected to grow fastest, by

TABLE 12.9

Eastern Europe: estimated future consumption of sawnwood, wood-based panels and paper and paperboard

	Real 1979-81 (million units)	Low			High		
		1990 (million units)	2000 (million units)	Average annual percentage growth	1990 (million units)	2000 (million units)	Average annual percentage growth
Sawnwood (m³)							
Coniferous...............	15.52	16.54	16.54	+0.3	17.33	18.66	+0.9
Non-coniferous............	5.45	5.63	5.63	+0.1	5.97	6.44	+0.8
Sleepers	0.78	0.78	0.78	–	0.78	0.78	–
Total.................	21.75	22.95	22.95	+0.3	24.08	25.88	+0.9
Wood-based panels (m³)							
Particle board.............	4.14	6.88	9.62	+4.3	7.28	10.43	+4.7
Plywood..................	0.84	0.84	0.84	–	0.84	0.84	–
Fibreboard...............	1.60	1.60	1.60	–	1.60	1.60	–
Veneer sheets.............	0.30	0.30	0.30	–	0.30	0.30	–
Total.................	6.88	9.62	12.36	+3.0	10.02	13.17	+3.3
Paper and paperboard (m.t.)							
Newsprint	0.51	0.51	0.51	–	0.54	0.57	+0.5
Printing and writing	0.85	0.90	0.90	+0.3	0.93	0.98	+0.7
Other	4.35	4.82	5.30	+1.0	5.25	6.31	+1.9
Total.................	5.71	6.23	6.71	+0.8	6.72	7.86	+1.6

about 3% a year as a group, while paper and paperboard consumption is expected to increase between 0.8% and 1.6% a year. Growth in sawnwood consumption is expected to be under 1% a year.

12.7 OUTLOOK FOR ENERGY WOOD

Energy is one of the major uses for wood: it is estimated that about 40% of wood removals in Europe is ultimately used as a source of energy. Furthermore the situation as regards demand for energy has changed radically since the mid-1970s. Certainly, the assumption used in ETTS III, that fuelwood consumption will continue to decline, is no longer tenable. Yet the statistical and economic understanding of the fundamental trends is in many respects still unsatisfactory, despite the great progress since the first "energy crisis". The whole question of demand for energy wood (as well as the interaction with the rest of the forest sector) is treated separately, at length, in chapters 18 and 19.

The method chosen was to request national correspondents to make their own forecasts, taking into account national circumstances and policies. A very good response was received, considering that the degree of uncertainty relating to these forecasts is quite high. Table 12.10 summarizes the sum of the trends forecast in the thirteen countries that provided complete estimates (see table 19.1 for further details). Total demand for energy wood is expected to rise by 2-3% a year, rather slower for residues of primary mechanical processing, rather faster for recycled wood and forest products.

However nearly half of wood used for energy comes from residues or recycled wood and is thus not a *direct* drain on the forest, although it must be taken into account

for raw material balances (section 19.5 examines the effect of demand for energy wood on wood raw material supply, notably as regards primary processing residues). In order to evaluate the demand on the forest resource it is necessary to have forecasts, for every country, for consumption of fuelwood. Several countries, which were not able to provide forecasts for all types of energy wood were able to provide forecasts for fuelwood. Estimates were made for the few remaining countries (see section 19.4 for estimation methods).

A problem arose concerning the energy scenarios. Correspondents were requested to make forecasts for energy supply and consumption according to two scenarios:

- Hypothesis A (Higher GDP growth, stable energy prices);

- Hypothesis B (Lower GDP growth, rising energy prices).

Most correspondents considered fuelwood removals would be higher under hypothesis B, but a significant minority thought the opposite. The discussion in chapter 19 will present correspondents' forecasts, as submitted. In this chapter however, and, above all, in chapter 20 on the material balance, it is necessary to consider how the fuelwood removals scenarios should be combined with the scenarios for total removals and for consumption of forest products. Hypothesis A for fuelwood is closer in its

TABLE 12.10

Summary of national forecasts for use of wood and bark as a source of energy in 13 selected countries, according to hypotheses A and B[a]

	1980 (million m³)	2000 (million m³)		Change 1980 to 2000			
				Volume (million m³)		Annual average percentage	
		A	B	A	B	A	B
Fuelwood	36.3	50.9	60.8	+ 14.6	+ 24.5	+ 1.7	+ 2.6
Residues of industry	25.6	39.6	40.6	+ 14.0	+ 15.0	+ 2.2	+ 2.3
of which:							
Primary mechanical	13.3	18.1	18.2	+ 4.8	+ 4.9	+ 1.5	+ 1.6
Secondary mechanical	8.7	12.8	13.6	+ 4.1	+ 4.9	+ 2.0	+ 2.3
Pulp and paper[b]	5.2	8.7	8.8	+ 3.5	+ 3.6	+ 2.6	+ 2.7
Recycled wood	6.7	13.7	14.0	+ 7.0	+ 7.3	+ 3.6	+ 3.8
Total	70.1	104.1	115.3	+ 34.0	+ 45.2	+ 2.0	+ 2.5

Note: See table 19.1 and annex tables to chapter 19 for country detail.

[a] For discussion and definition of hypotheses A and B, see text and chapter 19.

[b] Excluding pulping liquors, but including wood and bark residues used for energy by the pulp and paper industry.

background assumptions to the scenarios for *high* levels of total removals: this might indicate that lower fuelwood removals should go with high consumption of forest products and high total removals. Yet, in some countries at least, notably Sweden, it is considered that the high level of total removals is dependent on the achievement of high fuelwood removals. This question would seem to merit an in-depth discussion at national level, taking account of local factors and policies. This was unfortunately not possible within the time constraint for the preparation of ETTS IV.

The secretariat has chosen not to follow correspondents' estimates according to hypothesis A and B, but to sort them into a "low" and a "high" forecast. The low fuelwood demand forecast will be combined with the low removals forecasts and low demand for forest products; high fuelwood will be combined with high removals and forest products consumption. The possible validity of the other solution is recognised. However, in most countries, the difference between the low and high fuelwood demand forecasts is not large (or does not exist when only one forecast is made), so a change in the arrangement of scenarios would not significantly affect the material balance. In a few countries however, notably France, Sweden and Turkey, the difference between the low and high fuelwood demand forecasts is large. For these coun-

tries, both combinations will be calculated in the material balance and the results discussed in chapter 20.

The fuelwood demand estimates to be used in the material balance are set out in annex table 12.23.

These estimates were prepared in 1984/85, when oil prices were over $30/barrel. The sharp drop in 1985/86, to $10-15/barrel, was not foreseen by correspondents. If these low prices were to be maintained, it is unlikely that fuelwood demand would grow at the rate forecast. Many analysts believe however that the drop is temporary and that oil prices will recover the levels of the early 1980s. It is not possible to prepare a new set of forecasts in the final stages of preparing ETTS IV for publication; the oil price fall of 1985/86 however adds an extra element of uncertainty to the outlook for fuelwood.

It should be pointed out that for many countries, national estimates for fuelwood consumption in 1980 are different, occasionally very different, from recorded fuelwood consumption, which in some cases does not take auto-consumption fully into account. It is the *national estimates* which are presented in annex table 12.23 and used for the material balance in chapter 20, as they are the only historical data comparable with the forecasts. This will necessitate some adjustment of the historical data for the material balance which is presented in chapter 20.

12.8 OUTLOOK FOR PRODUCTS WHICH HAVE NOT BEEN MODELLED SEPARATELY

For a number of forest products, mostly of relatively minor importance, it was not considered necessary to undertake an in-depth quantified analysis of consumption trends. Nevertheless, some assumptions must be made about future levels of consumption of these products, if the scenarios for total consumption in the future are to be comparable with the data for the past.

This section briefly presents the assumptions used.

12.8.1 Sleepers

Between 1950 and 1970, annual European consumption of wooden sleepers declined from around 3.3 million m³ to about 2.1 million. It was assumed for a time that this decline would continue. However, as ETTS III pointed out (Chapter 4), hardwood sleepers have a number of technical advantages, which had not been exploited. It expected the use of steel and softwood sleepers to decline but considered that "the relative positions of hardwood and concrete sleepers would depend more on economic than technical considerations" and that consumption of sleepers would remain between 1 1/2 and 2 million m^3 in 1980 and possibly 1990. Since the mid-1970s sleeper consumption has been between 1.8 and 2.2 million m^3 (except 1981 when it dropped to 1.6 million m^3) and showed no marked downward trend.

It has therefore been assumed that sleeper consumption will remain unchanged at the 1979-81 level (see annex table 12.21).

12.8.2 Veneer sheets

According to the FAO/ECE definitions, the category "veneer sheets" excludes those veneer sheets used in the manufacture of plywood. Yet it is known that several countries are not able to provide data according to this definition. There is undoubtedly some misclassification and even double-counting, although the volumes involved are probably not large.

The wide range of influences in the rather specialized end-uses for veneer sheets, the small volumes concerned in many countries, together with the above-mentioned classification problems, make it difficult to identify any trends in the use of veneer sheets, mostly in the furniture and packaging sectors.

The statistical series available show European consumption of veneer sheets rising from about 1.5 million m^3 in the mid-1960s to a peak of 2.0 million in 1974. Since then, consumption has remained around 1.9 million m^3, of which about 1.1 million m3 in the EEC(9), notably Italy, which consumes around 0.5 million m^3.

It appears to be a reasonable assumption, for the purposes of this study, that consumption of veneer sheets will also remain constant at the 1979-81 level (see annex table 12.21).

12.8.3 Dissolving pulp

Consumption of dissolving pulp remained remarkably stable at around 1.8 million m.t. up to 1977. Between 1977 and 1981, however, it fell by over 10%, to just under 1.6 million m.t. in 1981. In a market which had previously been so stable, this might be an indication of structural change – possibly that the chemical products made from dissolving pulp were becoming less competitive. In view of this rather rapid decline, it does not appear prudent to assume, as in ETTS III, that consumption of dissolving pulp will remain stable.

It has therefore been assumed that consumption of dissolving pulp will decline by 10% per decade to 2000.

12.8.4 Pitprops and other industrial wood

The use of wooden pitprops in mines has been declining steadily as they are replaced by mechanical roof supports and more coal is obtained by open cast mining. European consumption, which was over 12.6 million m^3 in the mid-1960s had fallen to 6.5 million m^3 by 1979-81 – a faster fall than that foreseen by ETTS III, which forecast 8 million m^3 in 1980. There seems no reason to expect a reversal of this trend. It has therefore been assumed that consumption of pitprops will fall by 20% per decade. (This applies to round pitprops only: the significant quantities of sawnwood used in mines are classified as sawnwood.)

The uses of "other industrial wood" are very diverse and impossible to analyse in detail. Furthermore, there is also some misclassification of this assortment at the removals stage in many countries, and much wood which is ultimately used as pulpwood is classified as "other industrial wood" or *vice versa*. In most cases there is in fact little or no technical difference between the assortments so that those responsible for removals and trade are frequently not aware of the ultimate destination of the wood. It became clear during the data validation exercise undertaken in connection with ETTS IV that some of the volumes misclassified were quite significant.

Between 1969-71 and 1979-81, recorded European consumption of "other industrial wood" dropped from 21.7 million m^3 to 16.8 million m^3, a drop of nearly 25%, but it is unclear to what extent this was a real drop and to what extent a result of misclassification, especially as the drop took place abruptly between 1977 and 1979, and was concentrated in two countries only (Romania and Yugoslavia).

It has been assumed, therefore, for the purposes of ETTS IV, that the consumption of "other industrial roundwood" will remain constant, on the understanding that the analysis of the global position will take into account the statistical problems connected with roundwood classification.

12.9 CHOICE OF ETTS IV CONSUMPTION SCENARIOS

It has been pointed out above that different methods of projection or estimation are suitable for different circumstances and objectives. Yet, a single set of figures must be chosen, which may be compared with the results of other parts of the study, notably on roundwood supply.

The aim of this brief section is to outline the reasons underlying the choice of the figures on the outlook for consumption of forest products which will be presented in the next section.

As pointed out in chapter 11, econometric methods are

not really appropriate for *eastern European* conditions. For this reason, the estimates in section 12.6 for sawnwood, panels, paper and paperboard are used rather than the results of the econometric projections.

For *paper and paperboard*, the only projections available are those prepared by the GDP elasticities method, which will therefore be used. However for the material balance in chapter 20, two scenarios are needed, whereas there is only one base scenario for the GDP elasticities model. It was therefore decided to use for the "low" and "high" consumption scenarios the "low GDP" and "high GDP" scenarios developed for the sensitivity analysis (section 12.5.2). Although this leads to some inconsistency as regards the independent variables, the range obtained in this way appears quite satisfactory.

For *sawnwood and panels*, projections were available from two sources:

- GDP elasticities model (section 12.2) for the two product groups, all countries, one scenario;

- End-use elasticities model (section 12.3), for five products, eighteen countries, two scenarios.

The two sets of projections are based on the same assumptions regarding GDP and forest products prices, with additional assumptions (for residential investment and manufacturing production) for the end-use elasticities model.

As regards statistical measures of the quality of the models, the *coefficient of determination* (R^2) is very high (often over 0.98) for the GDP elasticities model, and lower for many of the equations in the end-use elasticities model.

This is in fact not surprising as the GDP elasticities model uses real values (i.e. with a very wide range between the highest and the lowest values), while the end-use elasticities model uses indices. The remarks in annex 11.1 paras 15-19 would seem to indicate that a better measure is the *standard error of the residuals*, which is not affected by the size of the variables. In the GDP elasticities model, the standard error ranges from 0.21 (other paper and paperboard) to 0.42 (wood-based panels). For the equations of the end-use elasticity model, the standard error of residuals is below 0.1 in 32 cases, between 0.1 and 0.2 in 32 cases and over 0.2 in only 5 cases of which four concern plywood (see annex table 11.2). This would seem to indicate that, despite the lower R^2 values, the end-use elasticity models are of at least as good quality as the GDP elasticity models. As the former provide a greater degree of detail in the projections, it was decided to incorporate the projections by the end-use elasticity model (using country group elasticities) into the outlook for sawnwood and panels, whenever possible. For those few western and southern European countries for which projections by the end-use elasticities model were not available (Cyprus, Iceland, Israel, Malta), the projections by the GDP elasticities model have been used, assuming the same percentage breakdown by product as in 1979-81.

For *fuelwood*, the results of the enquiry which is presented in chapters 18 and 19 are used, with the adjustments mentioned in section 12.7 above.

The methods for estimating consumption of *sleepers*, *veneer sheets*, *pitprops* and *other industrial wood* were presented in section 12.8.

12.10 PRESENTATION OF ETTS IV CONSUMPTION SCENARIOS

Following the criteria presented in section 12.9, preliminary consumption scenarios with data by country and product were prepared. These are set out in annex tables 12.13 to 12.23 and summarized in tables 12.11, 12.12 and 12.13.

These scenarios should be considered preliminary as they do not take into account supply factors (except for Eastern Europe). In chapter 20, these preliminary consumption scenarios will be confronted with scenarios for roundwood removals, trade and recycling, and may have to be modified as a consequence.

The scenarios foresee continued slow growth for sawnwood, between 0.8% and 1.6% a year, with significantly faster growth for sawn hardwood than for sawn softwood.

For wood-based panels however the growth rates proposed are much lower than those foreseen by earlier studies – about 2% a year. As in the past, faster growth is foreseen for particle board than for the other panels. The rate foreseen for particle board is 2.0-2.9%, as compared to

8% in ETTS III. The development of new panels, some of which combine the features of existing types, will probably blur the boundary lines between the different type of panel. Not too much importance should be attached to the forecasts for individual panels.

Paper and paperboard consumption is expected to grow at about the same rate as that of wood-based panels, between 1.6% and 3.2% a year. Within the group, printing and writing paper consumption is expected to grow much faster than that of other grades.

Fuelwood consumption is expected to halt its long-term decline and grow by 1-2% a year, with more rapid growth in the Nordic countries, as well as a few other countries, notably France (high scenario).

Growth in Southern Europe is expected for most products to be faster than elsewhere, notably because of the higher rates of GDP growth forecast. It should also be borne in mind that *per caput* consumption levels in this region are still significantly lower than in other regions.

TABLE 12.11

Summary of ETTS IV consumption scenarios for sawnwood and wood-based panels

	1979-81 (million m³)	Low scenario (million m³)		High scenario (million m³)		Average annual percentage change	
		1990	2000	1990	2000	Low	High
Sawnwood							
Nordic countries	10.75	10.39	10.97	10.80	12.19	+0.1	+0.6
EEC(9)	50.46	51.70	58.60	54.60	69.29	+0.8	+1.6
Central Europe	5.04	5.26	5.72	5.52	6.65	+0.7	+1.4
Southern Europe	14.32	16.30	20.79	17.91	26.83	+1.9	+3.2
Eastern Europe	21.75	22.95	22.95	24.08	25.88	+0.3	+0.9
Europe	102.32	106.60	119.03	112.91	140.84	+0.8	+1.6
of which:							
Sawn softwood	78.15	80.52	86.94	85.25	102.47	+0.5	+1.4
Sawn hardwood	22.37	24.29	30.30	25.87	36.58	+1.5	+2.5
Sleepers	1.80	1.80	1.80	1.80	1.80	–	–
Wood-based panels							
Nordic countries	2.57	2.68	3.08	2.92	3.78	+0.9	+1.9
EEC(9)	20.78	21.99	26.16	23.41	31.83	+1.2	+2.1
Central Europe	1.35	1.45	1.73	1.55	2.15	+1.2	+2.3
Southern Europe	4.02	4.65	6.28	4.94	7.52	+2.3	+3.2
Eastern Europe	6.88	9.62	12.36	10.02	13.17	+3.0	+3.3
Europe	35.60	40.40	49.63	42.85	58.46	+1.7	+2.5
of which:							
Particle board	23.82	28.36	36.13	30.16	42.52	+2.1	+2.9
Plywood	5.44	5.61	6.61	6.02	8.20	+1.0	+2.1
Fibreboard	4.45	4.53	4.99	4.77	5.84	+0.6	+1.4
Veneer sheets	1.90	1.90	1.90	1.90	1.90	–	–

Note: For country and product data see annex tables 12.13 to 12.17 and 12.21.

The ETTS IV consumption scenarios imply that *per caput* consumption will continue to rise, as shown in table 12.13. The levels of *per caput* consumption in 2000 would be higher than all those previously recorded, except for fuelwood where they are still considerably lower than those in earlier years, although higher than in 1979-81.

It is also apparent from fig. 12.1 that growth rates are expected to be higher in the 1990s than in the 1980s. This is due to the fact that the GDP growth rates foreseen for the 1980s take into account the very slow growth recorded in the early years of the decade. It was assumed when preparing the GDP projections that the GDP growth rates after 1985 would be rather higher than in the first half of the 1980s (see chapter 1).

TABLE 12.12

Summary of ETTS IV consumption scenarios for paper and paperboard and dissolving pulp

	1979-81 (million m.t.)	Low scenario (million m.t.)		High scenario (million m.t.)		Average annual percentage change	
		1990	2000	1990	2000	Low	High
Paper and paperboard							
Nordic countries	3.34	3.82	4.82	4.34	6.46	+1.9	+3.4
EEC(9)	32.75	35.81	44.07	41.30	60.82	+1.5	+3.1
Central Europe	1.94	2.10	2.46	2.38	3.27	+1.2	+2.6
Southern Europe	5.44	6.54	9.15	7.83	13.60	+2.6	+4.7
Eastern Europe	5.71	6.23	6.71	6.72	7.86	+0.8	+1.6
Europe	49.18	54.50	67.21	62.57	92.01	+1.6	+3.2
of which:							
Newsprint	6.47	7.23	8.72	8.00	11.00	+1.5	+2.7
Printing and writing	13.66	17.78	25.58	20.25	34.43	+3.2	+4.7
Other paper and paperboard	29.05	29.52	32.92	34.35	46.59	+0.6	+2.4
Dissolving pulp							
Europe	1.65	1.46	1.33	a	a	–1.1	a

Note: For country and product detail, see annex tables 12.18 to 12.21.

[a] Only one scenario for dissolving pulp.

TABLE 12.13

Summary of ETTS IV consumption scenarios for fuelwood and other wood used in the rough

	1979-81 (million m³)	Low scenario (million m³)		High scenario (million m³)		Average annual percentage change	
		1990	2000	1990	2000	Low	High
Fuelwood							
Nordic countries............	9.95	13.30	15.40	18.00	20.10	+2.2	+3.6
EEC(9).....................	18.12	20.15	22.06	24.09	31.24	+1.0	+2.8
Central Europe.............	2.22	2.36	2.47	3.51	4.80	+0.5	+3.9
Southern Europe...........	30.00	31.78	33.08	35.46	39.47	+0.5	+1.4
Eastern Europe.............	11.67	11.51	12.91	11.87	13.11	+0.5	+0.6
Europe	71.96	79.10	85.92	92.93	108.72	+0.9	+2.1
Other wood used in the rough							
Europe	23.33	22.02	20.99	a	a	−0.5	a
of which:							
Pitprops	6.50	5.20	4.17	a	a	−2.2	a
Other industrial wood	16.82	16.82	16.82	a	a	−	a
Total wood used in the rough							
Europe	95.29	101.12	106.91	114.95	129.71	+0.6	+1.6

Note: For country and product detail, see annex tables 12.23 and 12.24.

a Only one scenario prepared.

TABLE 12.14

Europe: past and projected levels of per caput consumption of forest products

	Unit	1949-51	1979-81	2000 Low	2000 High
Sawnwood	m³/1000 cap.	150	193	204	241
Wood-based panels	m³/1000 cap.	7	67	85	100
Paper & paperboard ....m.t./1000 cap.	25	93	115	158
Fuelwood	m³/1000 cap.	295	136	147	186

12.11 COMPARISON OF FORECASTS BETWEEN ETTS III AND ETTS IV

It is of interest to summarize the effect that the changes in methods, assumptions and outlook between the early 1970s when ETTS III was drafted and the early 1980s have had on the perception of the outlook for consumption of forest products (table 12.15). It will be seen that the ETTS IV forecasts are rather higher than those of ETTS III for sawnwood and very much higher for fuelwood, but considerably lower for paper and paperboard and, above all, for wood-based panels.

Most of these trends are also visible from a comparison of 1980 results with the ETTS III forecasts for that year.

On a rough calculation, the ETTS IV forecasts for 2000 are equivalent to 595-765 million m³EQ (equivalent volume of wood in the rough), while the adjusted forecasts of ETTS III are for 780 million m³EQ. Thus the ETTS IV low forecasts are significantly below the ETTS III (adjusted) forecasts, but the ETTS IV high forecasts are about the same.

TABLE 12.15

Europe: comparison of consumption forecasts and projections in ETTS III and ETTS IV
(Million units)

	Unit	1969-1971	1979-1981	ETTS III		ETTS IV, 2000		Difference, for 2000, between ETTS III and ETTS IV	
				1980 [a]	2000 [b]	Low	High	Low	High
Sawnwood	m³	93	102	99	113	119	141	+6	+28
Wood-based panels	m³	23	36	46	113	50	58	−63	−55
Paper and paperboard.........	m.t.	38	49	56	100	67	92	−33	−8
Fuelwood	m³	69[c]	72[c]	54[c]	35[c]	86[c]	109[c]	+33[c]	+56[c]
Other industrial roundwood....	m³	30	23	26	20	21	..	+1	..

[a] Low hypothesis.

[b] For sawnwood, panels, paper and paperboard adjusted forecasts from table 9/12.

[c] Fuelwood data: 1969-71 data are for recorded consumption, 1979-81 data are national estimates. ETTS III forecasts are comparable with 1969-71 data and ETTS IV forecasts with 1979-81 data. To correct this lack of comparability the figures for the "difference" have been adjusted by the difference between recorded and estimated consumption in 1980 − 18 million m³ .

12.12 APPRAISAL OF THE CONSUMPTION SCENARIOS

The previous sections have presented, in considerable detail, scenarios for the consumption of forest products. The outlook in general may be characterized as for rather slow growth, in the range of 1%-3% a year. It should be borne in mind that these quantitative forecasts are the end of a chain of reasoning involving some mathematical analysis and some (quantified) assumptions about a number of aspects. In addition to the mathematical analysis, the reader must consider the underlying assumptions. These may be divided into two groups:

— Assumptions about external factors which the forest and forest products sector cannot reasonably hope to influence, such as the rate of growth in GDP, residential investment, or manufacturing production, or global energy prices;

— Assumptions about factors which the forest and forest products sector can influence. The most important of these are the price of forest products and the "technological" factors. By the latter is meant factors determining consumption levels other than

end-use activity and price. A good example is product development.

The impact of these two types of assumption on the consumption scenarios is different. The first type introduces an additional element of *uncertainty* into the scenarios. The second enables the preparation of new scenarios, according to a *strategy for the forest and forest products sector*. If action is taken to control, even reduce, prices, and to develop products suited to the end-user, then consumption can be increased, even in stagnant demand conditions. The possible impact of this type of action has been explored in section 12.5 on sensitivity analysis.

The consumption scenarios, and the methodology used to produce them, thus have two objectives:

— To provide an input into the material balance in chapter 20;

— To provide the starting point for the development of strategies at the national, or product group level, to develop future markets for forest products.

FIGURE 12.1

Europe: ETTS IV consumption scenarios

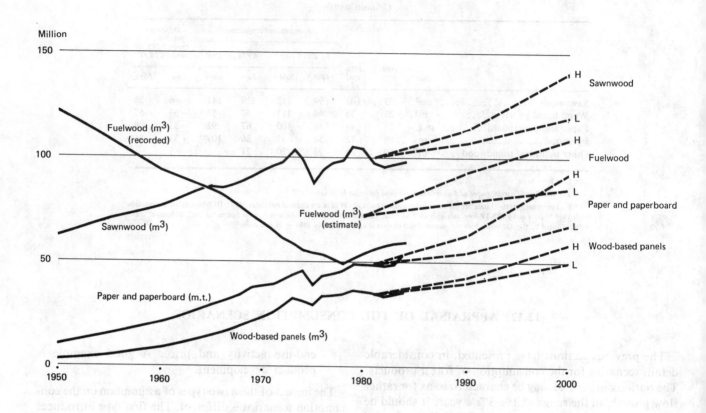

CHAPTER 13

Demand for forest products and non-wood goods and services of the forest beyond the year 2000

13.1 INTRODUCTION

A major problem for those who analyse the outlook for the forest and forest products sector is the difference in time-scale between developments for the forest and for forest products. A forest rotation is typically between 60 and 100 years in European conditions, although there are many exceptions on either side of this range. Therefore trees planted in the mid-1980s will mostly be ready for final harvesting around the middle of the twenty-first century. If the forester is to make his silvicultural decisions on a rational basis, he must have some idea, quantitative if possible, of the likely demand for the goods and services which his forest may produce over the whole rotation.

Yet a period of 10-15 years is often seen as a maximum for reasonably reliable demand forecasting. Beyond that period, the element of uncertainty increases until the forecasts can only be considered speculative. Some of the main elements of uncertainty are the following:

(a) *Broad macro-economic trends* are very difficult to foresee;

(b) *Technical changes* can be major for forest products themselves, for user sectors, for competing products and for society as a whole. Even when the technical breakthrough has already been made, it is often hard to foresee all its economic and social consequences (for instance, the world is undergoing profound changes in the 1980s because of technical breakthroughs in electronics and communication, many of which occurred in the 1960s or 1970s);

(c) *Society* can also change profoundly, sometimes in reaction to economic or technical developments, but sometimes for other reasons, requiring other skills to understand and analyse;

(d) Finally, the longer the period under consideration, the greater the likelihood that some *catastrophe*, natural or man-made, such as war or major climatic change, takes place, and completely changes the framework of analysis. It is not possible to take such an event explicitly into account in this study, but what follows, like everything else in the study, is subject to the assumption that no such catastrophe occurs.

Despite these problems, it appeared necessary to undertake a tentative exploration of the outlook for demand for forest products beyond 2000, in order to provide some indication of the environment in which the forestry forecasts (chapter 5) could be seen.

With shorter-term demand projections, the inertia of the situation plays an important role: changes, especially fundamental changes, cannot usually take place overnight, as the system can only change slowly. For longer-term analysis however, inertia plays a much smaller role: in the long run, all changes are possible, even the most fundamental. Many of the factors identified in this chapter are relevant already, but their full effect may not be felt until the twenty-first century.

Because of likely technical developments which could blur the traditional boundaries between forest products, this analysis is carried out according to very broad product areas:

– Sawnwood and wood-based panels;

– Paper and paperboard;

– Energy wood;

– New uses for wood.

13.2 SAWNWOOD AND WOOD-BASED PANELS

Construction, especially residential investment (with its associated activities), is by far the most important end-use for sawnwood and wood-based panels: future levels of consumption of these products will depend essentially on the answers to two questions:

(*a*) How much construction activity will there be?

191

(*b*) Will sawnwood and panels increase, decrease or maintain their share of the market for construction materials?

As pointed out in chapter 1, there is a limit to the amount of space each household can use, and in most countries the average level of housing quality is at present relatively satisfactory. There is also in many countries a limit on the amount of land which it is desirable to make available for housing. In these circumstances, it appears likely that more and more of the residential housing market will be limited to replacement, renovation or expansion of existing properties. This might imply a reduction in new residential investment, until a "steady state" level, fixed by the rate of replacement of existing dwellings, is reached. Before this level is reached however, it is possible that in several countries, there will be campaigns to replace much of the housing which was installed in haste in the two decades after the second World War. In several countries the quality of these buildings has proved to be unsatisfactory: in these circumstances, an effort might be made to replace a great part of the housing stock.

It is also likely that, as in recent years, a greater share of residential investment will be taken by repairs, maintenance and renovation of existing dwellings, including expansions and modifications to suit changed needs or fashions. This activity is a more intensive user of forest products than the construction of new dwellings, because of the unrivalled flexibility of sawnwood and wood-based panels and the fact that they can be easily worked, on site and, in more and more cases, by amateurs (Do-it-yourself).

A possible exception to this trend towards contraction and stability in the construction sector could be the closely related furniture sector, which in recent years has shown a tendency towards presenting furniture as a semi-durable good, i.e. with more fashion content and intended to be replaced much sooner than in earlier times. Thus, even if the number of households stays roughly constant, if each piece of furniture is replaced on average after 10 years instead of 20, consumption of furniture would double.

What will be the share of forest products in consumption of construction materials and of materials for furniture? In the short to medium term, traditional use patterns will change only slowly, because of the inertia in the construction sector due to conservatism in popular tastes, and rigidities in the legal framework (building codes). In the long term, however, forest products could either be replaced in practically all their uses, or could replace many other construction materials. Most architects, engineers or furniture designers do not have a special loyalty towards one type of material; they choose a design (e.g. for a house or a piece of furniture) which satisfies their technical and aesthetic requirements in the most convenient and economic way. Forest products will be chosen only if they meet these conditions.

One particularly important aspect is the standardization of forest products. The lack of homogeneity of wood, the variety of its species, forms, and strengths, while often making it attractive from an aesthetic point of view (especially sawnwood, plywood and veneers) can be a definite disadvantage from the technical point of view, notably to architects and designers who expect homogeneity in the materials they specify, and are used to selecting materials whose technical characteristics they can specify easily at the design stage. Efforts have been made, or are in hand, in many fields to improve the convenience (and safety) in use of forest products and of systems based on forest products: stress grading, finger-jointing, standardization of dimensions and prefabrication of building components such as roof trusses, windows and doors. Widespread introduction of these and other measures making the use of forest products more convenient and economic seems to be a pre-condition for the maintenance, or possibly the expansion, of the share of forest products in construction materials.

Forest products utilization should also be fully integrated into architectural and structural design courses. In some countries, especially those without a strong tradition of the use of forest products in construction, these courses do not always deal with the use of forest products as fully as with that of other materials.

Another important aspect is product development. In general, the forest industries spend a very small percentage of their turnover on R & D. If competing sectors develop attractive systems for using their materials, there is a real danger that the importance of forest products could steadily decline as they are replaced by products based on other materials which have been developed to satisfy more fully the needs of the ultimate consumer.

A prerequisite for effective product development is a much better understanding of the markets for forest products: what volumes are used in what end-uses, what are the technical requirements etc. It is clear from chapter 8 that knowledge in this area is far from satisfactory. Indeed it could be said, without very much exaggeration, that at present, most producers of sawnwood or wood-based panels cannot have a very clear idea of how their products will ultimately be used. They frequently are not well informed of how the product they sell will fit into a system and only rarely are whole wood-based systems put on the market. One reason for this is the fragmented nature of the industry and trade and the many different stages which the products (especially sawnwood) often pass through before reaching the consumer. Not only does this complicated structure increase costs, it separates producers from their markets and ensures that no individual link of the chain from forest to consumer has a *short-term* interest in undertaking product development. In the *long term*, of course, if improved market knowledge led to better products and thereby to larger markets, all concerned would benefit.

In most countries, end-uses outside the construction and furniture sectors are of marginal importance for sawnwood and wood-based panels. Nevertheless it seems likely that the same broad principles apply: forest products are not irreplaceable, but they are not doomed to extinction either. Their future will depend on whether the forest products sector is able to continue to produce, at a competitive price, products which suit users' needs. Their chances of doing this will be greater if the forest and forest industries sector invests in market research and product development, and if the structure of the trade and industry is rationalized, so that they are more market-oriented and sensitive to the needs of the consumer.

13.3   PAPER AND PAPERBOARD

The main uses for paper and paperboard can be classified into three groups:

– Cultural uses (newsprint, printing and writing);
– Wrapping and packaging;
– Household and sanitary uses.

In addition there exist very many rather specialized uses about which generalizations are difficult, if not impossible.

The major uncertainty around the long-term future of consumption of *cultural papers* is whether some or all of the means of communication based on paper will be replaced by electronic systems. The systems which could reduce the consumption of cultural papers already exist – electronic data storage, processing and transmission, teletext systems, electronic banking systems and so on. Frequently, their claimed advantages can only become evident after a long period of technical standardization, installation of systems (heavy investment costs), and, perhaps most important, modification of users' attitudes. In the medium-term therefore, say to the mid-1990s, it is likely that change will not be revolutionary, that consumption of some grades of paper (e.g. computer printouts) will actually be raised and that losses in consumption will be concentrated on a few vulnerable grades (e.g. paper for telephone directories which could well be replaced by on-line systems).

In the long run, however, those electronic systems which are both technically and economically viable and acceptable to users will be installed. It is already clear that some of these systems have very much to offer – speed and accuracy of reaction, ability to manage previously unheard of volumes of data, lower costs etc. However, probably not all the systems proposed at present will come into current use.

It is extremely difficult to foresee to what extent these new electronic communications systems will affect consumption of cultural papers. Some uses of paper will no doubt be totally replaced, while other uses will carry on unchanged: the widespread introduction of radio and television has not prevented people from reading books and magazines, as they have accepted an additional medium instead of replacing the old with the new. The new systems may be designed with or without a paper-based element. If users prefer paper, or if it is necessary for legal reasons (e.g. in electronic banking, a signed receipt on paper is still often necessary), the system will be designed accordingly. It is also clear that the cost of paper will play a role: if it remains relatively cheap and convenient, the new systems are more likely to incorporate an element of paper.

The level of advertising plays a major role at present in determining paper consumption, especially in market economies. The total volume of advertising is roughly determined by the level of economic activity, but the choice of medium – newspapers, magazines, posters, television, radio, cinema, direct mail, etc. – is determined by a wide range of technical and economic factors. In many cases the advantages of paper-borne media are their permanence and their more direct approach to selected "targets" (unlike, for instance, television). Paper-borne media are also often cheaper, increasing their attractiveness, especially to local advertisers. It is likely therefore, that even after the advent of new electronic communication systems, paper will have many attractions for the advertiser.

Thus the level of consumption of cultural papers in the twenty-first century will be determined by the choices taken during the installation of new electronic systems, which will be influenced not only by technical and economic considerations but also by consumers' preferences as to how the systems are presented to them. It must however be considered likely that after the mid-1990s consumption of cultural papers will grow slower than in earlier years – if it grows at all: a decline in absolute terms is quite conceivable.

Packaging is an extremely competitive end-use sector where well-informed consumers (the packaging companies) have a choice of a number of materials, including paper and paperboard, plastics, glass, and aluminium, which compete with each other on technical and cost grounds. The most direct competition for paper and paperboard is with plastic film, but there is also competition with other materials, for example for beverage packaging. The different materials are frequently combined (e.g. paperboard is lined with aluminium foil for orange juice containers). In this technical and competitive environment, the consumption of paper and paperboard will depend on the success of their manufacturers in developing and promoting their products and in keeping prices at a competitive level.

Consumption of packaging is linked to the level of production of goods. Assuming that paper and paperboard retain their share of the packaging market, it is still possible that consumption of paper and paperboard for packaging could grow slower than GNP as a whole, because services are expected to take a larger share of GNP, at the expense of production of material goods.

In *household and sanitary* uses, there are no obvious competitors to paper and pulp products, in those uses where they are now well established. However, most market sectors may now be considered saturated, so that expansion in consumption in Europe would only be in accordance with population. Only if new uses are identified and developed could consumption of these grades grow significantly faster than population. A possible exception is those countries which have low *per caput* consumption (e.g. Southern Europe) as there might be some narrowing of the gap between low and high *per caput* consumption countries.

13.4 WOOD FOR ENERGY

The outlook for the use of wood for energy will be discussed at length in chapter 19, which will also present the main factors which will affect this consumption for conventional and for new energy-linked uses for wood. Factors of particular importance for *conventional* uses are the world energy supply-demand balance and the resulting energy price, national policies to encourage domestic energy sources, the availability of satisfactory distribution systems and of reliable and convenient wood-burning equipment. For the *new* uses, important factors include progress in R & D programmes and the availability of sufficient wood and land to supply wood to the units which would produce ethanol, methanol, or electricity. Chapter 19 shows that national correspondents foresaw a rise in conventional uses of wood energy of 2-3% a year between 1980 and 2000. For the new uses, most countries foresaw little or no growth, but a group of countries with a high ratio of forest resource to energy needs are carrying on R & D programmes in this field. Some of these countries felt that if technical progress and global energy conditions were satisfactory, the new energy uses could expand rapidly after 2000 to play a significant role in national energy supply. However, if there is no severe prolonged worldwide energy crisis, necessitating the mobilization of all possible domestic energy sources, any large-scale expansion of new energy uses would be confined to a small group of countries (e.g. France and Sweden) where conditions are appropriate for this development.

The most important trend however, which is already apparent and is likely to continue after 2000 is towards more integrated systems which find the optimum balance between energy and raw material uses for wood (and bark), and a progressive elimination of all wastage, so that all material which is not used as raw material is used, effectively and economically, as a source of energy, in the forest industries, in rural households or in other wood-burning installations. The sources for this energy wood would be many: harvesting residues (subject to imperatives of preserving site fertility); wood processing residues from sawmills, wood-based panel mills, pulp mills, or from secondary processing industries such as furniture or joinery; recycled wood such as used pallets or wood from demolished buildings.

In short:

– Conventional uses of wood for energy may increase steadily, with a progressive elimination of wastage and the use of wood for energy becoming fairly widespread in rural areas;

– In most countries, large-scale processing of wood into "new" solid, liquid or gaseous fuels, or electricity generation from wood for the public grid is unlikely. In a few countries, in certain conditions, these uses could be developed. This could imply a rather radical re-arrangement of wood supply patterns in the areas concerned.

13.5 OTHER USES FOR WOOD

There is an enormous range of other products from wood ranging from traditional major uses like sleepers or pitprops to specialized uses such as the manufacture of violins, smoking pipes or specialized industrial filters. Some of these are of marginal importance, but others, while not consuming very large volumes, have stringent quality requirements and can generate a high percentage of value added. In a study as wide-ranging as ETTS IV, it is unfortunately not possible to examine the outlook for all these uses in detail. It is necessary, however, to ask whether there is a possibility that any use of wood other than those described in sections 13.2-13.4 could gain sufficiently in importance over the next 20-50 years to have

a significant impact on the overall supply/demand balance. There have been few, if any, significant new applications for forest products (as opposed to development of existing applications) over the last 20-30 years.

On many occasions this century, attention has been drawn to the potential of wood as a *feedstock for chemical processing*. So far, however, this potential has only been realized in a few isolated instances (if the chemical pulp industry is excepted). The main reason has been that other feedstocks, notably oil and natural gas, have proved more abundant, cheaper and easier to process. One important factor is that oil and gas are extracted from relatively concentrated sources and can be transported relatively cheap ly, enabling the construction of very large units. Units of a similar size which processed wood would require enormous raw material catchment areas, entailing high transport costs. Even the steep rises in the price of oil and natural gas in the 1970s did not bring about a shift from oil or gas to wood, or indeed to any other feedstock. One reason is probably the immense investment already made in the petrochemical industry, not only in plant, but also in R & D. There is also at present a severe world-wide overcapacity for many of the bulk chemicals.

In the circumstances, it is very unlikely that there will be any broad shift towards wood as a bulk chemical feedstock in the foreseeable future. Even if high prices and/or shortages caused some branches of the chemical industry to turn to feedstocks other than oil and gas, it is by no means certain that they would turn to wood, as other feedstocks, notably coal, are probably more attractive from an economic and technical point of view, and in abundant supply for the foreseeable future.

There are however a few chemical products whose molecular structure makes them easier to manufacture from wood than from oil — such as furfural, which is being produced from wood in several countries. It is likely that, where sufficient wood is available, at a suitable price, more of these chemical products could be manufactured from wood, especially where this processing can be in-

tegrated into a larger operation. It appears however, from the considerations above, that this use of wood as a chemical feedstock will neither increase radically nor have a significant effect on the overall supply-demand balance for wood.

A potential also exists for the use of wood (or residues of the forest products industries, or waste paper) as a source of *food*, usually for animal consumption (cellulose, the basic component of wood is not well suited to the human digestive system, and must therefore be broken down before it can be used for human consumption). Cattle feed has been prepared from needles and leaves ("muka"), and from waste paper. In addition, single cell proteins (SCP) for use as additives to cattle feed are grown on several mediums, including spent liquors from sulphite pulping (e.g. by the Pekilo process). In present conditions, however, these processes have only marginal importance and a major change does not appear likely as long as there are large agricultural surpluses in many parts of the developed world. The manufacture of animal feedstuffs is likely to remain a practical use for residues of various types in a few well-defined circumstances, when it is more profitable than the generation of energy (if a suitable market exists for the type of feedstuff produced), but it seems unlikely that it would ever develop into a major end-use sector for wood.

13.6 QUANTIFICATION OF ESTIMATES FOR WOOD CONSUMPTION AFTER 2000

It is clear from the above that great uncertainty surrounds the outlook for the distant future and it might seem rash to attempt any quantified forecast. Nevertheless, if a comparison is to be made with the forestry forecasts in chapter 5, it is necessary to dispose of at least some orders of magnitude.

It would be misleading to attempt projections based on sophisticated methodology such as that used in chapters 11-12 for at least two reasons:

- There is no certainty that the mathematical relationships established from data for 1964 to 1981 would still be valid after 2000. (Indeed, there is every likelihood that the relationships would weaken towards the end of the 1990s, if not before);

- There are no reliable estimates available for the independent variables (GDP, residential investment, manufacturing production) necessary to make projections.

Mathematical projections so far into the future would tend to confer spurious accuracy on the figures which would be put forward.

It did appear desirable to base the estimates on a relationship, however roughly calculated, with an in-

dependently forecast variable. Reasonably reliable forecasts for 2025 are only available for one variable — population. The method of estimation used in this chapter, unlike those used in earlier chapters, is based on scenarios for *per caput* consumption of forest products.

The estimates in this chapter are for 2025 as this is the year for which population forecasts are available, but they may be compared with forecasts for the forest in 2020 (chapter 5), because the speed of change is as uncertain as its direction and magnitude.

Data on population trends to 2025 are set out in chapter 1, with country data in annex table 1.1. They are summarized in table 13.1.

The population of the Nordic countries, the EEC(9) and Central Europe are all expected to decline slightly in the first quarter of the twenty-first century, that of Eastern Europe to expand slightly. The most noteworthy change is foreseen for Southern Europe whose population is expected to grow by 37 million between 1980 and 2000 and by a further 42 million between 2000 and 2025, taking the region from 24% of the European total to 33%. The population of Turkey is expected to increase by over 120% between 1980 and 2025, to reach 99 million, by far the largest in Europe.

TABLE 13.1

Population, 1980 and projections for 2000 and 2025

	Total population (millions)			Percentage share of European total		
	1980	*2000*	*2025*	*1980*	*2000*	*2025*
Nordic countries	11.4	17.6	16.9	3.3	3.0	2.7
EEC(9)	260.3	265.9	260.4	49.1	45.6	41.5
Central Europe	13.9	13.4	12.2	2.6	2.3	1.9
Southern Europe	128.5	165.4	207.2	24.2	28.4	33.0
Eastern Europe	109.8	121.1	131.2	20.7	20.8	20.9
Europe	529.9	583.4	627.9	100.0	100.0	100.0

Source: Annex table 1.1

Two *base scenarios* for 2025 have been prepared (see table 13.2). They are based on the conservative assumption that *the range of per caput consumption levels in 2025 is the same as that forecast for 2000 in chapter 12.* This would imply that:

– The basic pattern of consumption of forest products

would not change between 2000 and 2025, and radically new uses would not be developed;

– Forest products markets would mature around 2000 and thereafter change only in accordance with changes in population;

– Forest products would remain sufficiently competitive not to lose their traditional markets.

TABLE 13.2

Estimates of base scenarios for consumption of forest products in 2025

	1979-81 to 2000: scenarios from chapter 12						Estimates [a] (million units)	
	Per caput (unit/1000 cap)			Total consumption (million units)				
		2000			2000		2025	
	1979-81	*Low*	*High*	*1979-81*	*Low*	*High*	*Low*	*High*
A. *Sawnwood* (m³)								
Nordic countries..........	621	623	693	10.8	11.0	12.2	10	12
EEC(9).................	194	220	261	50.5	58.6	69.3	57	68
Central Europe...........	367	427	496	5.0	5.7	6.6	5	6
Southern Europe	111	126	162	14.3	20.8	26.8	26	34
Eastern Europe..........	198	190	214	21.7	23.0	25.9	25	28
Europe	193	204	241	102.3	119.0	140.8	123	148
B. *Wood-based panels* (m³)								
Nordic countries..........	149	175	215	2.6	3.1	3.8	3	4
EEC(9).................	80	98	120	20.8	26.2	31.8	26	31
Central Europe...........	101	129	160	1.4	1.7	2.2	2	2
Southern Europe	31	38	45	4.0	6.3	7.5	8	9
Eastern Europe...........	63	102	109	6.9	12.4	13.2	13	14
Europe	67	85	100	35.6	49.6	58.5	52	60
C. *Paper and paperboard* (m.t.)								
Nordic countries..........	190	274	367	3.3	4.8	6.5	5	6
EEC(9).................	126	166	229	32.8	44.1	60.8	43	60
Central Europe..........	137	184	244	1.9	2.5	3.3	2	3
Southern Europe	42	55	82	5.4	9.2	13.6	11	17
Eastern Europe...........	52	55	65	5.7	6.7	7.9	7	9
Europe	91	115	158	49.2	67.2	92.0	68	95
D. *Fuelwood* (m³)								
Nordic countries..........	575	875	1 142	10.0	15.4	20.1	15	19
EEC(9).................	70	83	117	18.1	22.1	31.2	22	30
Central Europe..........	158	187	358	2.2	2.5	4.8	2	4
Southern Europe	233	200	239	30.0	33.1	39.5	41	50
Eastern Europe..........	107	107	108	11.7	12.9	13.1	14	14
Europe	136	147	186	72.0	85.9	108.7	94	117

[a] Obtained by applying *per caput* consumption levels for 2000 to population data for 2025.

Although quantitative estimates for 2025 are presented in table 13.2, they are the result of a very simple calculation, totally different in nature from the methods used to prepare the scenarios for 2000. Their main purpose is to provide a rough order of magnitude for consumption which could be compared to the removals forecasts.

These estimates are for only small changes between 2000 and 2025 in consumption of forest products, with rises in Southern Europe compensating stagnation or slight declines elsewhere. This is due to trends in population.

With the extreme uncertainty inherent in any examination of the outlook forty years into the future, it is desirable to construct not only *base scenarios* based on a very conservative assumption, but also rough scenarios of *extreme cases, low and high*. Qualitative scenarios are presented below, followed by quantitative estimates.

A. SAWNWOOD

Low

As adequate accommodation is available for almost all residents, building land is scarce and costs are high, residential investment falls in absolute terms. There is however scope to improve the housing stock in Southern Europe. Furthermore, sawnwood loses ground to other building materials for combined technical and economic reasons (e.g. lack of product development and standardization). Use of pallets grows more slowly and sawnwood is replaced by other materials, including panels.

High

Effective demand for improved living conditions continues strong and residential investment expands. Sawnwood maintains its traditional uses in construction by increased standardization, technical development and marketing (possibly implying a more economic use pattern, i.e. *less* sawnwood consumed per volume built). Sawnwood maintains and improves its position as a prestige material for furniture and the use of sawnwood pallets expands.

B. WOOD-BASED PANELS

Low

Developments for residential investment as for sawnwood. The market penetration phase for wood-based panels is over. Other materials (metal, plastic, etc.) threaten the position of particle board in furniture manufacture. No significant new uses are developed. Panels lose their cost advantage over other materials in many sectors.

High

Developments for residential investment as for sawnwood. New panels and combinations of panels are developed and expand the intensity of use of panels in construction and furniture, as well as in new areas. Wood-based panels maintain their cost advantage over competing materials.

C. PAPER AND PAPERBOARD

Low

Many uses of printing and writing paper and newsprint are rendered obsolete by the development of electronic means of communication. Packaging grades lose their competitivity compared to other materials, notably plastic. *Per caput* consumption of household and sanitary papers remains stable.

High

Electronic communication continues to develop systems which incorporate a significant paper element (e.g. printouts). Traditional paper-based media (books, magazines, newspapers) remain dynamic. Packaging grades of paper and paperboard remain competitive compared to plastics. New uses are found for paper and paperboard, including in the household and sanitary paper sector. *Per caput* consumption levels of all grades of paper and paperboard in Eastern and Southern Europe rise, so that the gap between *per caput* levels of consumption in these countries and elsewhere diminishes.

D. FUELWOOD

Low

Conventional fuels (oil, coal, gas, nuclear) remain in abundant supply and renewable sources (solar, hydro, wind), are rapidly developed. The inconvenience of fuelwood use encourages its replacement by other fuels and the resumption of the long-term downward trend, interrupted between 1980 and 2000.

High

Conventional fuels become scarce and expensive, renewable energy sources prove difficult to develop. Governments strongly encourage the use of fuelwood as a domestic, renewable, decentralized source of energy. (In this scenario, there would be significant areas of energy plantations.)

What would be the results of these scenarios in volume terms? As stated above, sophisticated projection methods are not appropriate in the circumstances, but some, very rough, quantitative estimates would be useful for comparison with the removals forecasts in chapter 5. (This comparison will be carried out in chapter 21.)

There is enormous uncertainty as regards levels of residential investment, competitivity of building systems based on sawnwood and panels, the development of electronic communications and the broad energy supply/demand balance. It cannot therefore be ruled out that consumption levels in 2025 could be lower than those in 2000 – or even in 1980. On the other hand, in favourable circumstances, including retention or increase of market share, reasonably strong economic growth, and demand for wood-derived energy, levels of consumption could be well above the conservative assumptions of table 13.2.

On the basis of *subjective judgements*, the following very broad ranges of consumption in 2025 are proposed, as a foundation for discussion (million units):

	Base scenarios (table 13.2)	Extreme cases	
		Low	High
Sawnwood (m³)	123-148	90	160
Wood-based panels (m³)	52-60	40	70
Paper and paperboard (m.t.)	68-95	55	105
Fuelwood (m³)	94-117	50	130

FIGURE 13.1

Europe: estimates for consumption of forest products, to 2025

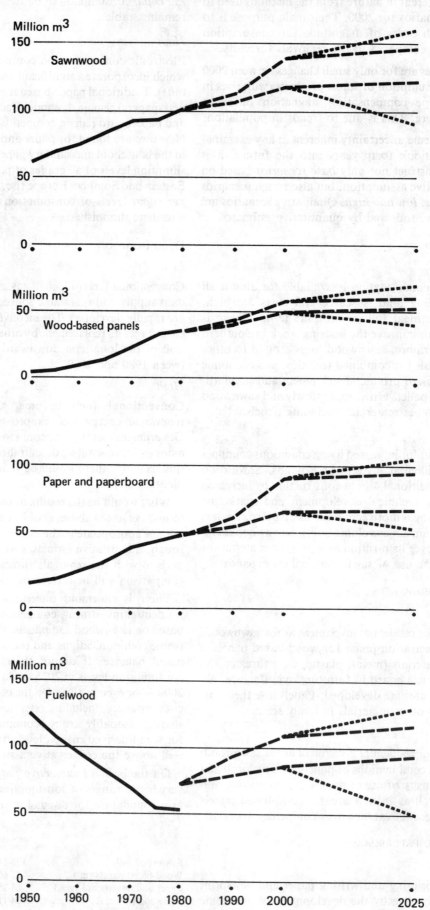

The main determining factors differ between forest products and are mostly not dependent on broad trends in economic growth: level of construction activity for sawnwood and panels, developments in electronic communication for printing and writing paper, energy price for fuelwood and so on. Thus, the "low" scenario for one product group may well co-exist with the "high" scenario for another.

13.7 DEMAND FOR THE NON-WOOD GOODS AND SERVICES OF THE FOREST

Chapter 4 has reviewed both the supply and demand of the non-wood goods and services of the forest. By the nature of the subject discussed, this review was largely non-quantitative in nature and did not apply rigidly to a defined period of time. It is therefore equally applicable to the period after 2000 as before.

It will be recalled that chapter 4 foresaw continuing growth in demand for the non-wood goods and services of the forest, particularly recreation, in a wide number of forms adapted to social circumstances and the nature of the forest in question.

In some parts of Europe, the contrast between the likely growth in demand for non-wood goods and services and the possible stagnation of demand for forest products may have important consequences for forest management and forest policy in general. This may become a central question of forest policy in the early twenty-first century.

13.8 CONCLUSIONS

The major factors affecting consumption of forest products identified above may be summarized as follows:

(a) Population growth;

(b) Macro-economic growth (which will certainly play a role, although the exact relationship between long-term GDP growth and consumption, especially of sawnwood and panels, needs further investigation);

(c) Level of residential investment;

(d) Technical and economic competitivity of sawnwood and panels in the building sector;

(e) Developments in electronic data processing and transmission and their acceptance by society;

(f) Technical and economic competitivity of paper and paperboard in the wrapping and packaging sector;

(g) The energy situation, especially the relative price and availability of different fuels;

(h) Changes in consumer habits.

While many of these − (a), (b), (c), (e) and (g) − are not susceptible to influence by the forest and forest products sector, which is of relatively minor importance in most countries, it should be stressed that the sector does have a strong influence over some others, notably the technical and economic competitivity of its products. It is to a large extent for the sector itself to decide whether consumption levels are to be allowed to drift down to the low "extreme case" for 2025 (or even lower) or whether the necessary measures will be taken to encourage higher consumption − control of costs, fundamental research, product development, standardization, marketing, etc.

CHAPTER 14

The European forest industries: structure and trends

14.1 INTRODUCTION

The forest industries[1] play a central role in the forest and forest products sector, as they are the crucial link, in physical and economic terms, between the forest, the source of wood raw material, and the consumers of forest products. Their function is to provide forest products with technical characteristics acceptable to the consumers (domestic or on export markets) from the available raw material. If they do not perform this function satisfactorily, consumers will substitute forest products with other materials or turn to imported forest products. In both cases, the domestic forest will become under-utilized, create less income and become therefore less well tended and frequently less able to provide non-wood benefits (these are probably often subsidised by the revenue from wood production). Furthermore, the forest industries provide employment, often in rural areas where there are few other employers, and thus may play an important role in the economic and social life of each country.

---

[1] This chapter does not address the questions of employment, value added, productivity, etc. in forestry (silviculture and harvesting) which are of a totally different nature and need a different methodological approach.

This chapter aims to present a brief overview of the situation of and trends in the European forest industries, in particular the following aspects:

- Geographical location;
- Trends in production and self-sufficiency;
- Size of manufacturing units;
- Capacity and capacity utilization;
- Place in the general economy;
- Trends in productivity.

Underlying this chapter is a concern for the economic health of the forest industries. If this health can be maintained and improved, then these industries can play their part in putting the European forests to the best possible use. If it is unsatisfactory, it is extremely unlikely that the physical potential of the European forests will be realized. It is acknowledged that precise statements concerning "economic health" can only be made at the level of companies, enterprises or small regions. Nevertheless, it is hoped that the analysis in this chapter will provide some broad indications of the situation and trends.

14.2 OVERVIEW OF GEOGRAPHICAL LOCATION AND RELATIVE SIZE OF THE INDUSTRIES

Although there are forest industries in every part of Europe, they are concentrated in the Nordic countries and the EEC(9). If consumption of raw material is taken as a rough indicator of industry size (table 14.1), it appears that in 1979-81 36% of the European total of those industries consuming wood raw material (i.e. sawmilling and manufacture of wood-based panels and woodpulp) were located in the Nordic countries and 26% in the EEC(9), a total of 62%. If paper and paperboard manufacture is considered, nearly half the European industry was in the EEC(9) and 26% in the Nordic countries. About a third of the European forest industries was in the EEC(9) and just under a third in the Nordic countries, with 16.2% in Eastern Europe, 11.6% in Southern Europe and 5.9% in Central Europe. The main trend since 1969-71 in the relative importance of the different groups has been an

increase in the share of Southern Europe and the EEC(9), at the expense of the Nordic countries, whose share fell from 37.8% to 32.2% between 1969-71 and 1979-81.

The data on raw material consumption per head, presented in table 14.1, show that the forest industries are much more important, relative to population, in the Nordic countries than elsewhere in Europe. The Nordic forest industries consume nearly 6m³/year of wood raw material for each inhabitant while the European average is just over 0.5m³/year. There is a similar picture for consumption of raw material for paper and paperboard. Raw material consumption per head of population in Central Europe, though much lower than in the Nordic countries is still significantly over the European average.

Table 14.2 shows that there are major differences between product groups in the distribution of capacity between regions. It appears that while sawmilling is distributed approximately in accordance with the volume of removals, over half the pulping capacity is located in the Nordic countries, which account for about one third of removals. On the other hand, nearly half the capacity of the European wood-based panel industries and of the paper and paperboard industries is located in the EEC(9) which accounts for only about 25% of removals. How is this possible, in raw material terms? In the first place there is specialisation, in that in the Nordic countries, a greater share of pulpwood (including sawmill residues) is directed to pulp, while in the EEC(9) it is directed more towards wood-based panels. For paper, the high relative importance of the EEC(9) is possible not so much because of its forest resource, but more because of pulp imports, notably from the Nordic countries, and waste paper recycling. The Nor-

dic share of paper capacity (27%) is significantly lower than its share of pulp capacity (54%).

The structural contrasts are not so great in other parts of Europe but it does appear that Central Europe is specialized in sawmilling, while Eastern Europe is specialized in sawmilling and wood-based panels. Central, Southern and Eastern Europe all have relatively low shares of the European pulp, paper and panels capacity, because of the concentration of these capacities in the Nordic countries and the EEC(9). Between them the latter two groups which account for 58% of removals of industrial wood, have about the same share of European capacity in sawmilling and wood-based panels, but over 70% in pulp and paper.

Sections 14.3 to 14.6 below present, for the four major industry sectors, the available data on structure and capacity and on trends in production and self-sufficiency.

TABLE 14.1

Consumption of raw material by the forest industries, 1969-71 and 1979-81

	Wood raw material [a]			Papermaking fibres [b]			Estimated percentage share of total raw material consumption [c]
	Volume			Volume			
	Total (million m³)	Per head (m³)	Per cent of total	Total (million m.t.)	Per head (m.t.)	Per cent of total	
1969-71							
Nordic countries	100.5	6.00	39.5	10.3	0.62	26.4	37.8
EEC(9)	70.5	0.28	27.7	20.6	0.08	52.9	31.1
Central Europe	14.1	1.36	5.5	1.8	0.18	4.5	5.4
Southern Europe	24.1	0.22	9.5	2.4	0.02	6.0	9.0
Eastern Europe	42.2	0.43	17.8	3.9	0.04	10.1	16.8
Europe	254.4	0.51	100.0	39.0	0.08	100.0	100.0
1979-81							
Nordic countries	101.2	5.80	36.0	13.4	0.76	26.0	32.2
EEC(9)	73.0	0.28	26.0	24.4	0.09	47.3	34.2
Central Europe	18.5	1.33	6.6	2.5	0.18	4.9	5.9
Southern Europe	35.5	0.28	12.7	5.0	0.04	9.8	11.6
Eastern Europe	52.6	0.47	18.7	6.2	0.06	12.1	16.2
Europe	280.8	0.53	100.0	51.5	0.10	100.0	100.0

[a] Apparent consumption of sawlogs and round pulpwood, plus net trade in residues, chips and particles.

[b] Apparent consumption of pulp and waster paper.

[c] Apparent consumption of wood raw material, plus papermaking fibres, converted to wood raw material equivalent. (Woodpulp is double counted as the objective is to estimate the size of the industries, not to establish a raw material balance.)

TABLE 14.2

Share of country groups in European total, for industry capacity and removals, around 1980

(Percentages)

	Sawnwood [a]	Wood-based panels	Pulp	Paper and paperboard	Removals of industrial wood
Nordic countries	25.4	12.6	54.3	26.7	33.6
EEC(9)	28.2	46.1	17.3	46.5	24.4
Central Europe	9.0	5.7	4.4	5.1	5.7
Southern Europe	14.6	15.4	11.2	10.9	12.9
Eastern Europe	22.8	20.2	12.8	10.8	23.4
Europe	100.0	100.0	100.0	100.0	100.0

[a] Share of production, as capacity data not complete.

TABLE 14.3

Self-sufficiency ratios, 1969-71 and 1979-81

(*Production as percentage of apparent consumption*)

	Sawnwood		Wood-based panels		Woodpulp		Paper and paperboard	
	1969-71	*1979-81*	*1969-71*	*1979-81*	*1969-71*	*1979-81*	*1969-71*	*1979-81*
Finland	276	303	317	241	146	133	590	521
Iceland	–	–	–	–	–	–	–	–
Norway	89	98	124	92	157	117	296	266
Sweden	231	215	154	154	184	155	283	373
Nordic countries	211	212	186	161	164	141	366	405
Belgium-Luxembourg	42	33	164	204	56	55	71	63
Denmark	42	43	54	56	74	60	39	32
France	89	81	95	93	62	55	86	84
Germany, Fed. Rep. of	71	70	92	88	53	47	72	798
Ireland	11	25	111	30	33	57	45	25
Italy	39	32	110	83	41	29	97	96
Netherlands	9	9	27	13	23	22	89	778
United Kingdom	12	20	17	20	13	12	67	53
EEC(9)	52	51	81	76	41	39	76	75
Austria	263	225	142	194	106	107	164	177
Switzerland	80	81	83	112	56	53	77	89
Central Europe	176	164	110	154	89	90	112	130
Cyprus	26	56	–	–	–	–	–	–
Greece	36	36	91	102	–	25	64	68
Israel	–	–	155	80	–	–	45	45
Malta	–	–	–	–	–	–	–	–
Portugal	129	171	196	140	318	224	81	123
Spain	70	71	104	128	69	91	90	96
Turkey	102	100	101	100	87	90	61	86
Yugoslavia	122	120	94	107	95	86	95	101
Southern Europe	93	97	104	114	91	100	83	92
Bulgaria	93	92	118	116	65	71	73	78
Czechoslovakia	116	126	93	96	92	91	99	110
German Dem. Rep.	59	64	81	76	83	82	88	90
Hungary	54	63	75	84	44	34	62	68
Poland	107	107	97	93	83	76	89	89
Romania	160	127	133	135	97	91	108	118
Eastern Europe	103	101	98	98	84	80	89	93
EUROPE	90	91	96	94	92	88	100	103

Annual data on production are published in the *Timber Bulletin*. Long-term series will also be published separately and are not therefore presented here.

National self-sufficiency ratios for the major product groups in 1970 and 1980 are presented in table 14.3 and commented on below.

14.3 THE EUROPEAN SAWMILLING INDUSTRY [2]

Data collected on the structure and capacity of the sawmilling industry are not yet comprehensive but from the data in annex table l4.5 it is possible to make some tentative estimates.

14.3.1 Number and size of mills

There are many very small sawmills in Europe. For the countries shown in annex table 14.5, nearly 90% of the

sawmills had a production capacity of less than 5,000 m³ of sawnwood a year. On the assumption that these countries are representative (they represent over 80% of European production and cover all country groups) it is possible to estimate the number of mills for Europe as a whole.

The only exception to the pattern is Eastern Europe where 56% of the mills have capacities of under 5,000 m³(s)/year. Yet this may be an underestimate (for

---

[2] The data in this section are taken from the latest enquiry on the structure of the sawmilling industry, published in 1985 (FAO/ECE *Timber Bulletin*, Volume XXXVIII, No. 3). Data collected for an earlier

survey were less complete and possibly not comparable with the later results. It has not, therefore, been possible to draw conclusions about trends over time.

Poland, only State-owned mills are considered, accounting for 80% of the production. It is likely that the Polish mills not included are the smaller ones).

	Number of mills	Percentage with production capacity of under 5 000 m³(s)/year
Nordic countries	9 789	92
EEC(9)	14 004	97
Central Europe	3 435	92
Southern Europe (5 countries)	5 824	90
Eastern Europe (3 countries) .	980	56
Europe (estimated)	42 000	90

It is likely that many of the smaller mills in Europe are not well equipped. It is known that in some countries they do not operate all the time, as they are brought into use only at specific seasons or when there is strong demand. They usually supply a local market and are often not capable of reaching the same quality levels as larger mills. Nonetheless, frequently their capital costs have long since been discounted and overhead costs are lower, so that they are able to remain economically viable. In addition they may have a better raw material input/output ratio, and their production may be well adapted to local needs.

14.3.2 Capacity of the sawmilling industry

Although 90% of the sawmills in Europe are small (under 5,000 m³(s)/year), most of the production capacity is in larger mills. For the countries which provided data on both number and capacity of mills, just under 7% of the mills − those with capacity over 5,000 m³(s)/year − accounted for over 65% of capacity. As the capacity utilization rates of larger mills are likely to be higher than those of small mills, because of higher fixed costs, it is likely that larger mills account for an even higher share of production than they do of capacity.

There is a very wide range in size within the group of "larger" mills: indeed, in many countries, mills of 5,000 m³(s)/year capacity would not be considered "large". Of the 3,236 "large" mills in annex table 14.5, 76 have a capacity of over 100,000 m³(s)/year. Of these, 36 are in Finland, 17 in Sweden and 6 in Austria. There are reported to be at least a further 134 mills with a capacity of between 50,000 and 100,000 m³(s)/year. Mills over 50,000 m³(s)/year accounted for 64% of production in Finland and 40% in Sweden.

Figures for European sawmilling capacity, as well as capacity utilization ratios at the national level are not very meaningful, as many countries cannot provide data on capacity or else show a very large theoretical overcapacity due to the many small under-used mills. Others have estimated capacity as the maximum recorded production level in a recent year. Cyclical factors also play a role.

It cannot be assumed that small mills are necessarily uneconomic. Indeed in some cases, a small to medium-size, family-run mill, with no new machines, can be very profitable, as the capital has been written off and the owners have great flexibility in operating the mill. Never-theless, it is likely that alongside this type of profitable small mill, there are many which are badly managed and unprofitable. It does appear from fragmentary data in the capacity surveys that a substantial number of mills, the majority of them small, have closed over the past 10-15 years. This trend could continue.

The crucial problem for the European sawmilling industry in the future is not to match a theoretical figure for "total capacity" with that for production, but rather to ensure that operating mills are efficient from an economic and technical point of view.

14.3.3 Trends in production and self-sufficiency of sawnwood

Production, like consumption, of sawnwood in the 1970s was dominated by cyclical factors, as is evident from fig. 14.1. The 10% difference between the total production data for 1969-71 and 1979-81 shown in table 14.4 cannot therefore be considered an indication of a trend to growth. A few structural trends may however be deduced from the table.

European production as a percentage of consumption (the self-sufficiency ratio) is around 90% and little change in this global ratio is visible over the decade at the regional level. This conceals some contrasting trends in self-sufficiency at the national level, as well as for individual products.

Countries with a production of sawnwood significantly higher than domestic consumption were Finland, Sweden, Austria, Portugal, Yugoslavia, Czechoslovakia and Romania. Changes in the relationship between production and consumption in these countries is very strongly affected by cyclical factors, as exports and therefore a part of production are much more volatile than domestic consumption.

Among net importers, self-sufficiency in sawnwood improved significantly in the 1970s in Ireland, the United Kingdom, Cyprus, Greece, the German Democratic Republic and Hungary, whereas this ratio declined markedly in Belgium-Luxembourg, France and Italy.

Production of sawn softwood, the major assortment, accounting for over 75% of European production, rose faster than that of sawn hardwood. There are strongly contrasting trends in the self-sufficiency ratios for the two products (see fig. 14.1): that for sawn softwood rose steadily from the mid-1960s to the early 1980s for a total rise of nearly 10 percentage points, while the movement for sawn hardwood was equally strongly downward, although the cyclical fluctuations were more marked. Among the reasons for these trends include the following:

Sawn softwood

− Rising domestic production by net importers;

− Rising exports to destinations outside Europe;

− USSR exports to Europe have not expanded in line with European consumption.

TABLE 14.4

Production of and self-sufficiency in sawnwood, 1969-71 and 1979-81

	Production (million m³)		Percentage share of European total		Self-sufficiency percentage of apparent consumption	
	1969-71	1979-81	1969-71	1979-81	1969-71	1979-81
Nordic countries	21.4	22.8	25	25	211	212
EEC(9)	25.1	25.8	30	28	52	51
Central Europe	6.9	8.3	8	9	176	164
Southern Europe	9.6	13.9	11	15	93	97
Eastern Europe	21.4	22.2	25	24	103	101
Europe	84.4	93.0	100	100	90	91
of which:						
Sawn softwood	64.2	72.1	76	78	89	92
Sawn hardwood	18.2	19.3	22	21	94	86
Sleepers	2.0	1.7	2	2	95	92

Sawn hardwood

- Increased market penetration by sawnwood imports from south-east Asia and to a lesser degree, the USA. Some of these sawnwood imports replaced imports of sawlogs for processing in Europe. The decline in self-sufficiency ratios for sawnwood as a whole has been marked in some countries which process large volumes of tropical logs (e.g. France, Italy, Belgium-Luxembourg).

Production of sawnwood in Southern Europe grew faster than elsewhere, causing this region to expand its share of the European total and to become a net exporter. Between the mid-1960s and the early 1980s, sawnwood production rose from 1.5 million m³ to 2.2 million m³ in Portugal, from 1.8 million m³ to 2.8 million m³ in Spain, from 1.3 to 4.3 million m³ in Turkey and 2.7 million m³ to 4.4 million m³ in Yugoslavia. Nevertheless, the regions with the highest production remained (in order) the EEC(9), the Nordic countries and Eastern Europe.

14.4 THE EUROPEAN WOOD-BASED PANELS INDUSTRIES[3]

14.4.1. Number and size of mills

In 1982 there were 389 recorded *particle board* mills in Europe, six fewer than in 1972. This overall change, however, concealed differences at the regional level: in the EEC(9) the number of mills dropped by 41 and in Central Europe by 10, while it increased by 28 in Eastern Europe and 18 in Southern Europe. 38% of European mills were in the EEC(9) and 27% in Eastern Europe.

These changes in numbers were part of a broader process, aimed at rationalizing the industry in the face of the structural overcapacity which developed in the mid-1970s, when overcapacity may have changed from being cyclical to being structural. A reduction in the number of mills was accompanied by a significant increase in their average size as some less efficient mills were closed. Capacity additions were mostly in the form of extensions to existing plants rather than totally new plants. In Central Europe, the average size of plants tripled over the decade and nearly doubled in the EEC(9) and Southern Europe. The average size of plants was largest in Central Europe and the EEC(9), followed by the Nordic countries. There were large inter-regional differences: the average size of a particle board plant in Central Europe was nearly four times the average for Eastern Europe.

The average size of *plywood* mills is much smaller (the European averages are 9900 m³/year for plywood and 74900 m³/year for particle board). The number of plywood mills dropped by a third (210 mills) between 1972 and 1982. This decline was concentrated in the EEC(9) where the number of plywood mills was halved (215 mills less). It was mostly the older smaller mills which closed. As a result, the average size of mills in the EEC(9) grew by 50%. There were mill closures elsewhere, but an increase of 25 mills in Southern Europe. This drastic restructuring was the result of the loss of market share to imports, notably from south-east Asia..

The *fibreboard* industry is also faced with stagnating demand (except for some new types of board such as MDF), but the changes in structure were less radical than for plywood. In the Nordic countries, the EEC(9) and Central Europe, the number of mills dropped and their average size increased. In Southern and Eastern Europe, however, the number of mills rose over the decade.

14.4.2 Trends in capacity and capacity utilization

In any industry where fixed costs, notably capital costs, account for a high share of total costs, it is very important to maximise the level of production, i.e. to keep the

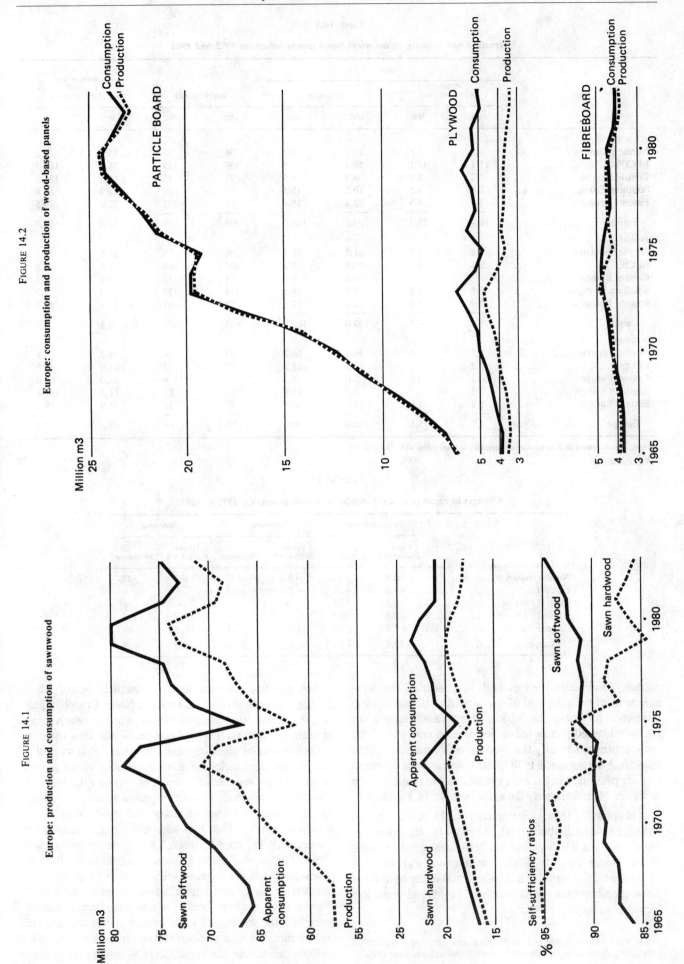

FIGURE 14.1

Europe: production and consumption of sawnwood

FIGURE 14.2

Europe: consumption and production of wood-based panels

TABLE 14.5

Structure and capacity of the wood-based panels industries 1972 and 1982

| | Capacity | | | | | | Average size of mill (1 000 m³/year) | |
| | Volume (million m³/year) | | Percentage | | Number of mills | | | |
	1972	1982	1972	1982	1972	1982	1972	1982
Particle board								
Nordic countries	2.0	2.6	10.7	8.9	36	35	55.6	74.6
EEC(9)	11.1	15.1	59.4	51.9	190	149	58.6	101.5
Central Europe	1.2	2.0	6.5	6.9	21	11	58.1	181.8
Southern Europe	1.9	4.5	10.2	15.5	71	89	27.0	50.8
Eastern Europe	2.5	4.9	13.1	16.8	77	105	32.0	46.5
Europe	18.7	29.1	100.0	100.0	395	389	47.4	74.9
Plywood								
Nordic countries	1.2	0.8	21.5	18.7	40	35	28.8	23.1
EEC(9)	2.4	1.7	44.5	39.2	403	188	5.9	9.0
Central Europe	–ᵃ	–ᵃ	0.8	1.0	11	10	3.9	4.3
Southern Europe	0.8	0.8	15.8	18.5	114	139	7.4	5.7
Eastern Europe	0.9	1.0	17.4	22.6	79	65	11.8	15.0
Europe	5.4	4.3	100.0	100.0	647	437	8.3	9.9
Fibreboard								
Nordic countries	1.7	1.5	36.7	26.2	23	18	75.2	80.8
EEC(9)	1.2	1.2	26.3	21.2	27	25	45.9	47.1
Central Europe	0.2	0.2	5.0	4.3	7	4	33.6	60.0
Southern Europe	0.3	0.7	7.3	11.9	11	21	31.5	31.6
Eastern Europe	1.2	2.0	24.6	36.4	27	36	42.9	56.2
Europe	4.7	5.6	100.0	100.0	95	104	49.6	53.5

Note: Annual country data on capacity are presented in annex tables 14.6, 14.7 and 14.8.

ᵃ 0.04 million m³ .

TABLE 14.6

Changes in capacity of the wood-based panels industries, 1972 to 1982

| | Particle board | | Plywood | | Fibreboard | |
	Volume (million m³)	Percentage	Volume (million m³)	Percentage	Volume (million m³)	Percentage
Nordic countries	+0.6	+30	–0.4	–33	–0.2	–12
EEC(9)	+4.0	+36	–0.7	–29	–	–
Central Europe	+0.8	+67	–	–	–	–
Southern Europe	+2.6	+137	–	–	+0.4	+133
Eastern Europe	+2.4	+96	+0.1	+11	+0.8	+67
Europe	+10.4	+56	–1.1	–20	+0.9	+19

capacity utilization level as high as possible. The measurement of capacity and of the capacity utilization ratio have therefore been the subject of increased interest for the wood-based panels industries over recent years. This section presents briefly the broad trends and main problems in this area since 1970. Year-by-year data on capacity are presented in the annex tables and in graphic form in figure 14.3. Summary data are in table 14.5 and 14.6.

It is apparent that while capacity in the particle board industry rose until about 1980, the trend for the other two panels over the whole period has been for stagnation, even decline, since the early 1970s. For all country groups and all products, the capacity utilization ratio declined and the difference between production and registered capacity[4] increased over the 1970s.

[4] In view of the problems in estimating and defining "production capacity", these measures should be considered approximate only.

The greatest changes were for particle board which during this period changed from a phase of rapid expansion in consumption and production to one of relative stagnation. In the mid-1970s, because of the lags inherent in the process of planning investments in capacity and of continuing expectation of a resumption of rapid growth in production, the rate of increase of capacity did not slow down by as much as growth in production. As a result, the difference between capacity and production at the European level, which had been around 2 million m³ in the early 1970s, grew to reach 5-6 million m³ in the early 1980s. This caused severe economic problems for the particle board industry – downward pressure on prices combined with high fixed capital costs. From around 1980, efforts were made to bring the situation under control, including the closing of some capacity, but there is still (in the mid-1980s) a structural overcapacity of around 4 million m³ at the European level. The biggest volume

TABLE 14.7

Capacity and capacity utilization in the wood-based panels industries

	Production (million m³)		Capacity (million m³)		Difference between production and capacity (million m³)		Capacity utilization ratio (percentage)	
	1972	1982	1972	1982	1972	1982	1972	1982
Nordic countries	4.1	3.6	4.9	4.9	0.8	1.3	84	73
EEC(9)	13.7	13.9	14.7	18.0	1.0	4.1	93	77
Central Europe	1.4	1.7	1.4	2.2	—	0.5	100	77
Southern Europe	2.5	4.3	3.0	6.0	0.5	1.7	83	72
Eastern Europe	4.2	6.6	4.6	7.9	0.4	1.3	91	84
Europe	25.9	30.2	28.8	39.0	2.9	8.8	90	77
of which:								
Particle board	16.8	22.9	18.7	29.1	1.9	6.2	90	79
Plywood	4.6	3.4	5.4	4.3	0.8	0.9	85	79
Fibreboard	4.5	3.9	4.7	5.6	0.2	1.7	96	70

increase in capacity between 1972 and 1982 was in the largest producing area, the EEC(9), but faster growth occurred in Southern and Eastern Europe which both increased their share of European capacity at the expense of the EEC(9) and the Nordic countries. The capacity utilization ratio declined from around 90% to around 80% between the early 1970s and the early 1980s.

By contrast, over the 1970s and 1980s in Europe both production and capacity of plywood decreased steadily and roughly in harmony with each other. The capacity utilization ratio declined less than for the other panels as capacity was adjusted to the production possibilities. Most of the decline in capacity occurred in the Nordic countries and the EEC(9), where capacity was reduced by about a third.

The European fibreboard industry also had to adjust to a situation of stagnating, even declining, production. In the early 1970s, production and capacity rose, but from the mid-1970s, capacity declined in the Nordic countries and in the EEC(9), although it rose in Eastern Europe. The net result was a severe drop in the average capacity utilization ratio.

For panels as a whole, capacity utilization in 1982 was rather higher than average in Eastern Europe and lower in the Nordic countries and Southern Europe.

14.4.3 Trends in production and self-sufficiency of wood-based panels

The 1970s saw sharply contrasting trends for production of the different panels: a decline for plywood, stagnation for fibreboard and veneer sheets, but a doubling for the dominant panel, particle board. Nevertheless, as mentioned elsewhere in this study, it appears that during the 1970s, the period of fast uninterrupted growth of production (and consumption) of particle board came to an end (see figure 14.2). There was a temporary halt in 1973-1975 followed by a resumption of growth, but at a slower rate, to 1980. Thereafter, it is difficult to discern a clear upward trend at all. As for sawnwood, production of wood-based panels grew faster in Southern Europe, which raised its share of European production from 9% to 14%. Central and Eastern Europe also slightly increased their share, while those of the Nordic countries and the EEC(9) fell. Nevertheless the EEC(9) remains by far

TABLE 14.8

Production of and self-sufficiency in wood-based panels, 1969-71 and 1979-81

	Production (million m³)		Percentage share of European total		Self-sufficiency (percentage of apparent consumption)	
	1969-71	1979-81	1969-71	1979-81	1969-71	1979-81
Nordic countries	3.5	4.2	16	12	186	161
EEC(9)	11.7	16.1	53	48	81	76
Central Europe	1.1	2.1	5	6	110	154
Southern Europe	2.0	4.6	9	14	104	114
Eastern Europe	4.0	6.8	18	20	98	98
Europe	22.2	33.6	100	100	96	94
of which:						
Particle board	12.5	24.1	56	72	100	100
Plywood	4.1	3.7	18	11	84	68
Fibreboard	4.2	4.3	19	13	102	98
Veneer sheets	1.4	1.5	6	4	98	82

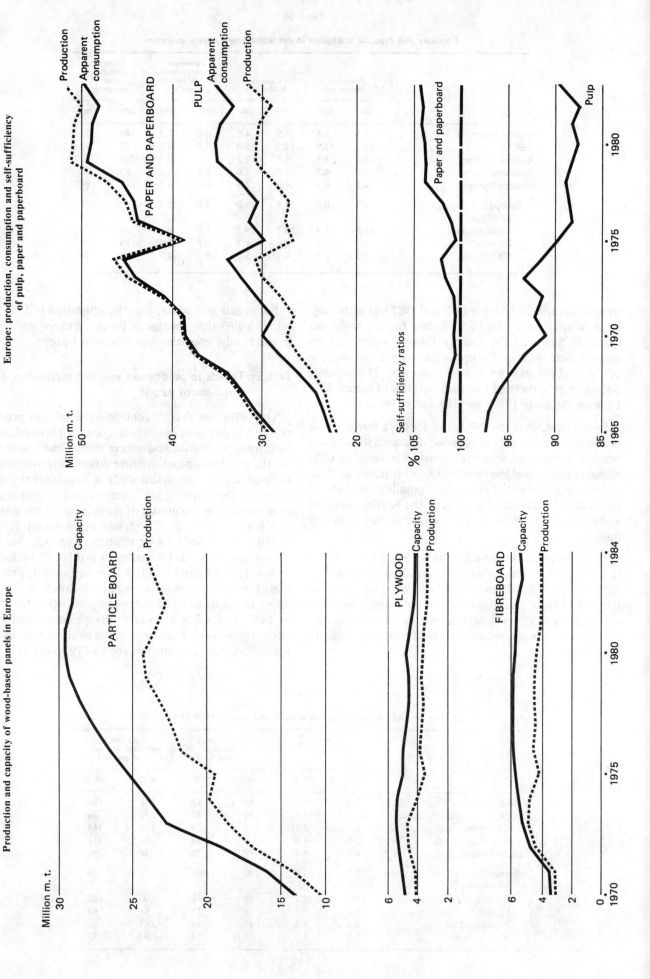

FIGURE 14.4

Europe: production, consumption and self-sufficiency of pulp, paper and paperboard

FIGURE 14.3

Production and capacity of wood-based panels in Europe

the most important European producing region (48% of the total) and one country, the Federal Republic of Germany, accounted for 21% of European wood-based panels production in 1979-81.

Europe is almost entirely self-sufficient in particle board and fibreboard, and as figure 14.2 shows, production and consumption move very closely together. For plywood and veneer sheets, however, self-sufficiency has dropped markedly as European production has stagnated or fallen, in the face of competition from imported panels, notably of tropical species. The fall in the panels self-sufficiency ratio for a number of European countries can be explained mainly by developments for plywood and veneer sheets. Examples are Finland, Norway, Italy, Portugal and Israel among countries which were net exporters in 1970 (some had become net importers of panels by 1980). Among net importers, drops in the self-sufficiency ratio can be at least partly explained by competition from tropical plywood and veneer sheets (France, the Federal Republic of Germany and the Netherlands). Countries more specialized in other panels saw their self-sufficiency ratios improve: Belgium-Luxembourg, the United Kingdom, Austria, Switzerland, Greece, and Spain.

14.5 THE EUROPEAN PULP AND PAPER INDUSTRIES

The economic health of the capital intensive pulp and paper industry with its large units is also dependent on maintaining a satisfactory capacity utilization rate. For this reason the quality and quantity of capacity data are far higher for this sector than for other forest industries. In particular, FAO, with the help of its Advisory Committee on Pulp and Paper, publishes medium-term forecasts for capacity which are updated annually. These surveys are the source of the information presented in this section. They do not, however, provide data on the number and average size of mills.

14.5.1 Trends in pulp and paper capacity

Data on production capacity by country group in 1970 and 1984 (the latest year for which historical data, as opposed to forecasts, are available) are presented in table 14.9 and a breakdown by grade of capacity in 1983 in table 14.10. Annual data by country, for total pulp and total paper are in annex tables 14.9 and 14.10. Figures 14.5 and 14.6 show the trends in graphic form.

More than half the European pulp capacity is situated in the Nordic countries, and the volume increase in capacity in these countries between 1970 and 1984 was the largest of all groups. Nevertheless, the share of the Nordic countries in the European total fell because of the very fast growth in capacity in Southern and Eastern Europe. Capacity in Eastern Europe increased over the period by over 60% and in Southern Europe more than doubled. Pulp capacity in the EEC(9) however decreased by 11%, and that region's share of the European total fell from 22% to 16%.

Pulp capacity did not grow steadily over the period: there are two distinct phases (see figure 14.5). Until 1978, total European pulp capacity increased at a fairly constant rate. From then to 1983 however, there was no growth in total capacity with a slight increase in 1984 and 1985. In the second period, after 1978, investments in capacity were concentrated on modernising or expanding existing machines, with the aim of limiting costs and increasing competitivity. Several uneconomic mills were closed altogether, and many others encountered severe financial problems. In particular several groundwood and sulphite pulp units were closed and investments concentrated on sulphate and high yield mechanical pulps.

In 1983, 60% of European pulp capacity was for chemical pulp (including semi-chemical) for paper-making and 33% for mechanical pulp. Pulp of other fibres and dissolving pulp (wood and non-wood) accounted for only 3% and 4% respectively. There are, however, quite large differences between country groups in the grade structure of the pulp industry. Mechanical pulp was more important, in relative terms, in the EEC(9) where it accounted for 47% of capacity, but less important in Southern and Eastern Europe (18-19% of capacity). Other fibre pulp was relatively more important than elsewhere in Southern Europe and the EEC(9) and dissolving pulp in Eastern Europe.

Chemical pulp capacity in the Nordic countries accounted for about 55% of European chemical pulp capacity and for a third of total European pulp capacity, demonstrating how important these grades and that region are for European pulp capacity. These plants produce the market pulp which is often processed elsewhere, notably in the EEC(9), as well as pulp for integrated operations.

Capacity for paper and paperboard also rose over the fourteen-year period, by 44%. Unlike pulp capacity it did not show stagnation or slow growth after 1978. Growth in paper and paperboard capacity was particularly rapid in Central, Southern and Eastern Europe. In Southern Europe, capacity nearly tripled. Growth in the EEC(9), however, was considerably slower, so that this group's share of the European total fell from 56% to 46%. The EEC(9) however still has by far the largest share of European paper and paperboard capacity (the next largest is the Nordic countries with 27%). Within the EEC(9) however there were differing trends. Paper and paperboard capacity in the Federal Republic of Germany grew between 1970 and 1983 by 45%, faster than the European average, by 29% in Italy and by 27% in France. In the United Kingdom, however, capacity dropped by 30%: in

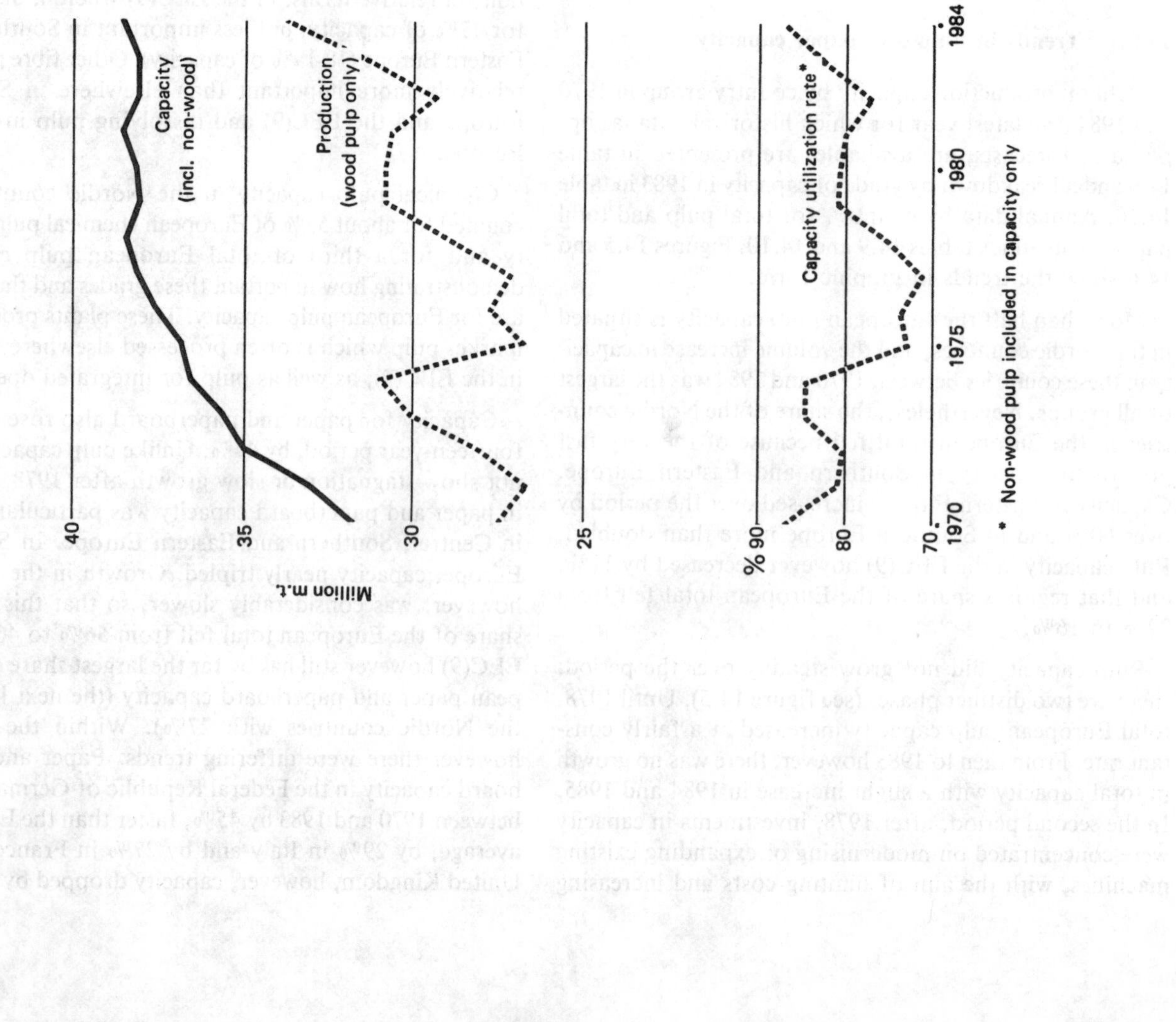

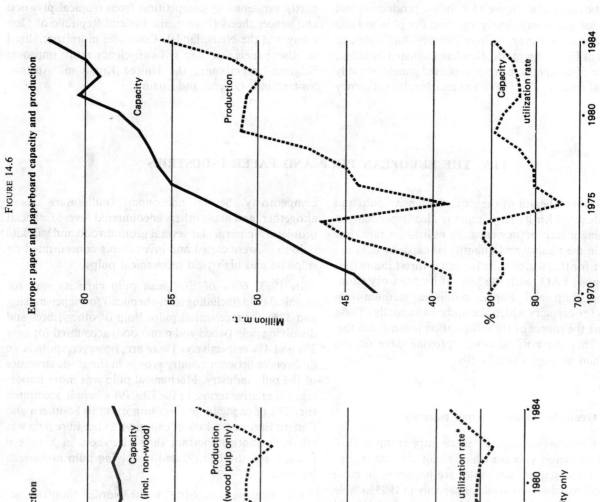

FIGURE 14.6

Europe: paper and paperboard capacity and production

Capacity

Production

Capacity

utilization rate

FIGURE 14.5

Europe: pulp capacity and production

Capacity
(incl. non-wood)

Production
(wood pulp only)

Capacity utilization rate*

* Non-wood pulp included in capacity only

FIGURE 14.7

European capacity for pulp and for paper and paperboard
(1970 = 100)

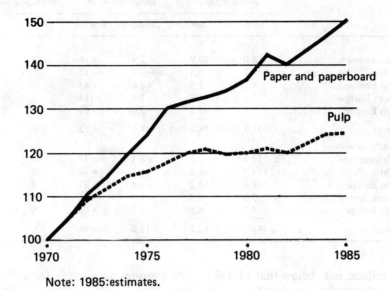

Note: 1985:estimates.

FIGURE 14.8

Pulp and paper capacity, 1983

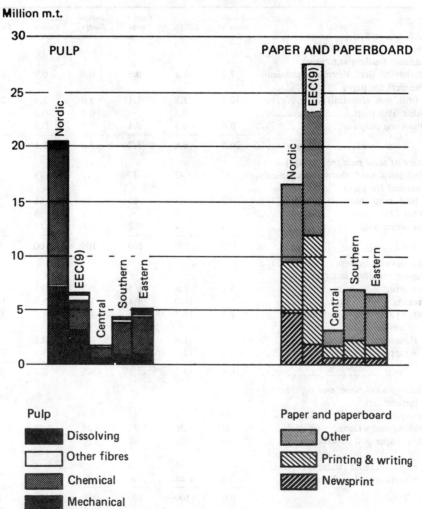

Pulp	Paper and paperboard
Dissolving	Other
Other fibres	Printing & writing
Chemical	Newsprint
Mechanical	

TABLE 14.9

Pulp and paper capacity 1970 and 1984

	Capacity (million m.t./year)		Change 1970 to 1984		Percentage of total Europe	
	1970	1984	Quantity	Percentage	1970	1984
Pulp						
Nordic countries...........	18.0	20.9	+2.9	+16	57	54
EEC(9)...................	7.1	6.4	−0.7	−11	22	16
Central Europe............	1.2	1.8	+0.6	+43	4	5
Southern Europe	2.1	4.5	+2.4	+114	7	12
Eastern Europe............	3.2	5.2	+2.0	+61	10	13
Europe	31.7	38.7	+7.0	+22	100	100
Paper and paperboard						
Nordic countries...........	10.8	17.1	+6.3	+57	25	27
EEC(9)...................	24.2	28.5	+4.3	+18	56	46
Central Europe............	1.8	3.2	+1.4	+83	4	5
Southern Europe	2.3	6.7	+4.4	+191	5	11
Eastern Europe............	4.0	6.5	+2.5	+63	9	11
Europe	43.1	62.1	+19.0	+44	100	100

1984 capacity was 1.6 million m.t. below that of 1970.

If paper and paperboard capacity is analysed by grade, it appears that wrapping and packaging is the largest product group, accounting for 45% of the European total, followed by printing and writing papers (31%) and newsprint (13%).

TABLE 14.10

Capacity of the European pulp and paper industry, by grade, 1983

	Nordic countries	EEC(9)	Central Europe	Southern Europe	Eastern Europe	Europe
PULP						
Capacity (million m.t./year)						
Mechanical (incl. thermo-mechanical).	7.2	3.1	0.5	0.8	0.9	12.4
Chemical for paper (incl. semi-chemical).............	12.7	2.8	1.1	3.0	3.4	23.0
Other fibre pulp	–	0.5	–	0.3	0.2	1.1
Dissolving pulp	0.6	0.3	0.1	0.2	0.5	1.7
Total	20.5	6.6	1.7	4.3	5.1	38.1
Share of total pulp (percentage)						
Mechanical (incl. thermo-mechanical).	35	47	27	18	19	33
Chemical for paper (incl. semi-chemical).............	62	43	65	70	66	60
Other fibre pulp	–	7	–	8	5	3
Dissolving pulp	3	4	8	4	10	4
Total	100	100	100	100	100	100
PAPER AND PAPERBOARD						
Capacity (million m.t./year)						
Newsprint......................	4.6	1.8	0.5	0.5	0.5	7.8
Printing and writing...............	4.8	10.0	1.1	1.7	1.2	18.9
Other paper and paperboard	7.1	15.7	1.5	4.5	4.7	33.5
of which:						
Household and sanitary...........	0.5	1.8	0.2	0.5	0.2	3.2
Wrapping and packaging	5.9	12.1	1.2	3.7	4.1	26.9
Total	16.5	37.5	3.1	6.8	6.4	60.2
Share of total paper and paperboard (percentage)						
Newsprint......................	29	6	15	8	7	13
Printing and writing...............	29	36	37	25	19	31
Other paper and paperboard	43	57	48	67	74	56
of which:						
Household and sanitary...........	3	7	6	7	4	5
Wrapping and packaging	36	44	40	55	63	45
Total	100	100	100	100	100	100

Figure 14.7 shows in index form the trends in European capacity for pulp and for paper and paperboard. Since 1972 paper and paperboard capacity has grown significantly faster than pulp capacity, which has more or less stagnated. A partial explanation of this is the growth in net imports of pulp (2.7 million m.t. in 1970, 4.2 million m.t. in 1982). The rise in use of waste paper and of coatings and fillers must, however, also be taken into account, especially as a similar difference between trends has been observed at the world level (where net trade clearly plays no role).

14.5.2 Capacity utilization

As mentioned above, the economic health of the pulp and paper industries with their high fixed costs is determined to a large degree by their rates of utilization of capacity. In particular, prolonged periods of low capacity utilization may be very damaging: losses in such periods can hinder necessary investments to increase productivity, develop new products and lower costs. It is not possible to analyse here trends in rates of utilization of capacity in greater detail, but the main lines for pulp and paper and paperboard are presented in figures 14.5 and 14.6.

The figures show clearly the negative effect of the recession in the mid-1970s. The steep drops in production of pulp and paper in 1975 were reflected in the capacity utilization rates which dropped to just over 70%. Thereafter production, and the utilization rate, recovered more quickly for paper and paperboard than for pulp. The latter did not recover the 80% level until 1979, whereas this was achieved in 1976 for paper and paperboard. The four years of low capacity utilization proved harmful to many pulp makers. The record capacity utilization rates of 1973 and 1974 were only approached in 1984.

There is a difference, between the medium-term trends for pulp and for paper in that production of the former seems to show hardly any upward trend at all, while that of the latter does move upwards (both lines are of course very much affected by cyclical fluctuations). These developments have been reflected, with a certain delay, in the trends for capacity. Total European pulp capacity did not increase between 1978 and 1983 (and only slightly in 1984 and 1985), while paper and paperboard capacity did.

In this context, two aspects should be mentioned:

(a) Frequently and for a number of different reasons, trends in capacity utilization vary according to the product and the country concerned;

(b) Data for total capacity do not distinguish between old and new machines. The total capacity may remain the same over a period when the efficiency and productivity of an industry are substantially improved as new efficient equipment replaces older and less efficient equipment. This has in fact been the case in recent years.

14.5.3 Trends in production and self-sufficiency of pulp and paper

Production of *woodpulp* in 1979-81 was about 15%

TABLE 14.11

Production of and self-sufficiency in woodpulp, paper and paperboard, 1969-71 and 1979-81

	Production (million m.t.)		Percentage share of European total		Self-sufficiency (percentage of apparent consumption)	
	1969-71	1979-81	1969-71	1979-81	1969-71	1979-81
Woodpulp						
Nordic countries	16.0	17.5	60	57	164	141
EEC(9)	5.4	5.4	20	18	41	39
Central Europe	1.2	1.5	4	5	89	90
Southern Europe	1.6	2.9	6	9	91	100
Eastern Europe	2.6	3.4	10	11	84	80
Europe	26.8	30.8	100	100	92	88
of which:						
Mechanical	8.0	9.4	30	31	99	100
Chemical (incl. semi-chemical and dissolving)	18.8	21.4	70	69	90	83
Paper and paperboard						
Nordic countries	9.9	13.5	26	27	366	405
EEC(9)	20.4	24.4	53	48	76	75
Central Europe	1.7	2.5	4	5	112	130
Southern Europe	2.5	5.0	7	10	83	92
Eastern Europe	3.9	5.3	10	10	89	93
Europe	38.4	50.8	100	100	100	103
of which:						
Newsprint	5.7	6.5	15	13	99	102
Printing and writing	10.2	15.0	27	30	110	110
Other paper and paperboard	22.4	29.3	58	58	96	105

higher than in 1969-71, but the decade was characterized above all by strong cyclical movements, notably the deep recession in the mid-1970s. Mechanical and chemical grades kept roughly the same share of European production, although there have no doubt been shifts between processes within the broad classification, e.g. towards thermo-mechanical pulping as opposed to more traditional mechanical pulping.

Pulp production grew faster in Southern Europe than elsewhere, although the Nordic countries continued to account for nearly 60% of European production.

There were however some structural changes, notably the reduction in the self-sufficiency ratio of the Nordic countries, which carried out their announced policies of increasing the share of pulp production which would be processed in these countries, thus adding value to local production. In the Nordic countries and elsewhere there was a trend towards increased integration between manufacture of pulp and of paper. In the mid-1960s, nearly half of Nordic pulp production was exported, but by the early 1980s, it was only about a quarter. In the EEC(9), the main importing region, however, the self-sufficiency ratio for pulp remained roughly stable, as pulp imports from elsewhere (outside Europe and Southern Europe) replaced Nordic supplies. The dependency on pulp imports of the paper-making industries increased between 1970 and 1980 in France and the Federal Republic of Germany. Where this dependency decreased in the EEC(9), this was generally not due to increased domestic supplies of pulp but to declines in paper production, leading to slower growth in pulp consumption or even to declines in absolute terms. This was the case, for example, in the United Kingdom. The pulp self-sufficiency ratio for Europe as a whole decreased steadily from the mid-1960s to the early 1980s, due to the increase in imports of pulp from outside the region.

Greece, Spain and Turkey increased their self-sufficiency in pulp by raising domestic production. For Portugal, the major pulp exporter in Southern Europe, the self-sufficiency ratio declined from 318% to 224%, despite increased pulp production. This was due to the installation of paper-making facilities and a consequent increase in pulp consumption, which more than tripled in 15 years.

For *paper and paperboard*, the only forest product group where Europe has a regional surplus of production over consumption, the 1970s were also marked by strong cyclical fluctuations, despite the growth in production between 1969-71 and 1979-81 of over 30%. The production surplus grew slightly over the period, largely because of higher European exports to markets outside the region.

Production grew particularly fast in Southern Europe, where it doubled between 1969-71 and 1979-81, due to large increases in most southern European countries. Portugal produced, in the early 1980s 15-30% more than it consumed; all countries in the region greatly improved their self-sufficiency ratio.

In contrast, growth was much slower in the major producing and consuming area, the EEC(9) (+20% between 1969-71 and 1979-81). The self-sufficiency ratio also deteriorated slightly for the EEC(9) as a whole. Drops in this ratio in the 1970s were recorded for all countries except the Federal Republic of Germany, whose self-sufficiency ratio rose from about 72% in 1969-71 to 79% in 1979-81.

Production also rose faster than average in the Nordic countries and the surplus for export of the group as a whole increased, especially in Sweden, although there are quite wide year-to-year variations due to swings in the export markets.

Production in Eastern Europe increased at about the average rate for Europe, but there were interesting trends in the self-sufficiency ratio: this was over 95% in the mid-1960s, declined to 87% in 1975, but thereafter started to rise again, reaching 96% in 1983. This trend may be attributed not only to rises in domestic production but to increasing constraints on imports.

14.6   THE FOREST INDUSTRIES IN THE ECONOMY

This section aims to present briefly the forest industries' place in the general economy and broad trends in employment and productivity. The international comparability of the available statistics is not good, so a detailed analysis could not be undertaken. Nevertheless the broad lines do not seem to be in doubt, nor the fact that trends for the forest industries have not been significantly different from those for manufacturing as a whole.

The source of information for the analysis in this section is the UN *Yearbook of Industrial Statistics*. There are some inconsistencies and gaps in the data, to which attention is drawn as appropriate. Where this source is used the forest industries as a whole are broken down into three main components, one of which has a significant sub-component. Each is identified by its UN International Standard Industrial Classification (ISIC) number as follows:

331 – Manufacture of wood, and wood and cork products, except furniture (wood and wood products)

332 – Manufacture of furniture and fixtures, except primarily of metal (furniture)

341 – Manufacture of paper and paper products (pulp, paper and paper products)

3411 – Manufacture of pulp, paper and paperboard (pulp and paper)

Data for Group 3411 are included in those for Major Group 341. It should be stressed that the coverage of these groups does not correspond exactly with the groupings by product used elsewhere in this study, notably by the inclusion of manufactures of wood and paper and of furniture. In this section the term "forest industries" refers to all three major groups.

The trends in industrial production[5] and employment for western European forest industries are presented in figures 14.7 and 14.8. Figure 14.7 shows that between 1970 and 1980 industrial production for wood products and furniture and for pulp and paper grew by just over 20%, a rate similar to that for the total of manufacturing industry. All sectors were strongly affected by the steep downturn in industrial production in the mid-1970s.

In western Europe (excluding Luxembourg and Switzerland for which no data were available) the forest industries accounted in 1980 for 6.3% of value added in manufacturing. This percentage was slightly lower than in 1970, although value added by the forest industries grew by about 3.5% in the decade. In North America, the industries accounted in 1980 for 7.5% of manufacturing value added and 9% of manufacturing employment.

TABLE 14.12

Europe and North America:
value added and employment in the forest industries
(ISIC 331 + 332 + 341)

	Western Europe [a]		North America	
	1970	1980	1970	1980
Value added				
Billion 1970 US dollars	17.1	17.7	23.4	31.4
Percentage of total				
manufacturing............	6.7	6.3	7.3	7.5
Employment[b]				
Number of employees				
(million)	3.1	3.0	1.7	1.9
Percentage of total				
in manufacturing	7.4	7.3	8.8	9.0

[a] Except Luxembourg and Switzerland for which no data were available.

[b] Includes data for Czechoslovakia, Hungary and Poland.

There are very large differences in forest industry size and structure between countries. From the data available, which are not fully comparable, it appears that by far the largest forest industry (ISIC 331 + 332 + 341) in Europe, by the criterion of value added, is that of the Federal Republic of Germany. Its value added in 1980 was roughly twice those of the United Kingdom and France, 2.5 times as large as those of Italy and Sweden and 4.5 times that of Finland. For comparison, while the value added of the Canadian forest industries is approximately the same size as that of the Federal Republic of Germany, in the United States it is nearly six times larger.

There is a considerable difference between the relative size of the different industries when measured by value

added according to the ISIC groups and that measured by levels of production of sawnwood, panels and paper in physical terms. The much greater importance of the industries of the EEC(9) by the value added criterion is due to the importance in those countries of the processing of semi-manufactures (e.g. sawnwood), some of which are imported.

If the forest industries are compared to the size of the economy as a whole there is a clear distinction between those countries where they have a "normal" share of value added in manufacturing, about 5%, and those countries where they play a particularly important role in the national economy: Finland (24.7% of value added in manufacturing accounted for by the forest industries in 1980), Sweden (16.7%), Norway (14.0%), Portugal (13.6%), Yugoslavia (10.3%), Spain (8.8%) and Austria (8.6%).

Over 3 million people were employed in the forest industries in Europe in 1980. (It is estimated, in very approximate terms, that a further 2 million were employed, full time or part time, in forestry, including related activities such as wood transport.) For the twenty countries for which data were available, employment was evenly distributed between the three sectors (ISIC 331, 332 and 341), if allowance is made for the fact that employment in furniture is included with that in wood and wood products for four major EEC countries. For most industry sectors and country groups, employment stagnated over the period: in Southern Europe, however, it increased quite strongly.

Between 1970 and 1981 employment in the western European forest industries fell, especially for pulp, paper and paper products, leading to a rise in labour productivity (see figure 14.8).

The increase in labour productivity and the reported increase in capital intensity indicate that the forest industries are seeking to substitute capital input for labour input in a rational response to the increasing cost of labour. Progress in this respect has been quite rapid, but there are some indications that future progress, at least in countries which already have high labour productivity, will be slower. It is possible that low labour productivity countries will, at least in part, catch up the leaders in this field, as modern technology spreads.

Capital, whose total availability is limited, will be attracted first to those sectors where it may expect the greatest return. Therefore, if capital productivity were to decline in the forest industries, they could have difficulty in attracting the capital which is necessary to continue to increase their labour productivity.

It is possible to speculate on the situation of the forest industries with regard to their use of labour and capital. In most countries of western Europe, there is at present a very high level of unemployment, which is unlikely to be totally absorbed in the foreseeable future. This might indicate that there would be little incentive for the forest industries to continue to substitute labour input by capital input (i.e. increasing labour productivity and capital in

---

[5] Shown as an index. "As recommended internationally, for most countries, the indexes relate to the value at constant prices of census value added". (Introduction to the UN Yearbook of Industrial Statistics).

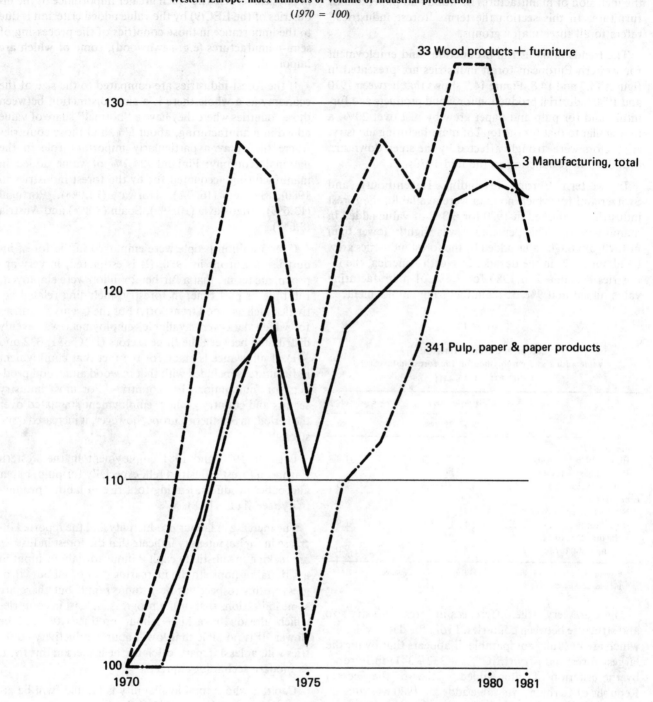

FIGURE 14.9

Western Europe: index numbers of volume of industrial production

(1970 = 100)

Source: UN, *Yearbook of Industrial Statistics.*

FIGURE 14.10

Western Europe: index numbers of volume of industrial production and number of employees
(*1970 = 100*)

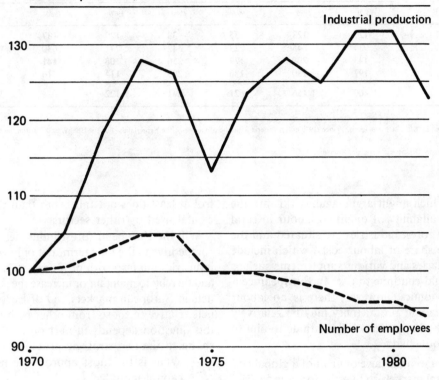

33 Wood products + furniture

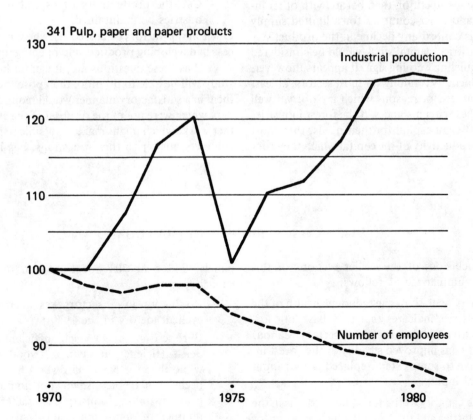

341 Pulp, paper and paper products

Source: UN, *Yearbook of Industrial Statistics.*

TABLE 14.13

Employment in the forest industries, 1970 and 1980

(*Thousand employees*)

	ISIC 331 Wood and wood products		ISIC 332. Furniture		ISIC 341. Paper and paper products		ISIC 331 + 332 + 341. Forest industries	
	1970	1980	1907	1980	1970	1980	1970	1980
Nordic countries	126	123	35	38	129	137	290	298
EEC(9)[a] plus Austria	934[b]	645[b]	223[b]	342[b]	771	630	1 928	1 617
Southern Europe	143	208	177	256	108	141	428	605
Eastern Europe[c]	197	180	186	205	114	116	497	501
Total (20 countries)	1 400	1 156	621	841	1 122	1 024	3 143	3 021

[a] Excluding Luxembourg.

[b] Data for ISIC 332 included in ISIC 331 for Belgium, France (1970 only), Federal Republic of Germany, the Netherlands. For the EEC(9) and Austria, the totals for ISIC 331 + 332 are 1970: 1157; 1980: 987.

[c] Excluding Bulgaria, Germany Democratic Republic, Romania.

tensity). Factors which might argue against this are the possibly limited availability of qualified labour in rural areas, where many forest industries are situated and the possibility that the share of labour costs, which include both wages and salaries and various supplements such as social benefits, could continue to rise. In European centrally planned economies, however, there is no labour surplus (although labour productivity may be relatively low), and it is likely that policies will continue to aim to increase labour productivity.

Recent years have seen the development of a global and rather volatile capital market, so that the forest industries are in more direct competition than before with other industrial sectors and other countries for a limited supply of capital. In this context any decline in the productivity of the forest industries' capital stock would be considered a potentially disquieting warning sign. It appears, however, that capital productivity is falling for most sectors, at least in market economies, for reasons which are not yet well understood. In these circumstances, if the forest industries are to obtain sufficient capital for their needs, they must ensure that the productivity of the capital which they con-

trol at least does not fall faster than the productivity of capital used by other sectors.

Will the European forest industries be able to remain competitive in the long term, not only with overseas forest industries but also with producers of non-wood products and thereby to maintain or increase the share of forest products in European markets and of European forest products relative to those from other regions? The answer to this question depends in part on how the industries respond to the two strategic problems which they face:

- What is the most appropriate mix of labour and capital input?
- Can the productivity of capital in the forest industries be maintained?

Other major factors discussed elsewhere include their success in developing products and opening up new markets.

Vital as these questions are, it cannot be said that this study will be able to do more than pose them. To answer them in a satisfactory manner would require resources and data which were not at the disposal of the secretariat. Further work, which might make it possible to provide a more complete answer to these questions, has been initiated.

14.7 CONCLUSIONS AND OUTLOOK

The main conclusions of the earlier sections of this chapter may be summarized as follows:

- A preliminary analysis of the economic health of the forest industries indicates that they have not lost ground relative to other industrial sectors. Labour productivity has improved steadily as the forest industries have to some extent replaced labour input with capital input;
- Most of the capacity of the forest industries is in the Nordic countries and the EEC(9) which account for about 55% of European removals, production of sawnwood and production capacity for wood-based

panels, and over 70% of European pulp and paper capacity;

- Production in all sectors was strongly affected by cyclical factors in the 1970s;
- Europe is about 90% self-sufficient in sawnwood. Self-sufficiency in sawn softwood has increased, while that for sawn hardwood has fallen, chiefly because of tropical sawnwood imports; the European sawmilling industry is characterized by a large number of small mills; most of the production capacity however is in larger mills;
- Over the 1970s, Europe particle board capacity in-

creased by about 55%, while that for plywood fell by 20%; the average size of particle board and fibreboard units increased significantly over the 1970s. Both are considerably larger on average than plywood mills. The number of plywood mills decreased, especially in the EEC(9); production capacity for plywood and fibreboard was adapted to declining or stagnating levels of production. For particle board, a major problem of over-capacity developed in the mid-1970s; European self-sufficiency in plywood and veneer sheets has decreased markedly, to a large extent because of imports from tropical regions;

- Europe is just under 90% self-sufficient in pulp, but is a net exporter of paper and paperboard (notably printing and writing grades). Total pulp capacity ceased to expand steadily around 1978 (although 1984 and 1985 did show increases). Investments in capacity were concentrated on improving efficiency and profitability. Total capacity for paper and paperboard, however, continued to expand over the 1970s and early 1980s;

- For wood-based panels, woodpulp, paper and paperboard, capacity increased at a faster than average rate in Eastern and, especially, Southern Europe.

What of the outlook? In the short – to medium-term, production cannot rise above the limits set by existing capacity. Furthermore, much plant in the forest industries has a rather long useful life. Functioning, economic units can be up to 40-60 years old, if they have been correctly maintained and periodically overhauled. Radical changes in the medium term are therefore unlikely: change will be of an evolutionary nature as the structure of the industries is adapted to changing conditions of demand and profitability.

In the long term however, the structure and capacity of the forest industries will be determined by more fundamental factors:

- The availability of raw materials, notably wood;
- The availability of other factors of production, including labour, capital, energy and land;
- The level of demand for forest products (determined by economic growth, activity in user sectors, technical and economic competitivity of forest products etc.);
- The competitivity of each country's forest industries (relative to other European forest industries, to forest industries outside Europe and to producers of non-wood alternative materials).

It is not possible here to combine these factors to produce a quantitative outlook for the forest industries. It is however possible to obtain some indications from other parts of this study:

- The outlook for *raw material* availability, in physical terms at least, is presented in chapters 5 (round wood) and 15 (wood residues and waste paper). It is important to bear in mind that for a specific mill,

the raw material supply must not only be sufficient in volume, but of the right species and quality, with a reasonable expectation of continuing supply over the lifetime of the mill (taking into account social factors, such as forest ownership), at an acceptable price. In view of all these considerations, it is becoming increasingly difficult to find sites in Europe for totally new ("green field") mills producing those products which require very large manufacturing units (e.g. chemical pulp), even if, as shown elsewhere in this study, there is a considerable difference between total increment and total drain, which indicates a theoretical possibility to expand capacity

- The outlook for *demand* for forest products, which has been presented in chapter 12, is characterized by slow growth. If over-capacity, of the type already encountered by the particle board industry, is to be avoided, capacity should not grow faster than demand. It therefore appears likely that investments will be concentrated less on increasing total capacity in volume terms but rather on increasing efficiency by reducing costs and raising yields, or towards adapting production more closely to the qualitative needs of the markets;

- It seems unlikely that the availability of the *other factors of production* (labour, capital, energy, land) will be a constraint on the expansion of the European forest industries. One possible exception is capital: in both market economy and planned economy countries, if the forest industries cannot offer a better return on investment than other potential users, they could have difficulties getting access to the limited overall investment funds available. This again highlights the necessity for the forest industries to ensure that they remain efficient and profitable in order to ensure a satisfactory future;

- The *competitivity* of an industry is determined by many factors, notably the quality of its products, its marketing ability, the prices of its products relative to competitors and its ability to limit costs. These vary widely between countries and industry sectors, even between individual companies or production units, and cannot therefore be analysed in quantitative terms by a study like ETTS IV.

The analysis attempted in section 14.6, of the economic health of the industries and trends in productivity, is of a type which has not been attempted in earlier timber trends studies. It has revealed two things:

- That a better understanding of the economic characteristics of the forest industries, trends in productivity and profitability etc. would be valuable to policy-makers;

- The statistical base for this analysis is at present quite inadequate.

Work has been started, as part of the follow-up to ETTS IV to investigate these questions further, and research at the national level, including improvement of the data base would seem to be justified.

CHAPTER 15

Trends in raw material consumption and the outlook for secondary raw materials (wood residues and waste paper)

15.1 INTRODUCTION

The forest industries take raw material (logs, round pulpwood, wood residues, pulp, waste paper) and transform it into products (sawnwood, panels, pulp and paper). This chapter will briefly review trends in the volumes of raw material consumed (notably the share of different types of raw material) and the raw material/product conversion factors. Thereafter, it will concentrate on the outlook for secondary raw materials (residues of wood processing, and waste paper) as the consistency analysis model, to be presented in chapter 20, requires specific scenarios for secondary raw materials in order to evaluate the consistency of the scenarios for supply and demand.

Secondary raw materials (residues of wood processing and waste paper) represent a significant and increasing part of the raw material input to the industries manufacturing paper and paperboard, particle board and fibreboard. Any study of the outlook for the forest industries must, therefore, examine how much waste paper and wood residues may be recovered and used in the future.

This chapter, therefore, has three objectives:

(a) To present broad trends in total raw material consumption;

(b) To present briefly trends for secondary raw materials since 1970;

(c) To prepare hypotheses for the future development of key ratios, notably the waste paper recovery rate and the share of residues generated which is used as raw material. These will be essential input to the material balance in chapter 20.

It should be pointed out that data quality for secondary raw materials is often quite low: apparent anomalies may appear and in some cases it is not advisable to draw firm conclusions from the data presented.

Chapter 7 of ETTS III analysed the trends for these materials up to the early 1970s and the factors underlying these trends. This analysis will not be repeated here as many of the observations made in ETTS III are still valid.

In this chapter, as elsewhere in the study, the term "pulpwood" includes pulpwood, round and split, coniferous and non-coniferous, as well as residues, chips and particles. The term "logs" is often used as an abbreviation for "sawlogs and veneer logs".

15.2 SITUATION AND TRENDS IN TOTAL RAW MATERIAL CONSUMPTION

Tables 15.1 to 15.5 present the consumption of logs, pulpwood, pulp and waste paper in 1969-71 and 1979-81 and analyse the changes over time.

TABLE 15.1

Consumption of logs and of pulpwood, 1969-71 and 1979-81

| | Apparent consumption (million m³) | | | | Percentage share of logs in total | |
| | Logs | | Pulpwood | | | |
	1969-71	1979-81	1969-71	1979-81	1969-71	1979-81
Nordic countries	42.8	47.4	70.7	73.4	37.7	39.2
EEC(9)	48.7	50.1	29.9	35.1	61.9	58.8
Central Europe	10.2	12.5	5.7	9.4	64.2	57.0
Southern Europe..............	17.4	22.5	7.5	15.4	70.0	59.4
Eastern Europe	34.6	37.5	14.5	21.5	70.5	63.6
Europe	153.7	170.0	128.3	154.8	54.5	52.3

220

TABLE 15.2

Consumption of coniferous and non-coniferous logs, 1969-71 and 1979-81

| | Apparent consumption (million m³) | | | | Percentage share of coniferous in total | |
| | Coniferous logs | | Non-coniferous logs | | | |
	1969-71	1979-81	1969-71	1979-81	1969-71	1979-81
Nordic countries	40.3	45.6	2.5	1.8	94.1	96.1
EEC(9)	26.3	29.2	22.4	20.8	54.0	58.4
Central Europe	9.4	11.1	0.8	1.4	92.5	88.6
Southern Europe.............	11.7	15.1	5.7	7.4	67.1	67.0
Eastern Europe	22.5	25.1	12.0	12.3	65.2	67.1
Europe	110.2	126.1	43.4	43.7	71.7	74.2

TABLE 15.3

Consumption of pulpwood, 1969-71 and 1979-81

| | Coniferous, round and split | | Non-coniferous, round and split | | Residues and chips | |
	1969-71	1979-81	1969-71	1979-81	1969-71	1979-81
Volume consumed (million m³)						
Nordic countries	47.6	42.8	9.4	9.7	13.7	20.9
EEC(9)	10.8	11.4	10.5	11.7	8.6	12.0
Central Europe	2.7	3.6	1.1	1.9	1.8	3.9
Southern Europe...........	3.7	7.3	3.0	5.9	0.8	2.1
Eastern Europe	8.6	10.2	2.1	5.2	3.8	6.1
Europe	73.4	75.3	26.1	34.4	28.7	45.0
Percentage share of total pulpwood consumption						
Nordic countries	67.4	58.3	13.3	13.2	19.4	28.5
EEC(9)	36.2	32.4	35.1	33.4	28.6	34.2
Central Europe	48.4	38.3	20.2	20.0	31.3	41.7
Southern Europe...........	49.5	47.6	40.2	38.6	10.3	13.8
Eastern Europe	59.3	47.4	14.5	24.2	26.2	28.4
Europe	59.0	48.6	20.4	22.2	22.3	29.1

TABLE 15.4

Consumption of paper-making fibres, 1969-71 and 1979-81

| | Total fibres (million m.t.) | | Pulp (million m.t.) | | Waste paper (million m.t.) | | Percentage share of pulp in total | |
	1969-71	1979-81	1969-71	1979-81	1969-71	1979-81	1969-71	1979-81
Nordic countries	10.3	13.4	9.8	12.4	0.5	1.0	94.9	92.8
EEC(9)	20.6	24.4	13.1	13.7	7.5	10.6	63.4	56.4
Central Europe	1.8	2.5	1.3	1.7	0.5	0.8	74.2	66.7
Southern Europe.............	2.5	5.0	1.8	2.9	0.7*	2.1	72.1	57.6
Eastern Europe	4.3	6.2	3.1	4.3	1.2*	1.9	72.3	69.0
Europe	39.5	51.5	29.1	35.0	10.4	16.4	73.6	68.0

Table 15.1 brings out clearly the wide difference in the pattern of raw material consumption[1] between the Nordic countries and other regions. In 1969-71 in the Nordic countries, logs accounted for under 38% of wood raw material input, while in all other regions this percentage was over 60%. This difference between regions diminished over the decade as the share of logs in the total diminished in all regions except the Nordic countries, where it increased. Except in the Nordic countries, pulpwood consumption increased considerably faster than consumption of logs over the ten-year period. The two trends almost counterbalanced each other, so that there was only a small (2.3 percentage point) decline in the share of logs at the European level to 52.3% in 1979-81.

Two regions, the Nordic countries and the EEC(9), accounted for over half of European consumption of logs and pulpwood combined (37% and 26% of the European total respectively in 1979-81). The Nordic countries consumed 28% of the European total for logs in 1979-81 and 47% for pulpwood.

---

[1] Conclusions about *supply* of raw material or raw material *balances* may *not* be drawn from these data as wood residues are double counted — as part of logs and of pulpwood.

TABLE 15.5

Changes in consumption of raw material between 1969-71 and 1979-81
(Percentages)

	Nordic countries	EEC(9)	Central Europe	Southern Europe	Eastern Europe	Europe
Total wood raw material	+ 6.4	+ 8.4	+ 37.7	+ 52.2	+ 20.2	+ 15.2
Sawlogs and veneer logs	+ 10.7	+ 2.7	+ 23.0	+ 29.3	+ 8.4	+ 10.6
Coniferous	+ 13.1	+ 10.0	+ 17.8	+ 29.2	+ 11.5	+ 14.4
Non-coniferous	− 27.1	− 7.0	+ 87.3	+ 29.7	+ 2.6	+ 1.0
Pulpwood	+ 3.9	+ 17.1	+ 66.2	+ 103.2	+ 48.4	+ 20.7
Coniferous, round and split	− 10.1	+ 5.6	+ 33.3	+ 97.3	+ 18.6	− 0.4
Non-coniferous, round and split	+ 3.2	+ 11.4	+ 72.7	+ 96.7	+ 147.6	+ 31.8
Residues and chips	+ 52.6	+ 39.5	+ 116.7	+ 162.5	+ 60.5	+ 57.3
Total papermaking fibres	+ 30.1	+ 18.4	+ 38.9	+ 100.0	+ 44.2	+ 30.4
Pulp	+ 26.5	+ 4.6	+ 30.8	+ 61.1	+ 38.7	+ 20.3
Waste paper	+ 100.0	+ 41.3	+ 60.0	+ 200.0	+ 58.3	+ 58.7

Nearly three quarters of the logs consumed in Europe are of coniferous species; this proportion has increased from just over 70% in 1969-71. Consumption of coniferous logs rose by nearly 15% in Europe over the ten-year period (even faster in Southern Europe) but the increase for non-coniferous was of only 1%. There was a drop of 7% in the consumption of non-coniferous logs in the most important consuming region, the EEC(9), as the self-sufficiency ratio for plywood and sawn hardwood fell, and the share of imports of these products, from tropical countries and North America, rose.

The main trend observable for pulpwood is the stagnation (or decline) in the consumption of the most important assortment, coniferous round pulpwood, and the sharp rise (+ 57% for Europe as a whole) in the consumption of residues, chips and particles, whose share of total pulpwood consumption went from 22% to 29% while that of coniferous round pulpwood fell from 59% to 49%. European consumption of non-coniferous round pulpwood also rose, by 32%, but the increase was relatively modest outside Eastern Europe where consumption more than doubled. It is possible that some of the apparent increase in the consumption of residues, chips and particles is due to better measurement and wider awareness of this assortment's importance, but most of the increase is real. The reasons for it are discussed below.

Structural shifts in the pattern of raw material consumption are also apparent for the paper and paperboard industries. Their consumption of fibres rose by 30% between 1969-71 and 1979-81, but consumption of pulp grew more slowly than that of waste paper (+20%, compared to +59%). Pulp's share of the European consumption of

paper-making fibres fell from 74% to 68%. The reasons for this trend are discussed briefly below. There are major differences between regions with regard to the structure of consumption of paper-making fibres. Pulp remains dominant in the Nordic countries, despite a doubling of waste paper consumption. Most paper produced in this region is exported and therefore not available for recycling while very large quantities of virgin fibre are produced and consumed. In the EEC(9) however, a pulp importing region with very high consumption and recycling of paper and paperboard, pulp and waste paper are of roughly equal importance as raw materials for the industries.

The main changes between 1969-71 and 1979-81, for Europe as a whole, may be summarized as follows (see table 15.5):

– Consumption of logs (+10.6%) grew more slowly than that of pulpwood (+ 20.7%);

– Consumption of coniferous logs grew steadily, (+14.4%) while that of non-coniferous logs fell in the Nordic countries and EEC(9), but rose in Central and Southern Europe;

– Among pulpwood assortments, consumption of coniferous round pulpwood stagnated, with a fall in the Nordic countries, while that of non-coniferous round pulpwood and especially of residues and chips grew strongly;

– Consumption of all papermaking fibres grew significantly over the decade (+30.4%). Growth was much faster for waste paper (+ 58.7%) than for pulp (+20.3%).

15.3 RAW MATERIAL/PRODUCT CONVERSION FACTORS

The volume of raw material consumed by an industry depends on:

(*a*) The number of units produced;

(*b*) The volume of raw material consumed per unit produced.

This section discusses the situation and outlook for (*b*).

TABLE 15.6

Raw material/product conversion factors, as supplied by countries
to the Joint FAO/ECE Working Party on Forest Economics and Statistics

Product and Unit	Range of national conversion factors			Common conversion factors [d] applied in ETTS III
	Around 1970 [a]	1979 [b]	1983 [c]	
Sawn softwood (m³/m³)	1.40-2.00	1.44-2.55	1.49-2.43	1.64
Sawn hardwood (m³/m³)..............	1.31-2.00	1.32-2.39	1.35-2.33	1.67
Sleepers (m³/m³)	1.43-2.00	1.39-2.00	1.40-3.12	1.75
Plywood (m³/m³)	1.95-2.78	1.63-2.94	1.65-3.40	2.28
Veneer sheets (m³/m³)	1.30-2.94	1.39-2.45	1.71-3.20	1.85
Particle board (m³/m³)	1.30-2.00	1.10-2.20	1.10-2.28	1.43
Fibreboard (m³/m³)	1.28-3.00	0.79-3.01	1.60-3.65	1.83
Mechanical woodpulp (m³/m.t.)	2.25-3.26	2.40-3.17	2.26-3.20	2.50
Semi-chemical woodpulp (m³/m.t.).....	2.00-3.90	2.04-3.30	2.37-3.30	3.00
Chemical woodpulp (paper) (m³/m.t.)				
Unbleached sulphite	3.31-5.65	4.00-5.76	4.35-5.93	⎤
Bleached sulphite	3.80-6.50	3.70-6.39	3.35-6.73	4.80
Unbleached sulphate	3.19-5.75	3.90-5.52	4.30-5.25	
Bleached sulphate	3.21-6.40	4.51-6.72	3.00-5.58	⎦
Dissolving pulp (m³/m.t.)	3.50-5.91	5.50-6.96	4.17-10.61	5.46

[a] Source: TIM/EFC/WP.2/R.4.

[b] Source: Supplement 12 to Volume XXXIV of the *Timber Bulletin for Europe*.

[c] Source: TIM/EFC/WP.2/R.74, provisional data.

[d] Calculated from weighted average of conversion factors for European countries around 1970.

There are two basic approaches to determining these conversion factors, one using technical or engineering criteria, the other a more statistical approach. The former is widely used for investment decisions and cost comparisons but is not appropriate for broader outlook studies as conditions vary widely between mills, so that it is not possible in general to draw conclusions at the national or regional level.

The statistical approach is to calculate national average conversion factors by dividing the total volume of raw material consumed by the total production. It should be pointed out that serious statistical problems can arise, chiefly because of difficulties in identifying the volume of raw material consumed for each type of product (problems of roundwood classification and of raw material which can be used for different production processes, such as pulpwood).[2]

The Joint FAO/ECE Working Party on Forest Economics and Statistics reviews trends in conversion factors at four-yearly intervals. The results of these surveys are summarized in table 15.6. The size of the ranges presented in this table brings out the wide differences between national circumstances.

What are the factors which have affected conversion factors in the past and will continue to influence them in the future?

- Other things being equal, *technical development of process equipment* would be expected to improve

yields of raw material. However, designers of equipment may pursue goals, such as fast throughput or recovery of certain types of by-product, which might not be compatible with higher raw material yield. *It does not therefore appear possible to assume,* a priori, *that technical development will improve yields*;

- The quality, in terms of both physical properties and dimension, of *raw material* plays an important role in determining the yields. For instance, sawnwood yields from small logs are usually lower than from large logs, even with the most advanced technology. It is expected that supplies of high quality logs, of the type found in mountain areas or the North of the Nordic countries, will diminish in the future. This will tend to lower sawnwood yields;

- *Product development* also plays a role. Probably the most significant trend in yields in recent years has been the development of new "high-yield pulps", notably thermo-mechanical (TMP) and chemi-thermo-mechanical (CTMP), which have raw material yields similar to those of conventional mechanical pulps but can wholly or partly replace chemical pulp in many uses, making possible significant savings of wood raw material in the final product;

- The price which can be obtained for *residues* will also affect yields. If this price is low, a producer will tend to maximise the yield of sawnwood or plywood. If it is high, or if demand for the "primary" product is weak, he will try to find the product/residue mix which will maximise his income. In certain market circumstances (low sawnwood demand, high

---

[2] For a discussion of this question, see annex II to Conversion Factors for Forest Products (supplement 12 to vol. XXXIV of the *Timber Bulletin for Europe*, ECE/FAO, Geneva, 1982).

TABLE 15.7

Raw material/product conversion factors used in consistency analysis (chapter 20)

Country	Sawnwood (m³/m³)			Wood-based panels (m³/m³)				Wood pulp (m³/m.t.)			
	Coniferous	Non-coniferous	Sleepers	Veneer sheets	Plywood	Particle board	Fibreboard	Mechanical	Semi-chemical	Chemical	Dissolving
Albania	1.64	1.67	1.75	1.85	2.28	1.43	1.83	2.50	3.00	4.80	5.46
Austria	1.48	1.31	1.82	1.67	2.55	1.75	1.97	2.50	3.00	5.00	5.50
Belgium-Luxembourg	1.67	1.82	1.82	2.00	2.30	1.00	1.28	2.50	3.00	4.70	5.50
Bulgaria	1.58	1.86	1.52	2.48	1.59	1.48	2.16	2.83	2.12	5.35	6.30
Cyprus	1.67	1.67	1.67	1.90	2.30	1.60	1.83	2.50	3.00	4.80	5.46
Czechoslovakia	1.54	1.40	1.43	1.30	2.23	1.46	2.29	2.68	3.00	5.00	5.61
Denmark	2.00	2.00	2.00	3.50	4.20	2.40	1.00	3.00	2.50	5.00	5.50
Finland	1.84	1.84	1.84	2.78	2.78	1.53	1.89	2.40	2.50	4.60	5.83
France	1.70	1.75	2.00	2.00	2.00	1.30	2.08	2.25	2.60	4.70	5.00
German Democratic Republic	1.67	1.82	1.82	1.90	2.30	1.60	2.69	2.50	3.00	4.70	5.50
Germany, Federal Republic of	1.40	1.40	1.70	1.80	2.30	1.40	2.40	2.70	2.80	4.70	5.50
Greece	1.69	1.69	1.71	1.90	2.30	1.30	1.90	2.50	2.50	4.80	5.50
Hungary	1.68	1.97	1.97	2.70	1.83	1.31	2.73	2.29	2.46	4.43	5.50
Iceland	1.80	1.80	1.82	1.90	2.30	1.60	1.75	2.50	3.00	4.80	5.46
Ireland	1.80	1.66	1.82	2.00	2.30	1.60	1.75	2.50	3.00	4.70	5.50
Italy	1.67	1.67	1.43	1.60	3.00	1.30	1.30	2.64	4.00	4.80	5.50
Malta	1.67	1.82	1.82	1.85	2.30	1.43	1.83	2.50	3.00	4.80	5.46
Netherlands	1.67	1.82	1.82	1.90	2.30	1.30	1.45	2.25	3.30	4.90	5.50
Norway	1.86	1.71	1.86	1.90	2.30	1.50	1.79	2.42	3.00	4.90	5.50
Poland	1.47	1.35	1.49	2.00	2.41	1.44	1.67	2.74	3.27	5.77	6.96
Portugal	1.94	1.48	1.67	1.58	2.47	2.00	2.75	2.50	3.00	3.50	5.50
Romania	1.49	1.73	1.76	2.94	2.37	1.62	2.33	3.20	3.50	5.50	5.91
Spain	1.67	2.00	2.00	1.50	2.20	1.30	1.94	2.30	3.00	4.70	5.50
Sweden	1.99	1.85	1.82	1.90	2.50	1.35	2.40	2.40	2.40	4.55	5.50
Switzerland	1.40	1.50	1.70	1.60	2.40	1.40	1.50	2.50	3.00	4.70	3.50
Turkey	1.67	1.82	1.82	1.90	2.30	1.60	2.68	2.50	3.30	4.90	5.50
United Kingdom	1.67	1.72	1.82	2.00	2.30	1.06	1.27	2.43	2.43	4.84	5.50
Yugoslavia	1.54	1.92	1.82	2.00	2.70	1.69	2.11	2.60	3.00	5.00	5.50

pulpwood demand) some North American sawmillers have in the past opted for 100% "residue" production by chipping whole sawlogs;

— Technical improvements directed towards *other objectives*, such as pollution control, energy conservation, or reduction in use of water may bring about changes in yield.

It is clear from the above discussion that many factors specific to particular products or sites affect the yield of individual mills. It does not seem possible to draw any general conclusions about broad trends in the past or the outlook for the future. One exception may be the move towards high-yield pulp, which is likely to improve the average yields of pulp. Furthermore, as mentioned above, there are wide differences in yields between countries.

How, then, should yields be taken into account in the future material balance, to be presented in chapter 20?

As the balance is calculated at a national rather than a regional or European level, it has been possible to use national data for conversion factors, supplied in the context of the regular reviews mentioned above. For non-responding countries the standard factors developed for ETTS III were used. These were circulated for checking to national correspondents. These conversion factors are set out in table 15.7. In view of the uncertainty about the direction and strength of future trends for yields, a conservative initial hypothesis was adopted, that the *conversion factors would not change from their 1979-81 level*. Changes in the conversion factors are one of the possibilities to be considered in analysing the results of this consistency analysis (see chapter 20).

15.4 OUTLOOK FOR WOOD RESIDUES, CHIPS AND PARTICLES

As seen from section 15.2, the use of wood residues, chips and particles in the manufacture of pulp, particle board and fibreboard has expanded strongly. Total European consumption of this raw material was estimated at only 5 million m³ in 1950 and 13 million m³ in 1960. In 1969-71, however, it was 28.6 million m³ and in 1979-81, 45.0 million, an increase of well over 50% in a decade. (See data by country group in table 15.8 and by country, for domestic supply, in annex table 15.1.)

The great majority of these residues come from the primary wood processing industries, notably sawmilling and plywood manufacture, but small volumes also come from secondary processing such as furniture manufacture and joinery. Chips made directly from roundwood in the forest may also be included in the statistics for "chips and particles" along with chips from processing residues, although, at least at present, the volumes concerned are certainly not significant on a regional level.

Tables 15.3 and 15.8 show very clearly that during the 1970s not only did the consumption of residues, chips and particles increase in absolute terms but they increased their share of pulpwood consumption, that a greater proportion of available residues was used as pulpwood and that the volume of trade increased while trade patterns changed somewhat.

The main reason for the increase in residue consumption seems to be economic: once technological advances had made possible the use of chips in most pulping and panel making operations, and residue collection circuits had been established, this raw material was often cheaper than roundwood.

It is also apparent from table 15.8 that more of the residues generated are being used as raw material for pulp and panels than before. This is in part a consequence of the fact that *demand* for residues, determined by production of pulp, particle board and fibreboard has in general

TABLE 15.8

Residues, chips and particles: supply, and trade

| | Domestic supply | | | | | |
| | Volume (million m³) | | Recovery ratio [a] (percentage) | | Net trade [b] (million m³) | |
	1969-71	1979-81	1969-71	1979-81	1969-71	1979-81
Nordic countries	13.0	19.7	68.2	88.7	− 0.7	− 1.2
EEC(9)	8.2	11.8	45.8	64.4	− 0.5	+ 0.1
Central Europe	1.8	3.4	54.5	86.8	−	− 0.5
Southern Europe	0.7	2.4	9.3	21.5	−	+ 0.2
Eastern Europe	3.8	5.9	28.9	42.5	−	+ 0.3
Europe........................	27.6	43.1	44.8	62.0	− 1.2	− 1.2

Note: These data for 1979-81 are taken from the "consistency analysis" data base (see chapter 2), and are not exactly comparable with those elsewhere in ETTS IV, although the differences are minor.

[a] Domestic supply of residues and chips as a percentage of estimated availability of primary processing residues (see text).

[b] + = net exports: − = net imports. Data on imports and exports are presented in chapter 16.

grown faster than *availability* of residues, determined by production of sawnwood and plywood. The "recovery ratio" for Europe as a whole rose during the 1970s from about 45% to over 60%. It should be stressed, however, that this ratio, which shows domestic supply of residues as a percentage of residues generated by primary processing industries, may not be entirely reliable for a number of reasons:

– The data on domestic supply of residues of a few countries show anomalies;

– The volume of residues generated has been estimated using conversion factors (raw material/product for sawnwood, plywood and veneers) which may not always be accurate;

– The residues generated by secondary processing could not be taken into account. ETTS III estimated that 12-19 million m^3 of residues were physically available from the furniture, joinery, building component and packaging industries and from reconversion of imported timber. This figure is 20-30% of the estimated volume of residues generated by primary processing around 1970. It is therefore quite conceivable that the "recovery ratio" could exceed 100%. (This is in fact the case for the Federal Republic of Germany in 1979-81.)

Nevertheless, the recorded changes in recovery ratio during the 1970s have been so great that there can be no doubt that they reflect a real movement and that an increasing part of the residues generated is being transferred to the pulp and panel industries. The percentage wasted or put to other uses (of which there are many) has therefore declined. An exception is probably the use of residues as a source of energy, which appears to have increased. However, the residues burnt are often those which were not suitable for use as raw material – bark, fines, contaminated material, etc. (see chapter 19).

There are major differences in recovery ratios between countries. In the Nordic countries, the Federal Republic of Germany, Central Europe and the German Democratic Republic, the recovery ratio is very high (over 70%), and it must be doubtful whether the percentage of primary processing residues transferred to pulp and papermaking can be significantly increased. The ratio is much lower in other countries, although in only eight countries is it below 40%. Specific factors with a negative effect on the recovery ratio may be a prevalence of hardwood residues, which were considered unsuited to several pulping processes, a fragmented sawmill sector or the absence of pulp mills able to use the type of residues available.

Trade in wood residues, chips and particles has also expanded, as European imports doubled between 1969-71 and 1979-81, while exports nearly tripled over the same period. Net imports, however, for Europe as a whole, were the same in 1979-81 as in 1969-71, although the pattern had changed somewhat: the deficit of the Nordic countries increased, partly as a response to domestic roundwood supply shortages and partly because of the industries' desire to profit from the availability of pulpwood for export in other countries; the EEC deficit was transformed into a surplus; and Central Europe became a significant net importer. Trade in residues and chips, like that in round pulpwood, is quite volatile and increases rapidly in periods of high demand or of roundwood supply shortages only to decrease equally rapidly when circumstances change. For instance, Nordic imports of wood residues, chips and particles which were around 1.2 million m^3 until 1972, doubled to 2.4 million m^3 in 1975, fell back to 1.3 million in 1978, then rose again to 2.9 million m^3 in 1981, to fall again thereafter. Most pulpwood using industries are unwilling to install capacity dependent on imported raw material to a large extent, but see imports as a temporary measure in special circumstances. An exception to this is the regular import by Finland of wood residues from the USSR, of over 1 million m^3 which in fact accounts for the bulk of European net imports.

What of the outlook to 2000 for the supply of wood residues to the pulp and panel industries? Resources were not available for a detailed analysis of the supply and demand for residues, nor is it certain that such an analysis would be fruitful, as local and technical factors are so important and the data base must still be considered weak. In the following paragraphs some hypotheses are proposed which provide the basis for the quantified scenarios for the recovery ratio:

– Residues will remain a cheaper raw material than roundwood;

– Technical progress will slowly widen the acceptability of residues for pulping;

– The use of residues as a source of energy will continue to expand but this will *not* affect the qualities suitable as raw material for pulp or panels;

– The "other" uses (e.g. for animal litter, chemicals, brick making, wood flour, etc.) will stay constant;

– There will be no major change in net imports, although the total volume of trade may well increase. There are likely to be considerable fluctuations in trade flows and volumes;

– The volume of residues actually wasted or merely incinerated will continue to diminish;

– Substantial differences between countries in the use patterns of residues will remain, although they may diminish. For instance, in some countries, residues are given or sold to employees or the local population as a source of energy; such customs may prove resistant to change, especially if the residues are of non-coniferous species and there is no strong local industrial demand for them.

The major elements determining residue *availability* are the level of production of sawmills, plywood mills and veneer mills and the raw material/product ratios in these industries, which were examined in section 15.3. Quantified hypotheses for domestic supply of residues must also take into account the "recovery ratio", which should be applied to the figure for "residues generated". It is suggested that the *recovery ratios will continue to rise as in the past*, but that the rate of increase will be lower in countries which already have a high recovery ratio. The following quantified hypotheses for the increase in recovery ratio over the two decades 1980-1990 and 1990-2000 are tentatively proposed:

Recovery ratio at beginning of decade	Increase in ratio (percentage points per decade)	
	Low	High
Below 40	10	30
40-60	10	20
60-80	5	10
80-100	1	5
Over 100	–	2

The low scenario assumes for "low ratio" countries the persistence of social, technical and economic barriers to expansion and for "higher ratio" countries increasing difficulties in expanding recovery as well as competition from other uses. The high scenario assumes that "low ratio" countries will rapidly mobilize their residue resources and that "high ratio" countries will continuously seek new sources, for instance in secondary processing industries (this factor is what makes possible a ratio of over 100%). The high scenario also assumes that the pulp and panel industries have priority over all other uses (including energy) in competition for wood raw material.

The resulting hypotheses for recovery ratios are shown in annex table 15.1. The resulting levels of residue supply and share of residues in raw material consumption will be presented and discussed in the context of the consistency analysis model (chapter 20).

15.5 OUTLOOK FOR WASTE PAPER

During the 1970s, the recovery and use of waste paper increased quite fast in most European countries. In 1972-74 (the first year for which nearly complete data are available), about 12.2 million m.t. of waste paper was recovered – nearly 28% of consumption of paper and paperboard. In 1979-81, the comparable figure was nearly 16 million m.t. (+31% in seven years) for a recovery rate[3] of 32.4%. The data for 1972-74 already represented a significant advance on those for earlier years (see annex table 15.2 and tables in ETTS III, chapter 7).

The following comparison shows the broad evolution for waste paper, for Europe as a whole, in the 1970s:

	1972-74	1979-81
Recovery (million m.t.)	12.18	15.97
Recovery rate (%)	27.9	32.4
Exports (million m.t.)	1.47	2.21
Imports (million m.t.)	1.72	2.73
Net imports (million m.t.)	0.25	0.52
Consumption (million m.t.)	12.43	16.49
Production of paper and paperboard (million m.t.)	44.10	50.77
Consumption rate (%)	28.2	32.5

Over the 1970s the importance of waste paper as a raw material increased in Europe in absolute and relative terms in almost every way. (It is possible that in a few countries part of the increase may have been due to improved statistical reporting, but the broad trend is quite clear.)

Over the period there were significant improvements in the technology of cleaning and de-inking waste paper. A recent OECD report[4] notes that:

There have been no real technological changes in the past few years, but existing processes (e.g. for the elimination of contaminants) have been improved. The result has been a marked improvement in the physical quality of recycled fibre, closing the gap with virgin fibre. In addition, there has been a favourable effect on the cost of recycled fibre because technology has allowed lower grades to be used – and therefore on the profitability of the industry.

Technological research by the paper-making industries has extended the uses to which recycled paper can be put, e.g. printing papers containing recycled fibre are now of a quality suitable for present methods of printing. The encouragement of the use of products containing a high proportion of recycled fibre may act as a further stimulus to the paper industry.

There are constraints, however, affecting the use of waste paper as a raw material by the paper industry (physical or technical problems relating to printability, stiffness of packaging products, purity and whiteness standards, etc.). [Summary and conclusions.]

There has also been a wider acceptance by buyers of some of the reductions in visual and technical qualities caused by replacing virgin with secondary fibre, where the essential purpose of the product was not affected. There were significant R and D programmes, a few of which were funded from public sources.

Some legal or fiscal measures were taken to encourage the use of waste paper, but such measures were not as strong or as widespread as had been expected.

The waste paper markets remained volatile but their characteristics appeared to be better understood and the quality of market information may have improved.

Uses for waste paper other than for paper making (e.g. for insulation) were perceived as less of a threat to their supplies by most papermakers. Indeed, it appeared to have

---

[3] In this study, the term "recovery rate" is used to indicate waste paper recovered for use by the paper industry as a percentage of apparent consumption of paper and paperboard and the term "consumption rate" for the consumption of waste paper for paper making as a percentage of production of paper and paperboard. It was not possible within the overall statistical framework of the study to use the more precise FAO definitions which take into account notably non-recoverable paper and paperboard, net trade in paper products and total fibre input to paper-making. The ratios shown in this study are derived from the ETTS IV data base and are not therefore exactly the same as those in other publications. The secretariat considers, however, that no distortion has resulted from this necessary simplification.

[4] *Future trends in consumption of waste paper by the pulp and paper industry*, OECD, Paris 1976.

become generally accepted that, in most cases, the best use for waste paper, where recovery was economic, was as raw material for paper making. An exception to this is waste paper which has already entered the "solid waste stream", i.e. become mixed with other types of refuse. Separation of secondary fibre from the waste stream was considered an expensive way of producing low quality raw material, only economically justifiable in exceptional circumstances. Furthermore, local authorities have been increasingly disposing of municipal refuse by incineration, with energy recovery, and therefore need the energy content of the waste paper, which accounts for 30-50% of the total. In some cases objections have been raised to separation of waste paper before it enters collection systems on the grounds that this procedure reduces the energy content of the municipal waste which will be incinerated.

The underlying reason for the expansion of waste paper recovery and use seems to be its economic advantages in many circumstances compared to virgin fibre. (Secondary fibre usually does not totally replace virgin fibre: costs may be reduced by increasing the percentage of secondary fibre used.) There is at present no reason to believe that this general trend will change, although there are, of course, certain limitations:

– A significant proportion of paper and paperboard consumption is not physically or economically available for recovery (paper in books, household and sanitary tissues, consumption in remote areas with prohibitive collection costs). It is difficult to estimate this percentage accurately on a comparable basis. Estimates reproduced in FAO's regular publications *Waste Paper Data* indicate that the *physically* non-recoverable waste paper amounts to 15-20% of paper and paperboard consumption in most European countries. The economic availability, which is closely linked to social and institutional factors, varies from place to place and cannot be estimated in any generalized way;

– Despite the processing improvements mentioned above, a certain proportion of virgin fibre is necessary for many grades of paper and paperboard, for technical reasons. Detailed analysis of the outlook for waste paper must try to match the supply of different grades of secondary fibre (old newspapers, container waste, pulp substitutes, mixed waste paper, etc.) with the technical requirements of production of different grades of paper and paperboard. Such detailed analysis is not possible in ETTS IV, but has been attempted by, among others, OECD (*Trends in Supply and Demand Pattern for Various Grades of Waste Paper*, 1978). These questions will also be addressed by the FAO Industry Working Party in their outlook work;

– Social and institutional factors specific to each country play a major role and cause large differences between neighbouring and apparently similar countries (see, for example, the differences in recovery rates between Finland and Sweden, Belgium and the Netherlands, or Austria and Switzerland). Thus, although most countries may move in the same direction, because of common influencing factors, significant differences between countries will remain, even if they are somewhat smaller than before.

The OECD study on future trends in consumption of waste paper presents forecasts by some OECD countries for future recovery and consumption of waste paper which indicate that the experts who prepared the forecasts expect recent trends to continue, i.e. that recovery and consumption of waste paper will expand, essentially for economic and technical reasons, even without large-scale official stimulus. Among European countries, the following percentage increases in recovery rate are forecast:

	1980	1990
Austria	75	78
Norway	28	33
Sweden	41	46
Switzerland	50	54

(N.B. These estimates are for recovery as a percentage of *theoretical availability*, taking into account estimated non-recoverable volumes: there are large national differences in methods of estimating this.)

On the basis of the above considerations, it appears a reasonable hypothesis for ETTS IV that recovery of waste paper will continue to expand, but without any explosive increase. Not only will recovery expand in absolute terms, but also in relation to consumption of paper and paperboard, i.e. the recovery rate will continue to rise. Differences between countries will remain. As the recovery rate rises above 40% of total consumption, further increases will become more difficult and slower as the potential for expansion is limited by technical and economic factors (lower quality, higher collection costs, difficulty of matching supply with demand in quality terms).

The *rate* of increase in the recovery rate is very difficult to estimate. In the past the increase was roughly as follows (in percentage points per decade):

	1960s	1970s
Nordic countries	−2	+ 6
EEC(9)	+1	+ 5
Central Europe	−	+ 6
Southern Europe	..	+10
Eastern Europe	..	+ 8

Some of the large gains in the 1970s may be unrepeatable as an under-used resource was rapidly exploited.

Very tentatively, the following hypotheses are suggested (for the two decades 1980-1990 and 1990-2000):

Percentage recovery rate at beginning of decade	Increase in rate (in percentage points per decade)	
	Low	High
Under 25	4	6
25 - 30	3	5
30 - 40	2	4
Over 40	1	3

The resulting recovery rates are presented in annex table 15.2 alongside the historical recovery rates. It is expected that by 2000 in most countries the recovery rate will be over 30%, with five countries over 45% (high scenario), but none over 50%.

Whereas the total volume of trade in waste paper may well continue to increase, as trade evens out imbalances in supply, it appears unlikely that there will be major shifts in net imports to Europe because of the low value/volume ratio of waste paper. The major supplier to the world waste paper market is the United States, which has a very considerable surplus, mostly transported to Canada, Mexico and the Far East, notably Korea and Taiwan (USA net exports in 1980 were 2.4 million m.t.). It is not clear to what extent waste paper recovery in the USA can in-

crease, whether the waste paper is likely to be processed domestically or what developments are likely in import demand by other countries. In the circumstances, it would not appear prudent to base an industry in Europe on the prospect of long-term supplies of imported waste paper.

In summary, therefore, although the waste paper market is likely to remain subject to cyclical fluctuations, long-term steady growth in recovery rates is foreseen. The volumes recovered will depend on the consumption of paper: these volumes, and the possible future role of waste paper in the supply of raw materials for papermaking will be examined in chapter 20.

CHAPTER 16

Structure and trends of European forest products trade

16.1 INTRODUCTION

This chapter is the first of two dealing with European forest products trade and describes the structure of European trade in forest products and broad trends and developments over the last two decades.

International trade is an important component in many countries' forest products balance, whether they be exporters, importers or, as in most cases, both. In Europe, the majority of countries import more than they export. But there are a number of exporters of major international importance which maintain the level of self-sufficiency of the region as a whole in forest products to about 88% in 1980. (Self-sufficiency is defined here as total domestic supply as a percentage of consumption.) This level has remained fairly steady in the past decade or so.

In 1980 the value of European countries' exports f.o.b. amounted to nearly US$ 25 billion and of its imports c.i.f. to $32 billion. The countries of the region are involved, either as exporters or imports, in nearly half of the total of world trade in forest products. This puts Europe far ahead of other regions in terms of trading importance. Compared with Europe's 44% share of total exports,

North America accounts for 31%; and Europe's 51% share of imports is followed by the Asia-Pacific region (including Japan) with 23% and North America with 13%.

Nevertheless, Europe is composed of a rather large number of countries, and the trade between these countries accounts for a substantial part of Europe's total trade in forest products, and even for around 30%, in volume terms expressed in equivalent volume of wood in the rough (m^3EQ), of the world total (table 16.2).

Trade is also a very important part of Europe's forest products supply/consumption balance. Of the total domestic supply in the region in 1980 of 424 million m^3EQ (removals, residue transfer and waste paper recycling), some 167 million m^3EQ, or 39%, were exported, mostly in processed form; and of total domestic requirements of around 480 million m^3EQ, 225 million m^3EQ or 47% were imported. Within the total figure of European countries' imports, 141 million m^3EQ came from other European countries (intra-European trade) and 84 million m^3EQ from other regions. Net imports, at around 58 million m^3EQ, accounted for about 12% of domestic requirements.

TABLE 16.1

Value of world trade in forest products, 1980 by product group, and Europe's share of it

	World		of which: Europe		
	Value [a] (billion US dollars)	Percentage of forest products, total	Value [a] (billion US dollars)	Percentage of forest products, total	Percentage of world trace in products
Exports					
Wood raw materials	8.8	15.7	1.4	5.6	15.8
Sawnwood	12.3	22.1	5.1	20.6	41.0
Wood-based panels	5.2	9.2	2.6	10.7	50.9
Woodpulp	9.5	17.1	3.3	13.5	34.8
Paper and paperboard	20.1	35.9	12.2	49.6	60.9
Total	55.9	100.0	24.6	100.0	44.1
Imports					
Wood raw materials	12.4	20.0	3.4	10.7	27.5
Sawnwood	14.0	22.4	8.3	25.9	59.4
Wood-based panels	5.2	8.4	3.4	10.6	65.0
Woodpulp	9.8	15.7	5.7	17.6	57.8
Paper and paperboard	20.8	33.5	11.3	35.2	54.2
Total	62.3	100.0	32.1	100.0	51.5

Source: FAO Yearbook of Forest Products, 1984.

[a] Exports f.o.b.; imports c.i.f.

TABLE 16.2

Europe's share of the volume of world trade in total forest products, in 1980

From: To:	Volume (million m³EQ)			Percentage of world total		
	World	Europe	Other	World	Europe	Other
World	474	167	307	100	35	65
Europe.........................	225	141	84	47	30	18
Other	249	26	223	53	5	47

Source: TIMTRADE data base.

These general statistics and remarks about Europe's place in world trade in forest products and the importance of trade in these products in the region's supply/consumption balance provide a backcloth to the description which follows of the main elements of European trade, in terms of structure, pattern and trends.

The chapter as a whole, while not attempting to address the complex question of the outlook for international trade, is intended to lead into the discussion of the actual and potential role of external suppliers to the European market in chapter 17 and to the discussion of the long-term supply/demand equilibrium in chapter 20.

The data cited in this chapter, and chapter 17, come from two separate data bases:
- The series for total imports and exports
- The TIMTRADE data base on trade flows, constructed from exporters' and importers' data (see annex 16.19 for an explanation of how the data base was constructed).

The data from these two sources do not always coincide exactly, although the discrepancies are not in most cases serious. When TIMTRADE data are used this is indicated. In particular, TIMTRADE data are for individual years only (e.g. 1980) whereas for total imports and exports, 3-year averages are often used (e.g. 1979-81).

16.2 EUROPE'S PLACE IN WORLD FOREST PRODUCTS TRADE

Tables 16.3 and 16.4 summarize world trade in 1980 in matrix form, in order to identify the volume of inter-regional trade flows and the main products concerned. For table 16.4, in the interests of clarity, minor flows, and all intra-regional trade flows have been omitted.

In tables 16.3 and 16.4, the Nordic countries are treated separately from the rest of Europe.

The volume of world trade in 1980 of four of the major product groups was 100 million m³EQ or more:

Sawnwood (125 million m³EQ);
Paper and paperboard (114 million m³EQ);
Wood raw material (107 million m³EQ);
Woodpulp (100 million m³EQ).

Trade in wood-based panels was equivalent to about 27 million m³EQ. Among individual products, the following were most heavily traded:

Sawn softwood (105 million m³EQ);
Chemical pulp (94 million m³EQ);
Paper and paperboard other than newsprint (72 million m³EQ);
Newsprint (42 million m³EQ);
Hardwood logs (42 million m³EQ).

By far the biggest importing region is Europe, the Nordic countries excluded (210 million m³EQ). If intra-regional trade flows are excluded, exports from other regions to non-Nordic Europe are 143 million m³EQ (30%

of total world trade). The next largest importers are the USA (84 million m³EQ) and Japan (71 million m³EQ). These three regions account for 63% of identified world trade flows.

With reference to the "regions" of table 16.3 (in fact sometimes large countries), the two largest inter-regional flows were the following:

Canada to USA (79 million m³EQ);
Nordic countries to the rest of Europe (65 million m³EQ).

All other inter-regional flows were much smaller (under 30 million m³EQ).

What is the size and assortment composition of Europe's trade with other regions? This is important background data to chapter 17 which will deal with the potential of countries outside Europe to supply forest products to Europe. The Nordic countries are treated separately from the rest of Europe.

In 1980, the *imports of "the rest of Europe"* from other regions were as follows:

Nordic countries: 65 million m³EQ, essentially of sawn softwood, woodpulp and all types of paper and paperboard;

Canada: 21 million m³EQ, essentially sawn softwood, plywood, woodpulp and newsprint;

TABLE 16.3

Estimated world trade in forest products, by origin and destination, 1980

(Million m³EQ)

To:	From World	Nordic countries	Rest of Europe	USSR	Canada	USA	Latin America	Africa	Japan	Other Asia/ Pacific	Unidenti-fied
World.....................	474	90	77	36	119	63	12	10	3	61	4
Nordic countries............	15	7	2	4	–	1	1	–	–	–	–
Rest of Europe.............	210	65	67	20	21	17	3	7	–	6	3
USSR	5	4	1	–	–	–	–	–	–	–	–
Canada...................	6	–	–	–	–	6	–	–	–	–	–
USA	84	1	–	–	79	–	1	–	–	2	–
Latin America..............	14	2	–	1	3	5	3	–	–	–	–
Africa....................	9	3	2	1	1	1	–	–	–	–	–
Japan	71	1	–	7	9	24	2	–	–	28	–
Other Asia/Pacific.........	41	4	1	1	3	8	1	–	1	20	–
Unidentified	20	4	3	2	2	1	1	2	1	3	–

Source: TIMTRADE data base (see annex 16.19). "Forest products" here includes roundwood, chips and residues, sawnwood, wood-based panels, pulp, paper and paperboard, converted to m³ EQ using the standard conversion factors.

USSR: 20 million m³EQ, essentially sawn softwood, woodpulp and paper and paperboard (the latter two product groups mostly to Eastern Europe);

USA: 17 million m³EQ, essentially sawn softwood and hardwood, woodpulp and paper and paperboard;

Africa: 7 million m³EQ, essentially non-coniferous logs and sawnwood;

Asia/Pacific: 6 million m³EQ, essentially non-coniferous sawnwood and plywood;

Latin America: 3 million m³EQ, essentially coniferous sawnwood and woodpulp (the latter trade has grown rapidly since 1980).

Exports by non-Nordic Europe to other regions were relatively minor (10 million m³EQ in total), consisting essentially of pulpwood to the Nordic countries (2 million m³EQ) and paper and paperboard to a wide variety of destinations.

The main destination of *Nordic exports* is the rest of Europe, as mentioned above, but these countries also export pulp, paper and paperboard to the USSR and paper and paperboard all over the world. They *import* wood raw material (pulpwood from the rest of Europe, logs and pulpwood for the USSR and chips from the USA). Within this broad structure, the volumes traded vary widely in accordance with market conditions.

16.3 TRENDS IN EUROPEAN TRADE IN FOREST PRODUCTS

TABLE 16.5

Europe: exports, imports and net trade of main groups of forest products, 1964-66 to 1979-81

	Million units				Change 1964-66 to 1979-81	
	1964-66	1969-71	1974-76	1979-81	Volume	Percentage
A. *Exports*						
Wood raw material (m³)	11.9	15.4	19.7	22.2	+ 10.3	+ 87
Sawnwood (m³)	19.2	22.0	21.0	24.8	+ 5.6	+ 29
Wood-based panels (m³)	3.3	4.9	6.5	8.0	+ 4.7	+142
Wood pulp (m.t.)	7.2	7.4	6.7	7.1	– 0.1	– 1
Paper and paperboard (m.t.)	7.3	10.9	13.3	18.4	+11.1	+152
Total (m³EQ)	106.6	131.1	140.6	171.5	+64.9	+ 61
B. *Imports*						
Wood raw material (m³)	25.9	31.8	37.2	38.8	+12.9	+ 50
Sawnwood (m³)	29.2	31.3	29.9	34.3	+ 5.0	+ 17
Wood-based panels (m³)	3.5	5.8	8.2	10.3	+ 6.8	+236
Wood pulp (m.t.)	7.9	9.7	10.0	11.4	+ 3.5	+ 44
Paper and paperboard (m.t.)	6.8	10.9	13.0	16.8	+ 9.9	+145
Total (m³EQ)	139.3	175.7	191.7	223.8	+84.5	+ 61
C. *Net trade*[a]						
Wood raw material (m³)	– 14.0	– 16.4	– 17.5	– 16.6	..	..
Sawnwood (m³)	– 10.0	– 9.3	– 9.0	– 9.5	..	..
Wood-based panels (m³)	– 0.2	– 0.9	– 1.8	– 2.3	..	..
Wood pulp (m.t.)	– 0.7	– 2.3	– 3.3	– 4.2	..	..
Paper and paperboard (m.t.)	+ 0.4	+ 0.0	+ 0.3	+ 1.6	..	..
Total (m³EQ)	– 32.7	– 44.6	– 51.1	– 52.3	..	..

Source: Tables 16.6 and 16.7.   [a] + = net exports; – = net imports.

TABLE 16.4

Product structure of inter-regional trade in forest products, by origin and destination, 1980

To: ► / From: ▼	World [a]	Nordic countries	Rest of Europe	USSR	Canada	USA	Latin America	Africa	Japan	Other Asia/Pacific
World [a]	474	90	77	36	119	63	12	10	3	61
Nordic countries	15		Pulpwood	Logs (C); Pulpwood		Chips				
Rest of Europe	210	Sawnwood (C); Wood-based panels; Wood pulp; Paper and paperboard		Sawnwood (C); Pulp; Paper and paperboard	Sawnwood (C); Plywood; Pulp; Newsprint	Sawnwood (C+NC); Pulp; Paper and paperboard	Sawnwood (C+NC); Pulp	Logs (NC); Sawnwood (NC)	Sawnwood (NC); Plywood	
USSR	5	Pulp; Paper and paperboard								
Canada	6					Logs (C)				Plywood
USA	84				Sawnwood (C); Pulp; Newsprint; Other paper and paperboard		Sawnwood (NC)			
Latin America	14				Newsprint	Pulp; Paper and paperboard				
Africa	9	Sawnwood (C); Paper and paperboard	Sawnwood							
Japan	71			Logs (C); Sawnwood (C)	Logs (C); Sawnwood (C); Pulp	Logs (C); Chips; Sawnwood (C); Pulp	Logs (C); Sawnwood (C); Pulp			Logs (C+NC); Chips; Sawnwood (C+NC); Pulp
Other Asia/Pacific	41	Paper and paperboard	Sawnwood (C+NC)							

Note: (C) Coniferous; (NC) Non-coniferous.
For volume of flows of each product see annex table 16.19.
For volume of flows in m³EQ, see table 16.3.
Minor flows are omitted.

[a] Exports and imports in million m³EQ.

FIGURE 16.1

Europe: trade in sawlogs and veneer logs

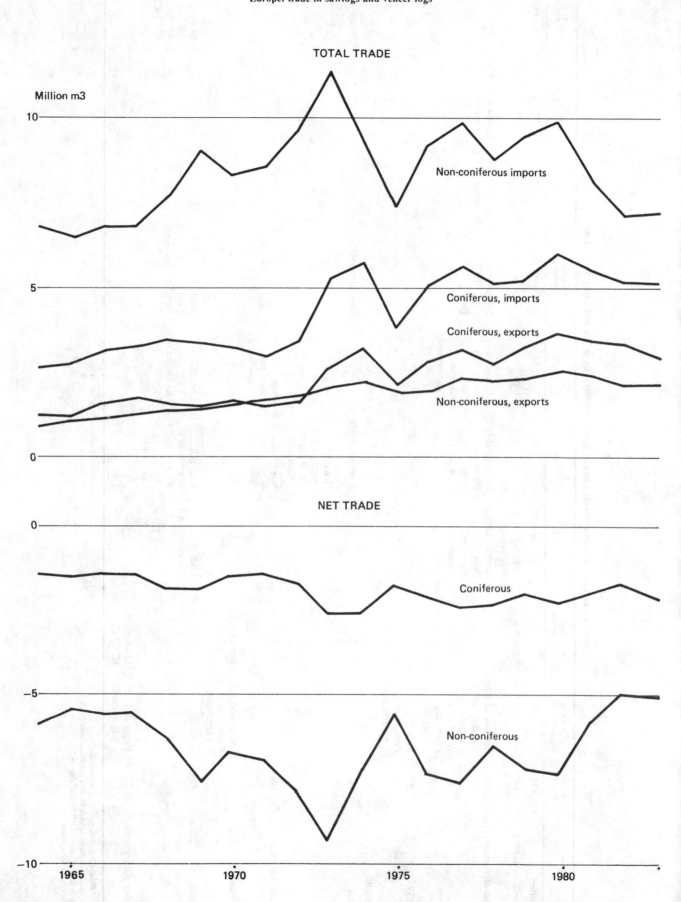

FIGURE 16.2

Europe: trade in pulpwood

FIGURE 16.3

Europe: trade in sawnwood

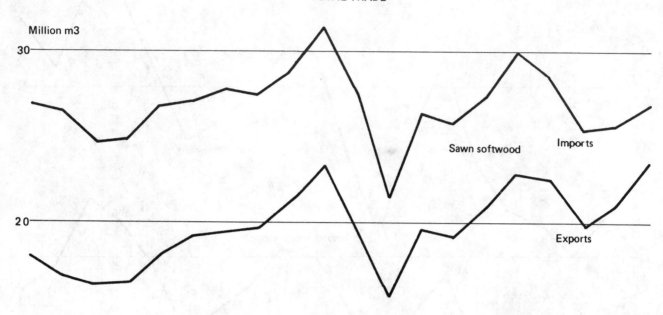

TOTAL TRADE

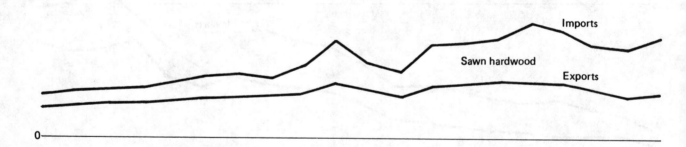

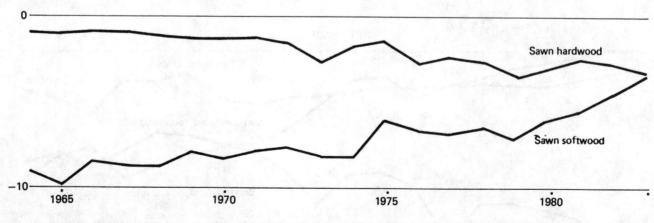

NET TRADE

FIGURE 16.4

Europe: trade in wood-based panels

TOTAL TRADE

Particle board and fibreboard imports*

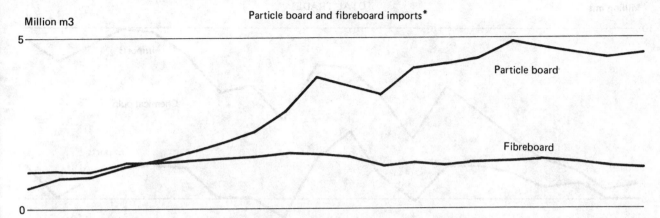

Million m3

* Exports of fibreboard and particle board not shown as they coincide almost exactly with imports

Plywood trade

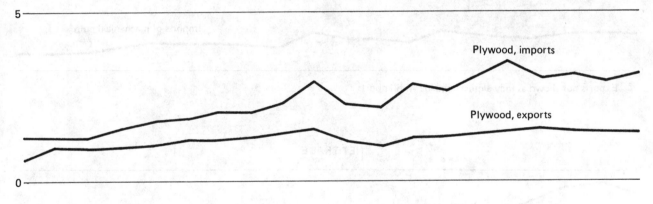

NET TRADE

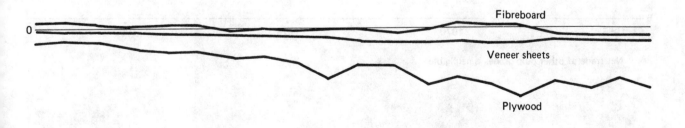

(net trade of particle board is negligible)

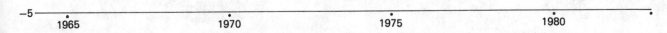

FIGURE 16.5

Trade in woodpulp

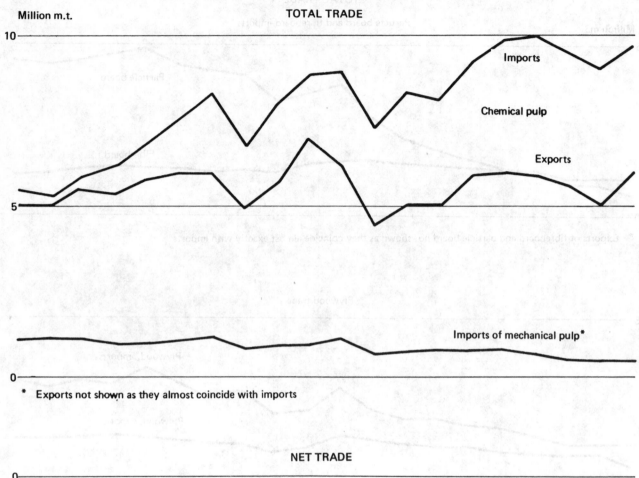

Million m.t.

TOTAL TRADE

Imports

Chemical pulp

Exports

Imports of mechanical pulp*

* Exports not shown as they almost coincide with imports

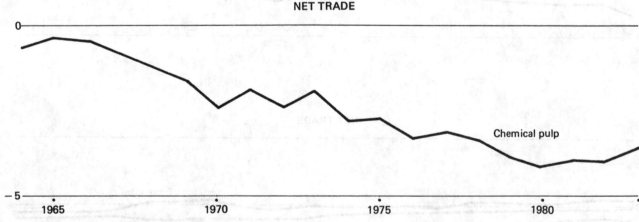

NET TRADE

Chemical pulp

Net trade of other pulp grades is negligible

FIGURE 16.6

Trade in paper and paperboard

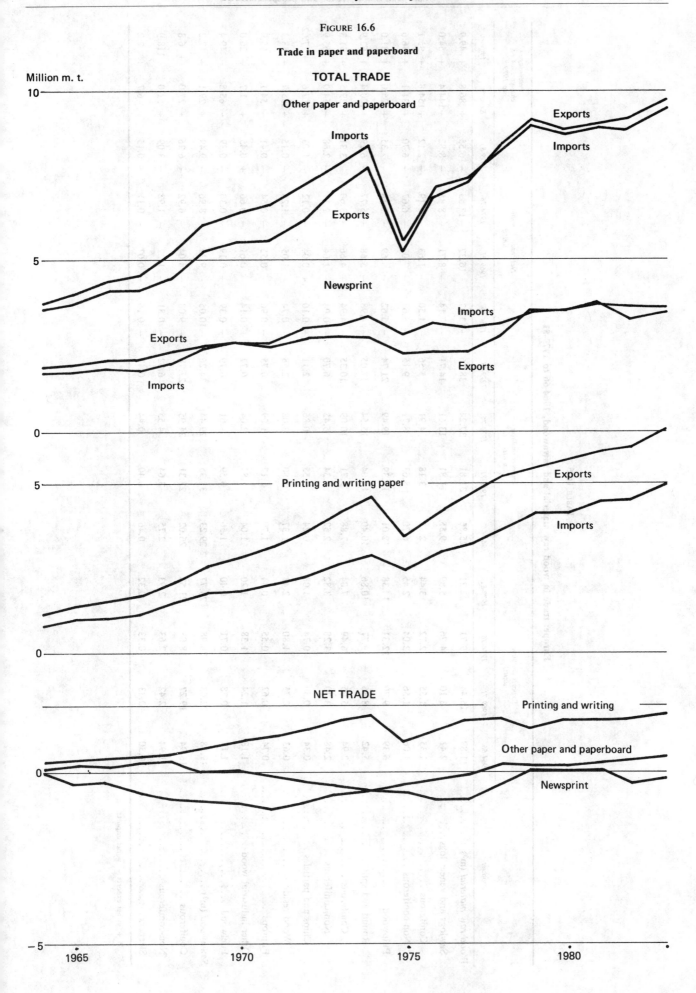

TABLE 16.6

Europe: trade in wood raw material and sawnwood, 1964-66 to 1979-81

Product	Exports (million units)				Imports (million units)				Net trade [a] (million units)				Percentage change 1964-66 to 1979-81	
	1964-66	1969-71	1974-76	1979-81	1964-66	1969-71	1974-76	1979-81	1964-66	1969-71	1974-76	1979-81	Exports	Imports
Wood raw material (m³)	11.89	15.44	19.71	22.21	25.86	31.81	37.24	38.76	− 13.97	− 16.37	− 17.53	− 16.55	+ 86.8	+ 49.9
Sawlogs and veneer logs	2.41	3.10	4.76	5.89	9.55	11.81	13.61	14.71	− 7.14	− 8.71	− 8.85	− 8.82	+144.4	+ 54.0
Coniferous	1.35	1.55	2.72	3.44	2.85	3.18	4.91	5.56	− 1.50	− 1.63	− 2.19	− 2.12	+154.8	+ 95.1
Non-coniferous	1.06	1.56	2.04	2.45	6.70	8.63	8.71	9.15	− 5.67	− 7.07	− 6.67	− 6.70	+131.1	+ 36.6
Pulpwood	6.39	9.70	12.31	14.38	12.01	16.39	19.49	21.74	− 5.62	− 6.69	− 7.18	− 7.35	+125.0	+ 81.0
Round and split	5.42	8.36	9.77	10.86	10.40	14.04	15.54	17.05	− 4.98	− 5.68	− 5.77	− 6.19	+100.4	+ 63.9
Coniferous	2.94	5.05	6.49	7.04	7.88	9.71	10.09	10.35	− 4.94	− 4.66	− 3.60	− 3.31	+139.5	+ 31.3
Non-coniferous	2.48	3.30	3.28	3.82	2.52	4.34	5.45	6.70	− 0.04	− 1.04	− 2.17	− 2.88	+ 54.0	+165.99
Chips and particles	0.34	0.59	0.74	1.08	0.24	0.55	0.86	2.11	+ 0.10	+ 0.04	− 0.12	− 1.03	+217.6	+779.2
Wood residues	0.63	0.75	1.80	2.43	1.37	1.80	3.08	2.58	− 0.74	− 1.05	− 1.28	− 0.15	+285.7	+ 88.3
Pitprops	0.74	0.62	0.55	0.34	1.70	1.17	1.09	0.75	− 0.96	− 0.55	− 0.54	− 0.41	− 54.1	− 55.9
Other industrial wood	1.18	1.24	1.38	1.20	1.04	1.85	2.04	0.77	+ 0.14	− 0.61	− 0.66	+ 0.45	+ 1.7	− 26.0
Fuelwood	1.17	0.78	0.71	0.40	1.56	0.59	1.01	0.79	− 0.39	+ 0.19	− 0.30	− 0.39	− 65.8	− 49.4
Sawnwood (m³)	19.18	22.03	20.98	24.77	29.23	31.35	29.44	34.26	− 10.05	− 9.32	− 8.96	− 9.49	+ 29.1	+ 17.2
Coniferous	16.94	19.27	18.03	21.52	26.02	27.31	24.95	27.78	− 9.08	− 8.04	− 6.92	− 6.26	+ 27.0	+ 6.8
Non-coniferous	1.94	2.42	2.63	2.93	2.75	3.64	4.55	6.02	− 0.81	− 1.22	− 1.92	− 3.09	+ 51.0	+118.9
Sleepers	0.30	0.33	0.33	0.32	0.46	0.40	0.44	0.47	− 0.16	− 0.07	− 0.11	− 0.15	+ 6.7	+ 2.2

[a] + = net exports − = net imports.

TABLE 16.7

Europe: trade in wood-based panels, wood pulp, paper and paperboard 1964-66 to 1979-81

Product	Exports (million units)				Imports (million units)				Net trade [a] (million units)				Percentage change 1964-66 to 1979-81	
	1964-66	1969-71	1974-76	1979-81	1964-66	1969-71	1974-76	1979-81	1964-66	1969-71	1974-76	1979-81	Exports	Imports
Wood-based panels (m³)	3.32	4.86	6.46	8.03	3.52	5.80	8.22	10.30	−0.20	−0.94	−1.76	−2.27	+141.9	+192.6
Veneer sheets	0.16	0.21	0.24	0.35	0.22	0.37	0.63	0.74	−0.06	−0.16	−0.39	−0.39	+118.7	+236.4
Plywood	0.90	1.20	1.14	1.49	1.25	1.97	2.40	3.25	−0.35	−0.77	−1.26	−1.76	+65.6	+160.0
Particle board	0.89	1.91	3.73	4.81	0.87	1.94	3.74	4.82	+0.02	−0.03	−0.01	−0.01	+440.4	+454.0
Fibreboard	1.38	1.55	1.35	1.39	1.19	1.51	1.45	1.49	+0.19	+0.04	−0.10	−0.10	+0.7	+25.2
Woodpulp (m³)	7.20	7.45	6.72	7.11	7.87	9.73	10.03	11.36	−0.67	−2.28	−3.31	−4.25	−1.2	+44.4
Mechanical	1.14	0.95	0.78	0.64	1.19	1.04	0.86	0.63	−0.05	−0.09	−0.08	+0.01	−43.9	−47.1
Semi-chemical	0.08	0.14	0.11	0.11	0.09	0.11	0.11	0.10	−0.01	+0.03	−	+0.01	+37.5	+11.1
Chemical	5.23	5.70	5.31	5.86	5.65	7.64	8.25	9.87	−0.42	−1.94	−2.94	−4.01	+12.0	+74.7
Dissolving grades	0.76	0.65	0.51	0.50	0.94	0.94	0.81	0.76	−0.18	−0.29	−0.30	−0.26	−34.2	−19.1
Paper and paperboard (m.t.)	7.29	10.94	13.27	18.38	6.84	10.89	12.98	16.76	+0.45	+0.05	+0.29	+1.62	+152.1	+145.0
Newsprint	2.04	2.58	2.48	3.62	1.84	2.58	3.13	3.60	+0.20	−	−0.65	+0.02	+77.4	+95.7
Printing and writing paper	1.37	2.82	4.09	5.70	0.96	1.89	2.79	4.25	+0.41	+0.93	+1.30	+1.45	316.1	+343.7
Other paper and paperboard	3.87	5.54	6.70	9.05	4.05	6.42	7.07	8.91	−0.18	−0.88	−0.37	+0.14	+133.8	+120.0

[a] + = net exports; − = net imports.

TABLE 16.8

Summary of world trade in major forest products, 1970 and 1980
(Million units)

| | Total | | From world — of which, from: | | | | | | | | | | | | | |
| | | | Europe | | USSR | | North America | | Latin America | | Africa | | Asia/Pacific | | Other [a] | |
	1970	1980	1970	1980	1970	1980	1970	1980	1970	1980	1970	1980	1970	1980	1970	1980
Coniferous logs (m³)	24.8	27.9	1.7	4.0	7.4	6.5	13.4	15.1	—	1.0	—	—	2.3	1.3	—	—
of which, to Europe	3.2	5.2	1.7	3.8	1.5	1.4	—	—	—	—	—	—	—	—	—	—
Hardwood logs (m³)	38.4	42.5	1.6	3.2	0.1	0.1	0.4	0.8	0.4	—	6.6	6.0	29.4	31.9	—	0.5
of which, to Europe	9.0	8.8	1.5	3.1	0.1	0.1	—	0.4	0.1	—	6.0	4.6	1.2	0.2	—	0.4
Round pulpwood, coniferous (m³)	12.9	10.6	5.8	6.1	4.4	3.1	2.7	1.4	—	—	—	—	—	—	—	—
of which, to Europe	10.7	9.1	5.8	6.1	4.0	2.5	1.0	0.5	—	—	—	—	—	—	—	—
Round pulpwood, non-coniferous (m³)	5.4	8.1	3.7	4.8	1.6	2.6	0.1	0.1	—	—	—	0.3	—	—	—	0.3
of which, to Europe	4.9	7.6	3.7	4.8	1.2	2.2	—	0.1	—	—	—	0.3	—	—	—	0.3
Wood residues (m³)	1.7	3.0	0.7	1.5	0.6	1.5	0.5	—	—	—	—	—	—	—	—	—
of which, to Europe	1.2	3.0	0.7	1.5	0.6	1.5	—	—	—	—	—	—	—	—	—	—
Chips and particles (m³)	5.1	17.2	0.6	1.2	—	0.5	4.5	8.0	—	—	—	—	—	7.4	—	0.3
of which, to Europe	0.7	1.7	0.6	1.2	—	—	0.1	0.5	—	—	—	—	—	—	—	—
Sawn softwood (m³)	49.0	64.0	19.3	21.5	7.7	7.2	20.1	33.6	1.3	1.4	—	—	—	—	—	—
of which, to Europe	27.8	28.5	18.0	18.2	6.7	5.6	2.7	4.1	0.4	0.4	—	—	—	—	—	—
Sawn hardwood (m³)	7.1	12.3	2.5	3.1	—	—	0.7	1.2	0.5	0.9	0.7	0.5	2.7	6.4	—	0.2
of which, to Europe	3.8	6.3	2.0	2.6	—	—	0.1	0.5	—	0.3	0.5	0.5	1.1	2.3	—	0.1
Plywood (m³)	4.5	6.6	1.2	1.5	0.3	0.3	0.5	0.8	—	0.3	0.1	—	2.4	3.7	—	0.1
of which, to Europe	2.1	3.4	1.0	1.3	0.2	0.3	0.4	0.6	—	0.1	0.1	—	0.3	1.0	—	0.1
Particle board (m³)	2.0	5.5	1.8	4.8	0.1	0.3	—	0.2	—	—	—	—	—	0.1	—	0.1
of which, to Europe	1.9	5.0	1.8	4.7	0.1	0.3	—	—	—	—	—	—	—	—	—	—
Fibreboard (m³)	2.0	2.2	1.5	1.4	0.1	0.3	0.2	0.3	—	0.2	0.1	—	—	—	—	—
of which, to Europe	1.5	1.4	1.3	1.1	0.1	0.2	—	—	—	0.1	—	—	—	—	—	—
Veneer sheets (m³)	0.7	1.3	0.2	0.4	—	—	0.2	0.5	—	0.1	0.1	0.1	0.2	0.2	—	—
of which, to Europe	0.3	0.7	0.2	0.3	—	—	—	0.2	—	—	0.1	0.1	—	—	—	—
Mechanical pulp (m.t.)	1.3	1.0	1.1	0.7	—	—	0.3	0.3	—	—	—	—	—	—	—	—
of which, to Europe	1.1	0.7	1.0	0.6	—	—	0.1	0.1	—	—	—	—	—	—	—	—
Chemical pulp (m.t.)	15.9	19.6	6.9	6.5	0.5	0.8	8.2	10.3	—	1.3	0.3	0.5	—	—	0.2	0.2
of which, to Europe	10.2	11.6	6.2	5.6	0.4	0.7	3.3	4.2	—	0.6	0.2	0.3	—	—	—	0.2
Newsprint (m.t.)	10.7	12.2	2.7	3.6	0.3	0.3	7.5	7.9	—	0.2	—	—	0.2	0.2	—	—
of which, to Europe	2.7	3.6	1.9	2.7	0.1	0.2	0.6	0.6	—	—	—	—	—	—	—	—
Other paper and paperboard (m.t.)	12.5	21.2	8.5	14.0	0.3	0.7	3.1	5.5	—	—	—	—	0.6	0.7	0.2	0.2
of which, to Europe	8.5	12.9	6.7	10.7	0.3	0.6	1.4	1.5	—	—	—	—	—	0.1	—	0.1

Source: TIMTRADE data base.

[a] These are flows whose origin was not specified in the original questionnaire, i.e. *not* the difference between identified flows and total trade (see annex 16.19).

TABLE 16.9

Share of world trade in major forest products by continent, 1970 and 1980
(Percentage of world total)

| | Total | | From world of which, from: | | | | | | | | | | | | | |
| | | | Europe | | USSR | | North America | | Latin America | | Africa | | Asia/Pacific | | Other [a] | |
	1970	1980	1970	1980	1970	1980	1970	1980	1970	1980	1970	1980	1970	1980	1970	1980
Coniferous logs	100.0	100.0	6.9	14.2	29.7	23.4	54.1	54.3	—	—	—	—	9.4	4.5	—	3.6
of which, to Europe	12.8	18.6	6.8	13.6	5.9	4.9	0.1	0.1	—	—	—	—	—	—	—	—
Hardwood logs	100.0	100.0	4.2	7.6	0.3	0.3	1.0	1.9	1.0	1.0	17.2	14.2	76.5	75.0	—	1.1
of which, to Europe	23.4	20.8	3.9	7.4	0.2	0.3	0.2	1.0	0.3	0.3	15.7	10.9	3.0	0.4	—	0.9
Round pulpwood, coniferous	100.0	100.0	44.6	57.6	34.3	29.6	21.1	12.8	—	—	—	—	—	—	—	—
of which, to Europe	82.9	85.4	44.6	57.4	30.6	23.4	7.8	4.6	—	—	—	—	—	—	—	—
Round pulpwood, non-coniferous	100.0	100.0	67.7	59.1	30.2	32.2	2.0	1.2	—	—	—	3.3	0.3	0.3	—	3.3
of which, to Europe	90.9	94.6	67.7	59.1	23.1	26.9	—	1.2	—	—	—	3.3	0.3	0.3	—	3.3
Wood residues	100.0	100.0	40.2	50.4	33.0	49.6	26.8	—	—	—	—	—	—	—	—	—
of which, to Europe	73.2	99.9	40.2	50.3	33.0	49.6	—	—	—	—	—	—	—	—	—	—
Chips and particles	100.0	100.0	12.2	7.0	—	3.2	87.7	46.5	—	—	—	—	—	43.3	—	—
of which, to Europe	14.1	9.7	12.2	7.0	—	—	1.9	2.7	—	—	—	—	—	—	—	—
Sawn softwood	100.0	100.0	39.4	33.6	16.3	11.2	40.9	52.5	2.7	0.6	—	—	0.7	—	0.5	2.2
of which, to Europe	56.7	44.5	36.6	28.4	13.7	8.7	6.3	6.3	0.9	0.6	—	—	—	—	—	—
Sawn hardwood	100.0	100.0	35.6	25.2	—	—	9.5	9.7	6.3	2.4	10.2	7.1	38.3	55.6	—	—
of which, to Europe	53.0	50.9	28.5	20.7	—	—	1.2	4.1	0.8	2.1	7.1	3.8	15.4	18.9	—	—
Plywood	100.0	100.0	26.5	23.4	6.3	5.1	10.8	12.2	0.6	0.8	3.2	2.8	55.9	55.6	—	1.2
of which, to Europe	45.8	52.0	21.6	19.7	5.3	3.9	8.9	9.9	0.5	0.7	2.8	2.8	6.7	15.0	—	—
Particle board	100.0	100.0	91.7	88.4	7.2	6.1	1.0	4.3	—	—	—	—	—	0.1	—	—
of which, to Europe	95.6	91.7	89.0	85.8	6.5	5.5	—	0.3	—	—	—	—	—	0.1	—	—
Fibreboard	100.0	100.0	75.9	64.1	6.9	14.0	9.3	13.5	2.2	2.7	3.5	2.2	2.3	2.7	—	—
of which, to Europe	73.0	66.7	61.8	52.0	6.6	10.9	1.6	1.1	0.7	—	2.2	2.2	—	—	—	—
Veneer sheets	100.0	100.0	30.2	28.3	—	0.5	26.4	34.9	5.2	7.9	17.4	11.0	20.8	2.7	—	—
of which, to Europe	41.3	57.1	24.9	24.5	0.4	0.4	4.4	18.3	1.1	1.1	10.2	10.6	0.3	0.9	—	—
Mechanical pulp	100.0	100.0	78.9	65.0	—	—	20.8	33.4	—	—	—	—	—	1.2	—	—
of which, to Europe	80.4	64.6	74.8	56.5	—	—	5.4	6.5	—	—	—	—	—	1.2	—	—
Chemical pulp	100.0	100.0	43.6	33.0	2.9	4.3	51.6	52.5	—	6.4	—	—	—	—	2.0	2.0
of which, to Europe	64.0	59.0	39.1	28.8	2.4	3.6	20.9	21.4	—	2.9	—	—	—	—	1.6	2.0
Newsprint	100.0	100.0	25.1	29.3	2.5	2.7	69.8	64.3	0.7	1.6	—	—	2.0	2.0	0.2	0.2
of which, to Europe	24.9	29.3	18.0	22.3	1.1	1.7	5.5	5.1	—	—	—	—	—	—	—	—
Other paper and paperboard	100.0	100.0	68.2	66.2	2.2	3.4	24.6	26.2	—	—	—	—	4.8	3.1	1.0	0.6
of which, to Europe	67.8	61.0	53.7	50.4	2.2	2.7	11.5	6.9	—	—	—	—	0.2	0.2	0.1	—

[a] These are flows whose origin was not specified in the original questionnaire, i.e. *not* the difference between identified flows and total trade (see annex 16.19).

Trends in European trade are summarized in tables 16.5 to 16.9 and figures 16.1 to 16.6. More detailed tables and summary tables of world trade matrices for 16 major products are shown in the annexes (annex tables 16.1-16.14 and annex 16.19). These tables and figures are mostly self-explanatory but are briefly summarized below. This section concentrates on trends for Europe as a whole, drawing attention to trade with other regions, where appropriate, as this aspect must be examined when evaluating potential supply of forest products for Europe.

Most European countries are net importers of forest products and of wood raw material. There are however a few exceptions: the Nordic countries and Austria which are major exporters of forest products (although importers of raw material) and other exporting countries including Czechoslovakia, Portugal, and Romania. The difference between the trade "profiles" of individual European country should be borne in mind when examining the broad trends for Europe as a whole.

16.3.1 Wood raw material

Since 1969-71, the volume of total trade in wood raw material has shown an upward trend, despite strong cyclical movements. Total net imports, however, have remained roughly stable at 16-17 million m³. There are significant differences in trends between assortments. Net imports of coniferous pulpwood, round and split, decreased, while those of non-coniferous pulpwood, round and split, increased, although net imports of both assortments remained around 3 million m³ after 1976. Imports of non-coniferous logs rose to a peak in 1973: this was followed by fluctuations in the late 1970s and a decline from 1980, nearly to the level of the 1960s. In 1980 the main external suppliers of wood raw material to Europe were the USSR (7.5 million m³ of pulpwood and of coniferous logs) and Africa (4.6 million m³ of non-coniferous logs). Within Europe, during the 1970s the Nordic countries became a major net importer of pulpwood, drawing on many sources of supply, inside and outside Europe.

16.3.2 Sawnwood

The volume of trade in sawnwood, especially sawn softwood, was heavily affected by the cyclical swings of the mid-1970s. Europe's net imports of sawnwood have been between 9 and 10 million m³ since the mid-1960s but this conceals differing trends for the two major assortments. Net imports of sawn softwood have declined steadily as imports from other regions (mostly the USSR and Canada) grew modestly or showed only cyclical fluctuations, while exports to non-European destinations, notably from the Nordic countries and Austria to Africa and Asia, increased. For sawn hardwood however, imports rose, chiefly due to higher imports from south-east Asia (as well as USA), while exports stagnated, leading to higher net imports.

Sawn softwood imports by the United Kingdom fell, although this country remained Europe's largest importer,

followed by Italy. Exports of sawn softwood by Portugal rose, but those of Romania fell by 60% in 15 years. Italy has expanded its sawn hardwood imports and is now Europe's leading importer of this product, followed by the Federal Republic of Germany and the United Kingdom, formerly the leading importer.

16.3.3 Wood-based panels

Trade in particle board is almost entirely intra-regional in nature: it expanded steadily to the end of the 1970s with increased specialization in manufacture and the resulting two-way trade, but stagnated thereafter at just under 5 million m³. For plywood, however, Europe imports significant volumes from other regions, including North America and the USSR and above all south-east Asia. Between the mid-1960s and the early 1980s, European exports, mostly to other European destinations, increased only moderately while imports more than doubled, leading to higher net imports. Finland accounted for 50% of European plywood exports in 1964-66, but for only 36% in 1979-81. The United Kingdom remained Europe's largest plywood importer (31% in 1979-81), but its share of European imports has halved since the mid-1960s. Trade in fibreboard and veneer sheets is largely intra-regional and accounts for relatively small volumes.

16.3.4 Woodpulp

European trade in chemical pulp (which accounts for the bulk of international trade in woodpulp), saw a major structural change between the mid-1960s and the early 1980s. In 1964-66 imports and exports were approximately equal at over 5 million m.t. Exports retained that level (with strong cyclical fluctuations) but imports had increased by over 4 million m.t. by 1979-81, leading to net imports of a similar amount. Europe's principal suppliers are Canada and the USA (4.2 million m.t. together in 1980), but Brazilian exports to Europe rose very fast in the early 1980s. The Nordic countries' share in European exports has fallen and that of other countries, notably Portugal, has risen.

16.3.5 Paper and paperboard

This product group is the only one for which Europe as a whole has a net surplus, as exports (to a wide range of destinations) are higher than imports, mostly from North America. This surplus has increased steadily since the early 1970s, largely because of developments for printing and writing grades and for other paper and paperboard. Trade in both these products rose steadily, despite strong cyclical fluctuations in the mid-1970s, especially for other paper and paperboard. In particular, exports from the Federal Republic of Germany grew from insignificant levels in the mid-1960s to reach 1.7 million m.t. in 1979-81. The volume of trade in newsprint also rose steadily, but imports and exports remained roughly in balance.

16.4 THE IMPORTANCE OF TRADE WITHIN THE EUROPEAN FOREST PRODUCTS ECONOMY

An analysis of forest products trade would not be complete without some indication of the importance of trade for the forest products economy as a whole, in particular as regards the shares which exports represent in production and imports in consumption of the main products. As a general rule, exporting countries show a relatively high ratio of exports to production, and importing countries a high ratio of imports to consumption. Relevant data for major products and countries for the years 1965, 1970, 1975 and 1980 are set out in table 16.10 for Europe as a whole. Annex tables 16.15 to 16.18 provide data for selected countries and products.

The data may be distorted by two main factors: first, there are stocks in exporting as well as importing countries which have not been taken into account in the calculations, due to a lack of relevant information, but which are known to be an important element and to show significant year-to-year fluctuations; second, in importing countries, re-exports of imported products result in double counting and thus to ratios overstating actual values. As data separating re-exports from exports are not generally available, the extent of the overstatement and thus the true ratio, cannot be assessed, although it is believed not to be significant.

Products where the share of trade showed a clearly rising trend during the period in question were coniferous and non-coniferous logs, non-coniferous sawnwood, plywood, particle board and veneer sheets, as well as all three major product groups of the paper and paperboard sector.

Several products showed contrasting trends between export and import trade, particularly coniferous and non-coniferous pulpwood, chemical woodpulp and, to a lesser extent, coniferous sawnwood and fibreboard.

For the majority of products shown in table 16.10, the percentage share of imports in consumption is higher than the corresponding share of exports in production. This reflects primarily the importance of imports from sources outside Europe.

Trade intensity is relatively low for roundwood products, not exceeding 20% except for imports of non-coniferous logs, primarily due to the European imports of such logs from sources outside Europe. It is generally higher for the products showing some degree of processing. The fact that shares of exports of pulp in total production declined markedly lies in the trend towards converting pulps, as intermediate products, to paper and paperboard, or even end-products for the consumer market.

TABLE 16.10

Europe: ratio of exports to production and imports to consumption of major forest products, 1965 to 1980

(Percentages)

	Exports/Production				Imports/Consumption			
	1965	1970	1975	1980	1965	1970	1975	1980
Coniferous logs	1.2	1.5	2.0	2.9	2.7	2.9	3.7	4.7
Non-coniferous logs	3.1	4.3	5.8	7.0	16.9	19.1	18.9	22.3
Coniferous pulpwood	5.7	8.0	9.9	10.3	14.1	14.7	13.9	14.9
Non-coniferous pulpwood	14.1	14.6	10.0	11.8	14.1	17.7	16.5	19.0
Coniferous sawnwood	29.1	30.0	24.9	30.1	39.5	38.0	31.5	35.3
Non-coniferous sawnwood	12.2	13.5	13.6	15.5	16.6	19.1	20.4	27.1
Plywood	26.1	29.5	28.5	41.1	32.9	42.0	45.0	58.4
Particle board	15.0	15.0	18.3	19.9	15.4	15.4	17.7	19.5
Fibreboard	35.9	36.1	27.2	31.2	32.4	36.6	29.8	33.7
Veneer sheets	11.1	14.4	13.9	24.8	15.2	23.1	31.9	39.8
Mechanical woodpulp	16.5	12.4	8.2	7.0	17.4	13.5	9.2	6.8
Chemical woodpulp	38.5	36.0	26.4	30.4	40.6	43.7	36.4	42.5
Newsprint	40.8	44.7	45.5	55.4	37.9	44.5	51.1	55.4
Printing and writing	20.7	26.8	33.1	37.6	15.6	19.5	26.5	30.6
Other paper and paperboard	21.3	24.8	22.7	30.6	22.3	27.5	23.9	30.4

16.5   TARIFF AND NON-TARIFF BARRIERS AND ENCOURAGEMENTS TO TRADE

Both exports and imports are generally subject to some control, by national or other public bodies, which frequently contains restrictive elements. Such control may serve a variety of purposes, such as security, environmental protection, fiscal, industrial, social and related policy aims. At the international level, the main forum for discussion and framework for negotiation is provided by the General Agreement on Tariffs and Trade (GATT).

The means employed to exert such control are generally divided into tariff and non-tariff measures. These may be of a temporary or a more permanent nature. Tariff and non-tariff measures may be combined, for instance, through quotas which relate to certain levels of tariff for a given product. As a general rule, tariff rates on imports tend to be low for raw materials, and progressively higher as the degree of processing or the content of added value of products increases.

Furthermore, different treatments may apply to the same product depending on its origin, i.e. whether or not it is favoured by special agreements with certain trading partner countries (e.g. inside free trade areas) or qualifies for Most-Favoured-Nation (MFN) duty treatment or for the Generalized System of Preferences (GSP), which is accorded to developing countries.

Non-tariff measures include a wide range of regulations,

procedures and practices. Such measures may be broadly classified under the following titles: government participation in trade and restrictive practices tolerated by governments; customs and administrative entry procedures; technical barriers to trade such as technical standards, testing and certification arrangements; specific limitations such as quantitative restrictions, licensing, and discrimination by bilateral agreements; charges on imports other than tariffs, such as import deposits and credit restrictions; and compensation trade. It is often difficult to assess accurately the effect of any one non-tariff barrier. There also exist measures to stimulate exports, such as export subsidies or guarantees, or to restrict them (e.g. by export bans). The stimulation of exports by means of aggressive pricing policies, may, however, be considered as dumping in the countries of destination and lead to counter-measures being taken; this is not infrequent, notably for wood-based panels.

Major trade negotiation rounds under the auspices of the GATT – of which the latest, completed in 1978 is referred to as the Tokyo Round, – have resulted in significant changes in the structure and level of tariffs and in the removal or alleviation of non-tariff obstacles to trade as a whole, including trade in forest products. There is discussion at present (1986) within GATT on the form and timing of the next round of negotiations.

TABLE 16.11

**Post-Tokyo round MFN (most favoured nation) import duties for forest products
of EEC, Japan and the USA**

(Duties shown in percent ad valorem *or ad valorem* equivalent*)*

CCCN code	Summary description	EEC	Japan	USA [a]
44.01	Fuelwood, wood waste	Free	Free	Free
44.02	Wood charcoal	Free	Free	Free
44.03	Wood in the rough (sawlogs etc.)	Free; ex 2.5	Free (Kiri 5)	Free
44.04	Wood roughly squared	Free	Free (Kiri 2.5)	Free
44.05	Wood, sawn, sliced, peeled, over 5 mm	Free; 3.8	Free; 2.5; 6.0; 10.0	Free
44.07	Sleepers	2.7; 3.8	Free	Free
44.07	Hoopwood etc., pulpwood (+ chips)	3.2; 4.4	Free; 5.0; 7.5	Free
44.11	Fibre building board	10.0	6.5	Free; 3.0-6.0
44.12	Wood wood and flour	3.8	2.5	5.1
44.13	Wood, planed, tongued, jointed	4.0	Free; 2.5; 10.0	Ex 5.0; free
44.14	Wood, sawn, peeled, sliced, under 5 mm, veneer sheets	Ex free; ex 6.0	Free; 15.0; 8.0	Free; 3.2; 4.0
44.15	Plywood, blockboard, etc.	Tariff quota 60,000 m³ free; 10.0	15.0; 17.0; 20.0[b]	(3.2); 4.1; 6.1; 6.9; 8.0; 11.0; 20.0
44.16	Cellular wood panels	3.8	10.0	4.0
44.17	"Improved" wood	3.0	7.0	5.8
44.18	Reconstituted wood board	10.0	10.0; 12.0	5.1; 4.0
44.19	Wooden beadings, mouldings	3.0	7.2	Free; 1.5; 4.5

Source: GATT. Tariff rates shown are final rates (i.e. those applicable as of 1 January 1987, unless implemented earlier). Most of the fixed rates shown are GATT bound rates. Temporary duty reductions are not shown. Where several percentage figures apply, this indicates rates for different subpositions within a four digit heading. Rates shown for the USA are allocated to CCCN headings on a provisional basis. All rates shown are for information only. For contractual rates refer to the GATT 1979 Geneva Protocol. "Ex" indicates the rate shown applies to only part of the heading.

[a] Approximate CCCN correspondence.

[b] A phased reduction of import duties is being drawn up.

Areas of preferential trade flows have been created among certain industrialized countries, as well as between developing countries in Asia, Africa and Latin America. A thorough analysis of all these various aspects could only be made on an item-by-item basis. An idea of prevailing tariff levels for forest products may, however, be gained from the data set out in table 16.11.

The structure of tariffs, as well as other measures (such as log export restrictions) may have a significant effect on the structure of international trade, notably on the question of where the raw material should be processed, e.g. for forest products, in the country with the forest resource, in the consuming country or in a third country.

Non-tariff measures are also commonly encountered in forest products trade; these are partly of a temporary nature, include phyto-sanitary restrictions on imports to prevent the spread of plant- or wood-attacking biological agents; import limitations in cases of sudden and severe forest damage (e.g. wind, snow); and quantitative restrictions on imports of processed products, in particular wood-based panels and paper products. Exports may either benefit from concessionary treatment or be restricted (e.g. log export tariffs). An example cited is technical specifications for forest products such as sawnwood and panels, notably in building construction, which may reflect traditions more than technical necessities, and are often referred to as obstacles to market access. It seems debatable, however, whether specifications of this type can truly be considered as non-tariff import measures, but rather as a lack of adaptation by exporters to the conditions prevailing in a specific market.

It is not possible within the scope of ETTS IV to analyse the effect these influences have had on trade in forest products or what influence possible changes in the future could have on trade flows. Nevertheless, when assessing the prospects for international trade, the possibility of changes in the structure of tariff and non-tariff barriers should be borne in mind.

16.6   TRADE IN MANUFACTURES OF WOOD AND WOOD-BASED MATERIALS[1]

Manufactures made of wood (such as mouldings, building joinery, wooden crates, boxes and similar packaging, wooden utensils, household articles, wooden furniture, etc.) and wood-based materials (such as paper bags, sanitary paper, paperboard boxes, stationery, books, etc.) are not covered by the present study. Yet trade in these manufactured products over the last 20 to 30 years has expanded in a very significant manner, not only between industrialized countries, but also with and between developing countries. While this trade is statistically well documented, in terms of weight or number and value of products treated, few attempts have been made to put it into the overall context of forest products trade or to assess its importance as part of total wood consumption. Therefore, in their present form, data on apparent consumption of forest products may not fully reflect the true extent of consumption by countries or regions.

One of the few European countries for which relevant information is available is the Federal Republic of Germany. This section sets out briefly the approach used in this work, and the findings, in particular as concerns the relative importance (in terms of m^3EQ) of trade in manufactures of wood and wood-based materials, and its development over time.

An external trade balance in value and volume terms was calculated for roundwood, semi-finished and finished wood products, where the volume is expressed in roundwood equivalent. Three aspects are of particular importance for the calculations: definition and coverage of finished products; the factors to be used for the conversion of volume or weight data into roundwood equivalent; and the level of aggregation for the detailed statistical information available, taking into account changes in product nomenclature. (It is not, of course, possible to take into account every single item having a wood component, such as some types of upholstered furniture.)

In order to ensure a degree of comparability with conventional classifications, such as that used in the present study, three sub-aggregate groups are used: wood in the rough (round, split or roughly squared, wood chips and residues); semi-finished products (sawnwood, other products of primary processing, pulp, waste paper, paper and paperboard); and finished products.

Trends between 1960 and 1980 in the Federal Republic of Germany's trade in these main forest groups (in terms of roundwood equivalent) are shown in table 16.12. Several interesting developments can be observed over that period. Exports as a whole rose much more strongly than imports; and roundwood exports, although considerably less important than those of the other two groups, grew considerably faster than those of either semi-finished or finished products. The share of the latter in total exports decreased quite markedly, while that of semi-finished products remained fairly stable at just over 50%. Over the 20-year period, imports of finished products grew most strongly and their share in total imports in 1980 was three times higher than in 1960. Nonetheless, the country remained a net exporter of finished products, but a net importer of wood in the rough and semi-finished products.

---

[1] For the preparation of this section, the secretariat used data prepared and made available by Mr. H. Ollmann (Federal Republic of Germany) to whom it expresses its profound gratitude.

TABLE 16.12

Federal Republic of Germany: trade of wood and products made of wood and wood-based materials, 1960, 1970 and 1980

(Roundwood equivalent)

	Volume (million m³ roundwood equivalent)			Percentage share		
	1960	1970	1980	1960	1970	1980
Exports						
Total	3.1	8.4	21.5	100	100	100
of which:						
Wood in the rough	0.2	0.9	3.6	6.5	10.7	16.7
Semi-finished products	1.7	4.4	11.6	54.8	52.4	54.0
Finished products	1.2	3.1	6.3	38.7	36.9	29.3
Imports						
Total	20.6	35.6	47.9	100	100	100
of which:						
Wood in the rough	4.9	5.0	3.9	23.8	14.0	8.1
Semi-finished products	15.0	28.6	38.7	72.8	80.4	80.8
Finished products	0.7	2.0	5.3	3.4	5.6	11.1
Net trade						
Total	− 17.5	− 27.2	− 26.4			
of which:						
Wood in the rough	− 4.7	− 4.1	− 0.3			
Semi-finished products	− 13.3	− 24.2	− 27.1			
Finished products	+ 0.5	+ 1.1	+ 1.0			

As the unit value of products normally rises with the degree of processing, the relative importance of finished products in total trade becomes even more evident when values are compared, as the following shares of total exports and imports of the Federal Republic in 1980 show (in percentages):

	Share of exports		Share of imports	
	By volume	By value	By volume	By value
Wood in the rough	17	4	8	5
Semi-finished products .	54	45	81	69
Finished products	29	51	11	26

Thus finished products accounted for over half of the exports by value, but for only 29% of exports by volume. Finished products were also the only product group to have a positive trade balance in value terms.

Similar calculations made for the external trade in wood and wood-based products in 1980 of the EEC(9) are shown in table 16.13. The share of finished products in total imports by the nine countries taken together was some 5.4%. On the other hand, the corresponding share in exports was much higher at 39%. These countries taken together were net exporters for only two categories of finished products – printed products and products made of dissolving pulp – and only to a very small degree in terms of total trade volume.

From this analysis, one may draw the following tentative conclusions for the EEC(9):

– Trade in finished products is not insignificant, although at the EEC(9) level the net trade effect is rather small;

– In general, and particularly for the Federal Republic of Germany, the trade balance in volume and value terms for finished products is markedly more positive than for wood in the rough and semi-finished products;

– Since 1960 trade in finished products has increased, along with trade in wood in the rough and semi-finished products.

At present sufficient data are not available for a more extensive analysis of trends in other country groups and of the outlook. While it does not at present appear that trade in finished products will significantly alter the outlook for the sector as a whole, this aspect should not be ignored; developments should be carefully monitored, as this factor has, in theory at least, the potential to distort raw material balances.

TABLE 16.13

EEC(9): external trade[a] of wood and wood-based materials, 1980

(Million m³ roundwood equivalent)

	Exports	Imports	Net trade [b]
TOTAL	19.8	145.6	− 125.8
of which:			
Wood in the rough	3.0	11.6	− 8.6
Semi-finished products	9.2	126.1	− 116.9
of which:			
Sawnwood	1.1	38.4	− 37.3
Wood-based panels	0.5	6.9	− 6.4
Other products of primary			
processing	0.6	4.3	− 3.7
Pulp and waste paper	2.2	40.5	− 38.3
Paper and paperboard	4.7	36.0	− 31.3
Finished products	7.7	7.9	− 0.2
of which:			
Wooden articles (including			
furniture)	2.2	3.9	− 1.7
Paper products	1.7	1.7	−
Printed products...............	1.6	0.9	+ 0.7
Dissolving woodpulp products....	2.1	1.4	+ 0.7

[a] Excluding trade between member States.

[b] − = net imports; + = net exports.

16.7 OUTLOOK FOR TRADE IN FOREST PRODUCTS

The structure and trends of trade in forest products are determined by a wide range of factors − levels of demand, costs of wood and other factors of production, transport costs, exchange rates, marketing strategies, consumer preferences, tariffs, policies for trade and for forest development and so on. Models of trade in forest products (other than those concentrating on a few well-defined and understood flows) should take all these into account, thus becoming effectively global forest sector models. Such a model has recently been constructed in the context of the Forest Sector Project of IIASA (International Institute for Applied Systems Analysis), but is not yet suitable for use as a decision-making tool or in the context of a study such as ETTS IV. It was not therefore possible to project, in the context of ETTS IV, on a scientific basis the levels and directions of future trade.

The approach chosen has been to concentrate on trade between Europe and other regions (notably imports from other regions), rather than on the levels of trade, i.e. to attach less importance to intra-regional trade. It is possible to assess the *potential* of countries outside Europe to supply forest products to Europe, drawing notably on assessments of their forest resource and their forest policies. The outlook for non-European importers is also relevant as developments there could affect availability for the European market. Chapter 17 will discuss the outlook for suppliers to Europe and non-European importers. This will provide valuable input for chapter 20 which will compare the forecasts of consumption with those for supply (including trade).

CHAPTER 17

Europe's trade with other regions: past trends and future potential

17.1 INTRODUCTION

European imports of forest products from other regions have a relatively small, but nonetheless essential, place in total supplies to the region's market. In 1980, these imports amounted to some 84 million m³EQ or 17% of consumption. Their importance lies partly in the fact that they often meet requirements for qualities and specifications which cannot be satisfied from domestic resources. To some extent also, they provide the competition to domestically-produced products which ensures that prices are kept attractive to the customer.

As seen in chapter 16, Europe is a net importer of forest products, but it is only one of several importing regions in the world, including parts of East Asia, of which Japan is the largest importer, the USA, Australia, North Africa and West Asia.

This chapter has as its principal objectives:

1. The analysis of the role which other regions have been playing as suppliers of forest products to Europe and an assessment, which cannot here be quantified precisely, of the role which they may play in future;

2. The analysis of import demand in the other main deficit regions, with a view to assessing, also in necessarily imprecise terms, how demand in these regions could affect availability for the European market.

The relative importance, in terms of m³EQ, of the four main external sources as suppliers to the European market in 1980 and changes since 1970 are shown in table 17.1.

The USSR was still the largest single external supplier of forest products to Europe in 1980 (30% of total shipments), although Canada and the USA in aggregate are larger (over 45%). The natural hardwood forests of the tropics in Africa, Asia and Latin America accounted for nearly 15% of Europe's imports from other regions by 1980 and imports from all other sources, whether tropical, sub-tropical or temperate-zone, for the remaining 11%. Between 1970 and 1980, the share of the USSR in European imports dropped from 34.2% to 30.0% and that of "other" suppliers (e.g. Brazil) rose from 3.2% to 10.6%.

There is an enormous difference between exporting countries as regards the *relative* importance of exports in the forest economy. Some areas, notably Canada and the Nordic countries, are strongly export-oriented, while for others, exports represent a relatively minor part of the output of the forest sector. Exports of forest products in total to other regions accounted for the following percentages of removals of industrial wood in 1980:

Nordic countries 86
Canada 77
Asia/Pacific exporters (Indonesia, Malaysia,
 Philippines, New Zealand, Papua New Guinea) 55
USA 19
USSR 12

TABLE 17.1

European imports of forest products from other regions, 1970 and 1980

	Volume (million m³EQ)		Change 1970 to 1980 (million m³EQ)	Percentage share of total	
	1970	1980		1970	1980
USSR......................	22.4	24.3	+ 1.9 (+ 8%)	34.2	30.0
North America....................	29.8	38.3	+ 8.5 (+ 29%)	45.5	45.6
Tropical hardwood sources	11.1	12.4	+ 1.3 (+ 11%)	16.9	14.8
Other[a]	2.1	8.9	+ 6.8 (+324%)	3.2	10.6
Total	65.5	83.9	+18.4 (+ 28%)	100.0	100.0

Source: TIMTRADE data base.

[a] Includes also flows whose origin is not specified.

Leaving aside the large part of total trade which is *intra*-regional (within Europe, within North America, etc.) Europe and Japan between them account for the bulk of *inter*-regional trade, as the following import figures show (1980 figures in million m³EQ).

Europe.. 84
Japan... 71
Other Asia/Pacific 21
North America................................. 5

The importance of Japan in inter-regional trade as an importer of forest products is apparent from the above figures, which show that its imports are nearly as great as those to all European countries together.

This chapter will deal in succession with the main trends and broad outlook for Europe's traditional overseas suppliers, the USSR, North America and the tropical regions, for Japan, which is the major importer outside Europe and thereby influences availablility for Europe, and for newer suppliers such as New Zealand and Chile, as well as for the interaction between the different trends.

17.2   THE USSR

In volume terms, the USSR remains the largest single outside supplier of forest products to Europe, while Europe is its main customer. Even so, because of the enormous area of the country and the size of its forest resource, which is far larger than that of any other country, forest products exports are relatively insignificant in comparison with the huge volumes of growing stock and roundwood removals, as well as of domestic production and consumption of forest products.

17.2.1   The USSR forest resource

According to the forest inventory data, the "forest fund" of the USSR covers an area of 1257 million ha. This is the area which comes under the responsibility of the forest authorities, but 328 million ha of this is non-forest land (farmland, sands, marshes, etc.), leaving about 930 million ha classifiable as forest and other wooded land. The details are as follows:

TABLE 17.2

USSR: land classification in 1978

	Area (million ha.)	Percentage of total
Total land area (excl. water)........	2227.5	100.0
Non-forest land	1297.9	58.3
Forest and other wooded land	929.6	41.7
of which:		
Closed forest	791.6	35.5
Exploitable	534.5	24.0
Unexploitable	257.1	11.5
Other wooded land	138.0	6.2

The concept of "exploitable" and "unexploitable" closed forest, as defined and applied by most European countries in the latest FAO/ECE forest resource assessment,[1] is difficult to apply to a country such as the USSR, where an appreciable part of the forest resource may be

theoretically exploitable (i.e. suitable for wood production), but economically and/or technically inaccessible for the foreseeable future. Much depends on development not primarily connected with the forest sector, such as the recent opening of the BAM (Baikal-Amur-Magistral) railway in Siberia to improve communications between eastern and western USSR and to allow the exploitation of vast mineral deposits in eastern Siberia. The bringing into production of forests is likely to follow on from the creation of the necessary infrastructure to open up such areas to other forms of exploitation. These are hard and expensive tasks given the climate, topography and previous absence of population and infrastructure.

A large proportion of the forests in the USSR are mature or over-mature and untouched, and, in consequence, the concepts of gross and net annual increment, as applied to managed, regularly exploited forests, are inappropriate. Natural losses in these conditions roughly correspond with increment, leaving the volume of growing stock more or less constant. Therefore, not too much weight should be given to the figure of annual increment (described as "mean annual increment" in the USSR reply to the FAO/ECE forest resource enquiry) of 750 million m³ overbark on exploitable closed forest or to the fact that it is nearly double the current level of annual removals. The bringing under regular management of the exploitable forest in the USSR could theoretically result in a level of net annual increment two to three times the figure shown in table 17.3 for mean annual increment, but to achieve this would also require a very large increase in removals, which would be difficult to realise in the foreseeable future.

Roughly three-quarters of the population of the USSR lives to the west of the Ural mountains but this area contains little more than one-fifth of the forest resources. Furthermore, there has reportedly been over-exploitation of the resource in the western parts of the country, which explains the policy of gradually shifting the centre of gravity of forestry and wood-processing eastwards into Siberia. Given the huge transport distances involved, this trend has implications for the long term supply possibilities from the USSR to the European market. Even if the volumes needed could be supplied from newly-opened-up forest

---

[1] The Forest Resources of the ECE Region (Europe, the USSR, North America), ECE/FAO, Geneva, 1985.

TABLE 17.3

USSR: composition of closed forest in 1978

				Closed forest		
				of which: *exploitable*		
	Total	*Conif-erous*	*Non-coniferous*	*Total*	*Conif-erous*	*Non-coniferous*
Area (million ha)	791.6	593.7	197.9	534.5	405.9	128.6
Percentage of total	100.0	75.0	25.0	67.5	51.3	16.2
Growing stock (million m³ o.b.)	84166	67336*	16830*	66996	54669	12327
Per hectare (m³ o.b.)	106	113	85	125	135	96
Percentage of total	100.0	80.0	20.0	79.6	65.0	14.6
Mean annual increment (million m³ o.b.)	970.0*	776.0*	194.0*	750.3	601.5	148.8
Per hectare (m³ o.b.)	1.2	1.3	1.0	1.4	1.5	1.2
Percentage of growing stock	1.2	1.2	1.2	1.1	1.1	1.2
Percentage of total	100.0	80.0	20.0	77.3	62.0	15.3

TABLE 17.4

**USSR: approximate species composition and geographical
distribution of the forest[a] in 1973**

(Percentage of total for USSR)

		By area		
			of which:	
	Total by volume	*Total*	*Siberia and Far East*	*European USSR*
Total	100	100	78	22
Coniferous	82	75	61	14
Larches	35	39	39	–
Scots pine (*P. sylvestris*)..........	19	17	9	8
Other pines ("cedars"[b])	9	6	5*	–*
Spruces	15	12	5	7
Firs............................	3	2	2	–
Others	1	1	1	–
Non-coniferous...................	18	25	17	8
Birches.........................	10	10	8*	2*
Others	8	15	9	6

Source: ETTS III.

[a] State forest fund only.

[b] Mainly *Pinus sibirica*, *P. koraiensis* and *P. cembra*.

areas, there would be cost implications of sending forest products to distant markets in the west. Other markets, including those on the western rim of the Pacific Ocean, such as Japan and China, could become more attractive than Europe.

Exports to Europe have been based mainly on Norway spruce and Scots pine, together with some birch. Recent information on the distribution of these and other important species is not available, but it is unlikely that the situation in the early 1960s as estimated for ETTS III will have changed significantly.

The relative importance of larch growing in Siberia and the Far East is striking – nearly two-fifths of total growing stock. Pine and spruce in the western part of the USSR accounted for 15% of the total, which may seem quite a modest share, but in volume terms it amounts to roughly 10 billion m³ or about the same as the total coniferous growing stock in Europe.

17.2.2 USSR removals

Roundwood removals in the USSR in 1980 amounted to 357 million m³. This was lower than in 1970 and even than in 1960 and reflects a downward trend which started after removals reached a peak of 395 million m³ in 1975. In recent years they have remained stable between 350 and 360 million m³. Between 1970 and 1980, the only assortment where removals rose was pulpwood (+ 4.8 million m³). Removals of fuelwood which accounted for some 30% of total removals in 1960 have decreased steadily since then, and were 29 million m³, or 27%, lower in 1980 than in 1960. In contrast, removals of industrial roundwood rose quite markedly until 1970 to reach nearly 300 million m³, 37 million m³ (14%) above the level reached in 1960. At the same time, the product pattern showed some changes, notably in respect to the share of pit-props and pulpwood. Problems with the statistical breakdown of the data, such as the inclusion of significant volumes of sawlogs and veneer logs in "other industrial roundwood", do not permit a more precise assessment of the shares of the different products. In 1980, however, production of industrial roundwood was again below the level of 1970, though still some way above that of 1960. The main categories of recorded removals are presented in table 17.5.

The figures in table 17.5 are in overbark measure, which overstates them by perhaps 15% in comparison with the underbark data used elsewhere in the study. On the other hand, they do not include certain categories of removals which are not recorded by the forestry authorities. According to the USSR response to the FAO/ECE forest resource enquiry, total removals in 1981 were 387.6 million m³ overbark; the officially reported figures for that year were 358.2 million m³ overbark.

Whatever the true level of removals, to which might be added harvesting losses, which are reported to be considerable, harvesting intensity in the USSR as a whole is low, whether measured by the ratio of fellings to increment or to growing stock. The major constraints on the fuller utilization of the forest resource have been mentioned earlier.

TABLE 17.5

USSR: recorded removals in 1960, 1970 and 1980

(*Million m³ overbark*)

	1960	1970	1980
Total	369.5	385.1	356.6
of which *by species:*			
Coniferous	297.1	320.6	297.7
Non-coniferous....................	72.4	64.5	58.9
of which *by assortment:*			
Total industrial wood	261.5	298.6	277.7
Sawlogs and veneer logs...........	174.5	167.0	152.0
Pulpwood	38.4	33.00	37.8
Other industrial wood[a]	48.6	98.6	87.9
Fuelwood.......................	108.0	86.5	78.9

[a] Includes a significant volume of sawlogs and veneer logs (see ETTS III, section 8.1).

17.2.3 Production and trade of forest products in the USSR

Production of sawnwood, by far the single most important product of the USSR timber industry, after rising steadily between 1960 and 1970, declined in the 1970s, and fell to below 100 million m³ in 1980, well below the level of 1960. This pattern was essentially the result of developments in the production of sawn softwood, which accounts for well over 80% of total sawnwood production. On the other hand, sawn hardwood production remained largely constant until the second half of the 1970s, when it too began to decline.

The other three main product groups, wood-based panels, woodpulp, and paper and paperboard, while as yet of minor importance compared to sawnwood, showed a marked upward trend in production, particularly for wood-based panels, with particle board recording the strongest expansion. Production of both the woodpulp

and paper and paperboard sectors, after doubling in volume between 1960 and 1970, grew more slowly in the 1970s.

The USSR is a major exporter of forest products. It must be borne in mind, however, that domestic consumption of forest products is high and, with the exception of sawnwood, expanding steadily. For all five main product groups, net exports in 1980 represented well below 10% of production (roughly 7% in the case of sawnwood, panels and woodpulp; less than 5% for industrial roundwood, and just over 1% for paper and paperboard).

Between 1970 and 1980 the level of the USSR's net exports of industrial roundwood did not change significantly, while those of sawnwood and paper and paperboard declined. On the other hand, net exports of wood-based panels (except veneer sheets) and woodpulp increased markedly.

In 1980, about three-quarters of the total export volume of coniferous logs went to Japan, with most of the remainder going to Europe, notably Finland. About 80% of the pulpwood was directed to Europe, half of it to Finland, whereas Japan accounted for less than 20% of the total. The bulk of exports of coniferous sawnwood was exported to Europe and to importing countries in North Africa and West Asia as well as Cuba. Panels and pulp products as well as newsprint were exported mainly to Europe, especially to Eastern Europe.

The USSR is a modest net importer of broadleaved sawnwood (0.2 million m³) and veneer sheets (0.1 million m³). It also imports a sizeable volume of paper and paperboard (0.9 million m.t in 1980) but is still a small net exporter of this commodity. Imports are not shown in table 17.6.

17.2.4 Prospects for exports of forest products from the USSR

The volume of timber permitted by the State to be cut in the accessible forests of the USSR in 1980 was some 638 million m³, but less than two thirds of this volume was actually removed. While this indicates that there is a considerable potential for the country to raise the volume of fellings, and thus the supply of roundwood for both the domestic and export markets, past developments in the forest and forest industries sector have highlighted a number of constraints to the expansion of roundwood supply. Since there does not seem to be much possibility of easing these constraints, it seems unlikely that present levels of volumes harvested will be raised to any significant extent for some time to come.

Since the 1960s, the rising raw material requirements of the rapidly expanding forest industries, notably the pulp and paper industries, as well as of the forest products export market, have had increasingly to be met from hitherto largely untouched forest resources located in the eastern parts of the country. Production has been basically of the extensive type, the utilization of the resource often remaining below its optimum level. Harvesting losses were

TABLE 17.6

USSR: production and trade of forest products 1960, 1970, 1980

(Million m³, except woodpulp and paper: million m.t.)

	Production			Exports		
	1960	1970	1980	1960	1970	1980
Roundwood, total	369.5	385.1	356.6	4.5	15.0	13.5
Fuelwood............................	108.0	86.5	78.9	..	..	..
Industrial wood	261.5	298.6	277.7	4.5	15.0	13.5
Total pulpwood	38.4ᵃ	33.0	37.8	1.6	6.0	5.7
Coniferous pulpwood	37.8	33.0	37.8	..	6.0	5.7
Broadleaved pulpwoodᵇ	0.6	..	..	..	..	..
Chips and particles	..	..	..	..	–	0.2
Wood residues	..	..	..	..	0.6	1.5
Coniferous logs	148.1	143.7	130.7	1.5	7.4	6.5
Broadleaved logs	26.1	23.3	21.3	–	0.1	
Pit-props	..	13.6	10.7	1.1	0.8	0.6
Other industrial wood..............	48.9	85.0	77.2	–	0.7	0.6
Sawnwood...........................	105.6	120.5	98.2	5.1	8.2	7.2
Coniferous sawnwood...............	89.8	101.9	86.0	5.0	7.9	6.9
Broadleaved sawnwood..............	15.8	14.6	12.2	–	–	0.2
Sleepers	..	4.1	..	0.1	0.3	0.1
Wood-based panels	1.7	5.4	10.1	0.1	0.5	0.9
Veneer sheets	..	..	..	..	..	..
Plywood	1.4	2.0	2.0	0.1	0.3	0.3
Particle board.....................	0.1	2.0	5.1	..	0.1	0.3
Fibreboard	0.2	1.4	3.0	..	0.1	0.3
Woodpulp...........................	3.2	6.7	8.8	0.2	0.4	0.8
Mechanical woodpulp...............	0.9	1.6	1.7	–	–	–
Semi-chemical woodpulp	..	0.2	0.3	..	–	–
Chemical woodpulp (total)...........	2.3ᶜ	4.9	6.8	0.2ᶜ	0.4	0.8
Dissolving grades	..	0.5	0.8	–	–	–
Chemical woodpulp (paper)........	..	4.4	6.0	..	0.4	0.8
Paper and paperboard	3.2	6.7	8.7	0.1	0.7	1.0
Newsprint	0.4	1.1	1.4	0.1	0.3	0.3
Printing and writing paper..........	0.7	0.9	1.1	–	–	–
Other paper and paperboard.........	2.1	4.7	6.2	–	0.4	0.7

ᵃ Including pit-props.

ᵇ It is known that significant volumes of broadleaved pulpwood are produced and exported, but precise data are not available.

ᶜ Including semi-chemical.

reportedly high, and the range of species, dimensions and qualities used, relatively limited. Thus, large areas needed to be worked each year in order to make the raw material available.

As forest activities continued to move east and north, as a result of the heavy cutting in the European part of the USSR, as well as parts of southern Siberia, the problems and difficulties encountered became more acute. Besides the changing resource pattern referred to earlier (lower representation of pine and spruce, the most sought after woods), the prevailing climatic and terrain conditions, limited labour resources, and problems in harvesting and processing technologies appeared increasingly as constraints on the further expansion of raw material supply. Furthermore the opening up of such vast areas will require correspondingly large investment: the forest sector will have to compete with other sectors for limited resources for investment.

Rather than extending the raw material base by increasing the area harvested, policies now emphasize the more intensive utilization of available resources and the reduction of waste at both the harvesting and processing levels.

This includes, in particular, the use of a broader range of species, in particular of hardwood, as well as of qualities and assortments, such as small-dimension wood. Furthermore, at the processing industry level, in order to increase yields, the utilization of raw material has to be improved and uses found for forest and industry residues; one means of achieving this would be by the further integration of industries. Further improvements are needed in the mechanization of forest work and in transport. This will require considerable adaptations and developments in technologies, and in labour and management skills. One important aspect is the intensification of reforestation activities in harvested areas.

The main conclusion to be drawn from the foregoing is that considerable scope exists to meet increased demand for wood products without having to expand the present area harvested to any major extent.

It is likely that domestic consumption of forest products, notably wood-based panels and paper and paperboard, will increase in the future, but it is not certain to what extent this would affect availability of forest products for export.

As far as exports of forest products are concerned, it would appear that there is little prospect of any significant increase in the foreseeable future, but that basically they will be maintained at present levels and follow the same patterns unless unforeseen events necessitate a significant change in policy priorities.

The main part of the USSR export trade in forest products – notably with its eastern European trading partners, but also with Finland, Japan and some other countries – is carried out within the framework of long-term trade agreements which frequently make provision for trading partners to be involved in some way in the development of the raw material resources and the industries on which they are based. Broader trade policy considerations may therefore have a direct effect on the level of forest product exports from the USSR.

17.3 NORTH AMERICA (CANADA AND THE UNITED STATES)

For the supply of forest products to the European market, North America plays an even more important role than the USSR, both in terms of volume and of value. While exports of the USSR to Europe were very roughly at the rate of 24 million m³EQ in 1980, those of Canada and the United States taken together amounted to over 38 million m³EQ, of which about two thirds were constituted of pulp and paper. Of the total volume of North American exports of forest products to Europe in 1980, roughly three fifths (in terms of m³EQ) was accounted for by Canada and two fifths by the United States. These two countries are the world's leading exporters of forest products; for both of them, Europe is only one of many markets, and for most products not the main market.

Canada and the United States taken together have a land area of 1 829 million ha, almost equally divided between the two countries. Forest and other wooded land account for some 735 million ha or slightly over 40% of the total area.

There are large differences between the circumstances of the two countries. Large areas of the north of Canada are not productive because of the harsh climate, and in general growing conditions are worse than in the USA. Canada's population and economy are both much smaller than those of the United States. The forest sector is very important in Canada, and is strongly export-oriented. The United States has a large population, the world's largest economy and the largest market for forest products. Despite being (with the USSR) the world's largest roundwood producer, the USA is also the largest importer of forest products, mostly from Canada. US exports are small relative to domestic production or consumption, but nevertheless account for significant volumes.

Both countries are nearing, or have reached the end of the phase of exploitation of old growth stands, and are going through a period of adjustment of forestry concepts and methods to the changing situation. Both countries are moving from an extensive to an intensive type of silviculture, with major regional differences.

TABLE 17.7

North America: land and forest area around 1980

	North America					
	Total		Canada		United States	
	(Million ha)	Percentage	(Million ha)	Percentage	(Million ha)	Percentage
Total land area (excl. water)	1829.2	100.0	916.7	100.0	912.5	100.0
Forest and other wooded land	734.5	40.2	436.4	47.6	298.1	32.7
of which:						
Coniferous	..	..	..[a]	..	157.9	17.3
Broadleaved	..	..	..[a]	..	140.2	15.4
Closed forest.....................	459.4	25.1	264.1	28.8	195.3[b]	21.4[b]
Exploitable closed forest	410.1	22.4	214.8	23.4	195.3[b]	21.4[b]
Unexploitable closed forest	..		49.3	5.4	..[c]	..[c]
Other wooded land	275.1	15.1	172.3	18.8	102.8[d]	11.3[d]
Non-forest land	1094.7	59.8	480.3	52.4	614.4	67.3

[a] Coniferous, 162.3 million ha; non-coniferous, 25.1 million ha; mixed, 53.5 million ha; undetermined, 101.5 million ha.

[b] Commercial timberland.

[c] Included under other wooded land.

[d] Non-commercial timberland, including areas of unexploitable forest.

TABLE 17.8

**North America: exploitable closed forest
by main ownership, around 1980**

(Million ha)

	North America		
	Total	Canada	United States
Exploitable closed forest,			
Total......................	415.2	219.9[a]	195.3[b]
Public, total	256.8	201.9	54.9
of which:			
State	242.0	201.8	40.2
Other ...,...............	14.7	0.1	14.7
Private, total................	158.3	18.0	140.3
of which:			
Forest industries	..	..	27.8
Farm.....................	..	..	46.9
Other		..	65.6

[a] Slightly different coverage to that in table 17.7.

[b] Commercial timberland.

TABLE 17.9

**North America: growing stock and annual increment
on exploitable closed forest**

(Million units)

	North America		
	Total	Canada	United States
A. GROWING STOCK (m³ o.b.)			
Total	46354	22958	23396
of which:			
Coniferous	33494	18310	15184
Non-coniferous...........	12860	4648	8212
B. ANNUAL INCREMENT (m³)			
Total			
Gross overbark............	..	..	840
Net overbark	1067	356	711
Net underbark	918	305	613
of which:			
Coniferous			
Gross overbark..........	..	..	486
Net overbark	676	267	409
Net underbark	575	227	348
Non-coniferous			
Gross overbark..........	..	..	54
Net overbark	391	89	302
Net underbark	335	78	266

TABLE 17.10

**North America: net annual increment, overbark,
on exploitable closed forest**

	Canada		United States	
	m³/ha	(%) [a]	m³/ha	(%) [a]
Total	1.7	1.6	3.6	3.0
of which:				
Coniferous	2.2[b]	1.5	4.9	2.7
Non-coniferous	1.0	1.9	2.7	3.7

[a] Percentage of growing stock.

[b] Natural stands, unmanaged.

Statistical data for the two countries are presented together in tables 17.7 to 17.10, but they are analysed separately below.

17.3.1 Canada

17.3.1.1 THE CANADIAN FOREST RESOURCE

Forest and other wooded land accounts for 436 million ha, 48% of the total land area of Canada. Most of this is coniferous: of the area whose species has been determined, 90% is coniferous or mixed. There also exist over 100 million ha whose species composition has not been determined. This is presumably largely remote and relatively unproductive land in the north of Canada. Closed forest accounts for 60% of total forest and other wooded land in Canada.

Nearly 215 million ha, or 81% of the area of closed forest in Canada is classified as exploitable, and of this area, some 89% is stocked. Just over 90% of Canada's closed forest is in public ownership, primarily that of the Provinces.

The volume of growing stock on exploitable closed forest in Canada is just under 23 thousand million m³, of which 80% is coniferous. Average growing stock per hectare of exploitable closed forest in Canada is 107 m³ o.b.

Net annual increment, over bark, on exploitable closed forest in Canada is 356 million m³, equivalent to 1.7 m³/ha or 1.6% of growing stock. This relatively low increment rate, which refers to mostly unmanaged stands, is chiefly due to the harsh growing conditions in much of Canada, as well as to the important natural losses occurring in these stands.

There are wide differences in the type of forest between the different parts of Canada. In general it can be said that in the east the trees are smaller, growing stock per hectare lower and harvesting conditions easier than in British Columbia, which has the largest forest resource, characterized by many old-growth stands, relatively fast growth rates and difficult logging conditions on the coast, with drier conditions in the interior. The shares of the different groups of provinces in the Canadian total are shown below (in %):

	Forest land	Productive forest land	Standing volume of timber
Maritimes	5.8	6.8	5.3
Quebec.................	21.5	32.1	21.4
Ontario	18.5	16.1	15.6
Prairies................	20.1	17.5	12.4
British Columbia	14.5	19.5	42.2
Yukon and North-West Territories	19.6	8.0	3.0
Canada..............	100.0	100.0	100.0

Source: G.M. Bonnor *Canada's Forest Inventory 1981*, Environment Canada, 1982.

The above data also point to the large differences in the standing volume per hectare existing in the country. The high standing volume per hectare of British Columbia is particularly striking: the province has less than 15% of Canada's forest land but over 40% of its standing volume. It contrasts sharply with the conditions prevailing for the north, in Yukon and North-West Territories, where nearly 20% of the forest land carries only 3% of the standing volume.

The Canadian forest is mostly coniferous, hardwood forest being confined to areas relatively near the US border with Quebec and Ontario and north of the wheat belt in the Prairies.

17.3.1.2 PRODUCTION AND TRADE OF FOREST PRODUCTS IN CANADA

Canada's large forest products industries are essentially export-oriented. The relative importance, in 1980, of Canadian exports compared to production was as follows:

	Volume of exports	Percentage of Canadian production	Percentage of world exports
Sawnwood	29.3 million m³	66	37
Sawn softwood	29.0 million m³	68	44
Wood-based panels	1.3 million m³	28	8
Plywood	0.5 million m³	23	8
Pulp	7.2 million m.t.	36	34
Chemical pulp (for paper)	6.7 million m.t.	57	37
Paper and paperboard	9.3 million m.t.	69	26
Newsprint	7.7 million m.t.	89	63

Canada has become a net importer of wood raw material, but this trade essentially concerns cross-boundary trade (in both directions) with the USA.

Table 17.11 shows production and trade of forest products in Canada in 1960, 1970 and 1980. When analysing these data, it should be borne in mind that there are considerable year-to-year fluctuations which may conceal underlying trends. Canada, as an exporting country, is particularly vulnerable to swings in the economies of its customer countries, and to changes in its own international competitivity (e.g. due to changes in currency parities).

Production of industrial roundwood showed a markedly rising trend between 1960 and 1980 with a volume in 1980 that was over 75% higher than in 1960. Most of the rise in the production of industrial roundwood in Canada concerned coniferous logs, production of which was one and a half times higher in 1980 than in 1960. On the other hand, production of coniferous pulpwood during the same period fluctuated between 33 and 38 million m³ (except in 1974, with nearly 43 million m³), and did not show a rising trend. Production of both broadleaved logs and pulpwood expanded, by 63% and 70% respectively, between 1960 and 1980, but their share in total production remained modest, (5% and 10% respectively in 1980). Furthermore, chips from sawmills account for a large part of Canada's pulpwood supply.

TABLE 17.11

Canada: production and trade of forest products, 1960, 1970, 1980

(Million m³, except woodpulp and paper: million m.t.)

	Production			Exports			Imports		
	1960	1970	1980	1960	1970	1980	1960	1970	1980
Roundwood, total	96.4	121.4	161.8	5.2	4.2	2.3	1.4	2.1	3.0
Fuelwood	6.9	4.1	4.5	–	–	–	–	–	–
Industrial wood	89.5	117.3	157.3	5.1	4.2	2.3	1.4	2.1	3.0
Round pulpwood	39.5ᵃ	40.4	41.0	2.8	2.6	1.1	0.6	0.3	0.6
Coniferous pulpwood	37.0	36.7	36.8	..	2.5	1.0	..	0.3	0.6
Broadleaved pulpwood	2.5	3.8	4.2	..	0.1	0.1	..	–	–
Chips and particles	..	..	..	2.0	1.0	1.5	..	–	0.1
Wood residues	..	..	..	..	0.5	0.4	..	–	–
Coniferous logs	45.6	71.3	109.1	–	1.2	1.0	0.8ᵇ	1.3	1.6
Broadleaved logs	3.3	4.4	5.3	0.1	0.1	0.1	..	0.3	0.4
Pit-props	..	0.1	–	–	–	–	–	–	–
Other industrial wood	1.1	1.2	1.8	0.2	0.3	0.1	–	0.2	0.4
Sawnwood	19.0	26.9	44.3	11.3	17.7	29.3	0.6	0.6	1.6
Coniferous sawnwood	17.9	25.4	42.9	10.9	17.3	29.0	0.3	0.4	0.8
Broadleaved sawnwood	1.0	1.3	1.4	0.3	0.4	0.3	0.3	0.2	0.7
Sleepers	0.3	0.2	–	0.1	–	0.1	–	–	0.2
Wood-based panels	1.6	3.3	4.8	0.2	0.6	1.3	0.1	0.2	0.2
Veneer sheets	..	0.2	0.5	0.1	0.1	0.2	–	–	–
Plywood	1.0	1.8	2.3	0.1	0.4	0.5	–	0.1	0.1
Particle board	–	0.3	1.3	..	–	0.5	..	–	–
Fibreboard	0.5	0.9	0.7	0.1	0.1	0.1	–	–	0.1
Woodpulp	10.4	16.6	19.9	2.4	5.1	7.2	0.1	–	0.1
Mechanical woodpulp	5.5	7.2	7.5	0.2	0.3	0.3	–	–	–
Semi-chemical woodpulp	0.1	0.3	0.3	–	–	–	..	–	–
Chemical woodpulp (total)	4.8	9.1	12.1	2.1	4.8	6.9	0.1	–	0.1
Dissolving grades	0.4	0.4	0.3	0.2	0.3	0.2	–	–	0.1
Chemical woodpulp (paper)	4.4	8.7	11.8	1.9	4.5	6.7	0.1	–	0.1
Paper and paperboard	7.9	11.3	13.4	5.8	8.1	9.3	0.1	0.2	0.3
Newsprint	6.1	8.0	8.6	5.6	7.3	7.7	–	–	–
Printing and writing paper	0.4	0.8	1.5	0.1	0.3	0.7	–	–	0.1
Other paper and paperboard	1.5	2.4	3.3	0.1	0.4	0.9	0.1	0.1	0.1

ᵃ Including pit-props.
ᵇ Including broadleaved logs.

There were also significant increases in production of the major products between 1960 and 1980:

	Percentage
Sawn softwood	+ 139
Plywood	+ 139
Particle board	+ 2900 (from a very low base)
Chemical pulp	+ 167
Newsprint	+ 42
Printing and writing paper	+ 313
Other paper and paperboard	+ 250

Canadian exports of products have also risen significantly, while imports are generally of minor importance, being mostly small transboundary flows or imports of special assortments, e.g. from the tropics. An exception is the trade in industrial wood: in 1960 Canada was a net exporter (3.7 million m^3), but exports fell and imports rose slightly, so that by 1980 Canada was a net importer of industrial wood (0.7 million m^3).

The main trade flows for Canadian exports can be identified as follows:

- Exports of coniferous sawnwood, the bulk of which (over 75% in 1980, and nearly as much in 1970) goes to the United States, over 10% to Europe and under 10% to Japan. The remainder goes mainly to Australia and countries in North Africa and West Asia;

- Exports of chemical woodpulp, of which nearly half goes to the United States, one-third to Europe, and 12% to Japan;

- Exports of newsprint, of which the bulk (81% in 1980) goes to the United States and less than 10% to Europe.

17.3.1.3   OUTLOOK FOR CANADA'S PRODUCTION AND TRADE OF FOREST PRODUCTS

Total production of coniferous roundwood (including fuelwood) reached an historical record of close to 150 million m^3 in 1979. Although this level was not maintained in the early 1980s, complete data for 1984 and 1985 will show new records. A little over half of this volume came from the forests of British Columbia. The total allowable annual cut in 1979 for this group of products was 205.1 million m^3, so that there was a theoretical under-utilization of the potential of over 55 million m^3. However, over 39.4 million m^3 of the potential was economically inaccessible and another 9.5 million m^3 not available for other reasons. Therefore, production remained only 6.4 million m^3 below the annual allowable cut and most of this unused volume was in areas outside the present main centres of production. In contrast to coniferous roundwood, harvests in 1976 of broadleaved roundwood amounted to less than one-quarter of the economically accessible annual cut (57.6 million m^3). These data show the existence of both unutilized and under-utilized resources. On the other hand, the coniferous resources in some areas of the country — in particular, old-growth timber — are already now used intensively and probably could not sustain present harvest levels indefinitely under the conditions of recent forest management programmes.

Authorities and forest-based industries have become increasingly aware of the developing imbalances in Canada and the uncertainty of their extent. There has been wide support for policies and programmes developed by the Provinces and designed to place Canadian forestry on a sustainable basis, with due attention given to protection of the forest (e.g. from insects and fire) and regeneration.

It was estimated in 1979 that the implementation of these policies and programmes could increase roundwood harvests by the year 2000 by over one-quarter and, in certain cases, to an even greater extent. It is also considered possible that the hardwood resource could become important for pulp and paper production (e.g. chemi-thermo-mechanical pulp). A policy target of 210 million m3 was set in a federal-provincial paper in 1980, based on the assumption that spending on forest renewal would double or triple. This has not yet happened. Moreover, insects and fire have worsened the situation. More recent analyses indicate that sawlog supply problems could occur in the next few years.

Levels of harvest could be further increased by the use of resources at present considered as economically inaccessible. As these include stands with a significant share of small diameter logs (although some of these logs may not be small by European standards), the development of systems permitting the processing of such logs would be an important element in broadening the industry's raw material base. Further technological developments may include increasing the use of lower quality raw material and improving the yield from the wood material available.

If the raw material base of Canada's forest industries can thus be expanded, there is undoubtedly a potential for the country's exports to grow further. Some concern has been expressed, however, over trends in production costs in Canada which are expected to rise more rapidly than those in other major countries which produce similar products. Among the reasons for this are said to be unfavourable trends in wood stumpage prices, harvesting costs (not only due to increasing transport distances, but also to lower stand densities in the interior and more northerly forest regions), as well as less attractive species mix and declining log grades. Without the expansion of the raw material base in the ways outlined above, it would hardly be possible for Canada's exports of forest products to rise much beyond present levels, at least in the longer-term and on a sustained basis.

A problem for Canadian (and United States) exports to Europe, in which pulp and paper products account for a major share, could also be more restrictive import regimes by European countries in respect of these particular products, of which there has been some evidence in recent years, notably with the coming into force of the free trade agreement between the EEC and EFTA.

The share of preferred species and of the better quality roundwood assortments, as produced particularly in the old-growth stands, will certainly decline as the reservoir of prime sawlogs is depleted. This will have an effect on the type of product which Canada will have available for export (e.g. on the width or length of sawnwood pieces).

At present Canada's exports are specialized in the so-called "bulk grades" – sawn softwood, chemical pulp and newsprint of standard quality produced in large volumes. It is intended to move towards a greater variety of products, notably by including higher value added products, thereby increasing the income from the same raw material input. It is also intended to reduce Canada's dependency on one market – the United States. In present circumstances, it could prove difficult to achieve these policy objectives.

17.3.2 The United States

17.3.2.1 THE FOREST RESOURCE OF THE UNITED STATES

The United States has just under 300 million ha of forest and other wooded land, a third of its total land area. Of this, two thirds, 195 million ha, is classified as "commercial timberland" which may be considered roughly equivalent to "exploitable closed forest", although the detailed criteria for classification are different. Just over 40% of commercial timber land (83 million ha) is classified as coniferous of which nearly all is stocked and a further 13% is mixed coniferous/non-coniferous. There are 106 million ha of stocked broadleaved commercial timber land.

In contrast to the situation in Canada, less than 30% of commercial timberland is publicly owned (55 million ha). Of the remaining 140 million ha, 28 million ha belong to the forest industries, 47 million to farmers and 66 million to other owners.

Growing stock in the USA on commercial timberland is just over 23 thousand million m^3 o.b., practically the same as in Canada. The net annual increment (overbark) is 711 million m^3. Nearly two thirds of the growing stock (15 thousand million m^3 o.b.) is coniferous, and about 58% of the increment (409 million m^3 o.b.). Net annual increment per hectare, underbark, is 3.1 m^3, just over 3% of growing stock.

The difference between regions as regards forest type and ownership patterns is even greater in the USA than in Canada. The USDA Forest Service in its analytical work divides the United States into four forest regions:

North (New England, Middle Atlantic, Lake States Central);
South (South Atlantic, East, Central and West Gulf);
Rocky Mountain (Northern and Southern);
Pacific Coast (North-west and South-west).

In addition, twenty forest types are identified. The division of US commercial timberland between the regions was in 1977 (in percentages of national total) as in table below.

It is not possible here to analyse these data in detail, but two aspects stand out:

(a) Nearly half the US coniferous growing stock is in the Pacific Coast region. These are old-growth forests, many publicly owned, some with a low road density;

(b) Nearly half the net annual increment and removals is in the South, where there are good growing conditions and low wood costs in the southern pine region.

17.3.2.2 PRODUCTION AND TRADE OF FOREST PRODUCTS IN THE USA

In 1980, the USA produced 327 million m^3 of industrial wood, 75 million m^3 of sawnwood, 26 million m^3 of wood-based panels, 46 million m.t. of woodpulp and 57 million m.t. of paper and paperboard, more than any other country in the world (except for sawnwood, where the USSR produces 30% more). The percentage changes in production since 1960 have been as follows:

	1960 to 1970	1970 to 1980
Sawnwood	+ 5	− 9
Wood-based panels	+ 76	+ 14
Woodpulp	+ 72	+ 17
Paper and paperboard	+ 55	+ 24

Production rose faster in the 1960s than in the 1970s, which were heavily affected by cyclical fluctuations. Sawnwood production in 1980 was actually lower than in 1960 and 1970.

Despite these very high production figures, the USA is a major net importer of most products, as the following data for net trade show (+ = net exports, − = net imports in million units):

	1960	1970	1980
Industrial wood (m^3)	+ 7.3	+ 10.1	+ 13.9
Sawnwood (m^3)	− 7.3	− 11.2	− 17.1
Wood-based panels (m^3)	− 0.7	− 2.2	− 1.2
Pulp (m.t.)	− 1.1	− 0.4	− 0.2
Paper and paperboard (m.t.)	− 4.2	− 4.0	− 3.7

The source of most of US imports is Canada with some imports of hardwoods from tropical regions or in-transit processing countries.

		Growing stock:				
	Area	Total	Softwoods	Hardwoods	Net annual increment	Removals
North	34.4	24.4	9.8	50.4	26.7	18.7
South	39.0	28.4	21.3	41.1	49.4	46.2
Rocky Mountain	12.0	14.0	20.8	1.9	7.8	5.9
Pacific Coast	14.6	33.2	48.1	6.6	16.1	29.2
USA	100.0	100.0	100.0	100.0	100.0	100.0

Source: *An Analysis of the Timber Situation in the United States 1952-2030*, USDA Forest Service, 1982.

TABLE 17.12

United States: production and trade of forest products, 1960, 1970, 1980

(*Million m³, except woodpulp and paper: million m.t.*)

	Production			Exports			Imports		
	1960	1970	1980	1960	1970	1980	1960	1970	1980
Roundwood, total	308.9	327.9	352.0	15.2	12.8	15.8	8.0	2.6	1.9
Fuelwood	42.5	15.3	24.9	–	–	–	0.3	–	–
Industrial wood	266.4	312.7	327.1	15.2	12.8	15.8	7.8	2.6	1.9
Round pulpwood.................	72.8ᵃ	109.6	124.3	0.3	0.2	0.4	4.8	1.1	0.5
Coniferous pulpwood..........	53.6	80.0	85.5	..	0.2	0.4	..	1.0	0.4
Broadleaved pulpwood.........	19.2	29.6	38.8	..	–	–	..	0.1	0.1
Chips and particles	..	..	..	..	3.5	6.5	2.0	0.9	1.0
Wood residues	..	..	..	..	–	–	..	0.3	0.1
Coniferous logs	142.8	156.5	154.1	1.0	12.2	14.1	0.1	0.5	0.5
Broadleaved logs	31.5	34.5	37.2	0.3	0.3	0.7	0.4	0.2	0.1
Pit-props	..	1.0	1.0	..	–	–	..	–	–
Other industrial wood	19.3	11.0	10.5	1.0	1.0	0.6	0.6	0.9	0.9
Sawnwood	67.7	81.8	74.7	2.0	3.0	5.8	9.3	14.3	22.9
Coniferous sawnwood	63.0	65.0	57.4	1.6	2.7	4.6	8.6	13.5	22.2
Broadleaved sawnwood	14.8	16.8	17.2	0.3	0.3	0.9	0.7	0.8	0.7
Sleepers	..	–	–	0.1	–	0.3	–	–	0.1
Wood-based panels	13.0	23.0	26.2	0.1	0.3	1.0	0.7	2.5	2.1
Veneer sheets	..	–	–	–	–	0.3	0.2	0.4	0.4
Plywood	7.9	14.1	14.9	–	0.1	0.2	0.4	1.8	1.0
Particle board	0.5	3.1	6.3	..	–	0.2	–	–	0.5
Fibreboard	4.7	5.9	5.1	0.1	0.1	0.2	0.1	0.3	0.2
Woodpulp	23.0	39.5	46.2	1.0	3.0	3.5	2.2	3.2	3.7
Mechanical woodpulp...........	4.1	5.9	4.2	–	–	–	0.2	0.2	0.1
Semi-chemical woodpulp	1.8	3.0	3.7	..	–	0.1	–	–	–
Chemical woodpulp (total)	17.1	30.6	38.4	1.0	2.8	3.4	1.9	3.0	3.5
Dissolving grades..............	1.0	1.5	1.4	0.4	0.8	0.7	0.2	0.3	0.2
Chemical woodpulp (paper)	16.0	29.1	37.0	0.7	2.0	2.7	1.7	2.8	3.3
Paper and paperboard	29.6	45.8	56.8	0.8	2.4	4.1	5.1	6.4	7.9
Newsprint	1.8	3.1	4.2	0.1	0.1	0.2	4.9	6.0	6.6
Printing and writing paper	5.9	9.9	14.1	0.1	0.1	0.2	–	0.3	0.7
Other paper and paperboard......	21.9	32.7	38.5	0.6	2.2	3.8	0.1	0.1	0.6

ᵃ Including pitprops.

The main US export flows can be identified as follows:

– Exports of industrial roundwood, primarily coniferous logs, and chips and particles. Exports of coniferous logs go mostly to Japan (mostly high quality logs from old growth stands), and the Republic of Korea. Well over a million m³ is exported annually to Canada. Of the total exports of chips and particles, the bulk goes to Japan, with smaller, but still appreciable volumes to a few other countries such as Sweden and Canada.

– Exports of other paper and paperboard (especially Kraftliner), of which Europe accounted for 28% in 1980, and Canada for 9%.

It is also of interest to draw attention to some other developments:

– Exports of coniferous sawnwood expanded appreciably between 1960 and 1980, with Europe accounting for 20-25% of the volume;

– United States exports of hardwood, both in the form of logs and sawnwood, have also grown strongly, particularly since 1970. Exports of sawn hardwood in 1980 were over three times those in 1970, and those of hardwood logs more than twice the 1970

volume. In 1970 Canada was still the most important market, both for hardwood logs and for sawn hardwood, but the situation was quite different in 1980. In that year, Europe's share of United States total exports of hardwood logs was 57%, against 29% for Canada, and of sawn hardwood 43% (Canada 44%).

17.3.2.3 OUTLOOK FOR EXPORTS OF FOREST PRODUCTS FROM THE USA

In 1982, a new study on long-term timber trends up to the year 2030[2] was published by the United States Forest Service. The study represents the latest in a series of analytical assessments of the current and prospective timber situation in the United States, conducted at regular intervals to serve as a basis for the formulation and direction of public and private timber policies and programmes. In very broad terms, the main findings of the study were summarized[3] as follows:

[2] *An Analysis of the Timber Situation in the United States 1952-2030*, United States Department of Agriculture, Forest Service; Forest Resource Report No. 23, December 1982, Washington.

[3] Sub-heading of relevant section of "Highlights", pages XX to XXV of the study.

1. Substantial growth is anticipated in population, economic activity and income.

2. Consumption of most timber products has been rising rapidly.

3. Projections show demand for most timber products continuing to rise rapidly.

4. Some increase in net imports of timber products is projected, but the increase is relatively small in comparison to the projected growth in demand.

5. Most of the projected growth in demand for timber will fall on domestic forests.

6. There is a large domestic timber resource — mostly in private ownership.

7. Trends in inventories, net annual growth, and harvests indicate the domestic timber situation has been improving in most regions.

8. The domestic timber resource in most regions can support larger timber harvests.

9. Projected timber demands on domestic forests are rising faster than supplies — rising prices and economic scarcity are in prospect.

10. A growing economic scarcity of timber will have significant and adverse effects on the economy, the environment, and general social well-being.

11. The adverse impacts which will result from a growing economic scarcity of timber are not inevitable — there are large opportunities to increase and extend timber supplies.

12. Moving forward to meet projected demands for timber products would require substantial investments, but these investments promise to be profitable.

With respect to exports, the study did not expect that the volume of exports of forest products (including roundwood) would change significantly from the 1976 level of about 42.5 million m³ roundwood equivalent. It expected, however, some diverging trends among major products. The authors of the study drew attention, however, to methodological problems in assessing the outlook for exports.

Sawn softwood exports were expected to rise till 1990, but subsequently to show a slow decline, due to decreased availability of the high quality clear sawnwood now produced from the old growth timber in the Pacific Northwest. Sawn hardwood exports would show a slow increase, reflecting in large part an improved hardwood timber supply situation in the United States. Exports of woodpulp and paper and paperboard were expected to show continued and fairly rapid growth through the projection period (i.e. until 2030), while pulpwood exports (including chips) would drop rapidly.

Exports of coniferous logs, most of which go to Japan, were projected to remain near present levels until 1990, then to show a continuous decline. Exports of broadleaved logs were expected to remain largely unchanged.

In the subsequent public discussion of the study, some doubts were expressed concerning the projected level of housing starts and the resulting projections of demand, but the assessment of the timber supply situation probably received the most attention. It was contended by some observers that the supply projections, especially for coniferous wood, may be too optimistic, as not having taken sufficiently into account a number of constraining factors. (Table 17.13 shows past and projected supplies of coniferous wood in the United States by main regions.)

TABLE 17.13

United States (contiguous states): past and projected (equilibrium level) production of coniferous roundwood 1952, 1976, 2000, 2020

(Million m³)

United States regions	Actual		Projected	
	1952	1976	2000	2020
Pacific coast	95.7	106.4	105.1	108.5
Rocky mountains.................	14.0	21.9	30.9	36.1
North..........................	16.9	18.0	29.2	34.9
South	86.3	120.1	176.5	204.9
Total	212.8	266.4	341.7	384.4

Source: USDA Forest Service *op. cit.*

The most uncertainty over the study's resource assessment concerned the forests of the south, which over the last two decades have shown spectacular growth in production under conditions allowing highly efficient operations. The study projected an increase of 56.4 million m³ in supplies of coniferous wood from the forests of the South between 1976 and 2000 (production had increased by 33.8 million m³ between 1952 and 1976).

However, some analysts consider the commercially useable forest area is declining more rapidly than was assumed in the study, as forest land, instead of being replanted, is being converted for agricultural purposes. Large areas of forest lands are said to have gone into the ownership of non-forest-based groups, whose future policies regarding land use may not necessarily include a commitment to continued forestry. Furthermore, due to age-class distribution, management regimes (notably the short rotation cycle) and species, the assortment pattern is shifting from sawlog production to pulpwood. Earlier estimates or assumptions are being questioned, notably those regarding increment and yield, especially on second rotation cycles, and in stands planted in areas offering less suitable management and natural conditions.

A reassessment of the forest resource of the southern region is at present being undertaken by the US Forest Service. Reports of preliminary findings of the study indicate that net annual growth in the South appears to be declining. The main factors isolated are a declining timberland base, low levels of regeneration after harvest, an increase in tree mortality and reduced growth per hectare. These factors are compensated to some extent by higher increment on intensively-managed commercial plantations.

In the early 1980s there has been a marked increase in the interest shown by industry and Government in expanding export markets for forest products, and several concrete steps were taken to encourage this trend. Among other things, the Forest Service is developing improved methods of modelling the outlook for international trade in forest products. No results of this work are yet available.

Another factor of uncertainty is the relationship between export and domestic markets, especially as the latter are so much larger than the former. An in-depth analysis of the question would have to compare the long-term wood supply potential of the USA and demand from domestic and export markets. Trade policy also plays a role. These factors will be taken into account in the Forest Service analysis mentioned above, but no results are at present available.

A major element of uncertainty for the outlook for US exports concerns the parity of the US dollar. If it were to recover the very high levels of early 1985, there would be little prospect of expanding exports of forest products: indeed increased imports from Europe would be more likely (as happened in 1984). The steep drop in the value of the US dollar in early 1986 improved the competitivity of US exporters.

17.4 TROPICAL REGIONS

17.4.1 Tropical forest resources

The following description is based on the survey undertaken between 1978 and 1981 by FAO and UNEP through the FAO/UNEP Tropical Forest Resources Assessment Project, with a view to reassessing the present situation and current development of the forest resources throughout the tropical world.

The survey covered countries in the tropical areas of America, Africa and Asia. Its findings were presented in a detailed technical report for each main region and synthesized in a fourth report for the tropical world as a whole[4].

In the following section, a brief summary is given of the main findings of the assessment of the tropical forest resources for the three tropical regions as a whole, and individually, by sub-regions and by main countries.

17.4.1.1 AREA

Table 17.14 gives an overview of forest areas and population in the tropical regions.

With an estimated area of 1,195 million ha in 1980, closed forests accounted for nearly two-thirds (62%) of all natural tree formations in the tropical regions, and for one-quarter of the total land area. Of the 1,195 million ha total area of closed forest, over half (679 million ha, or 57%), was situated in tropical America. Asia, with 300 million ha, accounted for one-quarter (25%) of total closed forest, and Africa, with 216 million ha, for 11%. Only a small part (3%) of total closed forest was predominantly of coniferous species, most of it in tropical America. There are considerable differences between continents in the ratios between the area of closed forests and that of tree formations as a whole. Closed forest represented a high percentage of the total area of tree formations (tree cover) in tropical Asia (89%) and tropical America (76%), but for only 31% in tropical Africa. In terms of total land area, closed forests account for 40% in tropical America, 32% in tropical Asia, but only 10% in tropical Africa.

Tropical closed forest is concentrated in a relatively small number of countries. In tropical America, 87% of closed, broadleaved forest was situated in the seven countries of tropical South America; in tropical Africa, seven countries in Central Africa accounted for 81%; and in tropical Asia the four countries comprising insular South-East Asia, for 49%.

For the whole of the 76 countries, the average area *per caput* of tree formations not yet affected by agriculture is just one hectare. This compares with an average of about 0.3 hectare of forest per person in Europe. However, differences between countries in the three tropical regions are enormous, and, especially in the present main producing and exporting countries, population density was well above the average and likely to increase further.

Of the relatively small area of *coniferous, closed natural forest* in the 76 tropical countries, some 34 million ha (75%) are to be found in tropical America, and the remaining 25% in tropical Asia. It must be noted, however, that over 24 million ha of these forests are located in the sub-temperate and temperate zones of Mexico (temperate pines); in southern Brazil (*Araucaria angustifolia*), and in the Indian subcontinent. The truly tropical coniferous forests (tropical pine stands in Mexico, Central America and the Caribbean; *Podocarpus* in the Andes and Africa; as well as *Pinus merkusii* and *Pinus kesiya* in South-East Asia), comprise not more than some 9.8 million ha overall.

Of particular importance in the context of the present study, due to the volumes of timber extracted from them for the world market, are the *closed, natural, tropical hardwood forests*. Table 17.15 sets out relevant data on their area and distribution, and their productive capacity, as well as the extent of their utilization.

The total area of all closed, natural broadleaved forest in the three tropical regions was estimated at 1160 million

---

[4] *Forest Resources of Tropical Asia*, FAO, Rome, 1981; *Forest Resources of Tropical Africa*, Part I: regional synthesis; Part II: country briefs, FAO, Rome, 1981; *Los Recursos Forestales de la America Tropical*, FAO, Rome, 1981. *Tropical Forest Resources*, by J.-P. Lanly, FAO, Rome, 1982, FAO Forestry Paper 30.

TABLE 17.14

Forest area and population in the tropical regions covered by the survey in 1980

Sub-region/region [a]	Total area (including inland water) (million ha)	Area of natural tree formations [b] (million ha)	Rate of tree cover (%)	Area of natural closed forest [b] (million ha) Broadleaved	Area of natural closed forest [b] (million ha) Coniferous	Bamboo forests (million ha)	Other natural tree formations [c] (million ha)	Population Total (million caput)	Population Area of natural tree formations per caput (ha/caput)
Central America and Mexico (7)	247.2	66.9	27.1	42.3	22.6	..	2.0	92.6	0.7
Caribbean (9)	70.0	46.7	66.7	45.4	0.5	..	0.7	26.6	1.8
Tropical south Latin America (7)....	1 362.4	782.1	57.4	566.2	1.6	..	214.3	202.6	3.9
Tropical America (23)	1 679.6	895.7	53.3	653.9	24.7	..	217.1	321.8	2.8
Northern savanna region (6)	423.6	43.7	10.3	0.8	–	..	42.9	29.6	1.5
West Africa (9)	212.1	55.7	26.2	17.9	–	..	37.8	113.8	0.5
Central Africa (7)	532.8	335.9	63.0	173.2	–	0.1	162.6	48.5	7.0
East Africa and Madagascar (13)....	881.1	216.9	24.6	22.5	1.1	1.0	192.3	149.8	1.4
Tropical south Africa (2)	139.9	51.0	36.4	–	–		51.0	1.8	28.3
Tropical Africa (37)	2 189.5	703.1	32.1	214.4	1.1	1.1	486.5	343.5	2.0
South Asia (6)	448.8	66.6	14.8	52.6	6.6	1.4	6.0	895.5	0.1
Continental south-east Asia (2)......	119.2	47.6	40.0	39.3	0.3	1.5	6.5	83.0	0.6
Centrally planned tropical Asia (3) ..	75.2	36.4	48.8	22.1	0.4	2.2	11.7	64.9	0.6
Insular south-east Asia (4).........	255.5	147.7	57.8	144.2	0.5	–	3.0	216.8	0.7
Papua New Guinea (1).............	46.2	38.2	82.7	33.7	0.5	–	4.0	3.1	12.3
Tropical Asia (16)	944.9	336.5	35.6	292.0	8.4	5.2	30.9	1 263.2	0.3
TOTAL (76 countries)	4 814.1	1 935.2	40.2	1 160.3	34.3	6.3	734.3	1 928.6	1.0

Source: FAO/UNEP (see text).

[a] In brackets: number of countries covered.

[b] Excluding plantation forest and forest fallow (i.e. formations affected by agriculture).

[c] Other than natural closed forest and bamboo forests.

ha – Brazil, Indonesia and Zaire – and their combined total of 575 million ha represented just half of the world total. Out of the total area of closed, natural broad-leaved forest, nearly three-quarters was considered as productive.

From the point of view of exports of tropical hardwood, the most important formations of closed forest are the moist evergreen forests, although some deciduous forest formations can be economically important for their valuable species (such as teak). Due to the great diversity of moist evergreen forest formations, only an approximate assessment of their extent can be made at present. Broadly speaking, it may be estimated that 90% (or about 590 million ha) of all closed, natural hardwood forest in tropical America is in lowland and hill areas and a possibly somewhat higher percentage (or about 200 million ha) in tropical Africa. In tropical Asia, lowland and hill forests represent a rather lower share of all closed, natural hardwood forest, possibly just under 50%, or about 135 million m^3, of which about two-thirds is in Indonesia.

Managed forest represented an insignificant part of the total area of closed, broadleaved forest in both tropical America and tropical Africa, but for 12.4% in tropical Asia, chiefly due to India where more than half of the closed hardwood forests are under management (table 17.18). The estimated 822 million ha of unmanaged productive forest accounted for just over 70% of all closed, broadleaved forest.

Of the 822 million ha of unmanaged, productive, closed, broadleaved forest, some 668 million ha, or 81%, were considered as undisturbed, while over 150 million ha had been logged, mainly during the last 20 to 30 years.

It is also apparent from table 17.15 that in the majority of the main present (and past) exporting countries, i.e. Indonesia, Malaysia and the Philippines in tropical Asia; Cameroon, Gabon and the Ivory Coast in tropical Africa, the area of logged-over forest is nearly as large as and, in some cases, even larger than that of forest recorded as being undisturbed.

The average volumes of timber (sawlogs and veneer logs) removed per hectare from undisturbed, closed, broadleaved forest at the time of first logging – the "volumes actually commercialized" – were estimated at 8.4 m^3 in tropical America, 13.5 m^3 in tropical Africa, and 31.3 m^3 in tropical Asia. The reason why the volume in Asia is higher than elsewhere lies in the greater homogeneity and richness of the forests of South-East Asia in commercial species of the Dipterocarp family. However, considerable differences in extracted volumes are found within the regions and between different localities, depending on the forest type. Especially in the more uniform types of forest, such as the Dipterocarp forests in South-East Asia, and in certain nearly pure broadleaved forests of edaphic origin in tropical America, volumes extracted can be as much as 100 m^3 per ha, the volumes felled or damaged in the process being considerably higher. On the other

TABLE 17.15

Production capacity, management and past logging of closed natural broadleaved forests, 1980

(Million ha)

Region/country	Productive				Unproductive			All closed natural broadleaved forest		Forest fallow [a]
	Unmanaged				For physical reasons	For legal reasons	Total	Total	Percentage of world total	
	Undisturbed	Logged over	Managed	Total						
TROPICAL AMERICA	452.98	53.49	0.01	506.48	133.54	13.91	147.45	653.93	56.36	99.34
of which:										
Mexico	12.28	0.30	–	12.58	13.78	0.21	13.99	26.57	4.06	17.45
Guyana	12.12	1.35	–	13.47	5.00	0.01	5.01	18.47	2.83	0.20
Brazil	288.63	12.00	–	300.63	51.00	4.65	55.65	356.28	54.48	46.42
Bolivia	17.76	12.09	–	29.85	14.16		14.16	44.01	6.73	1.10
Colombia	38.60	0.90	–	39.50	4.62	2.28	6.90	46.40	7.10	8.50
Ecuador	10.80	0.11	–	10.91	2.97	0.25	3.32	14.23	2.18	2.35
Peru	37.32	6.00	–	43.32	25.14	0.85	25.99	69.31	10.60	5.35
Venezuela	7.60	11.61	–	19.21	8.16	4.50	12.66	31.87	4.37	10.65
Surinam	12.08	0.42	–	12.50	1.76	0.58	2.34	14.83	2.27	0.27
Other	15.79	8.71	0.01	24.51	6.95	0.48	7.43	31.96	5.38	7.05
TROPICAL AFRICA	118.18	41.85	1.71	161.75	43.64	9.02	52.66	214.40	18.48	61.63
of which:										
Ivory Coast	0.20	3.09	–	3.30	0.52	0.65	1.16	4.46	2.08	8.40
Nigeria	0.38	2.59	–	2.97	2.98	–	2.98	5.95	2.78	7.75
Cameroon	7.00	9.94	–	16.94	0.98	–	0.98	17.92	8.36	4.90
Congo	10.33	3.36	–	13.69	7.52	0.13	7.65	21.34	9.95	1.10
Gabon	10.66	9.25	–	19.91	0.60	–	0.60	20.50	9.56	1.50
Zaire	79.74	0.38	–	80.12	19.84	5.69	25.53	105.65	49.28	7.80
Madagascar	1.60	5.07	–	6.67	2.70	0.03	3.63	10.30	4.80	3.50
Other	8.27	8.17	1.71	18.15	8.50	1.12	10.13	28.25	13.19	26.68
TROPICAL ASIA	97.26	58.42	36.19	191.87	83.62	16.46	100.08	291.95	25.16	67.25
of which:										
Burma	14.11	5.59	3.42	23.12	7.78	0.30	8.08	31.19	10.68	77.56
Thailand	3.92			3.92	2.04	2.18	4.22	8.14	2.79	0.80
Indonesia	38.92	34.62	0.04	73.58	34.57	5.43	40.00	113.58	38.90	13.46
Malaysia	7.53	5.52	2.50	15.55	4.48	0.96	5.44	21.00	7.19	4.82
Philippines	3.00	3.70	–	6.70	1.93	0.69	2.62	9.32	3.20	3.52
Papua New Guinea	13.82	0.22	–	14.04	19.62	0.06	19.68	33.71	11.55	1.25
Other	15.96	8.77	30.23	54.96	13.20	6.84	20.04	75.01	25.69	25.84
TOTAL (76 countries)	668.42	153.76	37.92	860.10	260.80	39.38	300.18	1 160.28	100.00	228.22

Source: FAO/UNEP, *op. cit.*

[a] All complexes of woody vegetation deriving from the clearing by shifting cultivation of closed broadleaved forests.

hand, the volume extracted and actually commercialized from the Amazonian forest was estimated at 5 m³/ha. Damage caused to the remaining trees in the stand during logging operations rises with the intensity of operations and the volume extracted, and under adverse conditions (e.g. terrain) can have long-lasting consequences. The area of undisturbed, productive, closed, broadleaved forest being logged annually in the early 1980s was estimated in the study at 4.34 million ha for the tropical countries in total, of which 1.96 million ha for tropical America; 0.64 million ha for tropical Africa; and 1.74 million ha for tropical Asia.

Besides logging activities, it is also important to consider the effects on closed, broadleaved forest of agricultural activities, in particular of shifting cultivation. Such *forest fallow*, constituted usually by a mosaic of various reconstitution stages, as well as of agricultural fields and patches of uncleared forest, can lead to the degradation of sites, particularly when site conditions are

unfavourable or when the fallow period is too short to allow full recovery of the site. Forest fallow, with an estimated 230 million ha, represented an area that corresponded to almost 20% of all tropical, closed, natural, broadleaved forest. In some countries, in particular in tropical Africa and Asia, the areas of once closed, mature forest thus affected by shifting cultivation, are extensive and, in fact, are larger than the area of remaining closed forest.

17.4.1.2 GROWING STOCK

The mean gross volume overbark[5] per hectare in unmanaged, productive, closed, broadleaved tropical forests was estimated to be as follows:

---

[5] The volume of the bole of living trees with a reference diameter greater than 10 cm (at 1.30 m height or above the buttresses or aerial roots) from stump or buttresses or aerial roots up to crown point or first main branch.

	Undisturbed forest (m³/ha)	Logged forest (m³/ha)
America (23 countries)	157	119
Africa (37 countries)	256	195
Asia (16 countries)	216	113
Total	183	137

The figures for average growing stock per hectare at the regional level and for the tropical world as a whole have only limited significance, because large variations exist between formations of the same category from one phase to another over a region. The data confirm, however, the commonly held view that average volumes per hectare of undisturbed, closed, broadleaved forests in tropical America are lower than those in tropical Africa and Asia. In particular, the Amazon forest seems to show greater heterogeneity, a smaller overall number of commercially accepted species, and a smaller number of large-sized trees than do forests in the other tropical regions. Yet, of the entire growing stock of productive, closed, broadleaved forests (unmanaged and managed) in the tropical world as a whole (average volume multiplied by area) estimated at the end of 1980 at 147 000 million m³, tropical America accounted for over half (53%), while Africa accounts for one quarter (26%) and Asia for one-fifth (21%).

17.4.1.3 Deterioration of tropical forests

Over the last thirty years or so the extent of alteration of tropical forests, essentially due to human needs and activities, has increased sharply. While some of these have already been referred to, notably logging for timber production, and shifting agriculture, it seems useful to consider them in a broader perspective.

Four main types of alterations of forests are identified. These are:

Deforestation
The complete clearing of tree formations (closed or open) and their replacement by other use of the land (alienation), such as permanent agriculture, human settlements (spontaneous migration and colonization programmes), industrial and communications infrastructure;

Degradation
Resulting from over-intensive use, for instance, for fuelwood, from grazing or from logging for sawlogs and veneer logs; natural causes can also be involved, for instance, repeated fires, cyclones, or plant parasites. Such degradation processes may in their final stages result in deforestation although, compared to clearing for other purposes, they account for only a very small part of annual deforestation;

Logging
Of unmanaged, closed forest, usually by selective logging of their more valuable timber species which results in the transformation of undisturbed forests into logged-over forests, but remains as a forest stand even if it may have been more or less altered. If the stand is subsequently clearcut and burnt for temporary or shifting agriculture, it falls under forest fallow, or, if clearcut for permanent new use, under deforestation;

Management
For productive or protective uses, for instance through silvicultural treatments.

Of these four different aspects, degradation processes are often the most difficult to detect and quantify, as many forms of degradation introduce progressive changes. Few precise data are therefore available on the losses in wood resources at the level of larger geographical entities.

While degradation processes and logging activities can, under certain circumstances, lead to deforestation, the major cause of deforestation in the tropical regions is clearing of forest (closed or open) for other uses of the land. Table 17.16 shows the estimated annual rates of deforestation of closed forest between 1976 and 1980.

The area of closed, broadleaved, tropical forest cleared (i.e. deforested) annually in the second half of the 1970s has been estimated by the FAO/UNEP project at close to 7 million ha, of which more than half in tropical America, nearly one-fifth in Africa, and one-quarter in

TABLE 17.16

Average annual deforestation of closed forest, 1976 to 1980
(1 000 ha per annum)

	Productive			Unproductive	Total
	Undisturbed	Logged over [a]	Total		
(a) Closed broadleaved forests					
Tropical America	1 135	1 684	2 819	988	3 807
Tropical Africa	220	1 036	1 256	63	1 319
Tropical Asia	483	1 174	1 657	110	1 767
Total	1 838	3 894	5 732	1 161	6 893
(b) Coniferous forests					
Tropical America	102	128	230	82	312
Tropical Africa	2	4	6	2	8
Tropical Asia	12	17	29	6	35
Total	116	149	265	90	355

Source: FAO/UNEP.
[a] Including managed production forests.

Asia. The rate of reduction in total closed forest area was thus 0.6% per year; this rate was very similar in all three tropical regions. The rate of reduction is slightly higher for productive forests, but significantly lower for unproductive forests. The annual percentage rate of reduction for the different forest types (world-wide) was as follows:

Closed broadleaved	0.60
of which:	
Undisturbed	2.06
Logged over	0.27
Closed coniferous	1.52
Open tree formation	0.52

17.4.1.4 PLANTATIONS

Although 33 forestry plantations for various purposes had already been established in the tropical world in the last century, more than 90% of the forest plantations existing in 1980 were established after 1951.

Almost two-thirds of all forest plantations by 1980 were industrial plantations (i.e. established totally or partly for the purpose of producing industrial wood assortments, such as sawlogs, pulpwood and pit-props). In recent years more effort has gone into the creation of non-industrial plantations, particularly for the production of fuelwood.

Table 17.17 shows the distribution of forestry plantations by main species over the three tropical regions.

Of the total area of *industrial* plantations, half was in tropical Asia, over one-third in tropical America and less than one-sixth in tropical Africa. The largest areas of industrial forestry plantation were to be found in Brazil,

followed by India and Indonesia; taken together these three countries accounted for 70% of industrial plantations. Of industrial plantations worldwide, 62% were of hardwood species. Two-thirds of all hardwood plantations were to be found in tropical Asia, while tropical America accounted for well over half (58%) of all softwood plantations (with pine as the most important species), with Brazil showing the largest area. Among the hardwood plantations, while the fast-growing species (*Eucalyptus spp, Gmelina* and some others) are mainly destined for chipping, others (e.g. teak) are mainly destined for the production of sawlogs. Of the *non-industrial* forestry plantations, hardwood species accounted for the bulk (94%) of the planted area, and among these, fast-growing hardwood species were by far the most important. Most of the *non-industrial plantations* in tropical Africa and Asia were for fuelwood and other smallwood production and for protection purposes. Nearly 36% of industrial plantations and 47% of non-industrial plantations existing in 1980 had been established after 1975.

17.4.1.5 CONCLUSIONS ON THE TROPICAL FOREST RESOURCE

A principal conclusion of the FAO/UNEP project was that the evolution of tropical forest resources observed in the recent past was likely to continue in the *near* to *medium-term* future. In particular:

- The closed forest of the 76 tropical countries overall would be deforested at a rate of about 7.5 million hectares per year;

TABLE 17.17

Areas of established plantations estimated at end 1980

(1 000 ha)

	Hardwood species						Softwood species		All species	
	Fast-growing		Other		Total					
	Total	1976-80 [a]	Total	1976-80 [a]	Total	1976-80 [a]	Total	1976-80 [a]	Total	1976-80 [a]
A. INDUSTRIAL										
Tropical America	868	346	129	37	997	383	1 571	662	2 568	1 045
of which:										
Brazil	675	280	66	13	741	293	1 232	496	1 973	789
Other	193	66	63	24	356	90	339	166	595	256
Tropical Africa	162	51	294	68	456	119	541	144	997	263
of which:										
Nigeria	62	32	82	34	144	63	2	2	146	65
Other	100	19	212	34	312	56	539	142	851	198
Tropical Asia	1 083	348	1 813	533	2 896	881	606	330	3 502	1 211
of which:										
India	941	265	537	144	1 478	409	58	15	1 536	424
Indonesia	15	5	1 001	282	1 016	287	430	255	1 446	542
Other	127	78	275	107	402	185	118	60	520	245
Total industrial (76 countries)	2 113	745	2 236	638	4 349	1 383	2 718	1 136	7 067	2 519
B. NON-INDUSTRIAL (including for fuelwood)										
Tropical America	1 583	722	419	257	2 002	979	50	26	2 052	1 005
Tropical Africa	483	102	294	98	777	200	6	3	783	203
Tropical Asia	1 220	608	163	93	1 383	701	226	183	1 609	884
Total non-industrial	3 286	1 343	876	448	4 162	1 880	282	212	4 444	2 092
OVERALL TOTAL (industrial and non-industrial)	5 399	2 088	3 112	1 086	8 511	3 263	3 000	1 348	11 511	4 611

Source: FAO/UNEP, *op. cit.*

[a] Plantations established between 1976 and 1980.

A further 4.4 million ha of undisturbed, closed forests of the 76 countries would be logged (for the first time) each year but not cleared;

- The open tree formations of the 76 countries in total would be cleared at the yearly rate of 3.8 million hectares, in particular in South America and Africa;
- Industrial and non-industrial plantations would increase by 1.1 million hectares per year.

In the *medium to long term*, a continuation at the same pace of present levels of reduction and logging of undisturbed, productive, closed forests would mean that by the year 2000 there would remain only about 540 million hectares of undisturbed, productive, closed forests (390 million ha in tropical America, 100 million ha in tropical Africa, and 50 million ha in tropical Asia). Adding to this the estimated area of 280 million ha of unproductive, closed forest remaining at that time, the total area of undisturbed, closed forest would be 820 million hectares.

However, the area of plantations will probably continue to increase at a pace that will be greater than the increase observed in the recent past. In addition, important reforestation programmes have begun to be implemented in a number of countries where destruction and degradation of the tropical forest cover have already reached a serious stage. These efforts will no doubt be pursued and expanded. At the same time, for both productive and protective purposes, a true, sustained management of forest plantations should be developed.

The result of these developments would be the partial replacement of the natural tropical resource by plantation silviculture, on a smaller area, but with much greater productivity and producing in most cases a totally different type of wood. It should be stressed, however, that great uncertainty surrounds many of the major factors which will determine the development of the tropical forest resource, including population growth, agricultural policy and demand for land, ecological factors and the techniques of tropical silviculture.

17.4.2 Production and trade of tropical hardwoods

Data on the production of tropical hardwood sawlogs and veneer logs, the raw material of the various sectors concerned with processing, distributing and using tropical hardwood products, are set out in table 17.18 for the three regions as a whole and in annex table 17.1 for individual countries. Due to problems in many countries with collecting statistics and probable underestimation of actual production volumes, these data should be considered approximations rather than accurate statistics in many cases. Nonetheless, the data show a number of remarkable developments.

Production of hardwood sawlogs and veneer logs in the 76 tropical countries nearly trebled between 1960 and 1980, reaching over 125 million m^3 in the latter year. Production in tropical America and Africa roughly doubled during that period, while in Asia its level was three-and-a-half times higher in 1980 than in 1960. Asia accounted in 1980 for nearly two-thirds of total production compared to just over half in 1960. The corresponding shares of tropical America and tropical Africa declined accordingly, from 27% to 21% and from 21% to 15% respectively.

The increasing concentration of the production of tropical hardwood logs in tropical Asia is accounted for essentially by three countries – Indonesia, Malaysia and the Philippines. In 1960, the volume of 15 million m^3 produced by these three countries represented 35% of total production for that year in the three tropical regions; in

TABLE 17.18

Production and exports of the major hardwood product by tropical regions
(Million m³)

	Production			Exports		
	1960	1970	1980	1960	1970	1980
Tropical America						
Sawlogs and veneer logs	11.5	15.3	26.0	0.3	0.4	0.1
Sawnwood	5.1	7.1	12.6	0.1	0.5	1.1
Plywood	0.2	0.7	1.4	–	0.1	0.2
Veneer sheets	..	0.1	0.4	–	0.1	0.1
Tropical Africa						
Sawlogs and veneer logs	9.2	14.7	19.5	4.4	6.8	6.1
Sawnwood	1.5	2.6	5.3	0.6	0.7	0.7
Plywood	0.1	0.2	0.4	0.1	0.1	–
Veneer sheets	..	0.2	0.3	–	0.2	0.2
Tropical Asia						
Sawlogs and veneer logs	22.7	52.2	79.7	6.1	29.0	32.1
Sawnwood	6.5	11.8	22.6	1.2	2.5	6.4
Plywood	0.4	1.3	3.0	0.1	0.5	1.7
Veneer sheets	..	0.4	0.9	0.1	0.2	0.2
Tropical regions, total						
Sawlogs and veneer logs	43.4	82.2	125.2	10.8	36.2	38.3
Sawnwood	13.1	21.5	40.5	1.9	3.7	8.2
Plywood	0.7	2.2	4.8	0.2	0.7	1.9
Veneer sheets	..	0.7	1.6	0.1	0.5	0.5

For country detail, see annex tables 17.1, 17.2 and 17.3.

1980, with nearly 60 million m³, it represented no less than 47%. This was due to the sharply expanding trend of production in Indonesia and Malaysia, while production in the Philippines, the main South-East Asian exporter of logs in the 1960s, declined in the course of the 1970s.

Compared to these countries, production in Brazil, the country with the single largest tropical forest area, appears modest, despite its marked rise between 1960 and 1980.

In tropical Africa, only a few countries – Cameroon, the Ivory Coast, Liberia and Nigeria – showed a marked rise in production between 1960 and 1980. In several others, production in 1980 was actually well below the level of 1960.

Exports of tropical hardwood logs in 1980 by the three regions taken together were some three-and-a-half times higher than in 1960. In 1980 they represented 30% of production, as against 25% in 1960. However, log exports in 1980 by the 76 countries taken together were not much higher than in 1970, and actually lower in the case of tropical America and Africa. There were several reasons for these developments. On the one hand, policies for the forestry sector in major exporting countries were being increasingly aimed at furthering the export of processed products, rather than logs, and developing forest products industries. Secondly, domestic consumption of forest products rose quite strongly, especially since the early 1970s, in many producing/exporting countries, and particularly where the economies benefited from rising oil prices. In many such cases, notably in Latin America and Africa, production capacities in the forest and forest products sector could not meet rising domestic and export demand. A third reason was that demand for tropical hardwoods did not, in the main industrialized importing countries, continue to rise with the same vigorous pace after the recession of 1974/75 as before it.

Exports of processed tropical hardwood products, essentially sawnwood, plywood and veneer sheets, although they may appear relatively minor compared to exports of logs, were already quite appreciable in terms of m³EQ in 1960. They increased quite strongly between 1960 and 1980, in particular in the 1970s as a result of industrialization policies in the forest sector. For 1960, exports of processed hardwood products, in terms of m³EQ, for the 76 tropical countries taken together, were equivalent to over 40% of their exports of logs, but in 1980 for nearly 60%. Trends in exports differed markedly between the three regions. Exports of the three processed products increased in terms of m³EQ in tropical Asia: from some 4.5 million m³EQ in 1960 to 22 million m³EQ in 1980 (+ 17.5 million m³EQ), although the percentage increase for tropical America was greater: tenfold from 0.3 to 3 million m³EQ. Exports from tropical African countries showed the least growth: from about 1.5 million m³EQ in 1960, they rose to close to 2.0 million m³EQ in 1980.

The bulk of imports of tropical hardwoods is accounted for by countries outside the tropical belt. The five main importing areas in order of importance in 1980 were: East Asia, including Japan; Europe; North America; West Asia, and Australia. The relevant data for the main importers – Japan, Europe and the USA – are set out in table 17.19. The salient features to emerge regarding trends in imports of these products between 1960 and 1980 were:

– The continued strong expansion of imports by the industrialized importing countries until the latter part of the 1970s;

– The marked change in the relative importance of different countries and areas;

– The emergence of new importing areas;

– The marked differences in the product pattern of imports by different countries and areas.

TABLE 17.19

Estimates of imports[a] of major hardwood forest products from tropical countries by Japan, Europe and the United States, 1960, 1970, 1980

(Million m³)

	1960	1970	1980
Japan			
Hardwood sawlogs and veneer logs	4.8	19.9	19.2
Sawn hardwood	–	0.3	0.6
Plywood	–	0.5	0.1
Veneer sheets	–	0.1	0.1
United States			
Hardwood sawlogs and veneer logs	0.3	0.1	–
Sawn hardwoood	0.4	0.4	0.5
Plywood	0.5	0.2	1.0
Veneer sheets	0.1	0.2	0.1
Europe			
Hardwood sawlogs and veneer logs	3.6	6.4	5.9
Sawn hardwood	1.0	1.5	3.2
Plywood	0.1	0.2	0.7
Veneer sheets	0.1	0.1	0.2

[a] Including imports based on tropical hardwoods from log importing countries in East Asia, which export processed products.

The strength of the growth in imports is illustrated by the fact that the combined volume of the four products, rose from about 13 million m³EQ in 1960, to an estimated 39 million m³EQ in 1970, and, to an estimated 51 million m³EQ in 1980, practically four times the level of twenty years earlier. This expanding trend was accompanied by marked changes in the pattern of imports and the relative share of different countries and areas. Thus, in 1960 Europe accounted for nearly half of the total volume of imports by the countries and areas concerned, North America for one-fifth; and East Asian countries for nearly one-third. By 1980, the East Asian countries as a whole accounted for well over half of the total volume of imports; Europe, despite a doubling of its imports, accounted for only just over one-quarter.

Throughout the period under consideration, there were wide differences between countries and areas as regards the pattern of products imported, despite a general trend towards a higher share of processed products compared to logs. In 1960 North American imports already consisted mainly of processed products. For Europe, the share of logs also decreased, but even in 1980 logs accounted for about two fifths of imports in terms of m³EQ, as compared to three fifths in 1960. The imports of the five East-Asian importing countries consisted almost entirely of logs. Of their imports in 1980, 87% were logs.

In contrast to the other importing areas under consideration, the East-Asian countries are also sizeable exporters of products processed from imported logs. Singapore is in addition a major in-transit centre.

Large export industries based on imported tropical hardwood logs have been built up, especially in the Taiwan Province of China and South Korea. The combined imports by these two areas of tropical hardwoods (nearly all logs) amounted to some 11.3 million m³EQ in 1980, compared to a mere 0.4 million m³EQ in 1960. Of this volume, some 4.3 million m³EQ or 38% was re-exported in the form of primary processed products, in particular plywood. Exports in the form of further processed products, such as building joinery, furniture and mouldings have also shown an increasing trend.

17.4.3   Prospects for trade in tropical hardwoods

In several of those countries which were the main producing and exporting countries of 15 to 20 years ago, intensive forest exploitation has resulted in a marked decline in timber production and exports, for instance in Ghana, the Ivory Coast, Nigeria and the Philippines. Many studies have been made over the years to assess the longer-term trends for supplies of tropical hardwood, in particular those to developed importing countries; they arrived at rather diverging conclusions. While earlier assessments sometimes saw in the tropical rain forests an inexhaustible source of wood, later studies,[6] using more detailed and comprehensive information, particularly as regards the forest resources, have tended to be more restrained, some even frankly pessimistic about longer-term prospects. Unofficial reports have tended to be even more pessimistic.

Therefore, rather than attempt to produce a synthesis, or another assessment, of future demand and supply of tropical hardwoods (and softwoods), which would in any case be beyond the scope of this study, it might be useful to review and consider the factors which appear likely to have an influence in this connexion.

Tropical hardwoods, in particular those intended for export to international markets, may be divided into two groups:

(a)   A selection of high quality raw material of a limited number of species, used predominantly in the form of sawnwood, plywood and veneer sheets;

(b)   Utility woods which account for most of the volume of tropical hardwood in international trade, and for which important criteria have been the consistency of quality, the ease of working, the high yield in processing, and the regular availability of large volumes, i.e. technical as well as economic factors.

It would seem that there is a certain potential for a wider use of the higher quality products, i.e. both of decorative tropical hardwood species, of which relatively few have found wider application in the importing countries, especially for furniture, and of special technical species.

The utility woods however may already have penetrated to a major extent the end-uses where their technical and particularly their economic performance was equal, if not superior to other timbers (temperate-zone hardwoods and softwoods) and competing non-timber materials. There seems to be little likelihood that major new industrial users will appear on the market for tropical hardwoods, as the manufacturers of industrial building joinery did in the past. Indeed there are indications that in some industries which have been using tropical hardwood extensively, the use of other timbers may be rising. In those areas where the utility timbers are used, the price of materials is a major cost element: the trend of price relationships during the latter part of the 1970s and the early 1980s has affected to some extent the competitivity of some important tropical hardwood species, especially in comparison to certain assortments of sawn softwood. A variety of factors has been involved in exerting upward pressure on producer prices, such as rising costs of labour, equipment and fuels; increasing transport costs as logging moves to more remote areas; and rising domestic demand for timber in many producing countries. To a certain extent it should be possible to contain the effect of such pressures by greater efficiency in the production of both logs and processed products, for which there is considerable scope in most producing countries.

Demand in importing countries, for utility tropical hardwoods in particular, could also be affected by the availability and technical and economic competitivity of domestic assortments (softwood and hardwood). For example, improved jointing techniques could improve the competitivity of some lower quality domestic timber assortments. In addition, it cannot be ruled out that consumers' concern about tropical deforestation, and unwillingness to be associated with it, could negatively effect demand for tropical hardwoods.

How will these factors effect demand for tropical hardwoods in the developed importing countries? A first approximation could be that demand may continue to grow, but at a considerably reduced pace compared to earlier years, with relative price and quality as determining factors. These aspects could take on added importance, particularly in the medium and longer term, if strains were to appear in the present supply pattern, a possibility which cannot be excluded.

In the producing countries, populations and economies will continue to grow, and pressure for land for settlements, agriculture and other uses, as well as for timber, will increase. Natural forest areas will inevitably continue to shrink, while logging activities will continue, with an increasing share of the logs being processed in their country of origin, and an increasing share of the processed products being consumed locally. It would seem that log production in the present main producing countries in

---

[6] For instance: *Agriculture Toward 2000*, FAO, Rome 1980; *World Forest Products Demand and Supply 1990 and 2000*, FAO Forestry Paper 29, Rome, 1982; *The Global 2000 Report to the President*, Washington, US Government Printing Office, 1980.

South-East Asia, taken together, may have passed its peak. The utilization of the forest resource is a major element of economic development and an important source of export revenue. The processing of logs in the country of origin, now the declared policy in the majority of exporting countries, is intended to add value to products for export so that the same (or even lower) production of logs, would bring higher revenues. A further reason for supposing that hardwood log production in South-East Asia is likely to decrease rather than increase is that, as the extent of logged-over areas rises further, forest exploitation policies tend to become more restrictive. While logging by itself does not destroy the forest resource, but only modifies it to a lesser or greater extent, depending on the intensity of logging and the success of regeneration, it does result in a certain depletion and makes the forest more vulnerable to influences causing degradation.[7] Without intensive management, logged-over forest does not normally revert to its earlier state, at least not within the normal 25- to 35-year interval between logging cycles in many tropical hardwood countries. Hence, volumes and qualities of the commercial species obtained in second logging cycles are usually considerably lower than those of the first logging cycle, although sometimes a second logging cycle may use additional species which have become marketable since the first cycle. If logged-over forests are subsequently used for shifting cultivation, forest regrowth is further retarded.

While the process outlined above, with its sequence of ascending and descending phases, has been typical of tropical forest development in the past, it seems likely that some form of management and silvicultural treatment will be introduced with a view to maintaining the forest under a sustained yield concept. It has been said that a first step could be closer control and refinement of logging operations, to ensure that the production potential of remaining trees is maintained and possibly further developed, and natural regeneration encouraged.

While there is considerable scope for making better use of the available wood raw material by including lesser-known species and lower qualities, as well as by reducing waste in both logging and processing activities, a gradual decline in the availability of wood and wood products for exports in the major, present day producing countries will

probably also be reflected in pressures on prices. This decline in availability might be compensated by increased supplies from forest-rich areas in other regions, e.g. in parts of West Africa, in Central Africa and in Latin America. The beginnings of this trend have already been observed, with the increase in imports of logs from Africa by East Asian countries, including Japan.

The development on any major scale of existing and new forest industry activities in these mostly remote areas will pose important problems in terms of investment, infrastructure and labour. Conceivably, log production capacity could be increased substantially under the impetus of a rising price trend for tropical hardwoods, but it is more doubtful whether corresponding processing capacities can be built up.

What general conclusions can we draw from this analysis of factors which may influence future demand and supply of tropical hardwood? Demand in importing countries, including those in West and East Asia with limited forest resources, seems likely to follow the trend of building construction activities, or possibly to grow more slowly if it becomes apparent that prices of tropical hardwoods would tend to rise faster than those of other forest products and materials.

From the supply side, the potential of tropical hardwood producing countries to meet increased demands from both domestic users and importing countries outside the tropical regions, exists well beyond the year 2000, although this would contribute to the further depletion of the natural forests. The growth potential of the forests for the species and qualities currently and potentially in demand could be seriously affected, and a sustained yield over the longer term could not be assured. The determining factor, therefore, seems to be the policies adopted by producing countries for the utilization of their forests. These policies take into account the various demands on the forests, and are based on their perception of the roles — economic, social, ecological and other — which forests should ultimately play in the national life and development of their country and in the well-being of its people.

A shift to a more restrictive stance in policies than in the past, for instance if conservation of resources was given higher priority, could lead to severe imbalances between demand and supply of the grades supplied by the natural tropical forests. Increased supplies from man-made plantation forests of fast-growing species in tropical countries could to a certain extent be an alternative to tropical hardwoods in terms of volume, though not of quality. This question is briefly discussed in the next section.

---

[7] It was noted in FAO document APRC/82/9 (para. 16) that "In general, logging operations are causing qualitative and quantitative degradation of the forest. The extent varies. It is dependent, both on intensity of logging as well as the skills in planning and execution of various operations". Sixteenth FAO Regional Conference for Asia and the Pacific, Jakarta, Indonesia, 1-12 June 1982: Regional Implications for a Development Strategy (Conservation, Utilization and Management) of Forest Resources.

17.5 OTHER PRESENT AND POTENTIAL SUPPLIERS OF FOREST PRODUCTS TO EUROPE

Although the USSR, Canada, the USA and the tropical hardwood producers are the major exporters of forest products to Europe, they are not the only ones. Furthermore, exports from some countries not mentioned above, such as Chile and New Zealand and Brazil (plantation forests) have gained in importance in recent years and could have a significant impact on the pattern of European supply in the future. This impact could take the form of direct exports to Europe or be in indirect form. Indirect effects could be of two kinds:

- Successful competition by new suppliers in non-European markets (e.g. Japan) with traditional suppliers to those markets, who might try to compensate these losses by increasing their sales to Europe;

- As the new suppliers mostly have rather low wood costs, they could affect the price for those forest products where it is possible to talk of a world market (e.g. chemical pulp, sawn softwood, newsprint, Kraftliner).

Which countries, other than traditional suppliers, are exporting forest products to Europe? Table 17.20 shows all European imports of forest products from outside the ECE region. ECE/FAO statistics on trade flows do not distinguish between hardwood of tropical and of temperate zone species. It may be assumed, however, that nearly all exports from Latin America, Africa and Asia/Pacific to Europe of hardwood logs, sawn hardwood and plywood are in fact of tropical species. This makes it possible to separate tropical hardwood products from other products.

Imports of the other products (i.e. assumed not to come from tropical hardwoods), which were equivalent to about 2.1 million m³EQ in 1970, nearly tripled to reach 5.7 million m³ in 1980. Imports of most of these products were insignificant, and those of sawn softwood from Latin America (0.4 million m³) did not change. There was however, a major increase in imports of chemical pulp from Latin America, which stood at 0.6 million m.t. in 1980. Most of this came from Brazil. In that country several large export-oriented pulp mills have been set up, mostly using hardwoods (e.g. eucalyptus, sometimes in combination with other species) as raw material. Low wood costs and modern equipment have enabled Brazilian hardwood pulp to claim a growing share of the world pulp markets. Brazil's total exports of chemical pulp grew from about 0.1 million m.t. in the early 1970s to nearly 1 million m.t. in 1983, i.e. from about 1% to over 5% of world exports (softwood and hardwood combined).

The increase in Brazilian production and exports of chemical pulp is only the most striking example of the potential of tropical and other plantations to produce low-cost wood or forest products for the world markets. Although this source of supply does not yet rival the traditional suppliers to Europe, its potential should be examined. Two different types of plantation are considered separately:

- Plantations in tropical countries;

- Other plantations, notably in Chile and New Zealand.

TABLE 17.20

Exports from areas outside the ECE region to Europe, 1970 and 1980

(*Million units*)

	Unit (million)	1970				1980			
		Latin America	Africa	Asia/ Pacific	Total	Latin America	Africa	Asia/ Pacific	Total
Exports to Europe									
Sawn softwood	m³	0.4	–	–	0.5	0.4	–	–	0.4
Hardwood logs	m³	0.1	6.0	1.2	7.3	–	4.6	0.2	4.8
Sawn hardwood	m³	0.1	0.5	1.1	1.7	0.3	0.5	2.3	3.1
Plywood	m³	–	0.1	0.3	0.4	–	0.1	1.0	1.1
Mechanical pulp	m.t.	–	–	–	–	–	–	–	–
Chemical pulp	m.t.	–	0.2	–	0.2	0.6	0.3	–	0.8
Newsprint	m.t.	–	–	–	–	–	–	–	–
Paper and paperboard (except newsprint)		–	–	–	–	–	–	0.1	0.1
Total	m³EQ	1.0	8.4	3.9	13.3	4.1	7.4	6.6	18.1
of which:									
Hardwood logs, sawnwood and plywood	m³EQ	0.3	7.2	3.7	11.1	0.6	5.5	6.3	12.4
Other	m³EQ	0.7	1.2	0.2	2.1	3.5	1.9	0.2	5.7
Exports to Europe as percentage of total exports									
Sawn softwood		33	–	–	28	25	–	–	25
Hardwood logs		30	92	4	20	–	76	1	12
Sawn hardwood		12	69	40	42	33	91	36	40
Plywood		–	85	13	18	–	100	27	29
Mechanical pulp		–	–	–	–	–	–	–	–
Chemical pulp		–	80	–	80	45	50	–	46
Newsprint		–	–	–	–	–	–	–	–
Paper and paperboard (except newsprint)		–	–	–	–	–	–	–	–
Total		27	86	9	24	41	70	12	23
of which:									
Hardwood logs, sawnwood and plywood		21	88	9	23	37	78	12	23
Other		29	79	6	29	42	51	6	35

Source: TIMTRADE data base. These data may not be exactly comparable with those for total exports and imports used elsewhere.

17.5.1 Outlook for tropical plantations

The area of industrial plantations in the 76 tropical countries covered by the FAO/UNEP project is as follows (in million ha) (see table 17.15 for detail):

	Hardwood			
	Fast-growing	Other	Softwood	Total
America..............	0.9	0.1	1.6	2.6
Africa	0.2	0.3	0.5	1.0
Asia	1.1	1.8	0.6	3.5
Total (76 countries)..	2.1	2.2	2.7	7.1

Of the 7.1 million ha, 2.5 million, or over a third were established between 1976 and 1980. Nearly 5 million ha (70% of the total) are in three countries: Brazil, India and Indonesia.

The theoretical potential of these plantations is enormous. (At Aracruz, in Brazil, growth rates of 70 m^3/ha/year have been achieved, equivalent to 18 m.t./ha/year of hardwood chemical pulp). If the very high rates of annual increment which are achievable in tropical growing conditions were to be attained, then about 170 million m^3 a year could be grown on the plantations already established, or over 10% of the world production of industrial wood (assuming 50 m^3/ha/year for fast-growing hardwood plantations, 10 m^3/ha/year for other hardwood plantations, and 15 m^3/ha/year for softwood plantations). Many technical, and possibly environmental factors will prevent this. On the other hand, there is the possibility of expanding the area under plantation.

In the first place, the techniques of intensive tropical silviculture are far from being fully developed and problems of choice of species, soil fertility and pest control have proved insuperable in some cases. In addition, the local inhabitants, accustomed to a natural forest, seen as the common good of the community, often do not easily accept the principles of ownership and management which accompany the setting up of intensive plantations for industrial wood. In particular there is a danger that the plantations, once established will not be properly tended. In addition, the establishment of large plantations in previously sparsely inhabited areas brings about enormous infrastructure costs, which can destroy a project's economic viability. Nevertheless these problems can be overcome, as proved by some notable success stories (e.g. Aracruz).

Finally, there is the question of markets. Some tropical plantations are established with the world market in mind, integrated from the start with a world-class processing facility. Others, however, are intended to satisfy the growing needs of the local population (or that of neighbouring countries) or to replace imported forest products. In the latter case, the influence on the international markets will be limited to some degree of import substitution, and the effect on availability of forest products for export to Europe will be insignificant.

In short, the influence of tropical plantations on wood supply to Europe is subject to two major uncertainties:

– Is it possible to achieve the desired production rates?

– Will the produce of the plantation be directed to the domestic or to the export market?

It is already clear that tropical plantation silviculture (of species like *Eucalyptus, Gmelina* or *Pinus spp.*) is making rapid progress and it is likely that in a few years most of the purely technical problems will be capable of solution. Also experience is being gained in carrying out plantation projects within the society and economy of developing countries. Thereafter, the fundamental advantages of these plantations, notably the good growing conditions, availability of land and labour, will become increasingly evident. Furthermore, because of the short rotations appropriate to tropical conditions, the lead time from planning to full production is very short by European standards – often less than a decade.

If forest products derived from tropical plantations can be produced and shipped at a price which is competitive with those on importers' markets, it seems likely that Governments or individuals in those developing countries with good growing conditions, and adequate land and infrastructure, will establish plantations as a commercial venture. They may well do this in partnership with enterprises from developed countries who could provide capital and expertise.

17.5.2 Chile and New Zealand as forest products exporters

Chile and New Zealand, both situated at similar, temperate, latitudes in the southern hemisphere, have undertaken extensive forestation, starting in the 1920s, but intensified since the late 1950s. Plantations of fast-growing, primarily coniferous species (Radiata pine in particular) have been yielding increasing harvests of different assortments, mainly for sawing and pulping purposes, with a significant part of the harvest being exported, mainly in unprocessed forms. Due to the relative importance which these two countries' exports have reached in several of the major importing markets, a brief look at their forest situation and export potential is of interest.

17.5.2.1 CHILE

The original natural forests of conifers (especially *Araucaria*) in Chile are largely depleted. However, by 1980, an area of nearly 800,000 ha had been planted with Radiata pine of which over half a million hectares since 1970. It is expected that by 1985 the planted area will have reached 1 million ha, which seems to be an initial target. Rotation is said to be 20 to 25 years, and the mean annual increment between 20 m^3 and 30 m^3 per ha. Due to the varying intensity of planting between 1960 and 1985 it is expected that harvests will reach an initial peak around the year 2000 with estimates ranging between 25 and 40 million m^3/year, depending on underlying assumptions of yield as well as on forestation activity through the latter half of the 1980s and the 1990s. After the year 2000, the volume of annual harvests is likely to decline somewhat. Two-thirds of the harvested volume is expected eventually to be sawlogs, and one-third pulpwood, mainly from thin-

TABLE 17.21

**Chile and New Zealand: production and exports of
selected forest products**

(1000 m³ (m.t. for wood pulp))

	Production			Exports		
	1960	*1970*	*1980*	*1960*	*1970*	*1980*
Chile						
Coniferous sawlogs	696	1 620	9 300	–	..	1 004
Coniferous pulpwood	694ᵃ	1 650	3 056	–	..	..
Coniferous sawnwood	458	700	1 833	11	172	1 258
Woodpulp	102	356	763	–	105	416
New Zealand						
Coniferous sawlogs	3 781	6 081	5 730	123	1 809	970
Coniferous pulpwood	720ᵃ	1 495	3 345	–	1ᵇ	35ᵇ
Chips and particles	..	..	..	..	183	400
Coniferous sawnwood	1 602	1 789	1 968	103	257	614
Woodpulp	266	576	1 122	71	98	475

ᵃ Including pit-props.
ᵇ All pulpwood (round and split).

nings, compared to some 45% in the early 1980s. In 1980, Chile's production of coniferous industrial roundwood was 12.5 million m³.

As the data in table 17.21 show, production of coniferous sawnwood increased rapidly, especially in the 1970s, and exports even more rapidly, in particular during the latter half of the 1970s. A significant part of the harvest is intended for export, mainly in the form of logs, sawnwood, woodpulp and, to some extent, paper. What the shares of the different products will eventually be will depend on market demand for specific products and available processing capacities. Some projections foresee exports of logs, now going mainly to China, Japan and the Republic of Korea, rising to 5 million m³ or even more, and those of sawnwood to an even higher level. Present markets for sawnwood are mainly in Latin America and West Asian countries. It should be noted, however, that, in terms of value, exports of woodpulp in 1983 accounted for half of the total exports of coniferous forest products (excluding non-wood forest products, such as mushrooms).

The intention is to encourage further development and expansion of forest-based industries, and detailed plans have been worked out for raising the installed processing capacity from the present figure of over 10 million m³ (roundwood consumption) to over 40 million m³ (assuming a continued expansion of the afforested area to over 2 million ha and regular reforestation of harvested areas), with a significant share of pulp production.

17.5.2.2 NEW ZEALAND

The present area of plantation forest in New Zealand, largely of Radiata pine, is comparable in size to that of Chile, about 1 million ha. Plantation activity, already quite intense between 1920 and 1940, but largely interrupted between 1940 and 1960, was later taken up again on a major scale. This is the main reason for the difference in round-wood production between New Zealand and Chile (table 17.21).

New Zealand exports a significant share of the products of its plantation forests (a relatively larger part than Chile) in the form of logs. The markets for logs are mainly Japan and China. In 1973, exports of logs, at 1.93 million m³, had reached a peak level, but have declined since then to only about half that volume, with marked year-to-year fluctuations. In contrast to logs, exports of sawnwood and other processed products continued to rise strongly during the 1970s.

It has been estimated that there are 1.5 to 2.0 million ha of land in New Zealand that could be used for plantation forest, and plans for continued planting activity exist, although at steadily declining intensity over the next few decades. To what extent these may be realized cannot be assessed with any degree of certainty. Among the factors that will have to be taken into account, are the prices that can be obtained on the world markets. In recent years, it has been claimed that these did not encourage investment in forestry in New Zealand. Furthermore, the cost factor, in particular for labour, tends to put New Zealand at a disadvantage compared to countries such as Chile. There has also been increasing environmental resistance to the clearance of native forest for exotic plantations.

One factor encouraging new plantation in New Zealand is the adjustment in world markets for agricultural products, to which New Zealand is a major exporter. As New Zealand's access to some markets, notably the EEC, is restricted by changing trade policies, a certain agricultural over-capacity is apparent. One solution to this problem would be to convert some of the lands used for agriculture to forestry, unless other markets for agricultural products could be developed.

It has been officially estimated (*The Forestry Sector in New Zealand,* New Zealand Forest Service, 1980) that total removals, at present below 10 million m³/year, would be over 20 million m³/year in 2000.

Finally, in this context, attention needs to be drawn to plantation activity in *Australia*, which since the 1960s has also seen the establishment of extensive pine plantations (Radiata and some other pine species). While these have the prime objective of meeting Australian domestic de-mand, a further substantial expansion of plantation forests from the present level of less than 1 million ha could make Australia an exporter of pine-based forest products. In any case, an increase in Australian domestic supply would have the effect of diverting a part of the New Zealand and North American exports at present directed towards Australia.

17.6 FOREST PRODUCTS IMPORTERS OUTSIDE THE ECE REGION

It is necessary to review briefly the situation and outlook for importers of forest products outside the ECE region, because, if there is any shortage of supply, demand in these regions could affect the availability of forest products for the European market.

Europe is by far the largest importer of forest products in the world, accounting for about 46% of world imports. North America accounts for 20%, Japan for about 17% and the Near East (see footnote *b* to table 17.22 for defini-tion of this grouping) for 4%, Latin America for 3%, and Africa and China for 2% each.

Countries other than Europe, North America, Japan and the USSR account for a relatively small share of world imports of forest products. Taking inter-regional trade only, these countries' imports were equivalent to about 38 million m³EQ, 10% of world inter-regional trade.

TABLE 17.22

Imports of forest products 1979-81

	Wood raw material (million m³)	Sawnwood (million m³)	Wood-based panels (million m³)	Woodpulp (million m.t.)	Paper and paperboard (million m.t.)	Total (million m³EQ)
Outside ECE region						
Africa	0.7	2.8	0.5	0.3	0.9	10.8
Latin America	0.2	2.6	0.4	0.8	2.1	16.2
Asiaᵃ	84.7	10.0	2.0	3.4	4.1	135.4
of which:						
China.................	7.0	0.1	0.1	0.4	0.5	11.0
Japan.................	67.3	4.9	0.2	1.9	0.6	86.9
Oceania..................	–	1.1	0.1	0.3	0.7	5.8
ECE region						
Europe....................	46.2	33.8	10.1	11.3	16.4	231.7
North America..............	7.7	25.7	2.9	3.9	8.0	101.7
USSR.....................	0.3	0.4	0.1	0.2	0.8	4.8
WORLD.....................	139.9	76.3	16.3	20.1	33.2	506.4
of which						
Near Eastᵇ	9.5	3.7	1.1	0.1	1.0	21.7

Source: FAO Yearbook of Forest Products.

Note: This table refers to total imports, not imports from other regions (i.e. intra-regional trade is included in the totals).

ᵃ Includes Cyprus and Turkey (FAO grouping).

ᵇ Africa: Egypt, Libya, Sudan, Asia: Afghanistan, Bahrain, Cyprus, Iran, Iraq, Jordan, Kuwait, Lebanon, Qatar, Kingdom of Saudi Arabia, Syria, Turkey, Yemen Democratic Republic. This grouping, from the FAO Yearbook of Forest Products, includes two countries covered by ETTS IV (Cyprus and Turkey).

17.6.1 Japan

This section will examine the outlook for Japanese im-ports, taking into account projections for consumption and domestic supply and then mention some factors which could affect the outlook for imports by other regions.

Japanese imports are dominated by wood raw material (roundwood, chips and residues), even though imports of processed products have shown a markedly rising trend in recent years. To put Japan's imports of all assortments of wood raw material into perspective, it may be said that their volume (67.3 million m³ in 1979-81) represented 57% of the world's total imports of wood raw material, and nearly double its domestic removals. In view of the heavy dependence on imports, which account for some two-thirds of total wood consumption, and the expected future changes in the pattern of world demand and supply of timber, the Japanese authorities (Cabinet Council) approved in May 1980 the "Basic Plan for Japan's Forest Resources" issued by the Ministry of Agriculture, Forestry and Fisheries of Japan.[8] The Plan's underlying assessment of the timber situation was stated in these terms:

The situation concerning timber demand and supply is as follows:

(1) The demand for timber is not expected to increase drastically because of the continuation of moderate economic growth and the change in the structure of demand with competition from non-wood materials.

(2) Although forestry production is sluggish due to resource restrictions, insufficient improvement of the productive base, such as forest roads, a decrease of forestry labour forces and competition from imported timber, timber production is considered to increase gradually because man-made forests will steadily reach the cutting period.

(3) Timber imports make up 68% of total supply in Japan. While the present import level will continue for the time being, it will become variable and begin to decrease in view of the tangible decrease of the supply of South-Sea timber, the shift to second-growth timber in America, and the transition to remote forest cutting sites in the Soviet Union.

Under such a situation, the demand for timber in the future, while its fundamental trend will continue for the short term, will change its aspect when the supply of foreign timber begins to level off or to decline. Furthermore, the structure of demand will change.

The ratio of domestic supply to total demand for timber in Japan will rise at that time when the supply of foreign timber begins to level off. The supply of domestic timber will increase gradually as the cutting period for young, man-made domestic forest arrives.

It is also expected that with the increasing scarcities in oil resources through the end of this century, forests, as renewable resources, will gain importance.

The Plan calls for the improvement of the quality and utilization of the country's forest resources and for the adoption and continuous implementation of a forest management system which should organically increase public benefits and at the same time improve timber production. The improvement of the forest resource is to be effected primarily through intensified management of existing forests rather than their extension in area (the total forest area of just over 25 million ha is expected in fact to decrease slightly). Low-stocked natural forests are to be converted to man-made forests and the stated goal is to increase the latter's share in total forest area from 38% in 1976 to over 50% by the mid-1990s. Growing stock is to be raised by over half to nearly 140 m³ per ha. This should result in a steady rise in the volume of annual fellings, which after the mid-1990s should be 50% higher than in 1976 (38 million m³). The Plan also sets goals for the various functions other than wood production and for management techniques for different functions.

---

[8] The document is reproduced in full in: *Review of Long-Term Trends and Prospects for the Forest and Forest Products Sector in Selected Countries Outside Europe:* Supplement 6 to Volume XXXIV of the *Timber Bulletin for Europe*, Geneva, November 1981. A new assessment of the situation by the authorities of Japan to take account of developments since 1980 is expected for 1986.

TABLE 17.23

Japan: trade in forest products by product groups, 1964-66 to 1979-81

	Unit (million)	1964-66	1969-71	1979-81
Exports				
Roundwood[a]	m³	–	–	–
Sawnwood	m³	0.3	0.2	0.1
Wood-based panels	m³	0.4	0.4	0.2
Woodpulp	m.t.	–	–	0.1
Paper and paperboard	m.t.	0.3	0.5	0.8
Imports				
Roundwood[a]	m³	17.3	42.3	52.5
Sawnwood	m³	1.1	2.4	4.9
Wood-based panels	m³	–	0.1	0.2
Woodpulp	m.t.	0.6	0.8	1.9
Paper and paperboard	m.t.	0.1	0.1	0.5
Net trade[b]				
Roundwood[a]	m³	– 17.3	– 42.3	– 52.4
Sawnwood	m³	– 0.7	– 2.2	– 4.8
Wood-based panels	m³	+ 0.4	+ 0.2	– 0.1
Woodpulp	m.t.	– 0.6	– 0.8	– 1.8
Paper and paperboard	m.t.	+ 0.2	+ 0.4	+ 0.2

[a] Including wood residues, chips and particles, and fuelwood.
[b] + = net exports; – = net imports.

In connection with the "Basic Plan for Japan's Forest Resources", long-range demand and supply projections up to the year 1996 were also made. (The authors assume timber shortages worldwide.) The main forecasts are set out in table 17.25.

TABLE 17.24

Japan: removals by main assortments, 1965 to 1980
(Million m³(r))

	1965	1970	1976 [a]	1980
TOTAL	59.2	49.8	35.9	34.4
Industrial wood	49.5	45.4	35.3	34.1
Sawlogs	34.7	28.1	22.0	21.5
Coniferous	28.3	22.2	17.9	17.8
Broadleaved	6.5	5.9	4.1	3.7
Pulpwood (round and split)	8.4	14.8	11.7	11.3
Coniferous	4.0	3.1	2.4	2.7
Broadleaved	4.5	11.7	9.3	8.5
Pit-props and other industrial wood	6.3	2.4	1.6	1.3
Coniferous	2.7	1.4	1.1	0.9
Broadleaved	3.6	0.9	0.4	0.5
Fuelwood	9.7	4.4	0.6	0.4
Coniferous	0.1	–	–	–
Broadleaved	9.6	4.4	0.6	0.4

Source: FAO. 1976 removals data not exactly comparable with those from Japanese sources in table 17.25.

[a] 1976 is shown for comparison with Plan forecasts.

TABLE 17.25

Japan: demand and supply projections for forest products
(Million m³)[a]

	1976 (actual)	1986 (projection)	1996
Timber demand			
Sawnwood	57.4	62.6	65.4
Pulpwood	29.6	36.1	44.9
Plywood, fibreboard and particle board	12.8	14.9	17.6
Other uses	4.6	4.8	5.3
Total	104.4	118.4	133.2
Timber supply			
Domestic supply	38.2	46.2	57.7
Imports	66.2	72.2	75.5
Total	104.4	118.4	133.2

Source: Basic Plan (see text). Data for 1976 removals not exactly comparable with FAO data in table 17.24.

[a] Equivalent to log volume; mill residues are not included in the figures.

17.6.2 Outlook for imports by other regions

Unfortunately, data of the type presented above for Japan (i.e. on the outlook for consumption, domestic removals and imports) are not yet available for other countries. Work carried out so far in the context of the FAO programme of outlook studies has resulted in projections of consumption of forest products, which are summarized in table 17.26 (for methodology of FAO study, see chapter 11).

TABLE 17.26

Projections of consumption of forest products in selected regions

			Change 1981 to 2000	
	1981	2000	Volume	Percentage per year
A. *Sawnwood* (million m³)				
Africa	7.2	12.4	+ 5.2	+2.8
Latin America	26.7	50.6	+23.9	+3.3
Near East	8.0	17.4	+ 9.4	+4.0
Far East (developing market economies)	25.7	57.7	+32.0	+4.1
Asia (centrally planned)	22.9	54.7	+31.8	+4.5
Oceania	6.2	7.6	+ 1.4	+1.0
B. *Wood-based panels* (million m³)				
Africa	1.0	3.5	+ 2.5	+6.5
Latin America	4.4	17.0	+12.6	+7.0
Near East	2.0	6.6	+ 4.6	+6.2
Far East (developing market economies)	3.4	20.3	+16.9	+9.4
Asia (centrally planned)	1.0	5.1	+ 4.1	+8.5
Oceania	1.2	2.3	+ 1.1	+3.3
C. *Paper and paperboard* (million m.t.)				
Africa	0.9	2.1	+ 1.2	+4.3
Latin America	9.2	21.4	+12.2	+4.3
Near East	1.8	4.4	+ 2.6	+4.6
Far East (developing market economies)	6.1	21.1	+15.0	+6.4
Asia (centrally planned)	6.4	20.2	+13.8	+5.9
Oceania	2.4	3.4	+ 1.0	+1.7

Source: FAO. In this table, "Africa" does *not* include the three African countries in the "Near East" grouping (Egypt, Libya, Sudan).

These projections, for almost all countries outside the ECE region, except Japan, are for significant increases:

Sawnwood	+ 103.7 million m³
Wood-based panels	+ 41.8 million m³
Paper and paperboard	+ 45.8 million m.t.

These increases are equivalent in total to about 400 million m³EQ. (World production of industrial roundwood in 1980 was 1,435 million m³, and production in the developing countries 324 million m³.) To satisfy this level of consumption would clearly require a very significant increase in production of roundwood world-wide and those countries with an export potential would endeavour to develop their resource to meet this demand.

There are however some factors which must be taken into consideration before accepting the above *projections* as a *forecast* of the future situation:

(a) The projections are based on assumptions regarding growth of population and GDP, which are of course tentative. Forecasting growth in GDP in developing countries is particularly difficult. In the last decade or so the economies of a number of developing countries have "taken off" and achieved high growth rates, but in other countries, real GDP has stagnated or even dropped. This type of uncertainty is inevitable with long-term consumption projections. These projections assume a GDP growth rate for all developing countries of 4.2% p.a. between 1980 and 2000;

(b) In addition, the projection method used (pooling of data from all countries, divided into four groups according to GDP per caput with dummy variables) assumes that there is a fairly constant relationship between national income (GDP) and consumption of forest products, applicable to all countries in the group. There is every likelihood that this relationship will change over a 20-year period in which the economies of many developing countries are expected to undergo radical changes. There is, unfortunately, no way at present of foreseeing what will be the direction of the change;

(c) Supply factors have not been taken into account. For many developing countries the high levels of consumption of forest products projected would imply either a large expansion of domestic supply or of imports. Either or both of these could imply a rise in prices, perhaps followed by substitution of forest products by other materials. Furthermore, higher imports imply the availability of foreign currency, which is severely limited in many developing countries.

Nevertheless, despite these uncertainties, the projections in table 17.26 may be considered a valid indication of *potential* demand, assuming steady growth in GDP, constant prices for forest products and no supply constraints.

From a European point of view, the most important aspect is not the level of consumption but of imports, i.e. to what extent will any increase in consumption be met from domestic sources and to what extent from imports? This question raises a number of subsidiary questions concerning the rate of depletion of the tropical forest, the possibility of establishing industrial wood plantations, the availability of currency to pay for imports, etc. It is not possible to examine these questions here, but only to draw attention to the very great uncertainty surrounding the important question of the future level of imports of forest products by developing countries, and to the major consequences for forest products trade of a significant increase in these imports.

17.7   CONCLUSIONS AND OUTLOOK

This concluding section summarizes the other sections and attempts to present the potential interactions between the developments foreseen above, and the possible implications for European supply.

The analysis will concentrate on supply *potential* and the balance outside Europe: it is intended as an input to chapter 20 which will analyse the raw material balance for Europe as a whole, where trade will be treated together with removals and recycling as part of Europe's supply of forest products.

In sections 17.7.1 and 17.7.2 the discussions in earlier sections are summarized. In section 17.7.3, possible interactions between the trends foreseen for different regions are considered. This discussion is expressed in qualitative terms because of the difficulties of preparing reliable quantitative estimates of future trade levels and patterns.

17.7.1   Outlook for exporters

In the *USSR* (section 17.2), there appears to be little prospect of a significant increase in exports for the foreseeable future, despite the large resource base, because of problems associated with developing remote parts of the country, notably Siberia. Even if removals were to increase, the resulting production might well be directed to satisfy domestic needs (*per caput* consumption levels of panels and paper and paperboard are relatively low). On the other hand, a significant reduction in exports is also unlikely, because of existing long-term supply agreements, (notably with Finland, Eastern Europe and Japan) and a continuing need for hard currency. There may well be a further shift in product mix from logs and sawnwood to more processed products and a shift in the mix of destinations towards a greater importance of Pacific Rim countries, which are better situated with regard to the Siberian resource.

For *Canada* (section 17.3.1), there appear to be two determining factors, the success or otherwise of the programmes being developed to expand sustainable timber supply, and the competitivity of the Canadian forest sector on world markets. It was noted above that, if the resource-oriented programmes are successful, Canadian removals could increase. If this occurs and the Canadian forest sector remains competitive, exports could rise. If not, it is unlikely that present levels could be surpassed. A reduction in exports is not inconceivable. There is another, very specific, possibility which must be taken into account − action by the United States, Canada's largest export market (79 million m³EQ) to restrict Canadian access to US markets. Although new markets could be found for some of the affected volumes, such action would certainly reduce Canadian total exports.

There is great uncertainty about the potential to increase exports from the *USA* (section 17.3.2). The forest industries of that country, supported by the Government, have stated their intention to expand exports, but many problems remain to be solved: developing new markets, maintaining competitivity (strongly influenced by exchange rates), finding a resource base adequate for domestic and export demand (there are doubts about the potential of the US south). In any case, exports will remain small compared to the domestic market. If demand from the latter is strong, there could be little incentive to develop exports; if it is weak, US producers could turn to overseas markets. Despite the wealth of information and analysis available for the USA, it is not possible to identify a likely trend for future exports. Resource constraints, a high dollar and strong domestic market could lead to a slight *decline* in exports, but low domestic consumption, due to low housing starts, a low dollar, vigorous export marketing and a satisfactory resource situation could lead to *expanded* exports. In recent years much more interest has been paid by both Government and industry to the opening up of overseas markets for forest products exports.

The *tropical hardwood* resource (section 17.4) seems ineluctably engaged in a process of deforestation similar to, but much more rapid than, the deforestation of Europe and many parts of North America in earlier millennia. To put this process into historical perspective, it will be recalled that the deforestation process in Europe lasted from about the fifth century B.C. to the nineteenth century A.D. In that time, forest went from being the dominant land cover (in some areas 80-90% of land) to an average of 15-20%. The negative consequences of the destruction of forest cover (erosion, landslides, soil loss, fuelwood and sawlog shortages, etc.) brought about a controlled reforestation, with a managed forest often quite different from the natural forest cover. There is, however, a major difference between historical conditions in Europe and present-day conditions in the tropics: many tropical ecosystems are very fragile and it may be much more difficult to re-establish a forest cover than it was in most parts of Europe. Some degree of further deforestation in the tropics appears inevitable until the growing populations of the developing countries can be said to have fully taken possession of their land as Europeans have taken possession of theirs. The supply of tropical hardwood for export can almost be seen as a by-product of this historical change in the relation between the populations of the countries concerned and their natural environment.

It appears therefore, as suggested in section 17.4.3 above, that the potential of the tropical hardwood producers to meet even increased demands from both domestic users and tropical hardwood imports, exists well beyond the year 2000, although not on a sustainable basis. The sources, species and qualities will change, as one resource is depleted and tropical hardwood importers turn to other sources. Governments of producing countries are also encouraging local processing in order to export sawnwood, plywood or further processed products instead of logs. The main factor determining when this "timber mining" will stop will be the decisions of Governments of producer countries as to what forest type and area they wish to maintain on a sustainable basis. When these decisions are taken and, equally important, made effective, the availability of tropical hardwoods will no doubt decrease abruptly; some countries have already reached this point: others will reach it in the twenty-first century. A temporary increase also cannot be ruled out. It is almost impossible, with the data available, to estimate the speed and extent of changes in exports from tropical hardwood sources.

Some *tropical plantations* (section 17.5.1) are in the same countries as the natural hardwood forests, but are different in all respects; they are intensively managed, very productive, and uniform as regards species and age composition. As mentioned above, their potential is enormous, although there is some doubt as to whether it will be achieved because of silvicultural, economic and social problems. To evaluate export potential, domestic demand must also be taken into account. However it is likely that *if economic conditions are right* (i.e. that demand exists and that wood can be grown, processed and transported to market at a competitive cost), then export-oriented plantations will be developed in suitable locations. Furthermore, it can take relatively little time (little more than a decade) to plan such plantations and to bring them into full production. The success of chemical pulp exports from Brazil, based on tropical plantations, is an example of a process which could be repeated in Brazil and elsewhere. This resource has the potential to respond reasonably quickly to increased demand if the economic, social, technical and ecological problems mentioned above (section 17.5.1) are overcome.

Chile and *New Zealand* are temperate-zone countries, with very productive plantations, which have already entered world markets and announced plans for increasing their exports. *Prima facie*, the national estimates presented above (section 17.5.2) appear reasonable.

There are other countries which have a similar potential for temperate or sub-tropical plantations, but for which no quantitative estimate is possible, notably Australia and Argentina. Australia is at present a net importer, notably from New Zealand, but aims at self-sufficiency, or even net exports by 2000. This would also have implication for exports from New Zealand.

In this review of world exporting regions, the *Nordic countries* should also be mentioned. They are technically very advanced, strong in marketing and near their main markets. Their major problems are wood costs and resource constraints. The national forecasts in chapter 5 indicate that they could increase their removals by 25 million m^3.

17.7.2 Outlook for importers

On the import side, trends in *Japan* are of the greatest importance. As indicated in section 17.6.1, the 1980 Japanese "Basic Plan" foresaw a rise in timber demand between 1976 and 1996 of about 30 million m^3, but an increase in imports of less than 10 million m^3, because of an increase in domestic removals. However, previous planning documents have also foreseen a rise in Japanese removals, although this has not in fact occurred. The possibility must therefore be envisaged that imports could rise by more than 10 million m^3. Demand could also grow more slowly than projected.

Trends in imports by the *USA* will also affect availability of forest products for export to Europe from Canada. If high domestic consumption in the USA attracts higher imports from Canada, supplies to the European market could be limited. On the other hand, if the USA seeks to limit Canadian penetration of its markets, as proposed at present (winter 1985/86), many Canadian exporters might actively develop new markets, including Europe. It is not possible to estimate at present what is the likely outcome of the debate being carried on in North America. The USDA Forest Service projected, before this debate, increases in imports of about 16% (in terms of roundwood equivalent) between 1980 and 2000.

Finally, it is necessary to assess the outlook for consumption and imports in the *developing countries* (section 17.6.2): consumption is likely to grow fast in some regions, under the pressure of population, increasing literacy and rapidly growing economies. The consumption projections presented in table 17.26 may be considered to be at the top end of a range: consumption could be lower because of slower economic growth, substitution by other materials, constraints due to availability of foreign currency etc. It should also be borne in mind that a rising population does not necessarily imply higher consumption of forests products, or of other goods: it is possible that per caput consumption may fall, due to supply constraints, even if it is already low by world standards. It must be admitted that there is an exceptional degree of uncertainty surrounding future consumption levels in developing countries, which inevitably affects the analysis of availability for export to Europe.

Among developing countries, specific attention should be paid to the *Near East* (defined in footnote *b* to table 17.25), the countries of the *Maghreb* and to *China*. In many countries of the Near East and the Maghreb, there has in recent years been a rapid increase in national income, mostly due to oil revenues. This has led to a construction boom in many places and sharp rises in imports of forest products (notably from Canada, Austria, southeast Asia and the Nordic countries). Although the outlook for oil revenues is now not as hopeful as before, it is likely that many countries in the Near East and the Maghreb, practically without forest resources of their own, will continue to expand their imports of forest products.

The situation of China is rather different: a huge population has created a strong latent demand for forest products which cannot be satisfied from the depleted local forest resource. Several analysts have foreseen a large increase in Chinese imports. However, this latent import demand can only become effective if the imports can be paid for (the Chinese authorities have traditionally taken a conservative attitude to overseas borrowing). It is not certain that forest products would have sufficient priority over other sectors to secure enough currency to expand imports significantly.

Thus, although it appears likely that imports by developing countries will increase, even a rough estimate of the maximum increase does not appear possible.

17.7.3 Outlook for import/export balance

It is not possible for ETTS IV to construct a "balance" of the outlook for world imports and exports, in order to give a quantitative answer to the question "Will sufficient supplies of forest products be available on world markets to satisfy any increased import demand by Europe?" It is however possible to make a rough assessment of the situation in qualitative terms.

The potential for *exports* may be summarized as follows:

- No dramatic changes, upward or downward in the export supply potential of traditional suppliers, USSR, Canada, natural tropical hardwood areas. (The Nordic countries, although inside Europe, may also be considered part of this group). Although supplies from the tropical forests will drop, this may not occur before the end of the century;

- Uncertainty for the USA;

- Significant expansion from Chile and New Zealand;

- Enormous potential from tropical plantations, in the right economic and social circumstances.

For the traditional suppliers, it should be pointed out that some of the uncertainty for individual countries or regions depends on the outlook for international competitivity. As competitivity is a relative concept, they cannot all become uncompetitive simultaneously, so that there is less uncertainty for the group as a whole than for its individual members.

The potential for *imports* may be summarized as follows:

- An increase in Japan, the size of which will depend on progress in developing the domestic resource;

- Uncertainty for the USA, with a sharp drop possible if protectionist action is taken;

- Enormous potential import demand from developing countries, with great uncertainty about the likelihood of this being converted to effective import demand.

This rough assessment does *not* appear to confirm the widespread opinion that a world wide "shortage" of forest products is imminent, as exporters appear to have the potential to satisfy even high import demand (if developing country imports are not at the top of the range).

The possibility of world-wide tension between forest products supply and demand would arise, however, if a strong effective import demand materialized from developing countries.

In such circumstances, there would be a strong incentive for existing exporters to increase their production beyond previously established limits and for fast-growing plantations to be widely established in tropical regions. In fact, if there were a very strong increase in *effective* demand from developing countries, it could be largely satisfied *only* from these tropical plantations; the supply possibilities of these plantations would constrain the import demand of developing countries, although there would be widespread consequences for traditional exporters and importers – notably a rise in real prices and probably substitution for forest products by other materials.

Finally, there will of course be no "shortage" or "gap" between imports and exports as total imports, world-wide, will always equal total exports. How would the adjustment of any tension which might arise be carried out? First, developing country imports are very vulnerable to supply factors. If there is difficulty for these countries in obtaining supplies in the world market, because of lack of foreign exchange, or because of competition for available supplies (e.g. from Japan or Europe), this could constitute a constraint on the level of their imports, which would not then attain the levels suggested. Second, some exporters, notably the tropical plantations, appear to have great flexibility to respond to higher levels of demand over relatively short time periods (5-10 years). One consequence of this reasoning is that there is an urgent need for assessments of future import demand for forest products in developing countries, taking into account their potential to satisfy it from local sources. If the effective import demand from developing countries were to become very strong, exports from other suppliers could probably also be increased by more intensive management, genetic improvement, etc., although the reaction time for non-tropical countries to increase their supplies is no doubt longer than for tropical plantations: this might lead to temporary market tensions.

In short, the secretariat believes the world forest products markets have the potential to be quite flexible in the medium to long term and to have the potential to adapt even to significantly changed situations. Whether this potential flexibility is realized depends on many economic and institutional factors, including the speed with which signals of impending change are understood and acted on.

These rough evaluations do not take quality, size or species of raw material into account, as there is practically no basis on which such a calculation, even in crude terms, could be founded. It is likely, however, that shortages and surpluses of particular products will occur in future, within the broad balance, despite the possibilities of changing the raw material base for individual products (i.e. sawing smaller logs, making pulp from different species etc.). In particular it is almost certain that high quality logs, whether from the north of the Nordic countries, the Pacific Coast of North America, old European hardwood stands or the natural tropical forest, will become increasingly rare and expensive. No substitute for these qualities is in sight in the short to medium-term, and renewable supply of these grades either demands very long rotations, or presents major silvicultural problems (e.g. managing tropical forests to produce logs equivalent to those being extracted at present from natural forests).

Chapter 20 will discuss Europe's potential demand for imports from other regions, and place it in the context of the world outlook, as briefly outlined above.

CHAPTER 18

Trends in the use of wood for energy, and the situation around 1980

18.1 INTRODUCTION

For over a quarter of a century, from shortly after the second world war until 1973, fuelwood was steadily replaced in Europe by other fuels: there seemed until then no reason to expect a change in this trend. The "energy crisis" of 1973-74 radically changed the situation and the general perception of the outlook for all forms of energy.

ETTS III, which was drafted shortly after the start of the "energy crisis", drew attention to these fundamental changes but did not have the statistical or analytical framework on which to base an analysis of the outlook for the use of wood for energy. In the absence of such a framework it assumed a continuation of pre-1973 trends but stressed the uncertainty on this point.

In the late 1970s and early 1980s, the Timber Committee undertook a number of activities to improve the understanding at the international level of the situation and prospects as regards the use of wood for energy. A seminar and a symposium[1] were held, and a team of experts was constituted, which helped to bring about major improvements in the international data base and ar-

ranged an intense exchange of information and experience on many aspects of wood and energy. As a result of the team's work, an *ad hoc* meeting was held in May 1983 on the impact of energy developments on the forestry and forest products sector. The objective of this meeting was to identify the main factors relating to wood/energy developments that needed to be taken into account in preparing ETTS IV.

The secretariat wishes to express its deep gratitude to the members of the team of specialists whose work provided the foundation for the analysis in chapters 18 and 19, as well as to the correspondents who prepared the replies to the enquiry which are presented below.

The work on chapters 18 and 19 was carried out with the valuable help of Mr G. A. Morin, generously put at the secretariat's disposal by the Government of France.

---

[1] Seminar on energy aspects of the forest industries, Udine (Italy), November 1978. Symposium on energy conservation and self-sufficiency in the sawmilling industry, Bonn (Federal Republic of Germany), September 1982.

18.2 OVERVIEW OF BROAD TRENDS IN ENERGY SUPPLY AND DEMAND

Until 1973, the price of oil had declined steadily in real terms for many years. This development, combined with the great convenience in use of oil and its products, had caused oil to reach a predominant place in the world energy market.

The political and economic events of 1973 led to a quadrupling of oil prices and temporary supply shortages. The shock which these events induced led all governments to develop new energy policies, aimed in most cases at reducing energy costs, encouraging conservation, diversifying energy sources, reducing energy imports and developing new and renewable sources of energy. Several major research and development programmes were initiated.

The global energy situation has in fact changed in many

important ways since the early 1970s, partly because of national adjustment to changed economic circumstances and partly because of the energy policies mentioned above. Energy conservation measures have been very successful (the GDP elasticity of total energy consumption, which was over 1.0 in most countries, has fallen to well below 1.0 in many); new oil resources especially in non-OPEC countries have been developed and come on stream; oil imports, especially from the OPEC countries, have fallen; other sources of energy such as coal, gas, or in a few countries, nuclear power, have been developed. A strong impetus was given to these developments by the second oil price rise at the end of the 1970s, which was in fact greater than the first, in dollar terms at least.

As regards new and renewable sources of energy, although many large research programmes and pilot pro-

TABLE 18.1

Consumption of fuelwood, 1949-51 to 1979-81, as reported to ECE/FAO[a]

	1949-51 (million m³)	1959-61 (million m³)	1969-71 (million m³)	1978 (million m³)	1979-81 (million m³)	Annual average percentage change 1950 to 1980
Nordic countries	26.4	19.8	11.8	5.6	6.1	−4.7
EEC(9)	34.8	23.4	11.5	10.1	11.8	−3.5
Central Europe	3.5	2.8	2.2	1.8	2.3	−1.4
Southern Europe	35.4	31.6	30.3	20.3	22.4	−1.5
Eastern Europe	21.6	15.9	13.4	11.2	11.1	−2.2
Europe.........................	121.8	93.6	69.3	48.9	53.8	−2.7

[a] Real consumption is significantly higher (see text and table 18.2), but no consistent time series are available.

jects have been undertaken, most of these sources have not yet been applied to a significant degree. Exceptions to this include hydro-electricity, passive solar heating, and wood.

The situation on world energy markets changed dramatically, however, in 1985 and 1986, when oil production overcapacity, together with successful conservation measures brought about a collapse of the crude oil markets. The price of oil fell below $US 10/barrel in spring 1986, as compared to over $US 30/barrel a year earlier. If this low price were to be maintained in the medium to long term the outlook for the energy sector would be radically different from that foreseen in the early 1980s, which assumed in most cases an increase in energy prices in the long term or at least stabilization of prices at the then prevailing levels. In particular the economic incentive for energy conservation and development of non-oil fuels would be nearly destroyed. Are the very low oil prices of early 1986 a temporary phenomenon? Or will the situation persist in the medium to long term? It is not possible yet to answer these important questions. The analysis which follows was prepared before the sharp drop in oil prices, and it has not been possible to modify it substantially at the final stages of revision. The possibility of lower oil prices, and the resulting disincentive to develop the use of wood for energy cannot however be excluded.

The long-term decline in recorded fuelwood removals slowed down during the 1970s and, from 1978, recorded European fuelwood removals started to rise again (see figure 18.1), without, in most cases, any significant fiscal stimulus or the support of large-scale R and D programmes (which mostly concentrated on larger-scale projects, which have not yet been put into practice).

However, it became apparent during the 1970s that recorded fuelwood removals are only a partial and possibly misleading indicator of how much wood is used as a source of energy.

Much of the wood harvested for fuel is done so in rural areas for autoconsumption, i.e. the farmer, forester or other inhabitant of a rural area who harvests the wood burns it himself. Nor does all the wood burnt come from the forest: wood from hedgerows, isolated trees, or fruit orchards, is also burnt. For obvious reasons, these volumes are usually not recorded. In many countries they were not even estimated until recently.

Energy is also derived from other forms of wood, mostly residues of manufacturing or processing. Sawmills, panel mills, furniture and joinery plants, among others, in many cases derive energy from their residues of wood and bark. A special case is that of chemical pulping, which separates the cellulose part of the wood, (which will become chemical pulp) from the lignin and hemi-cellulose. These latter, which, along with the pulping chemicals are suspended in water, are burnt in special boilers to provide process heat and to recover the expensive chemicals, which are then re-used. In fact, in many cases, chemical pulping provides a surplus of energy which can be used elsewhere, for instance in a paper-making operation. In addition, much wood is burnt for energy after its original use: this is notably the case for wooden packaging, including pallets, and wood waste from demolished buildings.

The ETTS IV correspondents were asked, in 1984, to provide estimates of the volumes of wood used as a source of energy in 1970 and 1980, as well as to estimate the outlook to the year 2000. Almost all correspondents were able to reply. Their estimates for 1980 are presented in table 18.2 (estimates of future trends are presented and discussed in chapter 19) and for all years (i.e. 1970, 1980, 1990 and 2000) in the annex tables to chapter 19. Due to the nature of the estimates, these data cannot be considered very precise but it is likely that the general picture is correct.

On the basis of the partial data available, which are summarized in tables 18.2 and 18.3, it is possible to make estimates for Europe as a whole and for individual country groups, in 1970 and 1980. These estimates are presented in table 18.4. Despite the need for estimation, these tables seem to represent a fairly realistic picture of developments in the 1970s.

It is estimated that 123 million m³ of wood was used as a source of energy in 1980. Of this estimated total less than 60% was accounted for by fuelwood removals, most of which were from non-coniferous species. Over 20 million m³ of residues from sawmilling and the manufacture of wood-based panels were used as a source of energy, as well as significant amounts of residues of secondary wood processing and pulp and paper manufacture, and recycled wood.

FIGURE 18.1

Europe: use of wood for energy

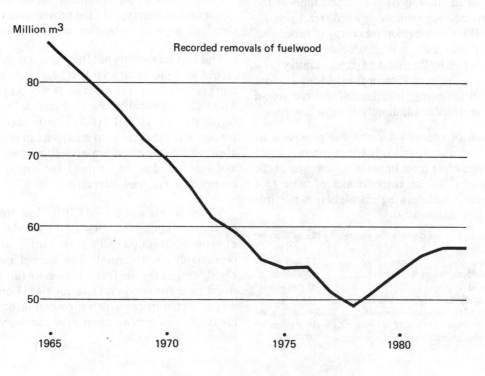

Recorded removals of fuelwood

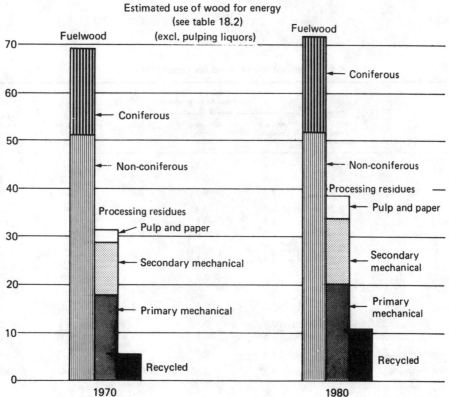

Estimated use of wood for energy
(see table 18.2)
(excl. pulping liquors)

There are significant differences between developments for the different assortments. Growth in European consumption was slowest for the most important wood energy assortment – non-coniferous fuelwood, which was only marginally higher at the end of the decade than at the beginning, after the long-term downward trend was reversed around 1978. Consumption for energy of processing residues grew by over 20% over the decade. The use of wood and bark as fuel by the pulp and paper industry grew particularly fast – by over 70% in the decade. The use of recycled wood for energy is estimated to have grown by 67% over the decade, admittedly from a low base.

The generation of energy was, with the provision of shelter, the earliest use of wood. It is not often realised that it remains one of its most important uses even in the developed countries. Taking the estimates of table 18.4 as a starting point, the following calculation is possible for Europe as a whole in 1980:

Wood (including bark) used directly for energy ... 123 million m³
of which:
    Fuelwood .... 72 million m³
    Industrial residues (wood and bark) .... 40 million m³
    Recycled wood .... 11 million m³
Total removals, underbark (adjusted to take account of unrecorded fuelwood removals) ...... 360 million m³
Total removals, overbark (estimate) .... 410 million m³
Percentage of removals used directly for energy .. 30

Estimated wood equivalent of pulping liquors .... 44 million m³
Total wood for energy (direct use plus liquors) ... 165 million m³
Percentage of removals used as a source of energy 41

If over 40% of the volume of wood removed is used as a source of energy, then the provision of energy remains the single most important use of wood, in volume terms.

This fact is probably not fully appreciated by many concerned with the forest sector. It bears out the opinion that there is very little real wastage of wood as almost all the wood (and, probably, bark) which is harvested either forms part of a final product or is used as an energy source. The importance of energy as an end-use for wood also justifies the attention now being paid, by ETTS IV and other studies, to demand for energy wood and to energy/raw material interactions.

Where is this wood used? It is clear from the above-mentioned enquiry that most of the wood used as a source of energy (around 60%) is burnt in households, presumably mostly rural. The second most important wood burners are the forest industries (under 30% of the wood used for energy). Data for the 17 countries (from all parts of Europe) which provided information on the breakdown of consumption of wood-derived energy are shown in table 18.5.

TABLE 18.2

Estimated use of wood for energy, 1980

| | Fuel-wood (million m³) | Residues of industry (wood and bark) (million m³) | | | Recycled wood (million m³) | Total wood and bark (million m³) | Pulping liquors [a] (1000 TJ) |
		Primary mecha-nical	Secondary mecha-nical	Pulp and paper			
Finland	3.7	1.6	0.4	1.2	–	6.9	50
Norway	1.8	0.6	0.1	0.2	0.2	2.8	3
Sweden	4.5	2.2	0.5	2.0	1.5	10.7	97
Denmark	0.2	0.3	0.4	–	–	1.0	..
France	10.0	1.5	0.8	0.6	3.0	15.9	14
Germany, Federal Republic of	3.6	2.0	5.0	0.4	0.5	11.4	11
Ireland	0.2	–	–	–	–	0.3	..
United Kingdom	0.2	0.5	0.1	–	–	0.8	..
Switzerland	0.8	0.2	0.2	–	1.0	2.2	1
Portugal	5.4	0.7	–	0.3	–	6.5	8
Spain	2.0	1.8	0.2	0.4	0.1	4.5	10
Bulgaria	1.9	0.3	0.1	–	–	2.4	1
Poland	1.9	1.5	0.8	0.1	0.3	4.6	7
13 countries with complete data	36.3	13.3	8.7	5.2	6.7	70.1	205*
7 countries with data for fuelwood[b]	29.7	3.9*	2.6*	0.4*	2.4*	39.0*	22*
8 countries which did not supply data[c]	6.0	3.3*	2.1*	0.5*	1.7*	13.6*	28*
Europe......................	72.0	20.5*	13.4*	6.1*	10.8*	122.8*	255*

[a] The questionnaire did not specify whether gross energy content of liquors or net (delivered) energy, taking account of boiler efficiencies, was required. Comparison of replies indicated that countries had interpreted the question differently. The data here have been adjusted by the secretariat to give estimated net values. The original data are reproduced in annex table 19.12.

[b] Czechoslovakia, Greece, Hungary, Israel, Italy, Turkey, Yugoslavia. Some of these also have data for some other types of wood energy (see annex tables).

[c] Austria, Belgium-Luxembourg, Cyprus, German Democratic Republic, Iceland, Malta, Netherlands, Romania.

Definitions: See note to table 18.3.

TABLE 18.3

Share of different types of energy from wood and bark, 1980
(Percentage of total, excluding liquors)

			Residues of industry (wood and bark)		
	Fuelwood	*Primary mechanical*	*Secondary mechanical*	*Pulp and paper*	*Recycled wood*
Finland	54	23	8	17	–
Norway	62	20	4	6	8
Sweden	42	21	5	19	14
Denmark	25	30	40	–	5
France	63	9	5	4	19
Germany, Federal Republic of	32	18	44	3	4
Ireland	75	23	–	–	1
United Kingdom	19	66	12	1	1
Switzerland	36	9	9	–	44
Portugal	84	11	–	5	–
Spain	44	40	4	9	2
Bulgaria	79	14	6	1	1
Poland	42	33	17	2	6
13 countries	52	19	12	7	10
Europe (estimate)	59	17	11	5	9

Source: Table 18.2.

DEFINITIONS:

Fuelwood: Wood in the rough, from the forest or elsewhere, used for energy.
Residues of industry: Wood and bark used as source of energy, arising from the operations of the following industries:

– Primary mechanical processing, defined as manufacture of sawnwood and wood-based panels;
– Secondary mechanical processing, defined as further processing of sawnwood or panels or manufacture of products from them, e.g. furniture or joinery;
– Pulp and paper industry, defined as the manufacture of all types of pulp and paper (pulping liquors are treated separately from wood and bark residues of the pulp and paper industry);

Recycled wood: Forest products used for energy after their original use, e.g. used pallets, wood from the demolition of buildings.
Pulping liquors: Lignin and hemi-cellulose separated from cellulose in chemical pulping and used for energy.

TABLE 18.4

Europe: estimated use of wood and bark for energy, 1970 and 1980 (excluding liquors)

	1970 (million m³)	1980 (million m³)	Change 1970 to 1980	
			Volume (million m³)	Percentage
Fuelwood	69.1	72.0	+ 2.9	+ 4
of which:				
Coniferous	17.8	20.0	+ 2.2	+12
Non-coniferous	51.3	52.0	+ 0.7	+ 1
Processing residues (wood and bark).	32.7	40.0	+ 7.3	+22
of which:				
Primary mechanical	17.8	20.5	+ 2.7	+15
Secondary mechanical	11.1	13.4	+ 2.3	+21
Pulp and paper	3.8	6.1	+ 2.3	+61
Recycled wood	5.8	10.8	+ 5.0	+86
Total	107.6	122.8	+ 15.2	+14

TABLE 18.5

Reported uses of energy from wood, 1970 and 1980

	Volume (million m³)		Percentage of total		Percentage change 1970 to 1980
	1970	1980	1970	1980	
Households	34.9	35.3	61.9	58.7	+ 1.1
Forest industries	15.2	17.8	26.9	29.6	+17.1
Other	6.3	7.0	11.2	11.6	+11.1
Total (17 countries with data)	56.5	60.0	100.0	100.0	+ 6.2

Source: Replies to enquiry.

Most of the "other" uses were accounted for by combustion in intermediate or large-scale units outside the forest industries (e.g. in schools, hospitals, barracks, or district heating plants), with minor volumes going to the production of charcoal. Very small volumes were used in a few European countries to generate electricity or to manufacture solid fuels (e.g. briquettes). No wood is used at present in the region to make synthetic liquid or gaseous fuels. Use of energy wood by the forest industries and "other" users has grown faster than use by households.

How important is energy from wood in national energy balances? The ECE/FAO medium-term survey of trends in the markets for pulpwood, wood for energy and miscellaneous roundwood[2] estimated (based on national replies) the contribution of energy from wood to final energy consumption, at the end of the 1970s (table 18.6).

In most countries, wood is a marginal source of energy, especially in those urbanized, industrialized economies which are heavy users of conventional forms of energy. In only a few countries does wood play a significant role; notably Turkey (23.3% of final energy consumption de-

---

[2] Supplement 15 to volume XXXIV of the *Timber Bulletin for Europe*, Geneva, 1982. See especially table 14. Because of the different source and the different period, these data are not fully comparable with those used elsewhere in this chapter.

TABLE 18.6

Share of wood-derived energy in final energy consumption, late 1970s

	Fuelwood and processing residues	Pulping liquors	Fuelwood, processing residues and pulping liquors
Nordic countries.	4.6	4.2	8.8
EEC(9)	0.7	0.1	0.8
Central Europe . .	1.2	1.2	2.4
Southern Europe	6.7	0.9	7.6
Eastern Europe . .	1.0	0.3	1.3
Europe	1.5	0.6	2.1

rived from wood, including pulping liquors); Finland (16.6%); Yugoslavia (8.3%); Sweden (7.2%); and Portugal (6.5%). There are major differences in the geographical and economic situation of these countries, but they mostly have major forest resources and no major domestic sources of energy. In the two Nordic countries mentioned (Finland and Sweden), use of wood residues for energy by the forest industries is very much more important than in the three southern European countries (Portugal, Turkey, Yugoslavia), where use of fuelwood by rural populations predominates.

This chapter has outlined the situation and trends for wood-derived energy, notably the volumes concerned, the importance of energy as an end-use for wood, and the contribution of wood to national energy balances. The outlook for the future is discussed in chapter 19.

CHAPTER 19

Outlook for the use of wood for energy

19.1 INTRODUCTION

It is clear from the account in the preceding chapter that:

- Developments in the use of wood for energy will affect the outlook for the forest products sector as a whole; and

- It is no longer possible to make simple assumptions for the future course of events, as circumstances have changed fundamentally over the past decade. Indeed, fundamental factors such as the price of oil, are in constant evolution.

How should the necessary analysis of the outlook for wood-derived energy be carried out? Despite the multiplicity of energy-linked studies which have been produced since 1974, no consensus has emerged on likely future developments for global energy supply and demand. Even if such a consensus on the broad outlook did exist, it is far from certain how this would affect the forest and forest products sector. Views expressed on the future for wood-derived energy in the past five years range from a conviction that the use of wood for energy will expand until the whole forest sector is changed beyond recognition, to the

belief that fuelwood use will soon resume its long-term decline towards insignificance.

The methodology and the data base were not considered sufficiently advanced to enable a modelling approach to be attempted, especially as many of the determining factors are highly local and could not therefore be integrated into a regional study.

The approach adopted is the following:

- To present the major factors which will determine future developments, and their likely evolution, while drawing attention to the high degree of uncertainty connected with some of them. The demand from conventional uses (combustion by rural households, forest industries, etc.) will be treated separately from demand for new types of wood-derived energy (e.g. methanol) and new sources of wood for energy, notably "energy plantations";

- To request national experts to evaluate the outlook for each country and to make estimates for future developments. The estimates made are presented below (section 19.4).

19.2 FACTORS AFFECTING THE OUTLOOK FOR CONVENTIONAL USES OF WOOD FOR ENERGY

For the purposes of this chapter, "conventional" uses of wood for energy are defined as those uses which account for significant volumes of wood-derived energy at present i.e. the combustion, for heat, of roundwood from forests or from trees outside the forest, of industrial wood residues and of recycled wood, in houses, forest industries and intermediate-scale (non-forest industry) units. For ease of combustion, the wood is usually either cut in short lengths and split or chipped. Compression into briquettes or pellets, which is a fairly simple mechanical process, may also be considered "conventional" although not at present widespread.

Wood energy use shows marked differences between regions due to the wide variations in forest resources and population density, and the fact that fuelwood is usually neither convenient nor economic to transport over long distances. There is a strong contrast between wood, whose

resources (the forest) are dispersed and of more concentrated energy sources, such as oil, gas or coal, which are usually obtained in large quantities from relatively few places, often transported very long distances and frequently burnt in large units such as electric power stations or heavy industry.

Perhaps the most important factor which will determine the future use of wood for energy is the development of the *overall balance between supply and demand of energy*. It was above all the easy availability and falling real prices of other forms of energy, notably oil, which caused fuelwood consumption to decline from 1950 to the mid-1970s, and the subsequent rise in energy prices which reversed this trend thereafter. For several years after the first energy crisis, most analysts foresaw a future characterized by increasingly scarce and expensive energy, due to the exhaustion of reserves of fossil fuels, especially oil and gas, and monopolistic behaviour by oil exporters. This view, however, underestimated a number of factors, notably the potential to conserve energy and to develop new sources of energy in a context of higher energy prices. Reserves of oil, gas or coal which had not been economic when the oil price was US$4 a barrel became very attractive when this price was US$30 a barrel or more, even if there was a need for enormous investments in the technology to extract oil in arctic conditions or from offshore wells. In addition, several countries which were heavily dependent on energy imports were prepared to pay a premium on their energy prices to develop domestic sources and to diversify sources of imported energy.

Conservation, the development of new energy sources and the slowdown in economic growth (attributed in part to the rise in energy prices) have led to a situation where the price of oil, still the world reference price for energy, has been dropping, in dollar terms at least, and the share of OPEC countries in world production and trade of oil has fallen sharply. In 1985 and 1986, the world oil price fell dramatically. In general, the world energy economy has proved more resilient and flexible than many expected.

What is the outlook for the future? There can, of course, be no certainty on this question; it has been the subject of numerous studies which have foreseen a wide range of different energy futures from a return to the cheap energy of the 1960s to a picture of dwindling resources, shortages and very expensive energy. However, the fact that fossil fuel resources are finite would seem to argue against the first of these extremes (notwithstanding the fact that the price of crude oil was between $10 and $15 a barrel as ETTS IV went to press) and the fact that the energy economy has proved itself very adaptable (in the medium term) against the second. Many analysts in recent years have foreseen a future of slowly rising energy prices after the decreases of the first half of the 1980s.

In particular it seems likely that there will be continuing advances in energy conservation which is immediately economically attractive to all energy consumers. The gains from conservation may however not be as rapid as in recent years, as the "easy" conservation measures have already been carried out.

In addition, demand will develop differently for different types of energy. Whereas there are many existing and potential sources of heat energy ranging from coal to solar, potential sources of energy for transport (especially road transport) are more limited. Much of the R & D effort has been devoted to developing transport fuels from sources other than oil e.g. by liquifying solid fuels such as coal but also wood. If economic or physical scarcity did occur, oil-based fuels would be increasingly reserved for these "premium" uses (as well as for petro-chemicals).

Most governments radically reviewed and revised their energy policies in the 1970s in the light of the changed situation. In most cases, they concluded that wood was for them a marginal source of energy and took no special measures to encourage or direct the generation of energy from wood, which was treated in the same way as other types of energy. A few countries, however, with a relatively high ratio of forest resources to energy needs, considered that wood was, in their particular circumstances, an important source of energy: the governments of these countries took specific measures concerning wood energy, such as the funding of significant R & D programmes. In several of these countries, a significant factor was the desire to reduce energy imports by encouraging a domestic energy source (even if this source was slightly more expensive than imported energy).

The overall energy situation, notably its price, is probably the major factor determining future demand for wood-derived energy, but there are a number of factors which affect specifically this sector. The future *availability* of wood for energy will be analysed together with the availability of wood for other uses, elsewhere in this study, but it should be borne in mind that, just as the consumption of fuelwood is often not recorded, so this material comes from sources which are not included in traditional forest inventories – parts of the tree other than the stem, trees outside the forest, hedgerows, woody agricultural residues (from fruit trees, vines, etc.). The results of the inventory of woody biomass (but not biomass from agricultural sources, such as fruit trees), have been summarized in chapter 3. Although there are no detailed data for future availability, estimates could be made by assuming a constant relationship between the total biomass growing stock and the inventoried growing stock. Such estimates would, however, be of limited validity for the following reasons:

(a) It is not known what proportion of fuelwood consumption comes from inventoried growing stock, from other forest biomass or from agricultural sources. Correspondents were requested to estimate these percentages but many were unable to do so and there was such a wide variation between those answers which were received that it is not possible to make even tentative global estimates;

(b) While national biomass inventories are a necessary part of energy resource planning, especially in the context of evaluating major biomass-based projects, conventional fuelwood availability is determined above all by local factors such as ownership régimes (communal or private), silvicultural practices, local practices, or seasonal employment patterns.

It is therefore proposed to compare changes in fuelwood demand with forecast changes in removals of small wood, while bearing in mind that part of the increased fuelwood demand could be satisfied by increased use of non-inventoried sources. Interpretation of the consistency analysis to be presented in chapter 20 will take this factor into account.

Although much of the fuelwood removed is for auto-consumption, a part is distributed through *commercial channels*. New fuelwood users are, in most cases, dependent on these for a supply of fuelwood, even in rural areas.

Certainly, no significant expansion of fuelwood use is possible, in European conditions, without a corresponding increase in harvesting and distribution services. However, the experience of the 1970s would seem to indicate that these services develop spontaneously, on a decentralized basis, if there is a demand. A technical aspect to which some governments have devoted attention is that of measurement of fuelwood and its calorific content (volume or mass, moisture content, degree of preparation, species) as uniform and rational measurement methods encourage effective and fair distribution networks. Whereas very small enterprises are usually sufficient to provide fuelwood for household use, the "intermediate" users require larger quantities on a continuous and reliable basis. Enterprises already active in a similar capacity in the forest products sector have in many cases undertaken this task, in some cases integrating it with their other harvesting activities e.g. by sorting out for energy use lower qualities or forest residues from harvesting operations for pulpwood or sawlogs.

The direct comparison of *fuel costs* between auto-consumed fuelwood and other fuels is complex because of the difficulty of evaluating the cost of harvesting and preparing the fuelwood, especially if this is done in a season when there are few other tasks for the farmer. In addition no taxes are paid on auto-consumed fuelwood, unlike other fuels (including fuelwood bought through commercial channels).

Another possible constraint on development of conventional wood energy applications is the availability of *wood-burning equipment*. The investment in appropriate equipment is often the major obstacle for households who wish to change their source of energy. In the initial stages of the recovery in fuelwood use, many users turned to old systems which had not been completely abandoned in the period of decline in fuelwood use. These, however, often proved inconvenient and inefficient from an energy conservation point of view. There was an enormous increase in the sale of modern, energy-efficient, wood-burning stoves during the 1970s. In most cases, however, these stoves (at least the smaller versions), still require hand stoking, which many users, accustomed to the convenience of modern oil- or gas-burning installations, are unwilling to accept. There has been progress in this field, however, and now completely automatic, chip-burning installations, which used to be restricted to larger units, are available for requirements as small as those of a medium-sized farm.

A related aspect is the emission of pollutants during combustion. From this point of view, wood is considered preferable to many other fuels, notably because of lower sulphur emissions. However, if combustion of wood takes place in unsatisfactory conditions (e.g. in older equipment which does not achieve full combustion) emission problems can also arise for wood-burning equipment, notably as regards particulates. Work is in hand in several countries in drawing up standards on emissions by wood-burning installations.

The development of commercial channels and of more convenient and effective wood-burning installations was a necessary condition for the expansion of fuelwood use, so that wood could compete with other fuels on its technical and economic merits. Developments over the last decade, in reaction to the "energy crisis" seem to indicate that the fuelwood sector is flexible enough to overcome this type of constraint, if the fundamental economic and resource situation is favourable to expansion in fuelwood use.

It should, however, be borne in mind that equipment constraints do impose a time lag in responding to relative price changes between fuels, as a change of fuel often implies a change of equipment. In some countries, there is a trend towards installing flexible equipment which can be rapidly converted from burning oil to burning solid fuels (wood or coal) and *vice versa*.

An important factor determining the economic accessibility of conventional fuelwood resources is the *cost of harvesting and transport* (except for auto-consumption, where economic criteria do not play a major role). These costs are higher, relative to the volume harvested, for small wood, such as fuelwood, than for logs, and even higher for the collection of logging residues. If wood is to remain competitive with other sources of energy, harvesting and transport costs (which are strongly influenced by the price of the oil-based fuels used for harvesting equipment) must not be allowed to rise excessively. In recent years, however, a number of systems have been developed to harvest small-sized wood economically — often systems which integrate fuelwood harvesting with the harvesting of other assortments.

In conclusion, consumption of conventional fuelwood could continue to rise, if the energy price also rises; harvesting, distribution and combustion are not expected to pose major problems. However, the rise is unlikely to be of such a magnitude as to transform completely the pattern of wood removals: wood is suitable as a fuel in rural areas, in decentralised systems. Larger systems imply longer transport distances and much higher costs, not to mention the problems of competition for raw material with the forest industries, which will be discussed below. Furthermore, in most countries, the wood resource is simply too small to provide more than a small part of total energy consumption. (For Europe, the entire tree biomass growing stock is equivalent to only four years' energy consumption: in the Nordic countries, it is equivalent to over 17 years, but in the EEC to only about 1.5 years.)

Most *forest industries* are now very conscious of their energy costs[1] and of the energy content of the residues of wood processing, and are therefore increasingly unwilling to allow any of their residues to be wasted — especially as the disposal of unused residues represents an additional cost. Some of the residues produced have no other rational use than as a source of energy — most bark, sander dust, material contaminated with glue or overlays, etc. The in-

---

[1] In Sweden, in 1980, purchased fuels and electricity accounted for 2.5% of the sales value for sawmills, 10.7% for pulp mills and 13.4% for the paper and paperboard industry.

dustries may, therefore, be expected to continue to review their energy balances and options and install wood-burning equipment as appropriate to the quantity and type of residues available and the specific energy needs of the industry.

Most of the uncontaminated wood residues, however, are used for the manufacture of pulp, particle board and fibreboard and form a major and increasing part of these industries' raw material input (see chapter 15). Whether any of the residues suitable as raw material are burnt by the producing industry will depend on a number of local factors, including the price offered by the potential raw material consumer, the cost and availability of other fuels, and the energy needs of the residue-producing industry. Moreover, decisions on these questions are often taken on a medium-term basis, e.g. when installing new energy equipment or drawing up raw material supply contracts; residue producers are normally not able or willing to let their decisions be influenced by short-term variations in price differentials, but will undoubtedly take these into account for their medium- and long-term planning. There is thus often an element of time lag between changes in differentials between prices for energy and raw material and changes in the distribution of the residues.

At present there has certainly been no major shift from using residues as raw material to using them for energy, although it is likely that energy prices have indirectly influenced the prices offered by buyers of residues as raw material. It seems likely that this situation will continue, unless there are very steep rises in energy prices: only marginal volumes of residues suitable for raw material (in areas where there are buyers) would actually be "diverted" for use as a source of energy, but there might well be a price effect with the "energy equivalent" price effectively setting a minimum price for wood residues.

It is also likely that, in the future, as at present, those who buy or sell wood raw material, whether from the forest or the forest industries, will take fully into account, in their economic calculations and negotiations, not only the physical product which can be derived from the material but also the energy which can be generated.

It has also been suggested that governments take measures to ensure that wood suitable for use as raw material is not used for energy, thus introducing an element of raw material allocation which is not normally found in market economies. No such measures have been taken, however, both because the problem has not been as acute as was feared and because governments in market-economy countries have preferred not to distort market forces. In some countries (e.g. Sweden) the legal framework has been modified to ensure that medium- to large-scale wood-burning enterprises are subject to the same planning regulations as all other wood-using plants over a certain capacity.

19.3 FACTORS AFFECTING THE OUTLOOK FOR NEW USES OF WOOD FOR ENERGY

The developments in conventional wood-derived energy described above may be considered normal reactions of the forest and energy economies to a higher price for energy. In most cases, it was not government subsidy or governmental directives which were the cause of increased wood energy use, and little was required in the way of large-scale R & D projects. Most technical developments were improvements on existing technology, not radical new departures.

However, in the ferment of new ideas which followed the first "energy crisis", and was sustained by the second, a number of proposals were made for radically new systems for deriving energy from wood or other biomass, (alongside proposals concerning other new and renewable sources of energy). Major programmes were set up by many governments to review and assess these ideas and to undertake the necessary research and development. At present, very few of the proposals in the field of energy from wood biomass have passed the pilot plant stage, but this is not surprising in view of the long lead times always associated with such ambitious projects. The object of this section is to review these proposals and some of the factors which will determine whether or not they are widely applied, and thereby result in radical changes in the demand for wood.

As mentioned above, much R & D work on energy is concentrated on finding replacement for a few fuels with special characteristics, notably liquid fuels for transport. Most of the new systems proposed for wood energy have in common that wood is not simply burnt to provide heat but processed, for instance into methanol, ethanol, gas, or electricity, to replace other forms of energy which, it is feared, will become in future scarce and/or expensive. There is a loss in energy content as well as extra costs, sometimes substantial, due to this processing, which, it is claimed, will be compensated by the desirable qualities of the new type of energy produced (and the premium price which is thereby made possible).

In almost all cases, the process being investigated, if carried out on a commercial scale, would require, for technical and economic reasons, that units be rather large. This implies a correspondingly large area from which wood must be transported to the processing plant, with a corresponding increase in the costs of wood delivered to the plant. Some of the units proposed would require similar volumes of wood to those used by a pulpmill and the same care would have to be taken as for a pulpmill to identify at an early planning stage a raw material resource which was technically and economically suitable, taking into account also the requirements of existing users.

The quality requirements for raw material of some of the processes are different from those of the existing forest industries (e.g. as regards moisture content, species, bark content). Indeed, in several cases wood is only one potential raw material alongside other types of biomass including special crops and agricultural residues.

Estimations of the economic viability of all these projects are, of course, strongly affected by the assumptions about future energy prices. In all cases the processing causes costs and energy losses to increase, so that to become economic these more ambitious proposals require higher energy prices than does direct combustion of wood (which is widely carried out at the present energy price). At the energy prices of the early 1980s, few, if any, of the processes proposed for the transformation of wood biomass into other types of fuel appear to be economically viable. This situation may be radically changed, either by technical progress (e.g. to lower processing costs) or by a rise in energy prices. Furthermore significant demand for wood for energy would probably stimulate a rise in the price for wood as other wood consumers defended their raw material supply. This would have negative consequences for the economies of the wood energy enterprises.

At present, governmental attitudes towards R & D projects in the "fuels from wood biomass" field appear to fall into two groups. The first, adopted by a majority of countries, considers that these projects are unlikely to be economically viable in the near future; furthermore, national forest resources are not large enough for wood-based fuels to make a significant contribution to national energy supply. These governments consider that the most effective way of deriving energy from wood is by simple combustion and have slowed down or abandoned R & D projects in the field of wood-derived energy. A few governments, however, in countries with a large forest resource, especially where they are without other domestic sources of energy, are continuing their programmes. This decision appears to be based partly on the expectation of an eventual energy price rise but also on a desire to dispose of a national "insurance policy" if energy supplies were interrupted. In these circumstances, a replacement for premium fuels, notably oil-based, would be available relatively quickly. Furthermore, it might be possible, if there was a demand for new wood-based fuels, to market to other countries the know-how obtained from these programmes.

Another major proposal concerned with wood energy, on the supply rather than the processing side, is for specially managed short rotation plantations of wood for energy (*energy plantations*). These are plantations of fast growing species, usually genetically improved, often of hardwoods such as willow, poplar or alder, but also of softwoods. These plantations are often coppiced, with major inputs of fertilizer and in general are treated more like agricultural crops than normal forest stands. Rotations are usually short — under ten years, sometimes only two to four years. Very high yields have been achieved on experimental plots — for instance about 20 m.t. dry mat-

ter/ha/year in Sweden. It is unlikely that such high yields would be achieved on average in Europe if energy plantations were widely introduced, in part because some of them were established on good quality agricultural land, which would not be widely available for forestry. The availability of wood from energy plantations would improve the economics of the large-scale, fuels-from-wood plants mentioned above, as the higher rate of biological production would make it possible to reduce transport distances, so that a constant supply of wood would be possible from a smaller area.

Like the fuels-from-wood projects, the future economic viability of energy plantations will be determined by the price of energy and by technical progress in increasing yields, reducing costs and integrating energy plantations into broader energy systems. For these plantations, however, there is another major factor — the availability of land. Even with the very high yields achieved on experimental plots, very large areas of land would be required to make a significant contribution to energy supply. It has been estimated[2] that to provide 1% of Europe's energy consumption in 1980, 1.4 million ha of land would be required for energy plantations (assuming the high yield of 20 m.t. dry matter/ha/year). This is 0.25% of Europe's land area. If more conservative yield estimates were made, the area required would of course be greater. For instance, if 10 m.t./ha/year is assumed, 0.5% of Europe's land area would be needed to provide 1% of Europe's 1980 energy consumption. The situation is different in areas with a more favourable ratio between forest area and energy requirements, such as the Nordic countries and Southern Europe, where to provide 1% of their energy needs, only 0.07% of the land area would be required (assuming 20 m.t./ha/year).

Are these quantities of land available? Could or should they be transferred to energy plantations? Several possibilities have been suggested — worked-over peat bogs, drained marshlands, and the millions of hectares of marginal agricultural land which exist in Europe, where agriculture has been abandoned (in some cases to be succeeded by the natural extension of unmanaged forest) or is maintained artificially by subsidies. It has even been suggested that these subsidies, which have led in many cases to enormous, expensive and economically irrational agricultural surpluses, should be partly redirected to support energy production on the same land which would thus enable energy imports to the countries concerned to be reduced. This is certainly a possible course of action if governments attach sufficient importance to energy self-sufficiency and reducing agricultural surpluses while maintaining rural employment. There are, however, several obstacles to such a course, notably the fact that experience

---

[2] This estimate, and others presented in this chapter, are taken from a paper prepared by a member of the secretariat entitled "How much biomass is available for energy in Europe?" and included in the proceedings of the Fifth Canadian Bioenergy R & D Seminar, edited by S. Hasnain., Elsevier Applied Science Publishers, London and New York, 1984.

with agricultural subsidy schemes has led many governments to be unwilling to subsidise projects which are not economic (at least in the long term) without subsidies. Great concern has been expressed about rising public expenditure and persistent budget deficits. Many governments are seeking ways to reduce expenditure and subsidies, not to find new recipients for them. Furthermore, most of the marginal agricultural land is so classified because of its low fertility, which would also reduce yields of wood biomass, and thereby the economic viability of plantations intended solely or primarily for energy wood production.

Another factor to be taken into account when assessing the outlook for energy plantations is the opposition from many people, including city dwellers who visit the countryside for recreation, to any changes – permanent or semi-permanent – in land use. The experience of the

United Kingdom with afforestation of bare uplands and of France with conversion of hardwood coppice to coniferous high forest ("*enrésinement*") has proved the strength of this type of opposition, which curiously totally ignores major changes in agricultural practice such as changes from arable land to pasture or *vice versa*. The establishment of millions of hectares of energy plantations, which are not very attractive from a landscape point of view, could be expected to encounter vigorous opposition on these grounds in many countries (even though some members of the same groups are advocates of renewable energy, including that from wood).

These questions of allocation of rural land, and policy for rural areas, are of great importance for the forest sector as a whole and will be discussed further in the conclusions of the study.

19.4   NATIONAL ESTIMATES OF OUTLOOK FOR USE OF WOOD FOR ENERGY

In view of the obstacles to modelling future demand for energy wood, notably the weakness of the statistical base and the importance of local factors, it was decided to rely on the judgement of those in contact with the situation at the national level. A rather elaborate enquiry was therefore circulated to ETTS IV correspondents requesting estimates for the present situation and the outlook to 2000 for the consumption and supply of wood-derived energy. For supply, estimates were requested on fuelwood (coniferous and non-coniferous), residues from primary and secondary mechanical wood processing, residues of the pulp and paper industries (pulping liquors, wood and bark) and recycled wood products. For consumption, estimates were requested for use by households, forest industries, intermediate-size users, and for electricity generation, solid wood-based fuels, and liquid and gaseous fuels. Correspondents were requested to check the coherence of their replies (notably that total supply was equal to total consumption) and to provide the reasoning on which their estimates were based. Replies were received from 22 European countries, as well as Canada and the USA. Most of the replies were very complete (national data are presented in annex tables 19.1 to 19.12). The secretariat wishes to express again its deep gratitude to the correspondents who undertook this difficult task.

Correspondents were requested to provide estimates for supply and consumption of wood-derived energy under two hypotheses for the general development in the energy sphere, which were defined as follows:

Hypothesis A (high demand scenario)

This scenario is based on an economic environment conducive to buoyant energy demand growth. Its basic assumption is that energy prices would maintain their real value in the long run, i.e. increase at the rate of inflation, between the mid-1980s and the end of the cen-

tury. Until 1985, however, real oil prices are assumed to decline. Economic growth would be relatively high (an average of about 3.2% in OECD countries from 1985 to 2000).

Hypothesis B (low demand scenario)

In this scenario energy demand growth would be dampened by gradually rising oil prices and subdued economic growth. The scenario assumes a 3% annual increase in the real energy price after 1985. For the immediate future, a price decline is assumed. Growth rates would be lower than in hypothesis A (an average of about 2.7% between 1985 and 2000 for OECD countries).

It is interesting to note that correspondents drew different conclusions as to the effect of these general circumstances on wood-derived energy. Whereas many expected consumption of wood-derived energy to be higher under hypothesis B, a significant minority foresaw an opposite development; other correspondents provided only one scenario. There were also differences of interpretation by product, frequently justified by detailed arguments.

It is unfortunately not possible to examine in detail in ETTS IV, for reasons of space, the reasoning behind the different scenarios, in different countries, although such a discussion would be of great interest. The existence of such a diversity of interpretations is further proof (if such proof were needed) that there is no consensus at present as to how general energy developments will affect wood-derived energy. In this chapter, no attempt has been made to impose a uniformity of approach and the data presented are as the correspondents supplied them.

The summary of these estimates in table 19.1 (including only those countries that provided complete forecasts)

TABLE 19.1

National forecasts for use of wood and bark as a source of energy in 13 selected countries

| | 1980 (million m³) | 2000 ª (million m³) | | Change 1980 to 2000 | | | |
| | | Hypo-thesis A | Hypo-thesis B | Volume (million m³) | | Annual average (percentage) | |
				A	B	A	B
Finland	6.9	7.9	9.3	+ 1.0	+ 2.4	+0.7	+1.5
Norway	2.8	3.5	3.7	+ 0.7	+ 1.0	+1.1	+1.5
Sweden	10.7	16.1	20.0	+ 5.4	+ 9.3	+2.1	+3.2
Denmark	1.0	0.8	1.5	− 0.2	+ 0.5	−0.8	+2.1
France	15.9	20.3	29.0	+ 4.4	+13.1	+1.2	+3.0
Germany, Federal Republic of	11.4	20.7	21.7	+ 9.3	+10.3	+3.0	+3.3
Ireland	0.3	0.6	0.7	+ 0.3	+ 0.4	+3.0	+4.0
United Kingdom	0.8	1.6	1.1	+ 0.8	+ 0.3	+3.7	+1.6
Switzerland	2.2	4.7	2.7	+ 2.5	+ 0.4	+3.8	+0.9
Portugal	6.5	11.2	10.7	+ 4.7	+ 4.3	+2.8	+2.6
Spain	4.5	8.6	7.5	+ 4.1	+ 3.0	+3.3	+2.6
Bulgaria	2.4	2.1	2.1	− 0.3	− 0.3	−0.7	−0.7
Poland	4.6	5.9	5.2	+ 1.4	+ 0.7	+1.3	+0.7
Total (13 countries) *of which:*	70.1	104.1	115.3	+34.0	+45.2	+2.0	+2.5
Fuelwood	36.3	50.9	60.8	+14.6	+24.5	+1.7	+2.6
Primary processing residues	13.3	18.1	18.2	+ 4.8	+ 4.9	+1.5	+1.6
Secondary processing residues	8.7	12.8	13.6	+ 4.1	+ 4.9	+2.0	+2.3
Pulp and paper residues[b]	5.2	8.7	8.8	+ 3.5	+ 3.6	+2.6	+2.7
Recycled wood	6.7	13.7	14.0	+ 7.0	+ 7.3	+3.6	+3.8

ª For explanation of hypotheses A and B see text.

[b] Excludes pulping liquors, but includes wood and bark used for energy by pulp and paper industry.

shows that sustained, but not explosive, growth in the use of wood for energy is expected by most countries, for most types of energy wood. The 13 countries, which between them accounted for about two-thirds of total European removals, expected the use of wood for energy (excluding pulping liquors) to be 30-45 million m³ higher in 2000 than in 1980, an annual average rate of growth of about 2.0 to 2.5%. This rate is, in fact, in the same range as the growth rates projected for consumption of other forest products (chapter 12), in contrast to earlier periods when fuelwood consumption was expected to decline while that of industrial forest products rose. For the 13 countries together, slightly lower growth rates were expected for residues of primary processing of wood than for other wood types, no doubt because it was considered that raw material uses for these residues would continue to prove more attractive than energy uses for residues which were technically suitable. Higher rates of growth were foreseen for burning of wood and bark by the pulp and paper industry and for burning of recycled wood products. The latter increase is essentially due to two countries: France foresaw a rise from 3.0 million m³ to 4.0-5.2 million m³ over the 20 year period and the Federal Republic of Germany from 0.5 to 5.0 million m³.

In volume terms, the largest increases are forecast by France (4.4-13.1 million m³), the Federal Republic of Germany (9.3-10.3 million m³), and Sweden (5.4-9.3 million m³). There are, however, interesting differences between these countries in the type of growth foreseen, as shown in table 19.2.

In Sweden, most of the increase is expected to be for fuelwood, in line with national policy to encourage the development of domestic fuels. In the Federal Republic of Germany, however, significant increases were expected from most sources, especially recycled wood products. France, too, foresaw fairly widespread increases, and indicated a particularly wide range between the "low" and the "high" hypotheses for fuelwood consumption with increases between 1.2 and 8.5 million m³. The high growth in fuelwood consumption under hypothesis B (low demand scenario) is assumed to result from a situation of high energy prices when strong Governmental action is taken to encourage the use of wood for energy. It is worth pointing out that in these circumstances growth in demand for forest products is considered unlikely to be high (due to low GDP growth as specified in hypothesis B). As a consequence high fuelwood consumption is unlikely to coincide, in France at least, with high consumption of forest products. The consequences of this line of reasoning for the overall material balance are discussed further in chapter 20. None of the three countries foresaw a significant increase in the burning of wood and bark residues by the pulp and paper industries, presumably as these residues are fully used at present. Indeed, at the European level, the increase for this category is due essentially to one country, Portugal, which expects an increase from 0.3 to 2.2 million m³, possibly due to an expansion of that country's pulp and paper industry, as well as to a more intense use for energy of available wood and bark residues.

TABLE 19.2

Changes in volume of wood used for energy, by source, in France,
the Federal Republic of Germany and Sweden, between 1980 and 2000

(Million m³)

	France		Federal Republic of Germany		Sweden	
	A	B	A	B	A	B
Fuelwood	+ 1.2	+ 8.5	+ 1.4	+ 1.4	+ 4.4	+ 8.5
Primary processing (mechanical)	+ 1.0	+ 1.0	+ 1.4	+ 1.4	+ 0.5	+ 0.3
Secondary processing (mechanical) ..	+ 0.9	+ 0.9	+ 2.0	+ 3.0	b	b
Pulp and paper industry[a]	+ 0.3	+ 0.3	–	–	+ 0.5	+ 0.5
Recycled wood	+ 1.0	+ 2.2	+ 4.5	+ 4.5	b	b
Total	+ 4.4	+ 13.1	+ 9.3	+ 10.3	+ 5.4	+ 9.3

[a] Excluding pulping liquors.   [b] No national forecast.

TABLE 19.3

Summary of national forecasts for consumption of wood for energy
in conventional uses, 1980 to 2000, in 17 selected countries[a]

	1970 *(million m³)*	1980 *(million m³)*	2000 *(million m³)*		Change 1980 to 2000			
					Volume *(million m³)*		Average annual percentage	
			A	B	A	B	A	B
Households	34.9	35.3	43.1	44.8	+ 7.8	+ 9.5	+ 1.0	+ 1.2
Forest industries	15.2	17.8	28.9	29.4	+ 11.1	+ 11.6	+ 2.5	+ 2.6
Intermediate and large-scale units[b] ..	7.0	7.6	17.4	17.9	+ 9.7	+ 12.0	+ 4.2	+ 4.4
Charcoal manufacture[c]	1.6	1.9	2.8	2.6	+ 0.9	+ 0.7	+ 2.0	+ 1.6
Total (17 countries)...........	58.7	62.6	92.2	94.7	+ 29.6	+ 32.1	+ 2.0	+ 2.1

[a] Bulgaria; Cyprus; Czechoslovakia; Denmark; Finland; the Federal Republic of Germany; Greece; Hungary; Ireland; Norway; Poland; Portugal; Spain; Sweden; Switzerland; the United Kingdom; Yugoslavia.

[b] Outside the forest industry.

[c] Where correspondents did not mention charcoal manufacture, it has been assumed that the volumes concerned are insignificant.

The estimates presented in this chapter have been used as the basis for the "ETTS IV consumption scenario" for *fuelwood*[3] in chapter 12 (see, in particular, annex table 12.23). Secretariat estimates were made for those countries which did not reply to the enquiry. For Austria, Belgium-Luxembourg, the German Democratic Republic, the Netherlands and Romania, it was assumed that fuelwood consumption would grow at the average rate of the other members of the respective country group. This procedure was not possible for Turkey, which is Europe's largest fuelwood user (over 20% of the total) and has a quite different social and economic situation from other countries covered by ETTS IV. In Turkey, there appear to be two conflicting trends in the fuelwood field. On the one hand, the forecast above-average rate of population growth could increase fuelwood demand; on the other, a modernization of economic structures and energy use patterns could reduce it. A rather wide range of estimates is therefore presented: from 15.7 million m³ in 1980, a fall to 13 million m³ or a rise to 18 million m³.

What are countries' forecasts for the consumption patterns of wood-derived energy? The replies are presented

in full in annex tables 19.7 to 19.11. A summary of the replies for conventional uses (as defined in chapter 18) is in table 19.3.

Consumption of wood for energy in all conventional uses, notably by direct combustion to produce heat, is expected to rise by around 2% a year between 1980 and 2000, rather more slowly in households; the potential for further expansion in wood-burning by rural families seems to be considered limited, and penetration of the urban household market, in competition with other fuels, unlikely. Consumption growth is expected to be stronger in the forest industries: unrecovered waste of wood and bark is expected to be steadily reduced as wood-burning equipment is installed. The fastest growth is expected in the area of intermediate and large-scale users outside the forest industries, which include schools, barracks, hospitals and residential complexes, as well as district heating units. When situated in rural areas, this type of heating installation is often well suited to wood-based heating, as they are not so large as to require inordinately high transport costs, but are large enough to justify long-term supply contracts, investments in handling and storage facilities, etc. In many cases the owners of the installations are also forest owners, making the whole system more attractive economically. Wood consumption by this type of unit is expected to grow by over 4% a year until 2000 with particularly large rises in Sweden (+ 1.6 to 2.9 million m³).

---

[3] For chapter 12, and the consistency analysis in chapter 20, only scenarios for fuelwood were needed, as scenarios for residues etc. would cause double counting. The effect of energy wood demand on availability of residues for raw material is discussed below.

The outlook is less clear for the "new" types of wood-derived energy, not included in table 19.3 These are defined as generation of electricity for the public grid (cogeneration of electricity by the forest industries is included with other wood-energy use by those industries), solid fuels (briquettes, pellets, etc.) and liquid or gaseous fuels (often seen as substitutes or additives to traditional fuels). National replies, reproduced in annex table 19.11, seem to indicate that very little expansion is expected. Nevertheless, in their comments, many correspondents drew attention to the very high degree of uncertainty about the future for these fuels, arising from possible technical progress.

Some of the correspondents' comments on the outlook for these new fuels are reproduced in annex 19.13 (along with replies from Canada and the USA, for comparison).

It is clear from these remarks that although most countries do not expect any growth in these "new" uses and all countries are extremely prudent, a significant increase cannot be entirely ruled out, in the right circumstances (technological progress, satisfactory wood supply, high prices for conventional energy sources). Most correspondents, however, expect little or no growth.

Only a few correspondents were able to provide estimates for future use of pulping liquors as a source of energy but these replies do come from countries which accounted for the great majority of European chemical pulp production. (These replies are presented in annex table 19.12). The slow growth rates foreseen (around 1% p.a.) reflect the expected growth in production of chemical pulp. However the absence of estimates for many countries does not have a significant effect on the wood balance which ETTS IV will draw up, as these liquors arise only in the process of chemical pulping and are in almost all cases entirely consumed in the same process. The quantity of energy produced is therefore entirely determined by the level and techniques of production of chemical pulp and does not affect other parts of the forest and forest products sector. It does however effect the calculation of the share of wood ultimately used as a source of energy (see estimate of over 40% for 1980 in chapter 18).

19.5   EFFECTS OF DEMAND FOR ENERGY WOOD ON SUPPLY OF WOOD RAW MATERIAL

In the period after the first "oil shock", considerable concern was expressed in many quarters that strong demand for energy wood would deprive the forest industries of part of their raw material supply. As mentioned above, this did not occur, although some price effects were noted. Would the increase in consumption of energy wood foreseen in the previous section have any negative effect on raw material availability? As the situation is different for each type of energy wood, they will be treated separately.

The effect of demand for *fuelwood* must be treated in the global context of wood supply, in chapter 20, as all wood can be burnt and it is necessary to compare total demand for wood with the total potential of the forest to supply it. There is one aspect, however, of particular relevance to fuelwood supply: a large part of fuelwood removals are taken from sources not covered in the inventory data on growing stock and increment – notably tops and branches, trees outside the forest, hedgerows, ligneous agricultural residues, etc. Although the forecasts for total removals (chapter 5) do include wood from outside the forest, it is difficult to make precise forecasts for wood arising in such a wide variety of places. This fact introduces an unavoidable element of uncertainty into the final balance calculation.

In fact the relationship between fuelwood demand and the level of supply of industrial wood is quite complex. On the one hand, increased demand for fuelwood may encourage thinning, bringing extra supplies of industrial wood onto the market and improving the quality of the forest resource and its ability to supply wood in the future. On the other, there can be competition between uses for available supplies of small wood, so that an increase in fuelwood demand reduces supplies of industrial wood. It appears that there are two types of fuelwood demand: the first, for auto-consumption, has little effect on supplies of industrial wood; the second, for fuelwood entering commercial channels, especially for the larger consumers (e.g. district heating) does interact with the industrial wood market. It is not possible to examine these questions in greater detail here, notably because of the importance of local factors.

Most concern has been expressed about competition for *primary processing residues*, especially the solid wood residues from sawmills, plywood mills and veneer mills, which represent a major and increasingly important source of raw material for the pulp and panel industries (see chapter 15, which also shows that an increasing share of these residues has in fact been transferred for use as raw material). This competition concerns essentially the solid wood portion of these residues, as bark is not usually attractive as a raw material for pulp or panels.[4]

It would be desirable to compare:

(a) The volume of solid wood primary processing residues generated;

(b) The volume of these residues required for the manufacture of pulp, particle board and fibreboard; and;

(c) The volume of these residues required as a source of energy.

---

[4] Bark does have several other uses, notably in horticulture, but these are of relatively local importance. In any case, most of these uses would be able to pay sufficiently high prices to prevent the bark being used for energy.

TABLE 19.4

**Outlook for availability of solid wood residues of primary processing
and demand for these residues for raw material and energy, 1980 and 2000**

(*Million m³*)

	Residues generated			Not used for raw material		
	1980 [a]	2000 low	2000 high	1980 [a]	2000 low	2000 high
Finland	9.0	9.4	9.9	1.2	1.0	0.3
Norway	2.1	2.1	2.3	0.2	0.2	–
Sweden	11.0	10.8	11.3	1.1	0.8	–
Denmark	0.9	1.1	1.2	0.7	0.5	0.5
Germany, Federal Republic of[a]	4.8	5.3	6.2	− 0.2	− 0.2	− 0.3
Ireland	0.1	0.1	0.2	0.1	–	–
United Kingdom	1.1	1.3	1.5	0.6	0.4	0.3
Switzerland	0.8	0.8	1.0	0.2	0.1	0.1
Greece	0.4	0.7	0.9	0.2	0.3	0.2
Spain	2.0	2.6	3.1	1.8	1.8	1.2
Bulgaria	1.0	1.1	1.3	0.7	0.5	0.2
Hungary	1.2	1.2	1.3	1.0	0.8	0.5
Poland	3.6	4.2	4.4	2.0	1.4	1.1

	Use for energy			Surplus (+) or deficit (−)		
	1980	2000 low	2000 high	1980	2000 low	2000 high
Finland	0.4	0.5	0.7	+ 0.8	+ 0.5	− 0.4
Norway	–	–	–	+ 0.2	+ 0.2	–
Sweden	0.7	0.8	0.9	+ 0.4	–	− 0.9
Denmark	0.3	0.2	0.5	+ 0.4	+ 0.5	–
Germany, Federal Republic of[a]	0.6	0.6	0.6	− 0.8	− 0.8	− 0.9
Ireland	0.1	0.2	0.3	–	− 0.1	− 0.2
United Kingdom	0.5	0.5	0.6	+ 0.1	− 0.1	− 0.3
Switzerland	0.1	0.1	0.1	+ 0.1	–	–
Greece	0.1	0.1	0.1	+ 0.1	+ 0.2	+ 0.1
Spain	1.1	2.0	2.1	+ 0.7	− 0.2	− 0.9
Bulgaria	0.1	0.1	0.1	+ 0.6	+ 0.4	+ 0.1
Hungary	0.4	0.5	0.6	+ 0.6	+ 0.3	–
Poland	0.6	0.7	0.7	+ 1.4	+ 0.7	+ 0.4

Sources: "Residues generated" and "residues not used for raw material", from consistency analysis model, assuming imports retain a constant share of consumption. See chapter 20. For "residue use for energy", see section 19.4.

[a] 1979-81 average, from consistency analysis model.

[b] Secondary processing residues also included on raw material consumption side (leading to negative figures for "not used for raw material"). See text.

The sum of (*b*) and (*c*) should not exceed (*a*). Estimates for (*c*) were presented in the previous section. (For country detail, see annex table 19.3, for solid wood only). It has been possible also to prepare estimates for (*a*) and (*b*) in the context of the "consistency analysis" which will be presented in chapter 20. This comparison is therefore presented in table 19.4, although it must be borne in mind that, because of the low quality of the basic statistics on residue availability and use, and the large number of assumptions made, this comparison should only be considered a very rough indicator of the situation. Nevertheless, a few facts are apparent:

(*a*) For 1980, in all cases except one (the Federal Republic of Germany, which will be discussed below), the figure for residues "not used for raw material" is higher, sometimes considerably higher, than the estimate of the use of these residues for energy. This indicates that it is likely that the comparisons undertaken are not too distorted;

(*b*) In 2000, with "low" consumption (of pulp, panels and sawnwood), and "low" use of primary processing residues for energy, there is usually a surplus. In this scenario, few, if any, problems of residue availability for energy or for raw material seem likely;

(*c*) In 2000, with the "high" scenario both for consumption and for energy use, many countries show a "deficit" or volumes of residues "not used for raw material" equivalent to the estimate for "residues for energy". This is an indication that in the "high" scenario, competition problems could well occur. Probably there would be no "deficit" in physical terms: it is more likely that the resulting tensions would cause prices of primary processing residues to rise;

(d) In the Federal Republic of Germany there is a "deficit" even in 1980 (also discussed in chapter 15). This indicates that part of the domestic supply of residues comes from the secondary wood processing sector (e.g. the sizeable planing industry) not taken into account in the consistency analysis (only the primary processing industries are taken into account when estimating "residues generated"). This artificial "deficit" does not increase significantly in the estimates for 2000. It is, of course, possible that the pulp and panels industries of other countries could also find a source of raw material in the residues of the secondary processing industries.

The situation for the other sources of wood-derived energy – *secondary processing residues, pulp and paper industry residues, recycled wood products* – is rather different. With a very few exceptions, this material is not at present used as raw material for pulp or panels: indeed, in many cases, it is technically not suitable for such use, as it has been contaminated with paint, glue, lacquer, nails, etc. Few reliable estimates have been prepared of the availability of this material and even fewer of the volumes in fact used as raw material by the pulp and panel industries. The consumption of this material for uses other than energy must, in most countries, be considered negligible at present. Furthermore, for reasons of expense and organization of collection circuits, it is unlikely that use for raw material will increase significantly, unless there arose a major raw material shortage for the pulp and panel industries. For these residues, therefore, few, if any, problems of competition between "raw material" and "energy" uses are foreseen.

19.6 OVERVIEW AND CONCLUSIONS

Chapter 18 established that wood is used as a source of energy in a number of forms: wood in the rough (fuelwood), residues of primary and secondary mechanical processing, and of the pulp and paper industries (wood and bark, as well as pulping liquors and recycled wood). It is estimated that 40% of the volume of wood removed in Europe is used as a source of energy. The decline in the consumption of fuelwood was reversed in the late 1970s, under the influence of the two "energy crises".

Chapter 19 examined the outlook for the use of wood for energy on the basis, among other things, of the replies to an enquiry. "Conventional" uses (i.e. combustion, usually by rural households or in the forest industries) were seen as being economically attractive when the study was drafted and were expected to continue to grow at a moderate speed (about 2-3% a year). "New" uses (e.g. the production of liquid or gaseous fuels from wood, electricity generation), were not expected to increase, but there was great uncertainty, notably about future technical developments and energy prices. Even with the most optimistic assumptions, wood could only provide a relatively small part of national energy demand (except in a few forest-rich countries).

Demand for energy uses was not expected to affect availability of wood residues for use as raw material, in a "low" demand scenario. In a high "demand" scenario, energy/raw material competition for these residues might occur.

How would the forecast increased demand for fuelwood affect the overall supply/demand balance? To answer this question, it is necessary to bring together, on a comparable basis, the various forecasts for consumption and supply of wood and forest products made in ETTS IV. This exercise is carried out in chapter 20.

CHAPTER 20

Consistency of forecasts for supply and demand and outlook for supply/demand balance to 2000

20.1 INTRODUCTION

Earlier chapters have presented the outlook for demand for forest products and for the different components of supply – domestic removals (the most important), transfer of wood residues, recycling of waste paper and international trade. These forecasts were arrived at independently of each other, by a variety of methods. They are summarized in table 20.1.

To analyse the outlook for the forest sector as a whole, it is necessary to compare the forecasts of supply and consumption with each other, in order to assess the outlook for the supply/demand balance. If there are significant differences between the forecasts for consumption and supply, it is also possible to suggest how this "gap" may be closed, since in reality there will be no gap and to analyse the implications for the sector as a whole of this "gap".

There are however major methodological difficulties in comparing the forecasts of supply and consumption, which are in different units (m³ and m.t.) and compare

different products (wood in the rough, sawnwood, panels, pulp, paper, etc.). The solution chosen by most studies has been to convert all the data into a common unit: in ETTS III this unit was the cubic metre equivalent wood in the rough (m³EQ). Apart from the inherent problems of obtaining satisfactory conversion factors, which were discussed in section 15.3, this method has a number of disadvantages:

- It may be misleading to assign, on the regional level, the same conversion factors to imported forest products as to domestically produced goods, as conditions in the area of origin may be quite different from those in the importing country;

- The factor used to convert waste paper to m³EQ will depend on the type of pulp replaced by the waste paper (it could vary between 2.5 and 4.5 m³EQ/m.t.);

- Account is not taken of losses when pulp or waste paper is converted;

TABLE 20.1

Europe: recapitulation of forecasts for consumption and supply of forest products

	Unit	1979-81 (million units)	2000 (million units)		Average annual percentage change 1979-81 to 2000	
			Low	High	Low	High
Consumption						
Sawnwood	m³	102.3	119.0	140.8	+0.8	+1.6
Wood-based panels	m³	35.6	49.6	58.5	+1.7	+2.5
Paper and paperboard	m.t.	49.2	67.2	92.0	+1.6	+3.2
Fuelwood	m³	72.0	85.9	108.7	+0.9	+2.1
Supply						
Domestic removals	m³	350.5[a]	390.8	438.1	+0.6	+1.1
Net trade[b]						
Sawnwood	m³	− 9.5	c	c	c	c
Wood-based panels	m³	− 2.3	c	c	c	c
Wood pulp	m.t.	− 4.2	c	c	c	c
Paper and paperboard	m.t.	+ 1.6	c	c	c	c
Wood raw material	m³	−16.6	c	c	c	c
Transfer of residues	m³	43.1	58.7[d]	83.5[d]	+1.6[d]	+3.4[d]
Recycling of waste paper	m.t.	16.0	24.6[e]	37.0[e]	+2.2[e]	+4.3[e]

[a] Base period around 1980. See chapter 5.

[b] + = net exports − = net imports.

[c] No forecast made for net trade.

[d] Estimated by consistency analysis model (see below).

[e] Calculated by applying scenarios for recovery rates (chapter 15) to forecasts for consumption of paper and paperboard.

– The treatment of residues is also difficult as many pulp and panels products can use either roundwood or residues as raw material. Supply of residues is also difficult to estimate without making estimates for production of sawnwood and plywood.

The use of computerized data bases made it possible to adopt a more ambitious approach: that of constructing a model of the material flows in each country including removals, trade in raw material, production, trade in products and consumption of forest products. It is possible to check the consistency of the different forecasts through this system. *This approach is not intended to provide a forecast of the future situation of the forest and forest products sector, but as a tool to analyse the interrelationship of the forecasts obtained earlier.*

It is believed that this approach, on the national level, using national conversion factors, and calibrated on real data for 1979-81, has made possible a significant increase in the accuracy of the analysis, compared to the conversion to m^3EQ, as well as the possibility to take account of specific national situations, which are often overlooked if analysis is carried out at the regional level.

Another approach, extremely attractive in theory, would have been the construction of a forest sector model, which would bring together the different parts of the sector in an integrated model. An experimental project of this nature has been, in fact, undertaken at the International Institute for Applied Systems Analysis (IIASA). However, neither the methodology nor the data necessary for this approach are yet at a level which would make it possible for it to be used for ETTS IV. It is to be hoped that the next European timber trends study, maybe in the mid-1990s, will be able to make use of the experience gained by IIASA and other organizations active in this field.

Needless to say, this type of analysis for a period 15-20 years in the future cannot be considered precise. What follows should therefore be considered an indication of the broad direction and magnitude of trends and potentials. For this reason, the data in the chapter are generally presented in millions of units. Those in the annex tables, however, are in thousands of units, essentially for reasons connected to the convenience of data processing. The annex tables must not, however, be interpreted as an attempt to forecast the future to the nearest thousand cubic metres.

Some of the analysis presented in this chapter is rather complex and may be difficult to follow. The secretariat recognises this and has done everything possible to simplify the presentation, while maintaining sufficient detail to support the argument. It considers that with a simpler presentation, it would have been necessary to omit some central parts of the analysis, thereby reducing the transparency of the argument.

20.2 STRUCTURE OF THE CONSISTENCY ANALYSIS FOR 1979-81, 1990 AND 2000

The principles used in designing the analytical system were the following:

– It must be compatible with the historical data base (1979-81 was taken as the starting point, and data for that period were used to verify the coherence of the structure);

– It must be able to take into account the scenarios to 2000 proposed elsewhere in this study;

– The assumptions which it may be necessary to make should be as simple as possible (i.e. the structure and reasoning should be transparent, and not a "black box" understood only by the devisers of the system);

– In order to gain the maximum of information, the structure should be as detailed as possible within the constraints set out above.

The structure chosen is presented graphically, in a slightly simplified form, in figure 20.1, and in a more precise formulation in annex 20.1. The underlying principle is that from forecasts for consumption, with certain assumptions concerning trade, raw material/product conversion factors and recycling, a figure for "derived removals" is calculated: this figure can be compared with national, independently prepared, forecasts for removals from chapter 5; conclusions can be drawn both from the difference between the two and from any shifts in the broad structure.

As a first stage, historical data, for the years 1976 to 1981, were used:[1] the resulting calculations were sent to national correspondents with the request that they check the data and the conversion factors, and, if possible, explain the difference between the figure for removals derived from consumption data, through the materials balance structure, and recorded removals. In most countries, a difference did remain, after correction of data series and conversion factors, which was attributed mainly to:

– Continuing problems with the accuracy of the conversion factors;

– The misclassification or mismeasurement of removals, which is perhaps the major source of error in national data systems.

This detailed exchange of information with correspondents, although it could not completely eliminate the discrepancies, did reduce them, significantly in a number of cases, and, most important, gave a precise indication of how large the discrepancies were. It was thus possible to take these statistical discrepancies into account when preparing the model for the years 1990 and 2000.

---

[1] The structure used in this initial "data validation" exercise was slightly different from that finally adopted, but the basic principles and methods were the same.

The secretariat wishes to express its deep gratitude for the devoted efforts of national correspondents in preparing and checking this model.

The procedure can be summarized graphically as follows (residues and waste paper omitted for simplicity):

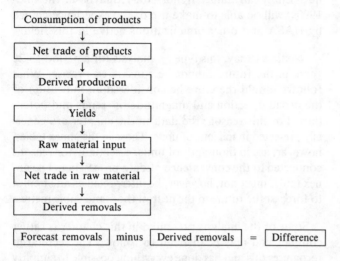

The main methods used were the following (only the most important totals are shown, as the calculations were carried out on a much more detailed basis. As far as possible the same order of presentation is used here as in the tables and the annex tables):

- The scenarios for *consumption* of products are those presented in chapter 12;

- *Derived production* is calculated from consumption and net trade (production = consumption minus imports plus exports);

- *Trade* of all products and wood raw material was assumed to remain constant at the level of 1979-81. (In order to save space, these data are not systematically presented; in particular they are not presented in the annex tables, although they are fully taken into account in the calculations.) The structure of the model made it necessary to make some assumption for trade. This assumption is *not* a forecast for trade but a means for calculating derived removals. The consequences for the analysis of the conservative assumption made for trade, in the absence of specific scenarios, are discussed below;

- *Waste paper* and *wood residue* recovery rates are those proposed in chapter 15; these rates are applied to forecasts for paper consumption and production of sawnwood, plywood and veneer to produce the quantitative data presented;

- *Derived consumption* of sawlogs and veneer logs, pulpwood (roundwood, residues and chips) and fibres for papermaking (pulp and waste paper) was calculated from the estimates for production, using the conversion factors valid for 1979-81. No change in conversion factors was assumed, although such changes are possible;

- *Derived removals* were calculated from derived consumption of sawlogs and veneer logs and smallwood

corrected for net trade. To ensure comparability with historical data, the derived removals also include "LOGDIFF" and "SMALLDIFF", the differences observed, in 1979-81, between the derived removals and recorded removals data, for logs and smallwood respectively. The reason for this is to ensure comparability with historical data and to minimize purely statistical discrepancies;

- *Forecast removals* are the national forecasts from chapter 5;

- The *difference* presented is that between derived and forecast removals. A positive difference indicates forecast removals higher than derived removals, a negative difference the opposite. The "difference" is calculated both for sawlogs and veneer logs and for smallwood.

The data for 1979-81, which are the starting point for the analysis of the situation in 1990 and 2000, are set out in table 20.2. For 1979-81, because of the use of SMALLDIFF and LOGDIFF, the difference between derived and forecast removals is negligible, and therefore not shown. In table 20.2, unlike most other tables in this chapter, imports and exports are shown.

There arose also two further problems of comparability between forecasts and the 1979-81 data base:

(a) The data for the "base period" for the removals forecasts were not always the same as those for 1979-81 removals (see chapter 5);

(b) The estimate for 1980 used as the basis for fuelwood removals forecasts (see chapter 19) did not coincide with data in the data base for fuelwood removals.

In both cases LOGDIFF and SMALLDIFF were adjusted in order to make derived removals in 1979-81 exactly equal to recorded removals (which are those given for the "base period", not those in the data base) and ensure that data for 1990 and 2000 are comparable with those for 1979-81. Because of this procedure, the results of the consistency analysis must be read as indicating *changes* in the situation, rather than as indications of absolute volumes. Furthermore, as a special data base, incorporating some modifications, was constructed for this analysis, data in chapter 20 do not always correspond exactly with those elsewhere in the study.

What is the significance of the "difference" or "gap" calculated by this model? Clearly there will be no "gap" in the future as consumption will exactly equal supply. The "difference" should rather be seen as an indication of the degree of consistency between the forecasts made in other chapters. No attempt has been made to reduce the differences by adjusting the forecasts. The differences, along with the secretariat comments on them, are intended as the starting point for a discussion at the national level of the implications of these forecasts not only for the national forest policy, but also for raw material, industrial and trade policy.

The "difference" could be adjusted by changing one or more of the following:

FIGURE 20.1

Simplified outline of consistency analysis

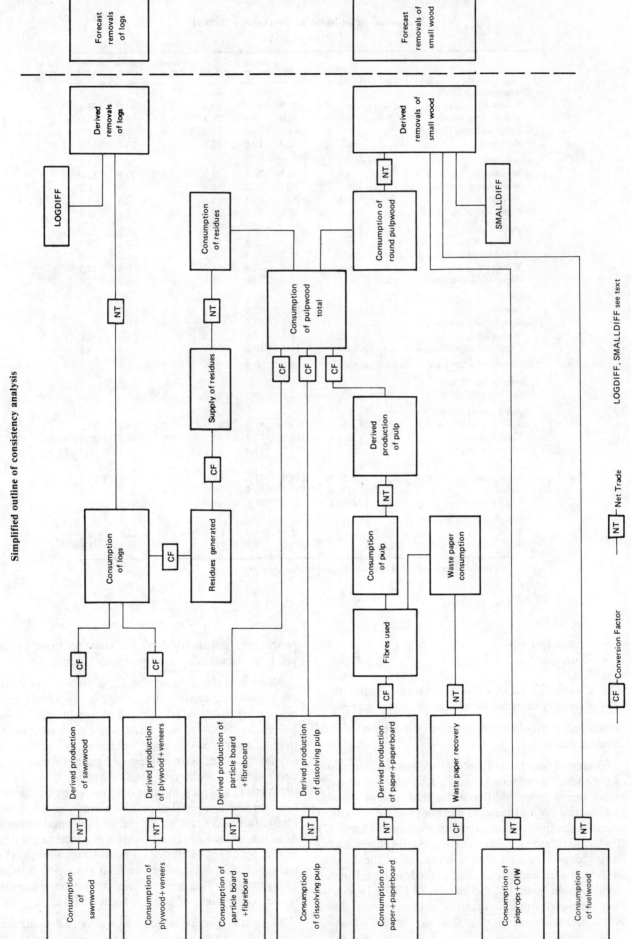

CF Conversion Factor

NT Net Trade

LOGDIFF, SMALLDIFF see text

TABLE 20.2

Summary of estimated material balance, 1979-81

(*Million units*)

	Unit	EUROPE	Nordic countries	EEC (9)	Central Europe	Southern Europe	Eastern Europe
Consumption							
Sawnwood	m³	101.9	10.8	50.4	5.1	14.0	21.8
Wood-based panels	m³	36.0	2.6	21.1	1.4	4.1	6.9
Paper and paperboard	m.t.	49.2	3.3	32.8	1.9	5.5	5.7
Fuelwood	m³	72.0	10.0	18.1	2.2	30.0	11.7
Production							
Sawnwood	m³	92.8	22.8	25.7	8.3	14.0	22.0
Wood-based panels	m³	33.7	4.2	16.0	2.1	4.7	6.7
Woodpulp	m.t.	29.4	17.1	5.1	1.4	2.9	3.0
Paper and paperboard	m.t.	50.8	13.5	24.4	2.5	5.1	5.3
Imports							
Sawnwood	m³	33.9	0.8	27.1	1.1	2.2	2.7
Wood-based panels	m³	10.3	0.4	8.5	0.3	0.2	1.0
Woodpulp	m.t.	10.6	0.4	8.3	0.4	0.6	0.8
Paper and paperboard	m.t.	16.6	0.4	13.6	0.6	1.0	1.1
Exports							
Sawnwood	m³	24.9	12.9	2.4	4.3	2.2	3.0
Wood-based panels	m³	8.0	2.0	3.5	1.0	0.7	0.8
Woodpulp	m.t.	6.6	5.1	0.5	0.3	0.7	0.1
Paper and paperboard	m.t.	18.3	10.6	5.3	1.2	0.5	0.7
Recovery of waste paper	m.t.	16.0	0.9	10.6	0.7	1.8	2.0
Domestic supply of residues and chips	m³	43.1	19.7	11.8	3.4	2.4	5.9
Derived consumption							
Sawlogs and veneer logs	m³u.b.	167.7	45.8	46.3	12.3	26.1	37.2
Total pulpwood	m³u.b.	164.7	73.7	37.7	9.9	18.4	25.2
Fibres for paper	m.t.	49.8	13.3	23.5	2.4	5.0	5.7
Removals							
Sawlogs and veneer logs	m³u.b.	159.9	47.3	44.0	10.1	21.2	37.3
Smallwood	m³u.b.	190.5	56.5	46.9	6.5	38.3	42.2
Total removals	m³u.b.	350.4	103.8	91.0	16.6	59.5	79.5
Adjustment factors							
LOGDIFF	m³	+ 1.3	+1.7	+3.9	−1.2	− 3.7	+0.5
SMALLDIFF	m³	− 18.88	−3.3	−0.1	−1.1	−10.5	−3.8

Note: These data are not exactly the same as those elsewhere in the study. See text.

- Consumption;
- Removals;
- Trade;
- Conversion factors (yields from raw material);
- Recovery rates for wood residues and waste paper.

Which factor should be adjusted will depend on national circumstances.

Whereas quantitative forecasts are available for three of the above factors (consumption, removals and recovery rates), the situation is different for trade and conversion factors. For both of these, it was not possible to provide scenarios for future developments (see chapters 15 and 17). It was therefore necessary for the consistency analysis model to assume that they will remain at the level of 1979-81. In no way should this procedure be misunderstood as indicating that the authors of ETTS IV believe that trade and conversion factors will not change. These assumptions are merely the first step in an analytical procedure, and the outlook for trade and raw material yields is discussed at greater length below.

As regards recycling, it should be pointed out that, in the consistency analysis, consumption of waste paper is calculated before that of pulp and consumption of residues before that of round pulpwood. The estimated consumption of pulp and of round pulpwood is therefore immediately and directly influenced by that of waste paper and residues. In other words, for a given level of paper production, the pulp input is determined by the assumptions on waste paper (i.e. pulp consumption is treated as a residual), and for a given level of pulpwood consumption, the volume of round pulpwood concerned is determined by the assumptions regarding consumption of residues (i.e. consumption of round pulpwood is treated as a residual). The extent to which such an assumption is acceptable may, of course, be debatable.

There is a specific aspect concerning fuelwood: many

fuelwood removals come from non-inventoried material – branches, trees outside the forest, etc. While in theory the forecasts for "total removals" in chapter 5 do include all removals from whatever source, there is a distinct possibility that they do not adequately take into account the unrecorded, unmeasured part of fuelwood removals. Attention is drawn to this problem below when it is necessary.

It should also be mentioned that, although in the analysis and tables summarizing the balances at the national level, data for the "minor" products such as sleepers, pitprops, other industrial wood or dissolving pulp are often not shown, they have been taken fully into account in the basic calculations on which the tables are based. The secretariat is willing to provide on request more detailed data on this and other aspects of the methods used.

The data base for this chapter was modified during the "data validation" exercise. Some of the data presented in this chapter, notably for 1979-81, are therefore not the same as those elsewhere in the study. In most cases, however, the differences are small.

The external sources of data for the material balance are as follows (all of these are by country):

Consumption . . . . Chapter 12 (annex tables 12.14 to 12.27)
Trade . . . . Real data for 1979-81
Raw material/product conversion
  factors . . . . 1979-81 data from "data validation" exercise (table 15.7)
Recovery ratio of wood residues
  chips and particles . . . . Chapter 15 (annex table 15.1)
Recovery rate of waste paper . . . . Chapter 15 (annex table 15.2)
Forecast removals . . . . Chapter 5 (table 5.6)

For consumption of forest products, recovery of residues and waste paper and for removals, "low" and "high" scenarios were proposed in the earlier chapters. In the consistency analysis, the "low" scenario contains all the "low" scenarios from other chapters and the "high" scenario all the "high" scenarios. It should be pointed out however that there is not necessarily a link between the scenarios in the other chapters: it is conceivable, for instance that "high" consumption could occur alongside "low" recovery rates or removals or *vice versa*. It would be possible to explore all the permutations of combinations of "low" and "high" scenarios, but this would result in an unmanageably high number of results.

20.3   CONSISTENCY ANALYSIS, 1990 AND 2000, BY COUNTRY AND COUNTRY GROUP

The difference between derived and forecast removals, calculated by the method described above is presented in table 20.3. Summary data, by country, are given in annex 20.2. It is stressed that the "difference" figures must not be interpreted as indicating simply a "surplus" or "deficit" of wood. They do however indicate a potential imbalance whose possible causes must be examined: in some cases, the "difference" may be an anomaly based on shortcomings in methodology, in others, it may indicate a real feature of the outlook for the forest sector of a particular country.

Removals forecasts were prepared by national correspondents but demand and recycling scenarios, as well as the consistency analysis model were prepared by the secretariat. The comments below are therefore not necessarily a reflection of the national forest policy of the countries concerned, and have not been submitted for official comment or approval.

20.3.1   Nordic countries

This group contains two of the major net exporters of forest products in Europe, *Sweden*[2] and *Finland*. For net exporters such as these, a positive "difference" figure implies that forecast removals are higher than derived

removals, assuming constant exports, i.e. that there is a potential to increase exports. This is indeed the case in the "high" scenario where in 2000 Sweden has a positive difference of 13.3 million m^3 and Finland of 3.4 million m^3. In the "low" scenario the potential for expansion is much more limited.

There are however, ways in which export availability of products could be increased, beyond the limits implied by the "difference" figures. These are:

– An increase in raw material yields;
– A recovery, wherever possible, of secondary processing residues (e.g. from furniture or joinery manufacture) for use as raw material. The example of the Federal Republic of Germany seems to show that it is feasible to use some at least of these residues (see chapter 15).

There is little scope in this region to expand further the use of secondary raw materials: nearly all available primary processing residues are already used as raw material, and domestic paper and board consumption (the source of waste paper) is relatively minor compared to the needs of the large export-oriented, virgin fibre based, paper and paperboard industries, even if quite high recovery rates are achieved.

Another course open to the industries, which has been established as a policy goal by many Nordic governments and associations, is not only to expand the volume of production but to increase the value added to products in the

---

[2] For Sweden the "difference" was adjusted to take account of the fact that a significant part of the increased fuelwood consumption is expected to come from non-inventoried sources, which were not included in the forecast removals.

TABLE 20.3

Difference between derived and forecast removals (total), 1990 and 2000, by country

(Million m³)

	Low scenario		High scenario	
	1990	*2000*	*1990*	*2000*
Finland	− 0.6	+ 0.7	+ 3.2	+ 3.4
Iceland	−	−	−	−
Norway........................	− 0.8	− 1.1	+ 0.4	−
Sweden	+ 1.6	+ 1.9	+ 17.0	+ 13.3
Nordic countries	+ 0.2	+ 1.5	+ 20.7	+ 16.6
Belgium-Luxembourg............	+ 0.4	− 0.6	−	− 2.1
Denmark	+ 0.2	− 0.5	−	− 1.6
France........................	− 1.6	− 12.5	+ 2.8	− 7.5
Germany, Federal Republic of.....	− 5.9	− 13.9	− 6.6	− 22.5
Ireland	+ 1.0	+ 2.2	+ 0.8	+ 1.5
Italy..........................	− 0.4	− 5.6	− 2.6	− 12.4
Netherlands	−	− 1.0	− 0.4	− 2.7
United Kingdom	+ 0.2	− 0.8	− 2.3	− 7.2
EEC(9)	− 6.1	− 32.6	− 8.3	− 54.5
Austria	+ 2.2	+ 1.6	+ 1.6	− 0.3
Switzerland	− 0.3	− 0.5	− 0.4	− 1.7
Central Europe	+ 1.9	+ 1.1	+ 1.2	− 2.0
Cyprus	−	− 0.1	−	− 0.1
Greece........................	−	− 0.3	−	− 0.7
Israel	− 0.2	− 0.6	− 0.2	− 0.6
Malta	−	− 0.1	−	− 0.1
Portugal	− 1.6	− 3.2	− 1.7	− 4.3
Spain.........................	+ 2.8	+ 3.8	+ 2.8	+ 0.4
Turkey	+ 0.3	− 0.5	− 1.9	− 7.0
Yugoslavia	+ 0.6	− 2.6	− 4.6	− 4.5
Southern Europe...............	+ 1.8	− 3.6	− 5.6	− 16.8
Bulgaria	− 0.3	− 0.4	−	− 0.2
Czechoslovakia ...., ...	− 1.8	− 3.0	− 1.0	− 2.4
German Democratic Republic	− 0.2	− 1.6	− 1.3	− 2.7
Hungary	−	0.1	+ 0.6	+ 0.8
Poland	− 2.1	− 0.6	− 0.9	+ 0.4
Romania	+ 0.5	+ 0.7	+ 0.6	+ 0.6
Eastern Europe	− 3.9	− 5.0	− 1.9	− 3.5
EUROPE	− 6.1	− 38.7	+ 6.1	− 60.1

Note: + = forecast higher than derived removals; − = forecast lower than derived removals. For national summary data of the calculations, see annex 20.2.

exporting countries, thus increasing the benefits to the Nordic forest industries, without increasing by the same extent the demands placed on the forest resource. In particular, it has been assumed that pulp exports will remain constant, although it is the intention of the Nordic industries steadily to reduce pulp exports and to process this pulp into paper and paperboard for export. A major problem with this policy is the understandable opposition it arouses from the processing industries in importing countries. This is clearly one of the major trade policy questions which will continue to be discussed in coming years, as it has been in the past.

20.3.2 European Economic Community (nine countries)

This region, where the largest forest industries are concentrated (see chapter 14), and where more forest products are consumed than anywhere else in Europe, is now, and will certainly remain, the largest net importing region in Europe. Yet despite the apparent homogeneity of the region, there are major differences between countries in their forest situation, policies and trends. The analysis will therefore be presented for individual countries or small groups of countries: the total could be misleading as opposing trends cancel each other out.

The smaller continental countries − *Belgium-Luxembourg, Denmark,* the *Netherlands* − all show for most years a negative difference indicating that the increases foreseen for removals, domestic supply of residues and waste paper recycling would not be sufficient to satisfy all the increased demand for forest products, although the volumes involved are not large. The resulting difference will most likely be absorbed by an increase in imports, which already account for a significant part of supply in these countries.

For *Italy,* where, in the absence of a national forecast, removals were forecast by the secretariat, the difference figure is also negative (5.6 to 12.4 million m³ in 2000) and very large compared to domestic removals (57 to 117%). The increase forecast for removals between 1979-81 and

2000 of 0.8 to 1.6 million m³ (9% to 17%) is not nearly enough to match the increases forecast for demand. Nor are the modest increases foreseen for residue supply or waste paper recovery in this traditionally import-dominated forest product economy sufficient to satisfy the expected rise in demand. The calculation carried out by the secretariat would seem to indicate that imports might further increase their share of consumption in Italy unless there were dramatic reductions in the growth of consumption of forest products or an equally dramatic increase in domestic removals. (Any foreseeable improvements in raw material yields could not compensate such a large "gap".)

The situation is quite different for *Ireland* and the *United Kingdom*, both of which have been carrying out major plantation programmes, whose production will come on stream during the period under review. Ireland has a positive difference, which would indicate that the volume of Irish imports could actually fall before 2000 by the equivalent of 1-2 million m³, simultaneously with a growth in consumption of forest products. The small negative figures for the United Kingdom indicate that even if imports increase slightly in absolute terms their share of consumption is likely to fall.

France too expects a sizable increase in its removals between 1980 and 2000 from 38.5 million m³ to 44 or 51 million m³, an increase of 6 to 12 million m³ (14-32%). Yet despite forecasts of moderate growth in demand for forest products and satisfactory use of secondary raw materials, the total difference is expected to be negative, by 7.5 to 12.5 million m³ in 2000. One reason is the expected level of fuelwood removals, which may be in the range (in 2000) of 11.2 to 18.5 million m³, 1.2 to 8.5 million m³ more than in 1980.

It was pointed out in chapter 19 that high demand for fuelwood (essentially determined by energy price levels) will not necessarily coincide with high demand for other forest products (essentially determined by the level of economic activity). In most countries, where the level of fuelwood removals is not great or where there is not a large difference between the high and low scenarios for fuelwood demand, the consistency analysis is not significantly affected by this question. In France however, as there are large fuelwood removals and 7 million m³ between the low and high fuelwood scenarios, it was decided to carry out the consistency analysis in such a way as to explore the consequences of different hypotheses in this matter. The "differences" obtained were as follows (in million m³):

	Total removals	Demand for products	Demand for fuelwood	Difference: 1990	Difference: 2000
1.	Low	Low	High	− 1.6	− 12.5
2.	High	High	Low	+ 2.8	− 7.5
3.	Low	Low	Low	+ 1.4	− 5.2
4.	High	High	High	− 0.2	− 14.8

Scenarios 1 and 2 above appeared slightly more plausible and have been incorporated into the European total, but the figures above show that the range of possibilities is wide. The "differences" shown above indicate that supply

and demand may be in balance for 1990 (without any increase in imports), but for 2000, demand might surpass supply. This imbalance could be resolved by an improvement in yields or by an increase in imports. In the latter case, it is possible that the *share* of imports in consumption would not increase. Another possibility is that part of the increase in fuelwood consumption would come from sources not fully taken into account in the forecasts for total removals. If this were the case for a significant volume, the difference would be considerably smaller.

For the *Federal Republic of Germany* the removals forecasts, although prepared in that country, are unofficial. The negative difference arising from the consistency analysis for the Federal Republic in 2000 is very large − 14 to 22 million m³ − and equivalent to 46-67% of total removals, despite the quite modest growth foreseen for demand of forest products and for fuelwood removals. There is little scope for increasing use of secondary raw materials, which are already very intensively used in the Federal Republic. Above all, however, the removals forecasts are for a decrease of 3 million m³ or stagnation around the 1980 level. If these forecasts (especially those for removals) are accepted, it is clear that practically all increases in consumption in the Federal Republic must be supplied from imports. A factor of uncertainty is the outlook for forest damage and possible sanitation fellings. If removals increased significantly because of sanitation fellings, imports would rise by less.

20.3.3 Central Europe

The situation of the two countries in this small group is very different. For *Austria*, a major exporter, the difference between forecast and derived removals in the low scenario is positive, indicating a slightly increased availability for export. Paradoxically, this disappears in the "high" scenario notably because the difference between high and low scenarios is greater for fuelwood consumption than for removals, indicating that fuelwood demand could constrain the potential to increase exports.

For *Switzerland*, derived and forecast removals are roughly in balance, with a negative difference in 2000 between 0.5 and 1.7 million m³, indicating that imports could increase by a corresponding amount.

20.3.4 Southern Europe

For *Cyprus, Greece, Israel* and *Malta*, the negative differences calculated are rather small, certainly well within the margin of error of this type of calculation. Cyprus, Israel and Malta are in any case almost totally import dependent.

For *Spain* however, the difference between forecast and derived removals is positive (3.8 million m³ in the low scenario for 2000, 0.4 million m³ for the high scenario). There could thus be a reduction in import volumes or an increase in exports, or both. What are the factors behind this? Only moderate increases (although often faster than in more northerly countries) are foreseen for demand of products and fuelwood. Removals however are expected to grow by 8 to 10 million m³ as a result of the major plantation programme which has been carried out over the past half century. Domestic supply of residues and of waste paper is also expected to increase (+0.7 to 2.0 million m³ and +0.6 to 1.6 million m.t. respectively over twenty years). If it proves possible both to increase removals and to produce internationally competitive products, it appears that Spain has the potential significantly to improve its forest products trade balance.

For *Portugal*, however, the difference was negative (between 3 and 4 million m³ in 2000), despite the quite moderate rises in demand for forest products and fuelwood and the base assumption that exports remain constant. A relatively minor cause of distortion may be found in pulpwood exports, which were assumed to remain at the exceptionally high 1979-81 level of 0.6 million m³, but this does not explain the broad trend. The explanation is probably to be sought in the rather prudent assumptions used by the secretariat, in the absence of complete national forecasts, when preparing the removals estimates; these were for a growth by 2000 of only 1.6-2.3 million m³. Nevertheless, this outcome of the balance calculation serves to emphasize the fact that a larger than foreseen expansion of Portuguese removals will be necessary if projected demand is to be satisfied without increasing imports' market share or diverting to the domestic market material intended for export.

For *Turkey*, for which removal forecasts were prepared by the secretariat in the absence of official national forecasts, and for *Yugoslavia*, the difference figure is quite significantly negative in the high scenario, which might be interpreted as an indication that imports could rise, if demand was at the high end of the range. In the low scenario the differences are not very large. Both these countries however have severe limitations on the availability of hard currency to pay for imports. It seems unlikely that imports of forest products would have the necessary priority with the governments of these countries to obtain the permission to devote scarce foreign currency to expanding imports of forest products – especially as both countries have major forest resources. How will the tension implied by the negative difference figure be resolved? There are three major possibilities:

- Consumption will be lower than projected in chapter 12;
- Removals will be expanded further (but Yugoslavia already foresees a rise of 5.5 to 6.2 million m³);
- A more efficient use of existing and forecast supply, by higher processing yields and more intensive use of secondary raw materials.

In any case, it appears that in these countries demand will be constrained by the supply factors mentioned above.

Some other factors may also play a role:

- For both countries the figures for LOGDIFF and SMALLDIFF are large and negative (the total of LOGDIFF + SMALLDIFF is minus 3.7 million m³ for Turkey and minus 4.6 million m³ for Yugoslavia) which might indicate that there are significant volumes of unrecorded removals, leading to underestimated removals forecasts;
- There is great uncertainty about the outlook for fuelwood removals in Turkey, as pointed out in chapter 19. If they fall, or if a significant part of the increase is not accounted for in the removals forecasts, derived and forecast removals would be in a more balanced relationship.

20.3.5 Eastern Europe

The removals forecasts for Bulgaria, Czechoslovakia, German Democratic Republic and Romania were prepared by the secretariat in the absence of official national forecasts. The foreign currency constraints described above for Turkey and Yugoslavia also apply to those Eastern European countries which show a significant negative difference – *Czechoslovakia* and the *German Democratic Republic*. The former is an exporter, but a fall or stagnation in removals was forecast in chapter 5. In these circumstances, despite the moderate increases forecast in demand for forest products and fuelwood, either supply must be increased (higher removals, if this is possible, or increased efficiency) or the volume of exports must be decreased. For the German Democratic Republic, an importing country, it appears unlikely that increases in imports are a viable option, either for foreign currency reasons, or in the context of CMEA agreements (see discussion in chapter 12), so demand will probably be constrained by supply factors.

For *Hungary*, the differences are mostly small, indicating that the rising demand will be satisfied by the increases in domestic supply due to the plantation programme.

Derived and forecast removals are roughly in balance for *Bulgaria* and *Romania*; those for Romania in fact show a modest positive difference.

For *Poland* derived removals are higher than forecast removals in 1990, as demand for products is expected to recover from the economic crisis of the early 1980s, but removals are expected to stagnate. Again, in the Polish situation, demand is likely to be constrained by supply factors. By 2000 however, derived and forecast removals are in balance or nearly so as removals increase, partly because of high levels of sanitation fellings. Indeed it should be recalled that the outlook for Poland as well as for some other neighbouring countries is particularly uncertain because of the possibility of air pollution damage. This question has been discussed in chapter 6.

20.4 CONSISTENCY ANALYSIS OF THE OUTLOOK FOR EUROPE AS A WHOLE

The preceding sections have analysed implications for individual countries of the difference between derived and forecast removals. Is it possible to make any more general statements, applicable to Europe as a whole? In particular, can anything be said on the basis of this consistency analysis about the outlook for the supply/demand balance as a whole?

There is, of course, no unified European market for forest products, so any analysis at the European level will conceal diverging trends at the national level and will thus be misleading. Nevertheless, there is enough communication between different sectors and countries, notably through trade, for analysis at the European level to be of some use.

The difference between derived and forecast removals for Europe as a whole in table 20.3 is quite small for 1990 (only 6 million m^3). No difficulty is foreseen in adjusting this difference, especially as the negative difference for Eastern Europe ($-$ 4.7 million m^3), arises from internal circumstances and will certainly not result in significantly increased imports to this region. It appears therefore that, for the period around 1990, supply and demand may be roughly in balance, without increased imports from other regions. If however extra-European suppliers were able to increase their exports to Europe, by successful marketing or competitive prices, there might even be some over supply. In any case, fairly competitive market conditions could be expected.

For the period around 2000, a rather different situation can be deduced from the consistency analysis. Demand (converted into derived removals) is expected to be between 39 and 60 million m^3 higher than supply (forecast removals). This should not however be misinterpreted as a forecast of a "shortage" of wood on international markets, for the following reasons:

- a number of countries, in Eastern and Southern Europe, are unlikely to increase their imports significantly, notably because of currency restraints. Adjustments in these countries, which account for 9-16 million m^3 of the negative difference, will probably not affect the *international* supply/demand balance

- it might be possible to improve raw material yields, notably through an increasing share of "high-yield" pulps in the paper furnish. This might reduce the "gap" by about 10 million m^3, assuming 5% improvement of yields.

These three factors together could reduce the "gap" at the European level by 20-25 million m^3. It does not appear realistic to propose a higher rate of residue transfer or waste paper recycling, as the high assumptions for these seem to represent a maximum. The remaining gap, of 20-35 million m^3 could be adjusted by one or more of the following:

- higher imports from outside Europe
- lower consumption of forest products (including fuelwood)
- higher European removals.

The scenarios for consumption and removals were based on analysis of trends and potentials, discussed at length elsewhere in the study. The scenario for trade however − that it remained constant at the 1979-81 level − was merely an assumption made to further the preparation of the consistency analysis model and is clearly *not* a likely future development. It appears reasonable, therefore, to assume that most of the remaining "gap" will be filled by increases in net imports. Thus, if the scenarios of the rest of this study, notably those for consumption of forest products and for European removals, are accepted, Europe's imports from other regions may be expected to increase by 20-35 million m^3 over their 1979-81 level, or Europe's exports to other regions to decline by a similar amount, or a combination of these two developments.

An increase in imports from other regions does not imply a tight supply/demand balance or "shortages", unless potential exporting regions have difficulty in supplying these volumes. Will potential exporters to Europe be able to expand their exports by 20-35 million m^3? Chapter 17 provided some indications, unfortunately not quantified, on this subject.

The main conclusions of chapter 17 on the outlook for trade in forest products outside Europe are reproduced below.

The potential for *exports* may be summarized as follows:

- No dramatic changes, upward or downward in the export supply potential of USSR, Canada, natural tropical hardwood. Although supplies from the tropical forests will drop, this may not occur before the end of the century;
- Uncertainty for the USA (with potential expansion for hardwoods);
- Significant expansion from Chile and New Zealand;
- Enormous potential from tropical and sub-tropical plantations.

The potential for *imports* may be summarized as follows:

- An increase in Japan, whose size will depend on progress in developing the domestic resource;
- Uncertainty for the USA, with a drop possible if protectionist action is taken;
- Enormous potential growth in import demand from developing countries, with great uncertainty about the likelihood of this being converted to effective import demand.

This rough assessment does *not* appear to confirm the widespread opinion that a world wide "shortage" of forest products is imminent before 2000, as exporters appear to have the potential to satisfy even high import demand (if developing country imports are not at the top of the range).

The possibility of world-wide tension between forest products supply and demand would arise however if a strong effective import demand materialized from developing countries.

In such circumstances, there would be a strong incentive for existing exporters to increase their production beyond previously established limits and for fast-growing plantations to be widely established in tropical and sub-tropical regions. In fact, if there were a very strong increase in *effective* demand from developing countries, it could be largely satisfied *only* from these tropical plantations; the supply possibilities of these plantations would constrain the import demand of developing countries, although there would be widespread consequences for traditional exporters and importers – notably a rise in real prices and probably substitution for forest products by other materials.

Chapter 17 concluded that the world forest products markets are *flexible in the medium to long term* and have the potential to adapt even to significantly changed situations. Thus it is likely that suppliers outside Europe would be physically able to supply an extra 20-35 million m^3 of forest products in 2000 to satisfy increased European demand. It is not possible however, with the analytical tools used in this study to say whether any economic scarcity, leading to higher prices, is likely or whether the low costs of new producers like Chile, New Zealand or Brazil will exert a downward pressure on prices.

It should be stressed that with the methodology and data used considerable uncertainty remains about the outlook. Two aspects in particular should be mentioned:

(*a*) It is conceivable that high levels of removals could occur with low levels of demand, or *vice versa*, as it is not certain to what extent correspondents took into account demand factors when preparing the removals scenarios. According to the enquiry circulated to correspondents (see chapter 5), the high removals scenario should assume high demand for forest products and satisfactory prices for roundwood; it is possible, however, that achievement of high removals may depend more on other factors (e.g. success in planting programmes, improvement of marketing arrangements) than on demand factors. In fact, forecast removals, high scenario, are nearly 5 million m^3 higher than the derived removals low scenario in 2000, indicating that if de-mand was at the low end of the proposed range, and European removals at the high end, there would be no need for net imports to increase. This is a rather unlikely combination, but cannot be ruled out altogether

(*b*) Fuelwood removals are expected to increase by between 14 and 33 million m^3. It is likely that part of this increase will come from non-inventoried sources – branches, fruit trees, forest residues, etc. or trees outside the forest. Correspondents were requested to estimate what percentage of fuelwood removals would come from non-inventoried sources, but most were unable to do so. This type of removal should be included in the forecasts for total removals made in chapter 5, but the nature of the material involved makes estimation of these quantities very difficult. It is possible that some removals of fuelwood from non-inventoried sources have been included in the forecasts of fuelwood demand, but not in the forecasts for total removals. The negative difference should be reduced by an amount corresponding to the increase in fuelwood from non-inventoried sources not taken into account in the removals forecasts. This is the case for Sweden. The Swedish "difference" has been adjusted to take account of this factor.

The negative difference for Europe as a whole is accounted for to a significant extent by two countries, the Federal Republic of Germany and Italy. The negative difference for the two countries together in 2000 is 19.5 million m^3 in the low scenario and 34.9 million m^3 in the high scenario (44% and 51% respectively of the European total). Both these countries are major importers, and foresee slow growth in removals (or no growth at all for the Federal Republic in the low scenario), leading to increasing "gaps", even though growth rates of demand are modest. The developments on the domestic markets of these countries will no doubt influence the supply/demand balance for Europe as a whole.

20.5 CONSISTENCY ANALYSIS BY ASSORTMENT (LOGS AND SMALLWOOD)

The different forest products have different requirements as regards raw material quality, which are often seen (in an over-simplified fashion) as a descending scale from veneer sheets, plywood and sawnwood, to pulp, to particle board and fibreboard, to fuelwood. Technically, it is possible to move down this scale but only to a very limited degree to move up it. For example, it is possible to burn sawlogs, but it is not usually possible to make sawnwood or plywood from fuelwood. This downward movement does not usually happen for economic reasons: sawlogs because of their technical characteristics command a higher price than fuelwood and are therefore too expensive to burn.

The quality requirements for each raw material assortment are not fixed immutably but change according to technical developments and economic pressures. The boundary between pulpwood and sawlogs is particularly important, but varies between places and over time as new techniques make it possible to saw smaller logs and the two sectors (pulpwood users and sawmillers) bid against each other for the material which is near the technical "frontier" between the two assortments. Market cycles also affect the relative strength of demand for each assortment.

For these reasons, as pointed out above, there are many problems of consistency regarding the historical data, and

TABLE 20.4

Difference between derived and forecast removals by assortment

(Million m³)

Countries	Logs				Smallwood			
	Low scenario		High scenario		Low scenario		High scenario	
	1990	2000	1990	2000	1990	2000	1990	2000
Finland	+0.3	−0.3	+0.7	+0.3	−0.8	+1.1	+2.5	+3.2
Iceland	−	−	−	−	−	−	−	−
Norway	−0.4	−0.4	+0.2	−0.2	−0.4	−0.7	+0.2	+0.1
Sweden	+6.8	+6.6	+8.4	+7.9	−5.2	−4.8	+8.6	+5.4
Nordic countries	+6.7	+6.0	+9.3	+8.0	−6.8	−4.5	+11.4	+8.6
Belgium-Luxembourg	+0.3	−0.2	+0.2	−0.9	+0.1	+0.4	+0.2	−1.2
Denmark	−	−0.6	−0.3	−1.3	+0.3	+0.2	+0.3	−0.3
France	+2.1	+0.2	+2.0	−2.5	−3.7	−12.7	+0.8	−5.0
Germany, Federal Republic of	−0.9	−3.3	−0.1	−5.5	−5.1	−10.7	−6.5	−17.0
Ireland	+0.5	+1.5	+0.4	+1.0	+0.5	+0.6	+0.4	+0.5
Italy	−0.1	−2.8	−0.9	−5.6	−0.3	−2.8	−1.6	−6.8
Netherlands	+0.2	−0.4	−0.1	−1.8	−0.2	−0.6	−0.4	−0.9
United Kingdom	−0.5	−1.1	−1.3	−3.2	+0.7	+0.4	−1.1	−4.0
EEC(9)	+1.6	−6.6	−0.1	−19.9	−7.7	−26.0	−8.2	−34.6
Austria	+0.9	+0.7	+0.9	+0.6	+1.4	+0.9	+0.6	−0.8
Switzerland	−0.4	−0.5	−0.3	−0.8	+0.1	−	−0.1	−0.9
Central Europe	+0.4	+0.2	+0.7	−0.2	+1.5	+0.9	+0.6	−1.8
Cyprus	−	−0.1	−	−0.1	−	−	−	−
Greece	−0.4	−0.9	−0.6	−1.5	+0.5	+0.6	+0.6	+0.8
Israel	−0.1	−0.4	−0.1	−0.4	−0.1	−0.3	−0.1	−0.3
Malta	−	−0.1	−	−0.1	−	−	−	−
Portugal	−0.3	−0.5	−0.4	−1.4	−1.3	−2.8	−1.3	−2.9
Spain	+0.3	+0.3	+0.1	−1.1	+2.4	+3.6	+2.7	+1.5
Turkey	−0.9	−3.4	−2.0	−7.7	+1.2	+2.8	+0.1	+0.7
Yugoslavia	+0.5	−1.0	+0.2	−3.1	−	−1.6	−4.9	−1.3
Southern Europe	−0.9	−5.9	−2.7	−15.3	+2.7	+2.2	−2.9	−1.5
Bulgaria	−0.1	−0.1	−0.1	−	−0.2	−0.3	+0.1	−0.2
Czechoslovakia	−1.3	−1.4	−1.1	−1.5	−0.5	−1.6	+0.1	−0.9
German Democratic Republic	+0.4	+0.3	−0.2	−0.6	−0.6	−1.9	−1.1	−2.0
Hungary	+0.1	+0.1	−	−0.1	−0.1	−0.2	+0.6	+1.0
Poland	−2.2	−1.5	−1.8	−2.0	+0.1	+0.8	+1.0	+2.4
Romania	+0.5	+1.0	+0.4	+0.6	−	−0.3	+0.3	−
Eastern Europe	−2.6	−1.6	−2.9	−3.7	−1.3	−3.4	+1.0	+0.3
EUROPE	+5.3	−8.0	+4.3	−31.1	−11.3	−30.7	+1.7	−29.0

not too much attention should be paid to forecasts for the separate assortments, as "shortages" for one can often be compensated by "surpluses" for the other. It is nevertheless of interest to carry out this analysis as it clarifies the broad problems described above and brings certain aspects into better focus.

Table 20.4 shows the difference between derived and forecast removals, with a breakdown into logs (i.e. sawlogs and veneer logs) and smallwood (all other wood i.e. pulpwood, pitprops, other industrial wood and fuelwood). In many countries, there is no significant difference between the trends for the two assortments. Where there is a significant difference, it is mentioned below.

For Finland, the difference for logs is small in all years indicating a rough balance between supply and demand, but a positive difference arises for smallwood, indicating a potential for increased exports of products based on smallwood.

Sweden shows a regular positive difference for logs of 6-8 million m³, indicating a potential for corresponding increases in sawnwood exports, while there are negative differences for smallwood in the low scenario arising chiefly from increased fuelwood demand (see discussion above).

For Italy, the differences are quite similar for the two assortments, but for the Federal Republic of Germany, the negative differences for smallwood are considerably larger than for logs (although the latter are not negligible). In France too, the difference for smallwood is more negative than for logs, notably because of demand for fuelwood. The positive differences of Ireland and Austria are roughly evenly distributed between the two assortments.

In Greece, Spain and Turkey, however, the difference for smallwood is positive (very strongly so in Spain), while that for logs is generally negative, possibly as young plantations are expected to come into production, for the time being producing mainly smallwood (thinnings).

In Poland also, the difference for smallwood is positive, nearly counterbalancing a negative difference for logs, while the opposite is true for Romania (although the volumes concerned are small).

It is possible to divide those countries where there is a divergence between the balances into two groups:

A. *Countries where difference for logs is more positive than for smallwood*

Federal Republic of Germany	France
Finland	Romania
	Sweden

B. *Countries where balance for smallwood is more positive than for logs*

Greece	Spain
Poland	Turkey

It is probable that in group B, some smallwood will be diverted to the higher value sector and used as raw material for sawnwood and possibly plywood. The opposite (e.g. pulping of small logs) could occur in group A and some upward price pressure due to competition between the sectors in the small logs/large pulpwood area is quite possible.

20.6   UNCERTAINTY SURROUNDING THE CONSISTENCY ANALYSIS

The discussion in this chapter has been based on statistical calculation and the conclusions have been expressed in quantitative terms. Yet this apparent precision should not be allowed to obscure the fact that considerable uncertainty surrounds the component elements of the calculation. In particular, the following elements of uncertainty should be mentioned:

 - Rate of macro-economic growth;
 - Level of residential investment;
 - Substitution for or by forest products in different market sectors (because of technical and cost factors);
 - Energy developments;
 - Technical developments in forest products industries which could, among other things, alter the raw material/product conversion factors;
 - Development in recycling;
 - Ability to mobilize the under-used European forest resource especially with reference to small, private holdings;
 - Competitivity of forest industries inside and outside Europe;

 - Government policies.

Most of these aspects have been discussed in earlier chapters of the study and reasons advanced for selecting the forecasts which have been brought together in this chapter. Readers of the study, therefore dispose of sufficient information to form their own judgement of the analysis presented in chapter 20.

Furthermore, the consistency analysis presented here has been constructed in such a way that it is relatively easy to replace any of its component parts with different forecasts. It is hoped that it will be possible, after the publication of ETTS IV, to undertake sensitivity analysis of this type and explore in more detail the consequences for the sector as a whole of various developments not included in this initial balance.[3]

Some of the main implications of the analysis in this chapter will be discussed in more detail in chapter 22.

---

[3] Researchers wishing to carry out consistency analysis (e.g. for their own country) with the same model but using different assumptions are invited to contact the secretariat.

CHAPTER 21

Outlook for supply and demand of forest products in the twenty-first century

21.1 INTRODUCTION

This chapter will analyse briefly the outlook for the supply and demand for forest products in the first part of the twenty-first century, comparing the forecasts for removals (chapter 5) and demand (chapter 13). A rather detailed quantitative comparison of supply and demand forecasts for 2000, was presented in chapter 20: a similar approach is not appropriate for 2025 because of the very great uncertainties affecting the outlook for demand. This chapter will therefore only make rough comparisons, at the European level.

There is a major difference in quality between the forecasts for supply, notably removals, and for demand. Because of the long rotations in most parts of Europe, changes in the forest sector will always be gradual. Trees already growing in the 1980s will provide practically all the removals in 2000, and the majority of those in 2025. Indeed it is because the forest ecosystem is slow to respond to changes in management that foresters need some

assessment, however rough, of the long-term outlook. The outlook for demand, however, as pointed out in chapter 13, is highly uncertain: the situation in 2025 could be radically different from that in the 1980s. This difference in quality between the supply and demand forecasts should be borne in mind throughout this chapter.

In this chapter, removals forecasts for 2020 are compared with demand forecasts for 2025 (based on population data for that year). In view of the major uncertainties mentioned above, it does not appear that this discrepancy will introduce significant distortions. For this reason the time reference for this rough comparison is labelled "the 2020s".

Earlier chapters did not examine the outlook for the period after 2000 for transfer of wood residues, recycling of waste paper, or international trade. Where necessary, rough estimates will be made in this chapter.

21.2 RECAPITULATION OF FORECASTS FOR REMOVALS AND DEMAND FOR FOREST PRODUCTS

Tables 21.1 and 21.2 recapitulate the forecasts for the 2020s of chapters 5 and 13.

National forecasts for removals are for continuing growth after 2000, with an increase between 2000 and 2020 of 40-52 million m³ u.b., to reach 431-490 million m³ u.b. As in the period before 2000, faster growth is expected in a small group of countries, mostly at present net importers, which are developing their forest resources (France, Ireland, Portugal, Spain, United Kingdom). The increase forecast by this group between 2000 and 2020 is 47% of the increase for Europe as a whole (low scenario) or 36% (high scenario). This group accounted for 18% of European removals around 1980, but its share is expected to rise to 21-22% in 2000 and 23-24% in 2020.

The speculative nature of the consumption estimates for 2025 in table 21.2 must be stressed again. The "base scenarios" are based on the simple assumption that *caput*

levels of consumption will not change between 2000 and 2025. The resulting scenarios are very slightly above those for 2000, obtained by econometric methods. However, the extent of possible change over 40 years is so great that it seemed only prudent to present "extreme" low and high scenarios, as an indication of what could happen. These extreme estimates are for:

(a) A return by 2025 to levels near or below those of the 1980s; or

(b) A further increase to new record levels, often about twice as high as those of the 1980s.

The levels of consumption of forest products are determined by two classes of factors:

(a) Those which the forest sector *cannot influence* in any significant way: overall economic growth, level

311

TABLE 21.1

Forecasts of total wood removals, 1980 to 2020

(Million m³u.b.)

	Base period [a]	Low scenario		High scenario		Change			
						Base - 2000		2000-2020	
		2000	2020	2000	2020	Low	High	Low	High
Nordic countries	104	110	119	130	140	+ 6	+ 26	+ 9	+ 10
EEC(9)	91	101	111	115	129	+ 10	+ 24	+ 10	+ 14
of which:									
France, Ireland, United Kingdom .	43	54	63	63	73	+ 11	+ 20	+ 9	+ 10
Central Europe	17	20	20	22	21	+ 3	+ 5	−	− 1
Southern Europe	60	77	92	82	99	+ 17	+ 22	+ 15	+ 17
of which:									
Iberian Peninsula................	21	30	40	33	42	+ 9	+ 12	+ 10	+ 9
Eastern Europe	80	83	88	89	100	+ 3	+ 9	+ 5	+ 11
Europe........................	350	391	431	438	490	+ 41	+ 88	+ 40	+ 52

Source: Chapter 5, especially, table 5.4.   [a] Around 1980. See chapter 5.

TABLE 21.2

Europe: scenarios for consumption of forest products, 1979-81 to 2025

(million units)

			Projections 2000 [a]		Estimates for 2025			
					Base scenarios [b]		Extreme scenarios [c]	
	Unit	1979-81	Low	High	Low	High	Low	High
Sawnwood	m³	102	119	141	123	148	90	160
Wood-based panels..............	m³	36	50	58	52	60	40	70
Paper and paperboard...........	m.t.	49	67	92	68	95	55	105
Fuelwood	m³	72	86	109	94	117	50	130

Source: Chapter 13, especially table 13.2.

[a] Econometric projections (see chapters 11 and 12).

[b] Based on constant levels of *per caput* consumption.

[c] Extreme scenarios, based on subjective judgement (see text of chapter 13).

of residential investment, technical development of electronics, energy prices etc.;

(b) Those which the forest sector *can influence*: product development (based on improved research),

cost control, standardization, improved marketing etc. Even if developments for the first group of factors are negative, if action is taken to maintain or improve the competitivity of forest products, levels of consumption can be improved.

21.3 COMPARISON OF SUPPLY AND DEMAND SCENARIOS FOR THE 2020s

Table 21.3 compares, in terms of m³EQ, the forecasts for removals and for consumption of forest products. To construct this table, it was necessary to make a number of assumptions:

— That consumption of dissolving pulp and wood used in the rough, other than fuelwood, would stay constant after 2000;

— That the volume of residues generated and trans-

ferred for use as raw material would grow at the same rate as removals. (The volumes for 2000 are taken from the consistency analysis model in chapter 20, assuming imports retain a constant share of consumption);

— That the waste paper recovery rate would not rise above the (high) levels forecast for 2000;

— As a starting point for analysis, that net imports would not change from the level, judged necessary

TABLE 21.3

Europe: comparison of forecasts for consumption and supply in 2000 and in the 2020s
(million m³EQ)

		2000		The 2020s Base scenarios		The 2020s Extreme scenarios	
	1979-81	Low	High	Low	High	Low	High
Consumption							
Sawnwood	169	196	232	203	244	148	264
Wood-based panels	58	79	94	83	96	64	112
Paper and paperboard	167	228	312	230	322	186	356
Fuelwood	72	86	109	94	117	50	130
Other (wood in the rough, dissolving pulp)	32	28	28	28*	28*	28*	28*
Total	498	617	775	638	807	476	890
Supply							
European removals	350	391	438	431	490	..	..
Transfer of residues	43	56ᵃ	75ᵃ	62ᶜ	84ᶜ	..	..
Waste paper	40	62ᵃ	92ᵃ	63ᵈ	95ᵈ	..	..
Net imports	58	78ᵇ	93ᵇ	78ᵉ	93ᵉ	..	..
Total	491	587	698	634	762	..	..
*Difference*ᶠ	−7	−30	−77	−4	−45		

Note: For the 2020s, consumption estimates are for 2025 and removals forecasts for 2020. See text.

ᵃ Assuming imports retain a constant share of consumption.

ᵇ Assuming an increase in net imports of 25-45 million m³ EQ (see chapter 20) between 1979-81 and 2000.

ᶜ Assuming growth at same rate as removals.

ᵈ Assuming same recovery rates as in 2000.

ᵉ Assuming no change from 2000 level.

ᶠ Arises partly from inaccuracies in conversion factors. See text for interpretation of this difference.

to balance supply and demand in 2000 (see chapter 20).

Comparison of the totals for supply and demand provides a "difference", which must be interpreted carefully. The "difference", of 7 million m³EQ, for 1979-81 (when supply and consumption were, of course, exactly equal) indicates that the conversion factors used to obtain the data in m³EQ are not completely accurate. For 2000, the data are derived from the consistency analysis model in chapter 20, but they are *not* the same as the difference in that chapter, which were measured at the removals stage.

Chapter 20 (section 20.4), using the same basic data as table 21.3 considered that there would be no "shortage" of forest products in 2000. The apparent deficit would be adjusted in several ways, including increased net imports (taken into account in table 21.3), improved raw material yields and, in certain countries, lower consumption than projected. The latter two factors are not included in table 21.3. It may be concluded from the analysis in section 20.4 that the deficit of 30-77 million m³EQ in 2000 apparent in table 21.3 would not lead to any real shortages.

For the 2020s, if the supply forecasts are compared to the base scenarios for consumption, the difference is either negligible (low scenario) or minus 45 million m³EQ. This is less than the difference calculated for 2000, which, it was considered, did not imply supply problems. From this, it may be concluded that *if per caput consumption in the 2020s remains at the levels projected for 2000, forecast European removals would be able to satisfy the increased*

demand with no increase in net imports over the levels forecast for 2000.

This situation changes however if the "extreme" scenarios for demand are considered. If activity in user sectors was low and forest products lost markets, leading to the extreme low scenario in table 21.3, there might even be an oversupply of forest products, with negative consequences for the forest sector as a whole (e.g. falling prices, closure of capacity, neglect of silviculture). In such circumstances, however, the system would tend to adjust itself to a certain extent, e.g. by not expanding residue transfer or waste paper recycling, possibly by reducing imports from other regions, and, above all, by developing new markets and products.

On the other hand, if demand for forest products was very high (extreme high scenario), the European forest resource would have considerable difficulty meeting demand, net imports would increase (if supplies were available in other regions) all wastage would be reduced and recycling encouraged, new fibre sources (e.g. biomass) developed and raw material yields improved. Furthermore, the inevitable rise in prices would brake the increase in consumption.

Demand for the three main product groups (sawnwood and panels, paper and paperboard, fuelwood) is determined by quite different factors, as pointed out in earlier chapters. It is unlikely that all product groups would experience extreme high (or extreme low) scenarios simultaneously. For instance, if extreme low fuelwood demand coincided with extreme high demand for sawnwood, panels and paper, and *vice versa*, the range of total consumption would be narrowed substantially (by nearly 40%), from 476-890 million m³EQ to 556-810 million m³EQ. This would reduce the seriousness of any imbalances which might occur.

Unfortunately, at present, there is no reliable way of assessing demand forty years ahead, and uncertainty will remain, as neither the extreme low nor the extreme high scenario can be entirely ruled out. In this sense, forestry will remain, as it always has been, an expression of confidence in the future.

Furthermore, there will continue to be wide differences between national situations. Europe is by no means a unified market for forest products and is unlikely to become one, even if levels of trade continue to rise. It is therefore quite conceivable that a surplus of forest products at the European level could co-exist with continuing efforts to improve national self-sufficiency in some countries, or for some forest resources to lack outlets for their products even in a situation of region-wide strong demand.

CHAPTER 22

Conclusions

22.1 INTRODUCTION

This chapter aims to bring together the main conclusions of the study, so as to draw attention to the most important points.

There are some major differences between the outlook for 2000 foreseen by ETTS IV and ETTS III (table 22.1). For consumption, ETTS IV foresees slower growth for panels, paper and paperboard, similar growth for sawnwood but a significant increase for fuelwood, as compared to the decrease forecast by ETTS III. The supply side forecasts of the two studies are close to each other, the only exception being waste paper, where lower volumes are forecast by ETTS IV because of the lower forecast for consumption of paper.

22.2 NATURE OF THE FORECASTS

The main aim of this study is to analyse trends in the past and, on that basis, to examine the outlook for the future, first for individual parts of the forest and forest products sector and then for the sector as a whole. To build up the broad picture, it has sometimes been necessary to make rough estimates or subjective judgements for aspects which could not be forecast in a more scientific way. There is therefore a wide difference in quality between the various forecasts in the study.

Each forecast is only as good as the data, assumptions and methodology which have been used to prepare it. For this reason the way in which each forecast was prepared was explained, so that readers may make their own judgement of its quality. In some cases it has been possible to present the data and methods in such a way that researchers could construct their own forecasts on the basis of different assumptions.

TABLE 22.1

Europe: comparison of selected forecasts for change between 1980 and 2000 in ETTS III and ETTS IV

(Million units)

	Unit	ETTS III [a] Low	ETTS III [a] High	ETTS IV Low	ETTS IV High
Consumption					
Sawnwood	m³	+ 11	+ 39	+ 17	+ 39
Wood-based panels	m³	+ 84	+ 96	+ 14	+ 23
Paper and paperboard	m.t.	+ 58	+ 99	+ 18	+ 43
Fuelwood and other wood in the rough	m³	− 25	− 25	+ 12	+ 34
Supply					
Removals	m³	+ 30	+ 80	+ 40	+ 88
Transfer of residues	m³	+ 20	+ 40	+ 16	+ 40
Recovery of waste paper	m.t.	+ 22	+ 52	+ 9	+ 21
Net imports	m³EQ	+ 30	+ 50	+ 20[b]	+ 35[b]

[a] From tables 1/12, 5/12, 6/12 and 7/12 of ETTS III. These are the unadjusted forecasts i.e. the total of increases forecast for supply are smaller than those for consumption.

[b] Estimate based on difference between derived and forecast removals. See text.

315

22.3 THE EUROPEAN FOREST

Forest and other wooded land covers 35% of the land area of Europe. There are enormous differences between forests in different parts of Europe, for instance as regards climatic and site conditions, ownership structure, or management objectives and levels. Over the past 30-40 years, there has been significant investment in forestry in almost all countries, a small increase in area and a larger increase in growing stock and increment. Yet there are indications that the forest is still not used to the maximum possible, as drain is still well below increment as well as below potential cut. The forecasts, by national experts, are for a continuation of these trends and higher levels of growing stock, increment and removals. The most dramatic increases forecast are concentrated in a group of countries which have adopted a policy of forest improvement, notably France, Ireland, Hungary, Spain and the United Kingdom. Elsewhere a more stable situation is foreseen.

The non-wood benefits of the forest include the provision of soil protection, leisure and recreational facilities and non-wood products. Demand for these, especially recreation in all its forms, has risen, and is expected to rise further, probably faster than the demand for wood. In many forests, especially those near large centres of population or in tourist areas, this demand could put forest managers and forest policy makers before difficult choices and possibly lead to changes in forest policy and management.

Developments in the agriculture sector, notably the large surpluses of agricultural products built up in certain countries, may lead to pressure to convert agricultural land to other uses, including forestry. There is a need for the development of integrated policies for rural land use. In some western European countries, there are very many small forest holdings belonging to private, sometimes absentee, owners, many of whom are not able to manage their forests efficiently or whose management objectives may be different from those of larger forest owners and national objectives. This factor may be a brake on achieving the full potential of the forest, for the provision of wood or non-wood benefits.

It is likely that the supply of high quality, slow grown sawlogs, of the type found in northern Scandinavia or the mountains of Central Europe for softwoods, or in Normandy or Slavonia for hardwoods will become increasingly rare. There does not seem any possibility to reverse this trend, except in the very long term.

In the early 1980s, great concern was expressed about the widespread damage observed in European forests and attributed to air pollution. If this damage increased, it would have serious consequences for the whole of the forest sector. Chapter 6 presented a number of non-quantitative scenarios for the effects of this damage on the forest, forest industries, consumption and trade of forest products. It is not possible, however, with the present state of knowledge, to make any firm conclusions about the future development of damage, or on the likely consequences for the forest sector.

22.4 CONSUMPTION OF FOREST PRODUCTS

Econometric models were used to project the consumption of the major forest products. The major assumptions used for the projections were:
- GDP growth at 2.6-3.3% p.a.;
- Residential investment stable (low scenario) or rising at half the rate of GDP (high scenario);
- Constant real prices of forest products;
- The market penetration phase for particle board has passed.

On these assumptions, moderate growth was foreseen for 1980-2000:

	Percentage per annum
Sawnwood	0.8-1.6
Wood-based panels	1.7-2.5
Paper and paperboard	1.6-3.2

These forecasts depend on the assumptions made, both about the external factors and about the success or failure of the forest sector in developing and marketing its products and in maintaining or improving its competitivity. The scenarios summarized above were based on the conservative assumption that forest products' technical and economic competitivity would not change from its 1980 level. The sensitivity analysis in chapter 12 showed that with different assumptions about technical and marketing development or prices, consumption could be significantly lower or higher than the base scenarios.

Thus the forest sector itself will have a determining influence on the future level of consumption of sawnwood and panels. Strategies should be developed to achieve these ends.

The most effective use will have to be made of the available wood raw material, notably by:

(a) Making the best use of the physical and other properties of the wood and bringing out to the maximum the advantages of wood;

(*b*) Minimizing the negative effects of potentially less positive properties, such as biodegradability, flammability, dimensional instability;

(*c*) Improving surface finishing, both for durability and appearance.

It will not be sufficient, however, to produce a better product. It has also to be brought efficiently to the market. To this end, the product will in many cases need to be promoted as part of a solution to a problem, for instance as part of a building component, wall unit etc. i.e. as part of a system. Research, development and technical innovation will take on added importance in the wood-processing and wood-using industries. In this connexion, prefabrication may well gain further in importance as new techniques (computer numerical control, CNC, and computer aided design, CAD) permit mass production in combination with considerable flexibility.

A further important point is that with technically more sophisticated products, great emphasis needs to be put on their correct application, still frequently a source of problems for sawnwood and panels, especially in construction.

To meet the challenge of other materials – which will no doubt be strong – the wood-processing industries will need to look beyond their own sector to anticipate changing conditions in other sectors, especially as regards emerging new technologies, in order to perceive areas where the use of forest products might be affected, as well as new opportunities and outlets for forest products that might arise. Sawnwood and panels can be substituted by other materials and products in virtually all their applications.

Such action should be based on good knowledge of the end-uses for forest products. It is clear from chapter 8 that this knowledge is not available in most countries. As a starting point for action, it is necessary to know what volume of product is used in each application, and what are the major technical and cost factors affecting competition with other materials. There is a need for major research projects in order radically to improve knowledge on end-uses. Detailed work at the national (or even local) level will be necessary before any international overview is possible.

For newsprint and printing and writing papers, the major factor of uncertainty concerns development of new electronic means of communication. These may not affect total levels of consumption significantly before the mid-1990s, but thereafter major changes are possible. The new systems may however include an element of paper, if this is economically and technically possible, and if it is felt that this is in accordance with consumers' preferences. For other paper and paperboard, especially packaging grades, there will be stiff competition on the technical and economic level, for which these grades are well equipped. For household and sanitary uses, the European market may well be approaching saturation (except, possibly, Southern Europe) leading to slow growth in future.

Between 1980 and 2000 the share of the products of sawlogs and veneer logs in total consumption (in terms of m³EQ) is expected to fall from 37.1% to 33-35%.

For the period after 2000, the same basic factors will continue to determine the level of consumption: activity in user sectors, itself influenced by GDP, and technical and economic competitivity of the products themselves. However, the degree of uncertainty for a period 40-50 years in the future is so great that it is only possible to make rough estimates, with a very wide range between the lowest and highest scenarios (see chapter 13).

22.5 THE EUROPEAN FOREST INDUSTRIES

The pattern of raw material input to the industries has changed, with increases in the share of pulpwood in wood raw material, of wood residues in pulpwood, and of waste paper in paper-making fibres. The scenarios for future supply of residues and waste paper in chapter 15 are for increased supply of these raw materials, relative to availability and for them to account for a larger share of raw material input. Using the consistency analysis model (assuming imports retain a constant share of consumption) the following scenarios were calculated:

	1979-81	2000
Consumption of residues (million m³)	44	57-77
Share of residues in pulpwood consumption	27%	30-37%
Waste paper recovery (million m.t.)	16	25-37
Share of waste paper in papermaking fibres	33%	41-48%

This increasing use of residues and recycled material, together with the widespread use of wood and bark as a source of energy in the forest industries reinforce the status of these industries as leading exponents of low- and non-waste technology. They are, as such, well suited to a world which will be ever more reluctant to accept the wastage of materials and determined to encourage the use of renewable materials, including wood.

The future of the forest sector as a whole depends in part on the economic health of the forest industries. If they are uncompetitive and unprofitable, and capacity is shut down, less income will be available to carry out necessary silvicultural work, leading to a parallel decline of forests and forest industries. ETTS IV has attempted to examine the economic health of the industries (chapter 14), but because of problems with data and methodology was not able to draw conclusions. This is an aspect where further research is necessary to understand better the forest industries, and evaluate their economic health with particular reference to trends in productivity and profitability and their place in the economy.

Raw material/product conversion factors are a vital part of balance calculations but depend to a great extent on local circumstances. It was not therefore possible to provide specific scenarios for changes in these factors from the 1980s level. It is likely, however, that they will improve in at least one major sector – the increasing use of "high-yield pulps", such as thermo-mechanical pulp (TMP) and chemi-thermo-mechanical pulp (CTMP), with factors of 2.5-3.0 m^3 wood/m.t. pulp, which can sometimes replace chemical pulp, with a factor of about 4.8 m^3/m.t.

What of the outlook for the capacity of the forest industries, a subject on which it was not possible to reach a conclusion in chapter 14? The consistency analysis model (chapter 20) was used to provide scenarios for production in 1990 and 2000 (derived production) and compare these with 1980 capacity levels. To obtain a more realistic scenario for production it was assumed that imports would retain a constant share of consumption. This produces a rough balance between supply and demand at the European level. These data are interpreted below. It should be borne in mind that definitions of capacity are often uncertain, so any conclusions must be tentative.

In 1979-81, the wood-based panels and paper and paper board industries were on average estimated to be working at 84% and 87% of capacity.[1] The sawmilling industry's capacity utilization rate was almost certainly much lower (see chapter 14).

---

[1] It is usually not possible to run at 100% capacity for long periods so capacity utilization rates around 90-95% may be considered full utilization.

For 1990, the derived sawnwood production levels are all below 105 million m^3, probably less than estimated capacity in 1979-81. Likewise for panels, the estimates for derived production are below the estimated capacity in 1979-81. Thus for sawnwood and panels in 1990 it appears that there is no need significantly to expand total production capacity over the 1979-81 level. Investment is likely to be concentrated on improving the competitivity of existing capacity. For paper and paperboard, the low scenario shows a similar picture as for sawnwood and panels in 1990 (derived production below 1979-81 capacity), but for the high consumption scenario, a capacity expansion of about 3 million m.t. might be necessary.

For 2000, however, the scenarios imply the following increases of capacity over the 1979-81 level, if the levels of derived production in the scenarios are to be attained:

Sawnwood .... 0-10 million m^3 (assuming that around 1980 sawmilling capacity was about 105-110 million m^3)

Wood-based panels 5-10 million m^3
Paper and
  paperboard ..... 5-20 million m.t.

(It is possible to obtain data for derived production of woodpulp, but as woodpulp is an intermediate product, and the consistency analysis model is not designed to analyse production trends, these data are not meaningful).

These are estimates for expansion of capacity and do not cover any investments which may be made to modernize or replace existing capacity. These capacity expansions would probably take place in areas where capacity is inadequate for the existing raw material supply, notably in areas where removals are expected to increase. It is likely that the process of structural rationalization in all sectors will continue. In particular a few large sawmills may replace many small ones.

TABLE 22.2

Europe: production capacity in 1979-81 and scenarios for derived production in 1990 and 2000

(Million units)

| | 1979-81 | | Derived production | | | |
| | | | Low scenario | | High scenario | |
	Production	Capacity	1990	2000	1990	2000
Sawnwood (m^3)	92.8	a	96.7	105.3	101.1	120.3
Wood-based panels (m^3)	33.7	39.9^b	37.4	44.5	39.2	51.0
Paper and paperboard (m.t.)	50.8	58.4	54.6	63.3	61.3	79.5

^{a} No European total available for sawmilling capacity (see chapter 14).
^{b} Excluding veneer sheets.

22.6   TRADE IN FOREST PRODUCTS

Trade in forest products has grown faster than either production or consumption. Intra-European trade has expanded, in particular two-way trade, with growing specialization in production. There is every reason to believe these trends will continue.

In the late 1970s, Europe's exports to other regions,

notably in North Africa and the Middle East, expanded. This trade ceased to expand in the mid-1980s, partly because of financial problems in the overseas importing countries. Will this lull be temporary or have these exports reached a maximum level?

However, the volume of Europe's imports from other regions is three times greater than its exports to other regions. If Europe were to need extra supplies from other regions, would they be able to provide them? To answer this question it is necessary to review the outlook at the global level.

The potential for *exports* from other regions may be summarized as follows:

- No dramatic changes, upward or downward, in the export supply potential of the USSR, Canada and natural tropical hardwood areas. Although supplies from the tropical forests will drop, this may not occur before the end of the century;

- Uncertainty for the USA;

- Significant expansion from Chile and New Zealand, but their role as suppliers to Europe is in doubt;

- Enormous potential from tropical plantations, in the right economic and social circumstances.

The potential for *imports* by other regions may be summarized as follows:
- An increase in Japan, the size of which will depend on progress in developing the domestic resource;

- Uncertainty for the USA, with a drop possible if action is taken to protect domestic markets;

- Enormous potential import demand in developing countries, but with great uncertainty about the likelihood of this being converted to effective import demand. Effective import demand from China is particularly uncertain.

This rough assessment does *not* appear to confirm the opinion that a worldwide shortage of forest products is imminent, as exporters appear to have the potential to satisfy even high import demand (if developing country imports are not at the top of the range). The possibility of world-wide tension between forest products supply and demand would arise, however, if a strong effective import demand materialized from developing countries.

In such circumstances, there would be a strong incentive for existing exporters to increase their production beyond previously established limits and for fast-growing plantations to be widely established in tropical and subtropical regions. In fact, if there were a very strong increase in *effective* demand from developing countries, it could be largely satisfied *only* from these plantations; their supply possibilities would constrain the import demand of developing countries, although there would be widespread consequences for traditional exporters and importers – notably a rise in real prices and probably substitution for forest products by other materials.

In short, the world forest products markets seems to have the potential to be quite flexible in the medium to long term and to have the potential to adapt even to significantly changed situations. Whether this potential flexibility is realized depends on many economic and institutional factors, including the speed with which signals of impending change are understood and acted on.

The direction and levels of trade flows will be affected by the competitivity of the various exporters as well as by a number of other factors, including governmental trade policy and exchange rates, which are not predictable in the long term.

A few data were also presented (section 16.6) on trade in finished products (e.g. building elements, manufactures of paper and paperboard), which may affect the overall forest products balance. For the EEC(9), the trade balance for these products is more positive than for wood raw material and for semi-finished products (i.e. sawnwood, panels, paper and paperboard).

22.7 WOOD AND ENERGY

The consumption of fuelwood fell steadily between the 1950s and the mid-1970s. With the successive "energy crises" and the rise in the general energy price, this trend was reversed around 1978. Research has shown that if all types of energy derived from wood (i.e. energy from fuelwood, industrial wood residues, pulping liquors and recycled wood products) are taken into account, about 40% of the volume of wood (with bark) removed in Europe is used as a source of energy, making the provision of energy the single most important end-use for wood in volume terms.

The results of an enquiry show that the use of wood for energy is expected to rise at 2.0-2.5% per year between 1980 and 2000 in conventional uses (combustion in rural households, the forest industries and in medium-scale units e.g. for district heating). The potential for "non-conventional" uses (e.g. manufacture of liquid or gaseous fuels, electricity generation) is very uncertain and dependent on rises in energy prices. No significant development of these non-conventional energy sources, or of the energy plantations which might accompany them, was foreseen although a number of forest-rich countries, poor in domestic energy sources, notably France and Sweden, are maintaining their R and D programmes in the wood energy field, as a further rise in the price of energy and a corresponding increase in non-conventional wood energy cannot be excluded.

The moderate rise foreseen for energy from wood does not appear to pose a serious threat to availability of wood as raw material for the forest industries, although some price competition for lower grades of pulpwood cannot be ruled out.

In little more than a decade, the price of crude oil rose from about $10/barrel to over $30/barrel, causing structural changes in the world economy, and has recently fallen again to just over $10/barrel. In mid-1986 the medium-term outlook for oil prices − and therefore for energy prices in general − is highly uncertain, as are the possible consequences of a continued low price. The forecasts for future wood energy use were based on the assumption that oil prices would rise moderately from $30/barrel (the price when the enquiry was circulated). Will oil prices recover rapidly the levels of the early 1980s? Should the forecasts of chapter 19 be modified if they do not? Or are the changes in the pattern of wood energy use since the mid-1970s unlikely to be reversed by changes in general energy prices? These questions, which arose only while the study was being finalized, could not, unfortunately, be answered. They constitute, therefore, another factor of uncertainty for the outlook.

22.8 OUTLOOK FOR THE SECTOR AS A WHOLE

In chapter 20 the forecasts for consumption are compared with those for supply, by the use of a "consistency analysis" model. It is stressed that the data generated by this model must be interpreted only with reference to the model's structure and assumptions (see section 20.2) and *not taken out of context*.

There is a wide range of national situations, which are discussed in chapter 20 but cannot be summarized here. On the European level, however, it appears that the forecasts for supply and demand are roughly in balance in 1990, with no increase in net imports or improvement in raw material yields. For 2000, however, the forecast for supply is below that for consumption. This should *not* however be interpreted as a forecast of a "shortage" of wood on international markets. After some technical adjustments to take account of national situations, supply and consumption may be brought into balance by:
  − An improvement of raw material yields (e.g. by more use of "high-yield pulps");
  − An increase in net imports of 20-35 million m³EQ (which, according to chapter 17, is probably feasible from the point of view of the supplying countries).
The likely developments between 1980 and 2000 may thus include moderate growth in consumption of forest products and fuelwood, increased European removals, higher rates of residue transfer and waste paper recycling, improved yields and an increase in European net imports, leading to a balance in 2000 at a higher level of consumption than in 1980.

For the early decades of the twenty-first century, the outlook is much less certain; European removals are expected to continue to increase, but there are no reliable ways to project levels of consumption. The wide range of scenarios proposed for consumption, when compared to forecasts of supply, can be interpreted as leading either to shortages or to oversupply. Levels of consumption, however, can be influenced by the forest sector itself, which therefore has the power if it chooses to exert it, to determine whether there are sufficient outlets for the likely increased supply of forest products from European forests in the twenty-first century.

What of the outlook for prices for forest products, an element of major importance for all investment decisions? Chapter 10 has shown that past trends provide no basis for forecasting future price movements. The analytical tools used to examine the future supply/demand balance do not take price specifically into account (although the demand models do, assuming that prices would remain constant), so there is unfortunately no way of providing quantified price scenarios, and only general conclusions can be drawn from the overall supply/demand balance. Chapter 20 concluded that supply and demand would be roughly in balance for 1990, without improved yields or increased net imports, and that in 2000, the apparent surplus of demand over supply could be adjusted, notably by increased imports from other regions. This would seem to imply a rather competitive overall situation and a slight downward pressure on prices, at least up to 1990.

However, it must be stressed that most markets, for wood in the rough or for processed products, are strongly affected both by local factors and by factors specific to particular grades or qualities, which play at least as large a role as the broader supply/demand balance for forest products. There are a few products, notably chemical pulp, newsprint and Kraftliner and, to a lesser extent sawn softwood, where trading takes place in what is practically a world market. For these, the price level will continue to be strongly influenced by the lowest cost producers, which are often not in Europe, but in the USA or Brazil. Some new producers, such as Chile or New Zealand, may aspire to become the lowest cost producers.

For other products, however, price movements will be determined above all by market conditions in small market sectors, oriented to particular users. There is a potential here for prices to be raised if products are developed and marketed which are well fitted to that particular market, e.g. because they offer savings to the user, are accompanied by advisory services, or have short delivery times. To a great extent the future of forest products markets (especially those for sawnwood and panels), both as regards volume and price, depends on developments in the competition between products in a great number of small sectors.

22.9 FINAL CONCLUSIONS

This study has identified a number of important areas where there is a need for *improved data and analysis*:

– The behaviour of small-scale private forest owners and their forest management objectives are as yet inadequately understood;

– There is a need to continue to monitor the extent of forest damage, and its possible effects on round-wood supply and forest products markets;

– Improvements are needed in data on the structure and capacity of the sawmilling and wood-based panels industries;

– The economic health of the forest industries (e.g. trends in productivity and profitability, if possible international comparisons) should receive more attention;

– The outlook for construction is not at all clear;

– The distribution of consumption of sawnwood and panels between different end-use sectors, and the technical and economic factors which determine the levels of consumption, need thorough investigation at the national level;

– Work is in hand on assessing the effect of new electronic media on consumption of paper, but much uncertainty remains;

– The outlook for the forest sector in the USSR, Canada, the USA and Japan is under continual review at the national level. It is hoped to bring together the results of this work at the international level before too long;

– There is a need to monitor the situation and outlook for tropical and subtropical plantations (area, species, growing stock, rotation, and increment), whether directed to domestic or export markets;

– The results of the FAO outlook studies on levels of consumption in developing countries and development of domestic resources, will also be of great interest for Europe, in as much as they will provide an indication of possible future import demand for these countries which could compete with that of European countries;

– It will also be necessary to monitor global energy trends, notably for energy prices, and to adjust forecasts for wood energy accordingly.

Finally it will be necessary to compare, on a continuing basis, real trends with those foreseen in ETTS IV, in order to obtain early warning of deviations from the expected trend lines.

A number of important questions of *policy* for governments, forest owners, forest industries and trade have been raised, which are summarized below:

(a) ETTS IV has pointed to the large and increasing importance of the demand for non-wood benefits of the forest. Yet in many cases, the provision of these benefits generates costs for the forest owner but no income, to the extent that forestry sometimes appears an economically unattractive undertaking. Should the provision of these benefits continue to be considered part of the forest owner's normal obligations as a member of society? Or should he receive compensation for them in some way? If the latter, how should this be organized? Should there be further changes in forest management practice in the light of the expected growth in the relative importance of non-wood benefits?

(b) In several countries there have been long-term intensive efforts to enlarge and improve the forest resource and raise levels of removals. These policies have often had several goals, notably improving the trade balance and supporting the rural economy, but have sometimes been based on the belief that there would be a "shortage" of forest products in the future. ETTS IV is suggesting a more balanced outlook for supply and demand, at least up to 2000. The question arises as to the appropriate policy response to this outlook, which differs in some important respects from those given in earlier studies. For example, instead of seeking to maximize production of wood fibre, the emphasis might be on the selection of species and qualities that would offer more flexibility in future supply and at the same time contribute more to the multiple-use management of the forest. Each country's situation with regard to growing conditions, trade balance, population pressure and other factors, as well as the relative importance attached to different policy objectives, will determine the response to the new findings. Net importing countries will have to weigh the benefits of reducing their import dependence against the costs of developing their forest resource, a complex equation involving not only the direct costs connected with the supply and trade of forest products but also the costs of present agricultural policies, the availability and quality of marginal agricultural land, trade policies, policies for domestic industries, for rural development and so on.

(c) Seen from the national policy-making point of view, fragmentation of ownership and management is a serious handicap in many countries to raising productivity and efficiency, both in forestry and certain forest industries, such as sawmilling. What can and should be done about this, within the existing social and political constraints?

(*d*) The present agricultural surpluses in many European countries are leading to a general reassessment of rural land-use policies and of the possibility of transferring some agricultural land to other uses, including forestry. What should be the future role of forestry in rural land use patterns?

(*e*) Doubts have been expressed in some quarters about the dynamism and profitability of the forest industries, although little detailed analysis has been done at the international level. If these doubts are justified, what would be appropriate strategies for their revitalization? What should be the role of governments and of the industries themselves? How should these policies be co-ordinated with forest policies and trade policies? The economic health of the forest industries is considered an essential contribution to the health of the forest and forest products sector as a whole.

(*f*) ETTS IV foresees a rather balanced supply/demand outlook to the year 2000, which suggests a continuation of the competitive market environment experienced in the 1980s. There appears to be a need for action to improve forest products' competitivity. What measures should be taken? Should more funds be devoted to research and development, to which the forest and forest products sector has devoted only a small part of its turnover in the past (with the possible exception of the capital intensive pulp and paper industry)? Should the traditionally production-oriented industries become more market-oriented, and how could this be achieved?

(*g*) The future of the forest sector is linked to a considerable extent with three major end-use sectors: construction, packaging and communications. The first has been falling in relative importance within most national economies, the second has been and will remain intensely competitive, the third is undergoing enormous structural changes as a result of the electronics revolution. Should the forest sector seek to diversify the outlets for its products? If so, how?

(*h*) Should the use of wood for energy be encouraged by governments? If so, should there be safeguards for the raw material supply of the forest industries? Should there be greater co-ordination between policies for energy, for raw materials, and for forestry?

(*i*) There are many different stages between planting a tree and the use of a forest product in its final form, including silviculture, harvesting, transport, processing, trade (sometimes international), marketing and final use. Most of the participants in this "wood chain" (*filière bois*) seek to maximize their own returns, sometimes at the expense of other participants in the chain. Yet all participants clearly have a common interest in optimizing the functioning of the chain, with the ultimate aim of raising levels of consumption and thereby the returns for all. Is there scope for improving the situation in this respect, for instance by facilitating the flow of information along the chain, in both directions (from the supplier to the customer and *vice versa*), or even modifying the structure of the chain?

All outlook studies are vulnerable, to shortcomings in data, methodology or analysis, or to unforeseen changes in the sector studied or outside it. These problems are magnified if the time span for the outlook is very long. Yet all decision-makers, and the analysts who advise them, need to have a mental picture of the future, as an indispensable element of the decision-making process. Furthermore, over the long term, there may be an interaction between the outlook studies and real trends (influenced by decision-makers who have read the studies), so that forecasts, when they indicate an unwelcome future, may be in a way self-invalidating. One example of this interaction is the increase in European removals foreseen by ETTS IV, which is at least partly the result of action taken in response to the analysis of earlier outlook studies, including the predecessors of ETTS IV, which stressed the necessity of developing the European resource. It is hoped that in a similar way, action will be taken to invalidate some of the negative aspects foreseen by ETTS IV.

Relatively little has been said in ETTS IV about future threats to European forests, notably from forest fires in the south and from the damage attributed to air-borne pollution in northern continental Europe. The reason for this neglect lies in the impossibility of constructing scenarios for the future, on the basis of the information currently available, and not in any underestimate of the gravity of the threats. The question of forest conservation and protection is one of the gravest issues facing policy-makers in Europe in the closing years of the twentieth century. The forest is not just a source of wood and non-wood benefits, but has a special place in the hearts of millions of people. This fact, along with the multiplicity of sources of air pollution and the long distances travelled by pollutants, has brought forests into the centre of national and international policy discussions. It would be deeply ironic if the remarkable recovery of the European forest in the nineteenth and twentieth centuries from the centuries of destruction which culminated in the early phase of the industrial revolution, should be jeopardized by the negative side-effects of recent technical progress. This is a great challenge facing not only the forest and forest products sector but European society as a whole, in the last decades of the twentieth century.

What, finally, of the twenty-first century? After all, forest policy decisions taken in the 1980s will not bear fruit, in most cases, until well after 2000. European supply of wood is expected to continue to expand in the twenty-first century, yet European demand, for wood or for non-wood goods and services of the forest, is highly uncertain. The outlook for the supply/demand balance at the world level in the twenty-first century is, if anything, more uncertain: populations and standards of living may be expected to raise demand, yet there is also great potential for higher supply. Will demand for forest products increase with standards of living? Or will other products replace forest products? How much could tropical and sub-tropical plantations supply in conditions of strong demand and at what price? These, and other, important questions cannot find definitive answers at present. Yet decisions must be taken, on the basis of imperfect information, and take into account all features of the national and international situation, and the priority attached to the different objectives of forest policy. In view of the uncertainty surrounding the outlook for the global supply/demand balance in the twenty-first century, more importance may come to be attached to:

(*a*) Security of supply of forest products at the national or regional level;

(*b*) Flexibility in the range of wood and non-wood goods and services which the forest may supply in the future.

This study will have served its purpose if it has provided some signposts which may be useful for those facing an uncertain future. This lack of certainty about the future is nothing new for the forest sector. Frequently, forests planted with one objective in mind have served different, equally valuable, purposes, when they reached maturity. An example is the forests established to provide wood for ships, which have in fact provided high quality sawnwood and veneers for use in construction. Forestry will remain, as always, an expression of confidence in the future.